AF614983

AUTOMATED ASSEMBLY

Jack D. Lane
Editor

Judy D. Stranahan
Senior Publications Administrator

Published by

Society of Manufacturing Engineers
Publications Development Department
Marketing Services Division
One SME Drive
P.O. Box 930
Dearborn, Michigan 48121

AUTOMATED ASSEMBLY

Second Edition

First Printing

Library of Congress Catalog Card Number: 86-061736

International Standard Book Number: 0-87263-246-6

Manufactured in the United States of America

SME wishes to express its acknowledgement and appreciation to the following contributors for supplying the various articles reprinted within the contents of this book Appreciation is also extended to the authors of papers presented at SME conferences or programs as well as to the authors who generously allowed publication of their private work.

Annals of the CIRP
Hallwag AG
Nordring 4
CH-3001 Berne
Switzerland

Assembly Automation
IFS (Publications) Limited
35-39 High Street
Kempston, Bedford MK42 7BT
England

Assembly Engineering
Hitchcock Publishing Company
Hitchcock Building
Wheaton, Illinois 60188

Robert L. Douglas
Gilman Engineering and Manufacturing Company
305 W. Delavan Drive
P.O. Box 1367
Janesville, Wisconsin 53547

Electronic Design
Hayden Publishing Company, Inc.
10 Mulholland Drive
Hasbrouck Heights, New Jersey 07604

Electronic Packaging & Production
A Cahners Publication
Cahners Plaza
1350 East Touhy Avenue
P.O. Box 5080
Des Plaines, Illinois 60018

Industrial Engineering
Institute of Industrial Engineers
Institute Headquarters
25 Technology Park/Atlanta
Norcross, Georgia 30092

Journal of Manufacturing Systems
Society of Manufacturing Engineers
One SME Drive
P.O. Box 930
Dearborn, Michigan 48121

B. Kuttner
Computer Tool and Die Systems, Inc.
Atrium Office Center
900 Victors Way
Ann Arbor, Michigan 48104

Machine Design
Penton/IPC Publication, Inc.
Penton Plaza
Cleveland, Ohio 44114

Manufacturing Engineering
Society of Manufacturing Engineers
One SME Drive
P.O. Box 930
Dearborn, Michigan 48121

Gary P. Maul
Ohio State University
1971 Neil Avenue
Columbus, Ohio 43210

Modern Machine Shop
Gardner Publications, Inc.
6600 Clough Pike
Cincinnati, Ohio 45244

Modern Materials Handling
Cahners Publishing Company, Inc.
Division of Reed Holdings, Inc.
221 Columbus Avenue
Boston, Massachusetts 02116

RCA Corporation
13 Roszel Road
P.O. Box 432
Princeton, New Jersey 08540

Tower Conference Management Company
331 West Wesley Street
Wheaton, Illinois 60187

Cover photo courtesy of *Modern Machine Shop* magazine

PREFACE

Manufacturing managers and engineers are constantly seeking new and improved techniques for producing a finished assembled product. With increased manufacturing competition on both a national and international level, attention is being focused more than ever on reducing the costs and increasing the quality level of assembled products. Major emphasis is being placed on adopting, modifying, and evaluating new technologies, techniques, and processes for achieving low cost assembly.

It has long been recognized by manufacturing personnel that the most costly operation encountered in production is that of assembly of manufactured parts into a final assembled product. Independent of the method of assembly; whether manual, automatic, or a combination of both, the assembly operations being performed are diverse in nature and require detailed analysis before a system is conceived, designed, built, and placed into production for successful operation.

Assembly, though used extensively by all industries producing a finished product, is probably the least understood and most poorly documented technology applied in the manufacturing community. Very few colleges and/or universities offer courses or curriculums on this subject, due mainly to a lack of documented information available in the literature. Assembly is still thought of as being an "art" with very little documented scientific evidence to support any conclusions to the contrary.

Prior to the last five years, very little new information evolved in the field of assembly and/or assembly processes. Manual assembly is predominant over hard automatic assembly techniques due to inherent assembly problems and the fact that these assembly machines lack the innate intelligence of a human operator and lack sufficient flexibility to changeover when product designs and market demands change. However, in the past five years, more and more emphasis and attention has been focused on assembly. Although many reasons are probably responsible for providing this impetus, new concepts, industrial robots, and the vision technology have predominated in terms of offering new approaches and capabilities for improvements in assembly operations. As a result, the introduction of this book with current and timely discussions not only on robots and vision, but also on other updated subject areas, has initiated the need to document the technical information that is being applied to current assembly processing and on future assembly trends.

This book has been compiled through detailed and comprehensive manual and computer based literature searches on assembly and assembly related topics. The book has been organized into the following separable chapters:

Planning of automatic assembly systems involves detailed analysis of many different concepts, techniques, and methodologies with many diverse factors requiring consideration. Final determination of the most suitable assembly system involves detailed economical analysis. Chapter 1, "Planning Automatic Assembly Systems", presents information on several approaches which need to be considered in the planning phase and offers new insights which should benefit the reader.

The concept of design for assembly has taken on greater emphasis in the past few years for facilitating the reduction in product and assembly costs. Whether assembly is performed manually, automatically, or with robots, it has been demonstrated that substantial cost savings can be achieved by employing various rules to accommodate assembly operations. Chapter 2, "Design for Assembly", contains papers which reveal what and how product design for assembly can result in substantial savings.

Chapter 3, "Feeding for Assembly", provides the reader with an update on feeders and feeder concepts which are currently being used and being considered for the future.

Flexible assembly systems are gaining prominance for the assembly of a family of products where production volumes and product mix can be automatically accommodated. Different approaches to achieving this flexibility are being practiced. The fourth chapter, "Flexible Assembly Systems", reveals some of the approaches being taken using automatic guided vehicles and robots.

The subject of Chapter 5, "Compliance Devices in Assembly", provides insight into what compliance devices are and how they can be beneficial for improving assembly operations. The different design concepts embodied in various compliance devices are presented to enlighten the reader to the concepts utilized. Assembly applications employing compliance concepts and devices are presented in the papers in this chapter to familiarize the reader with the potential benefits which can be achieved.

Robots are being employed more and more in assembly operations to offer flexibility and programmability to the process. The book's sixth chapter "Robots in Assembly", presents the reader with varied applications utilizing robots in different ways to achieve the assembly of both small and large parts (automotive body assembly). System design concepts and hardware components are discussed with regard to specific requirements of the robotic assembly process.

Chapter 7, "Lasers in Assembly", introduces the reader to the use of lasers and laser systems for assembly applications which are unique and require the precision operating characteristics of lasers.

Vision systems are gaining in prominance for assisting in automatic assembly operations since they provide the ability to "see" which is lost when manual operations are replaced. The title of Chapter 8, "Vision Systems in Assembly", contains papers which present a varied assortment of assembly applications which are assisted through the use of vision systems.

Automatic gaging and inspection is an important operation which is essential for any type of assembly operation. Concepts and hardware required for the automatic inspection of components and assemblies for in-process, final inspection, and selective assembly is presented in the book's ninth chapter entitled "Automatic Gaging and Inspection".

The automated factory of the future is incorporating automatic assembly into its design. The subject of the final chapter, "Assembly in an Automated Factory", reveals how total assembly automation can be achieved through a discussion of the concepts and methodology necessary for successful integration. The first major unattended factory built in Japan is presented in this chapter to illustrate the extent to which this has been accomplished and to reveal thinking which was utilized in its creation.

I wish to thank all the companies, organizations, publishers, and authors who gave permission to have their articles reprinted in this volume. Thanks also to the Publications Development Department staff at SME for their assistance in the research and development required in making this book possible.

Jack D. Lane, CMfgE, P.E.
R.I.S.E. Inc.

ABOUT THE EDITOR

Jack D. Lane is president of R.I.S.E., Inc. located in Flint, Michigan. Prior to founding R.I.S.E., he was a professor and assistant department head of Mechanical Engineering and director of the robotics center at GMI Engineering & Management Institute in Flint, Michigan, where he taught courses in manufacturing engineering and robotics. He obtained a B.S. in Metallurgical Engineering from Wayne State University and an M.S. in Manufacturing Engineering from the University of Detroit. Mr. Lane also worked as a development engineer for General Motors Corp. Manufacturing Development, a test and development engineer for Chrysler Missile Division, and a plant metallurgist for Revere Copper and Brass.

Mr. Lane has authored many papers on various subjects associated with automation, assembly processing, and robotics, and has also done considerable consulting in these areas. He has also presented many seminars on these subjects, both in the United States and Europe, for professional societies, General Motors, and other private corporations. Since 1967, Mr. Lane has been actively involved as an engineer, teacher, and now full-time consultant on numerous projects and activities associated with automatic processing, robotics, and assembly systems.

While with the General Motors Technical Center as a development engineer, Mr. Lane was involved with the design and building of several automatic assembly systems for various applications and General Motors plants. While teaching various manufacturing engineering courses at GMI Engineering and Management Institute, he developed and taught several courses associated with assembly processing, automation, and robotics and was responsible for the development of laboratory facilities and equipment used for instructional and development purposes. In addition to his activities in assembly, Mr. Lane has been actively involved in robotics and is currently editing a book on robotic arc welding.

Mr. Lane is a Certified Manufacturing Engineer and a Registered Professional Manufacturing Engineer. He is a charter member of Robotics International of SME (RI) and the Computer and Automated Systems Association of SME (CASA) in addition to being a member of the Society of Manufacturing Engineers, the American Welding Society, and the Machine Vision Association of SME (MVA). He is the past chairman of the Assembly Division of SME and the past founding chairman of the Education and Training Division of Robotics International. Currently, Mr. Lane is a Technical Vice-President of Robotics International of SME.

SME

The informative volumes of the Manufacturing Update Series are part of the Society of Manufacturing Engineers' many faceted efforts to provide the latest information and developments in engineering.

Technology is constantly evolving. To be successful, today's engineers must keep pace with the torrent of information that appears each day. To meet this need, SME provides, in addition to the Manufacturing Update Series, many opportunities in continuing education for its members.

These opportunities provide:

- Monthly meetings through five associations and their more than 300 chapters and 130 student chapters worldwide to provide a forum for membership participation and involvement.

- Educational programs including seminars, clinics, programmed learning courses, as well as videotapes and films.

- Conferences and expositions which enable engineers and managers to examine the latest manufacturing concepts and technology.

- Publications including the periodicals *Manufacturing Engineering*, *Robotics Today*, and *CIM Technology*, the *SME Newsletter*, the *Technical Digest*, *Journal of Manufacturing Systems*, and a wide variety of text and reference books covering everything from the basics to manufacturing trends.

- Information on Technology in Manufacturing Engineering database containing technical papers and publication articles in abstracted form. Other databases are also accessible through SME.

The SME Manufacturing Engineering Certification Institute formally recognizes manufacturing engineers and technologists for their technical expertise and knowledge acquired through experience and education.

The Manufacturing Engineering Education Foundation was created by SME to improve productivity through education. The foundation provides financial support for equipment development, laboratory instruction, fellowships, library expansion, and research.

SME is an international technical society dedicated to advancing scientific knowledge in the field of manufacturing. SME has more than 80,000 members in 70 countries and serves as a forum for engineers and managers to share ideas, information, and accomplishments.

The society works continuously with organizations such as the American National Standards Institute, the International Organization for Standardization, and others, to establish and maintain the highest professional standards.

As a leader among professional societies, SME assesses industry trends, then interprets and disseminates the information. SME members have discovered that their membership broadens their knowledge and experience throughout their careers. The Society of Manufacturing Engineers is truly industry's partner in productivity.

MANUFACTURING UPDATE SERIES

Published by the Society of Manufacturing Engineers and its affiliated societies, the Manufacturing Update Series provides significant up-to-date information on a variety of topics relating to manufacturing. This series is intended for engineers working in the field, technical and research libraries, and also as reference material for educational institutions.

The information contained in this volume doesn't stop at merely providing the basic data to solve practical shop problems. It also can provide the fundamental concepts for engineers who are reviewing a subject for the first time to discover the state of the art before undertaking new research or applications. Each volume of this series is a gathering of journal articles, technical papers and reports that have been reprinted with expressed permission from the various authors, publishers, or companies identified within the book. Educators, engineers, and managers working within industry are responsible for the selection of material in this series.

We sincerely hope that the information collected in this publication will be of value to you and your company. If you feel there is a shortage of technical information on a specific manufacturing area, please let us know. Send your thoughts to the Manager, Publications Development Department, Marketing Division at SME. Your request will be considered for possible publication by SME or its affiliated societies.

TABLE OF CONTENTS

CHAPTERS

6 ROBOTS IN ASSEMBLY

7 LASERS IN ASSEMBLY

8 VISION SYSTEMS IN ASSEMBLY

CHAPTER 1

PLANNING AUTOMATIC ASSEMBLY SYSTEMS

Presented at the SME 13th ISIR/Robots 7 Conference, April 1983

Choosing Manufacturing Systems Based on Unit Cost

by Richard E. Gustavson
Charles Stark Laboratory, Inc.

Introduction

When considering the various possible means for performing a manufacturing task, the types of capable resources available and their corresponding costs must be determined. For most situations, manual labor is a definite candidate. Various types of machinery that can be used for the particular set of tasks to be performed will also need to be investigated. All of the potential resource types can be evaluated economically on a unit cost (fixed and variable) basis. The goal is to determine what type of system is most economical for particular production quantities because it is often hoped that the required volume is increasing. Since variable costs are also increasing, a year-by-year set of snapshots must be put together in order to establish the appropriate relationships. Work by Lynch (1) and Boothroyd (2,3) was revised and added to.

The extent of the conditions to be included in the model must be established. Figure 1 exhibits the "worlds" of possible system models. Those costs external to the system "world" can be neglected or included in all comparisons.

Nomenclature

A	subscript denoting Alternative system.
B	subscript denoting Base system.
C	total unit cost (\$/unit).
C_F	fixed unit cost (\$/unit).
C_V	variable unit cost (\$/unit).
D	number of working days per year.
f	annualized cost factor (also known as Capital Recovery factor).
H	investment horizon (years).
$\bar{L}_H$	average loaded labor rate (\$/hr).
m_s	maximum number of stations per worker.
n	number of tasks to be performed.
O_H	total operating/maintenance rate (\$/hr.)
P_A	total annualized, installed system price (\$).
P_{R_i}	hardware price of resource i. (\$).
$\bar{P}_R$	average hardware price for resources in a system. (\$).
P_{T_j}	hardware price of tool/mat'l. handling eqpt. j. (\$).
P_T	average tool/mat'l. handling hardware price. (\$).
q_H	hourly production rate (unit/hr).
Q_Y	yearly production quantity (unit/yr).
r	minimum attractive Rate of Return (%/100%).
S	shifts actually used (not necessarily integer).
S_B	shift value of Base system at which a cross-over point occurs.
S_{max}	number of shifts available (probably integer).
$\bar{t}$	average task cycle time (second/units).
v	book value (portion of the system cost at end of year H. A function of depreciation type).

$\bar{V}_H$ average operating/maintenance rate per resource ($/hr).
w number of workers in the system.
ε system efficiency (%/100%).
ρ ratio of total installed cost to hardware price. (includes engineering, system design, debug, etc.).
σ number of stations (resources) in the system.
T theoretical system cycle time (seconds/unit).
(T/ε) actual system cycle time (seconds/unit).

General Characteristics

There are two cost factors to be considered:

1. The write-off of capital expenditures which do not vary from year-to-year on an aggregate basis. This expense will be called fixed cost; it is characteristically shown in Figure 2a as a constant function of production quantity.
2. Variable costs include all items that can be attributed to the manufacturing system being investigated. For present purposes, these costs are lumped into two classes: loaded labor cost and operating/maintenance cost. Figure 2b shows the increasing characteristic as a function of production quantity.

While total fixed and variable cost behavior is interesting, it has been found much more useful to deal with unit costs. The characteristic for fixed cost/unit as a decreasing function of production quantity is shown in Figure 3a while the constant nature of variable cost/unit is shown in Figure 3b.

It is worthwhile noting that manufacturing systems with heavy capital requirements, such as fixed automation, will have unit cost behavior similar to that shown in Figure 3a. At the opposite extreme, a basically manual labor system (especially one which has had fixed costs written off) will have unit costs like that shown in Figure 3b. Most real systems will have both cost components; this is especially true of adaptable, programmable systems.

Variable Unit Cost

The costs which vary on an overall production basis but can be determined to be constant on a per-unit basis fall into two general categories:

1. Loaded labor cost - composed of wages/salary and benefits/overhead. The true cost of labor to the company.
2. Operating/maintenance cost - made up of numerous cost items such as space, power, lubricants, rework, scrap, excess inventory, insurance on equipment and downtime. All costs considered appropriate should be included.

The total labor cost is the sum of the loaded cost of all workers required in a system. For present purposes, this total cost is expressed as the number of workers multiplied by their average loaded labor rate. $(w*\bar{L}_H)$

Variable unit cost is the sum of labor and operating/maintenance costs and can be expressed as:

$$C_V = \frac{w * \bar{L}_H + O_H}{q_H} \quad \left(\frac{\$}{\text{unit}}\right) \qquad (1)$$

It should be possible to define the total system operating cost (O_H) for a presently used or well-defined proposed system. When performing initial system design however, it will be more useful to define the operating cost in terms of each resource which could be used.

$$O_H = \sigma * \bar{V}_H \tag{2}$$

At this time, only systems composed of one resource type can be modelled. More complex systems design methods are under development.

The hourly production rate (q_H) is a function of the efficiency (percentage of up-time) and the system cycle time. This relationship is:

$$q_H = \frac{3600}{(T/\varepsilon)} \quad \left(\frac{\text{unit}}{\text{hr}}\right) \tag{3}$$

System cycle time (T/ε) is known for an existing system but must be predicted for proposed systems. When timing studies are used, well-defined system designs will yield good values for (T/ε). For a totally new system, particularly of the adaptable, programmable type, the cycle time can be expressed as:

$$T = \left(\frac{n}{\sigma}\right) \bar{t} \quad \left(\frac{\text{sec}}{\text{unit}}\right) \tag{4}$$

Here the number of tasks with an average cycle time are distributed over the number of stations (one resource per station) in the system. To be sure, Equation (4) strictly applies to only a perfectly balanced system; it does however, provide a very useful first approximation to expected cycle time especially when general purpose resources (such as robots) are to be used.

<u>Fixed Unit Cost</u>

All manufacturing systems require a capital investment of some magnitude. For cost accounting purposes, this expense is charged to the product being created in a formal manner, often called annualized cost. When dividing such a cost by the annual production volume, the fixed unit cost becomes:

$$C_F = \frac{P_A}{Q_Y} \quad \left(\frac{\$}{\text{unit}}\right) \tag{5}$$

The annualized cost of a manufacturing system is composed of numerous elements. The cost of each resource, the cost of a tool and material handling device for each task, the annualized cost factor and an installed-to-hardware cost multiple can be combined as:

$$P_A = \rho \left[\sum_{i=1}^{\sigma} P_{R_i} + \sum_{j=1}^{n} P_{T_j} \right] f \quad \left(\frac{\$}{\text{yr}}\right) \tag{6}$$

Equation (6) defines the annualized cost of any system. For system creation purposes, it is desirable to use average resource cost and average tool/material handling cost as in the following:

$$P_A = \rho\ (\sigma * \overline{P}_R + n * \overline{P}_T)\ f \quad \left(\frac{\$}{yr}\right) \tag{7}$$

Most real manufacturing systems can be readily approximated by (7). Use of average costs allows considerable variation in the actual resources required in a system. It remains possible (and desirable in this case) to categorize systems in terms of the average cost.

The factor, ρ, which defines the actual creation and installation cost (to get the system performing useful work) to hardware cost ratio is an interesting variable. For known technology, ρ is approximately 1.5. When technology being developed is to be used in a system, this factor is higher, often considerably so. Data for programmable assembly systems is shown in Figure 4 based on the author's estimates at the time of its creation. The point here is that true system cost for new systems will be a multiple of the hardware cost.

The annual cost factor, f, comes from the world of accounting. It requires an estimate of the minimum attractive rate-of-return as well as specification of the investment horizon, the economic life and the depreciation method. These parameters are combined as:

$$f = \left[1 - \frac{v}{(1+r)^H}\right]\left[\frac{r\ (1+r)^H}{(1+r)^H - 1}\right] \tag{8}$$

Annual cost factor (also called the capital recovery factor) establishes the proportion of a capital investment which can be charged to a product on a yearly basis. It thus allows direct comparison of equipment cost to variable cost for a system or when choosing among competing systems.

Yearly production volume, Q_y, is a multiple of hourly production rate. The number of working days per year and the (not necessarily integer) number of shifts that will be required are used as follows:

$$Q_y = s\ (8D)\ q_H \quad \left(\frac{unit}{yr}\right) \tag{9}$$

Substitution of (3) into (9) results in:

$$Q_y = s\ (8D) \left[\frac{3600}{(T/\varepsilon)}\right] \left(\frac{unit}{yr}\right) \tag{10}$$

If (4) is substituted into (10), the yearly production quantity for a balanced system is found to be:

$$Q_y = s\ (8D) \left[\frac{3600}{\left(\frac{n}{\sigma}\right)\overline{t}/\varepsilon}\right] \left(\frac{unit}{yr}\right) \tag{11}$$

It should be noted that ($0 \leq s \leq s_{max}$) for any system. s_{max} is likely to be 1, 2, or 3 in practical applications.

An often asked question concerns the allocation of fixed costs when a product requires only part of a year to manufacture. If one substitutes (9) into (5), the result is:

$$C_F = \frac{P_A}{s\ (8D)\ q_H} \qquad (12)$$

Here P_A is based on a year of work consisting of D days since it is an annualized cost. In other words (P_A/D) is a constant; if one factor is reduced, the other is automatically also. The unit fixed cost is thus independent of the portion of a year actually required to manufacture the product.

Total Unit Cost

Total cost per unit is the sum of the fixed unit cost and the variable unit cost:

$$C = C_F + C_V \qquad (13)$$

Substitution of (1) and (12) into (13) results in:

$$C = \frac{1}{q_H}\left[\frac{P_A}{s\ (8D)} + w*\bar{L}_H + O_H\right] \qquad (14)$$

If (3) is now substituted into (14), the total unit cost relationship is seen to be:

$$C = \frac{(T/\varepsilon)}{3600}\left[\frac{P_A}{s\ (8D)} + w*\bar{L}_H + O_H\right]\left(\frac{\$}{\text{unit}}\right) \qquad (15)$$

While Equation (15) describes the unit cost for any system, it is most appropriate for presently used or well-defined proposed systems.

When designing new systems, more detail will be required. Substitution of (2) (4) and (7) into (15) results in:

$$C = \left[\frac{\left(\frac{n}{\sigma}\right)\bar{t}/\varepsilon}{3600}\right]\left[\frac{\rho\ \left(\sigma*\bar{P}_R + n*\bar{P}_T\right) f}{s\ (8D)} + w*\bar{L}_H + \sigma*\bar{V}_H\right]\left(\frac{\$}{\text{unit}}\right) \qquad (16)$$

The number of workers, w, which will be required in a system is a function of the complexity therein. For present purposes:

$$w = \text{Integer Value of} \left|\frac{m_s + \sigma - 1}{m_s}\right| \qquad (17)$$

The maximum number of stations that a worker can handle and the actual number of stations used thus determines the number of workers. Only integer workers are used in the present model; even though their time might not be fully utilized, it is fully charged. This appears to be a potential penalty for an automated system. Since it is only a small portion of the total unit cost, no significant deficiency occurs.

General Technology Data

The fundamental idea is to develop general technology data that will be useful in system design. Table I exhibits presently used information for assembly systems. This data is intended to be descriptive but not necessarily precise. Values for $\bar{P}_R$, $\bar{P}_T$, ε, $\bar{t}$, $\bar{L}_H$, $\bar{V}_H$, m_s and ρ to be used in (17) and (16) are presented. Table I is a reasonable starting point for unit cost decisions on assembly systems.

Typical Results

The simple unit cost Equation (15) has been programmed in BASIC for the NOVA and MINC mini-computers, the APPLE-II personal computer as well as on the TI-59 Programmable Calculator. The complex unit cost Equation (16) has been programmed in BASIC for the NOVA mini-computer.

Various assembly system combinations are displayed in a series of figures 5 through 9. Only a few of the resource types shown (those where the technology is applicable) can be properly compared for a particular manufacturing system. For this data set, the assumptions are:

1. 20 parts in the assembly.
2. 240 working days per year.
3. 5 year economic life.
4. 2 year investment horizon.
5. ACRS depreciation.
6. 35% minimum Attractive Rate of Return.
7. 0 End of economic life salvage value.

Using (8), the annual cost factor is calculated to be:

$$f = \left[1 - \frac{.63}{(1.35)^2}\right]\left[\frac{.35\ (1.35)^2}{(1.35)^2 - 1}\right] = 0.5074$$

which says that slightly more than half the installed cost must be charged in each of the two years.

Manual assembly system costs are shown in Figure 5 for labor only and in Figure 6 for total (fixed plus variable) cost. Some companies want to compare a proposed new system to labor cost only instead of comparing to total cost. When a manual system has been in use for some time, the fixed cost portion can be considered zero. However, the operating/maintenance costs will never be zero and should not be neglected.

Fixed automation assembly system unit costs are shown in Figure 7. Most of the cost for these systems is due to the capital investment. Since they are special purpose, cycle time is often very low which leads to very large production volume capacity.

Both the Manual and Fixed Automation systems can be unit cost analyzed using Equation (15) or (16). Programmable Assembly Systems require Equation (16) to establish unit cost. For these systems, some very interesting conditions arise. Figure 8 displays unit costs for various programmable devices (defined in Table I). The characteristic "saw tooth" shape results when a system of σ stations (resources) reaches its capacity and one more resource is added. The tasks are redistributed among the resources. There will be an instantaneous rise in unit cost which will again decay as production quantity is increased to capacity.

The P15 curve in Figure 8 does not exhibit the same behavior. By looking at the unit cost vs. production quantity for 1 resource and 2 resources (Figure 9), it is seen that the least cost condition requires an additional resource before current system capacity is exceeded. This is certainly a surprising result! Mathematically, it is a special case of the cross-over point.

Cross-Over Point

If a base system is denoted by subscript B, its unit cost is (see (16)):

$$C_B = \left(\frac{n\bar{t}_B}{3600\varepsilon_B}\right)\left[\frac{\rho_B\, f\left(\bar{P}_{R_B} + \frac{n}{\sigma_B}\bar{P}_{T_B}\right)}{s_B\ (8D)} + \left(\frac{w_B}{\sigma_B}\right)\bar{L}_H + \bar{V}_{H_B}\right] \qquad (18)$$

The corresponding alternative system (presumably faster & probably more expensive) is denoted by subscript A and has unit cost:

$$C_A = \left(\frac{n\bar{t}_A}{3600\varepsilon_A}\right)\left[\frac{\rho_A\, f\left(\bar{P}_{R_A} + \frac{n}{\sigma A}\bar{P}_{T_A}\right)}{s_A\ (8D)} + \left(\frac{w_A}{\sigma_A}\right)\bar{L}_H + \bar{V}_{H_A}\right] \qquad (19)$$

By equating (18) and (19), the cross-over point for the two systems is established. The result of considerable algebraic manipulation is:

$$s_B = \frac{f\left[\rho_A\left(\sigma_A\bar{P}_{R_A} + n\bar{P}_{T_A}\right) - \rho_B\left(\sigma_B\bar{P}_{R_B} + n\bar{P}_{T_B}\right)\right]}{(8D)\ \sigma_B\left[\left(\frac{w_B}{\sigma_B}\bar{L}_H + \bar{V}_{H_B}\right) - \left(\frac{\bar{t}_A/\varepsilon_A}{\bar{t}_B/\varepsilon_B}\right)\left(\frac{w_A}{\sigma_A}\bar{L}_H + \bar{V}_{H_A}\right)\right]} \qquad (20)$$

A special case for this general cross-over occurs when two systems of the same type (see Figure 9) are compared. In that case $\bar{P}_{R_A} = \bar{P}_{R_B}$, $\bar{P}_{T_A} = \bar{P}_{T_B}$, ρ_A, ρ_B, $V_{H_B} = V_{H_A}$, and $(\bar{t}_A/\varepsilon_A) = (\bar{t}_B/\varepsilon_B)$. Equation (20) can be significantly reduced to:

$$s_B = \frac{f\ \rho\ \bar{P}_R}{(8D)\ \bar{L}_H}\left(\frac{\sigma_A - \sigma_B}{w_B - \frac{\sigma_B}{\sigma_A}w_A}\right) \qquad (21)$$

where w_j is defined by (17).

When using either cross-over point calculation, (20) or (21), the result must be less than the maximum value of s (usually defined as 1, 2, or 3 - the number of shifts available). For the P15 curve in Figure 8 (or Figure 9), s_B is less than 3 when σ=1 or 2.

If one were to overlay the curves of Figure 8 and Figure 7, it would be seen that the P15 (σ=3) curve A intersects the F30 (σ=10) curve B as shown in Figure 10. Substituting the appropriate values from TABLE I into (20) results in:

$$s_B = 0.4897$$

from which the yearly production quantity (11) is found to be:

$$Q_Y = 902,575$$

Below that production volume, it is most economical to use the P15 system, above to use the F30 system.

Conclusion

Use of these techniques allows one to make rational type-of-system choices. For any production volume, the corresponding minimum unit cost to produce can be determined. The system which produces that minimum cost may be manual, fixed automation or programmable (flexible) automation; all have their appropriate situations. The methods shown here allow a rapid, sensible choice.

Bibliography

1. Lynch, P.M., Economic-Technological Modeling and Design Criteria for Programmable Assembly Machines, Ph.D. Thesis, MIT Mechanical Engineering Dept., June, 1976.

2. Boothroyd, G. "The Economics of Robot Assembly Applications," SME Paper AD77-720, presented at Autofact I, Detroit, MI., November, 1977.

3. Boothroyd, G., C. Poli, & L. Muech, Automatic Assembly, New York, Dekker, 1982.

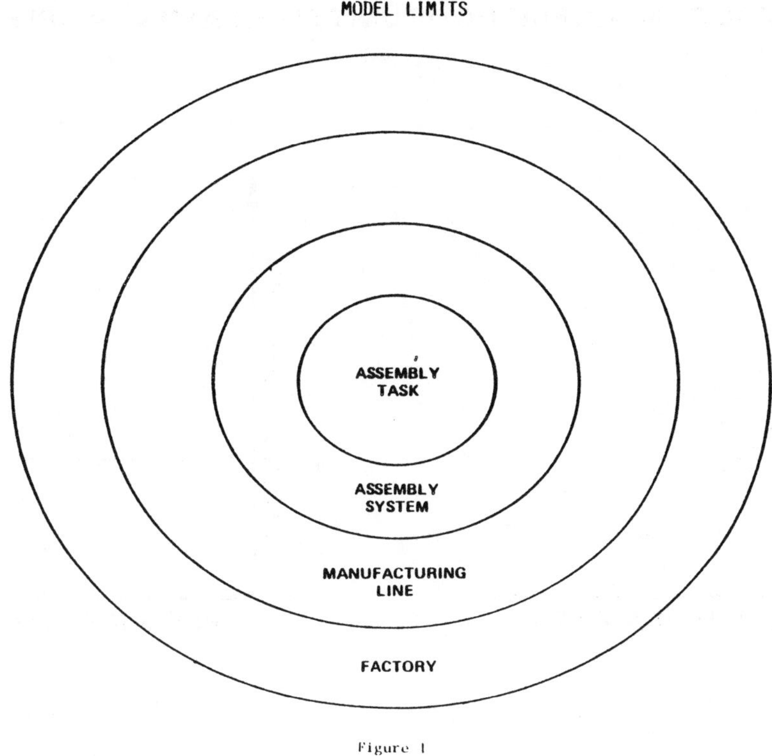

Figure 1

GENERAL CHARACTERISTIC OF TOTAL FIXED AND VARIABLE COSTS

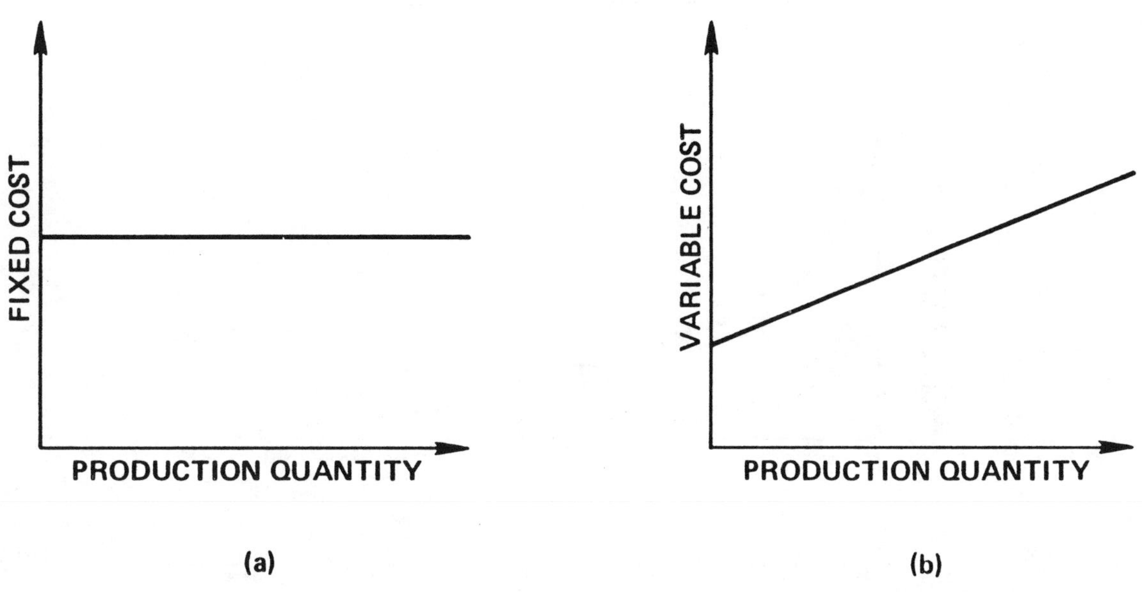

(a)

(b)

Figure 2

GENERAL CHARACTERISTIC OF UNIT FIXED AND VARIABLE COSTS

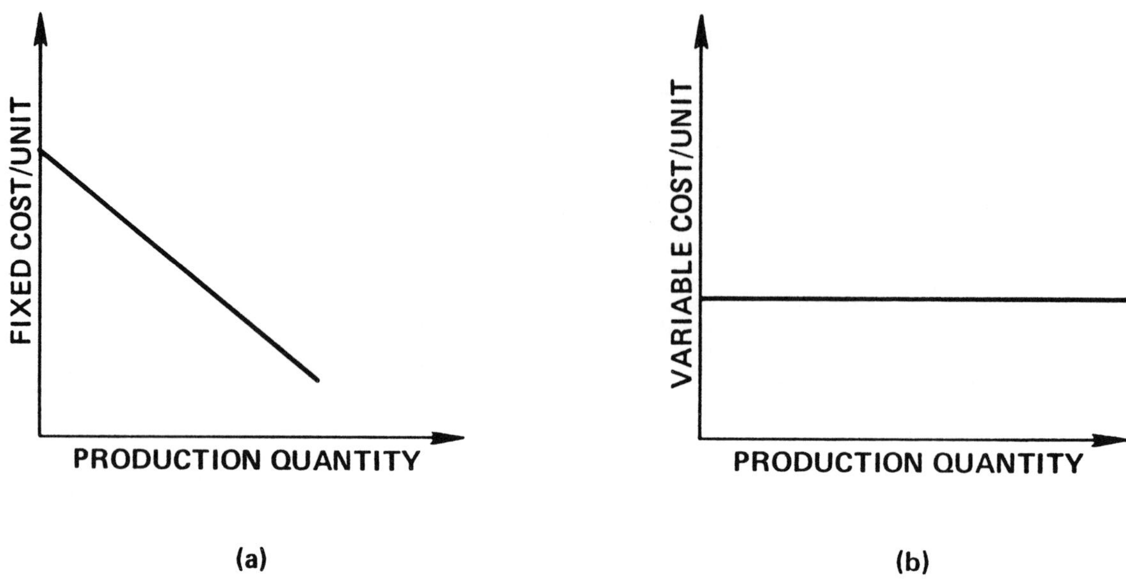

Figure 3

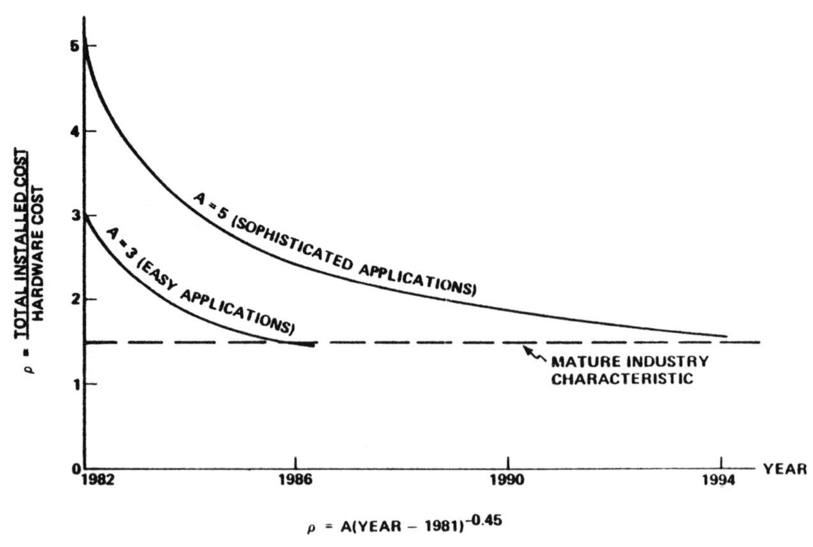

Figure 4

ASSEMBLY SYSTEM OPTIONS – GENERAL TECHNOLOGY DATA									
	Symbolic Name	Average Resource Price	Average Tool & Material Handling Price (per part)	System Efficiency	Average Task Cycle Time (seconds)	Average Loaded Labor Rate ($/hour)	Oper./Maint. Rate ($/hour) For each resource used	Maximum Number of Stations per Worker	Installed Cost ÷ Hardware Cost ρ
Manual	M05 M10 M15 M20	$ 200	$ 2,000	80%	5	5 10 15 20	0.5	0.833	1.0
Program-mable	P07 P15 P35 P45 P70	$ 7,500 $15,000 $35,000 $45,000 $70,000	$ 3,500 $ 5,000 $ 7,500 $ 8,500 $12,000	80%	2 2 3 4 7	12.5	1 2 3 3.5 5	8 7 6 5 4	2.0 2.5 3.5 4.0 5.0
Fixed Automa-tion	F30 F60 F90	$30,000 $60,000 $90,000	0	80%	1.5 5.0 10.0	12.5	1 2 3	6 4 2	1.5
		$\bar{P}_R$	$\bar{P}_T$	E	$\bar{t}$	$\bar{L}_H$	$\bar{V}_H$	M_S	ρ

TABLE I

MANUAL SYSTEMS – LABOR ONLY
20 PARTS

5-yr ECONOMIC LIFE
2-yr INVESTMENT HORIZON
35% RATE OF RETURN
240 WORKING days/yr
2 SHIFTS AVAILABLE
M20
M15
M10
M05
UNIT COST ($)
100
10
1.0
0.1
10,000
100,000
1,000,000
10,000,000
YEARLY PRODUCTION QUANTITY

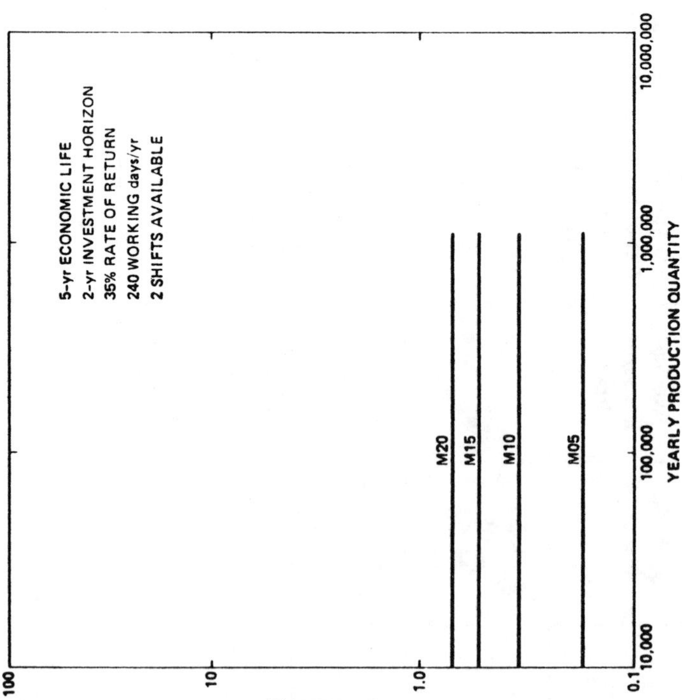

Figure 5

MANUAL SYSTEMS
20 PARTS

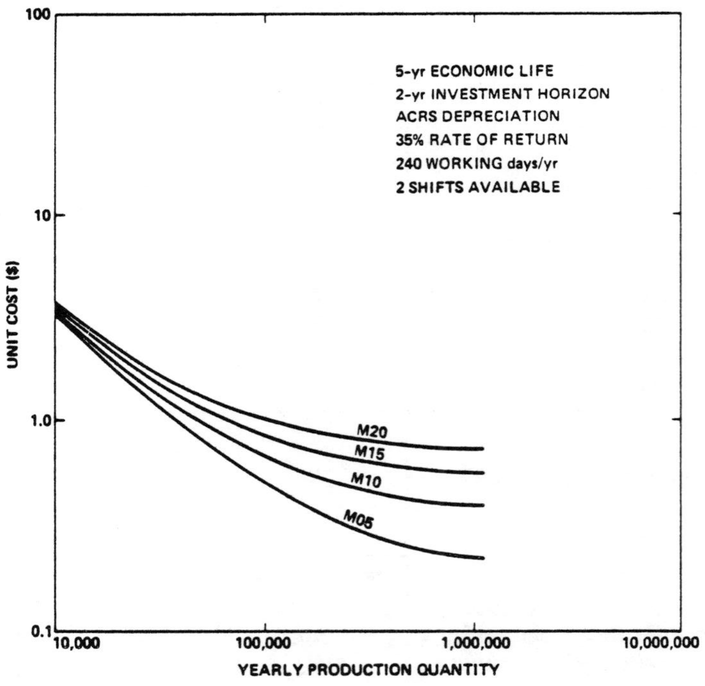

5-yr ECONOMIC LIFE
2-yr INVESTMENT HORIZON
ACRS DEPRECIATION
35% RATE OF RETURN
240 WORKING days/yr
2 SHIFTS AVAILABLE
M20
M15
M10
M05
UNIT COST ($)
100
10
1.0
0.1
10,000
100,000
1,000,000
10,000,000
YEARLY PRODUCTION QUANTITY

Figure 6

FIXED AUTOMATION SYSTEMS
20 PARTS

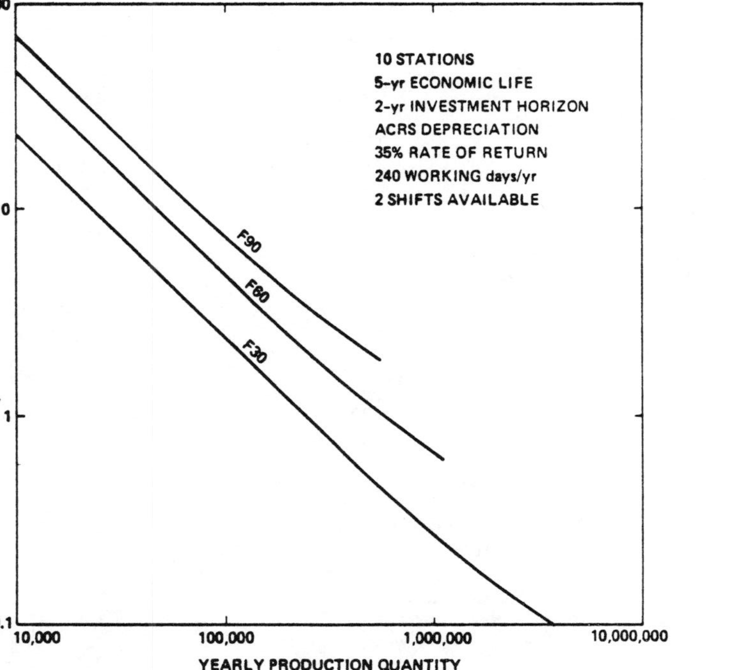

Figure 7

PROGRAMMABLE SYSTEMS
20 PARTS

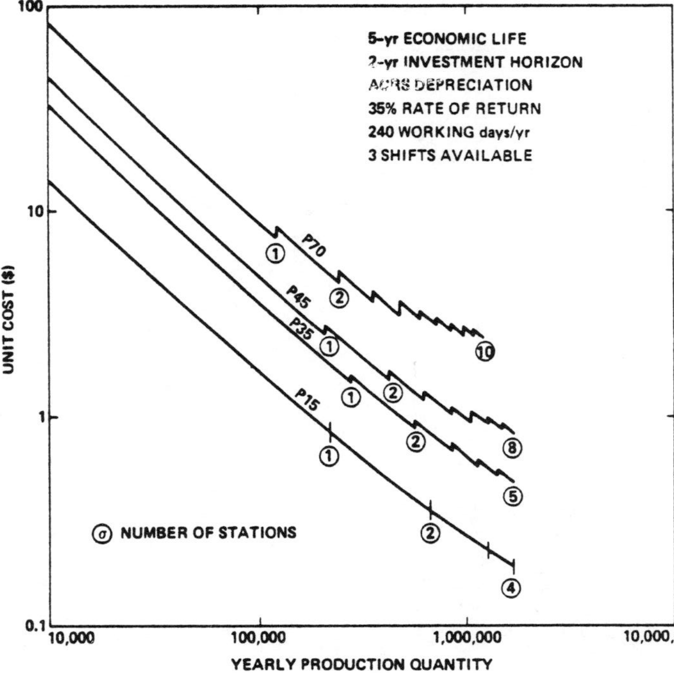

Figure 8

P15 SYSTEMS
20 PARTS

10
1.0
0.1
UNIT COST ($)
10,000
100,000
1,000,000
YEARLY PRODUCTION QUANTITY
5-yr ECONOMIC LIFE
2-yr INVESTMENT HORIZON
ACRS DEPRECIATION
35% RATE OF RETURN
240 WORKING days/yr
3 SHIFTS AVAILABLE
2 STATIONS
1 STATION
3 STATIONS

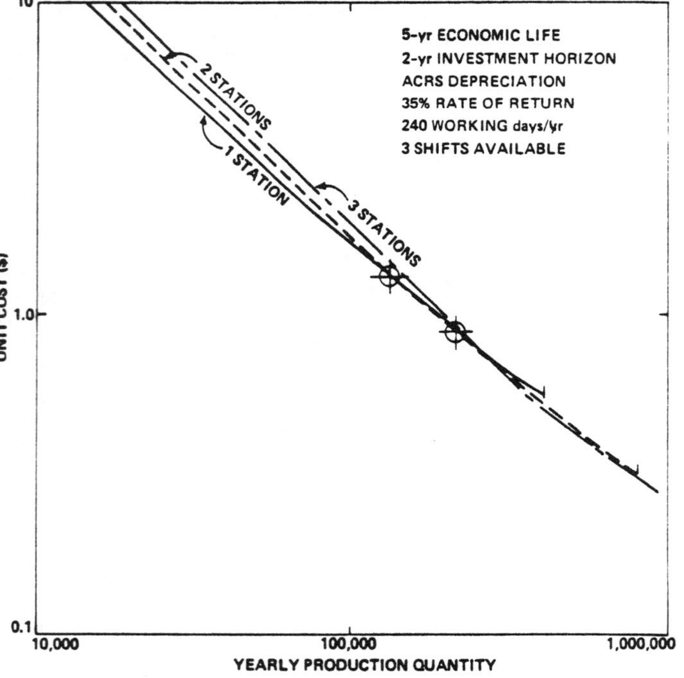

Figure 9

CROSSOVER POINT EXAMPLE
20 PARTS

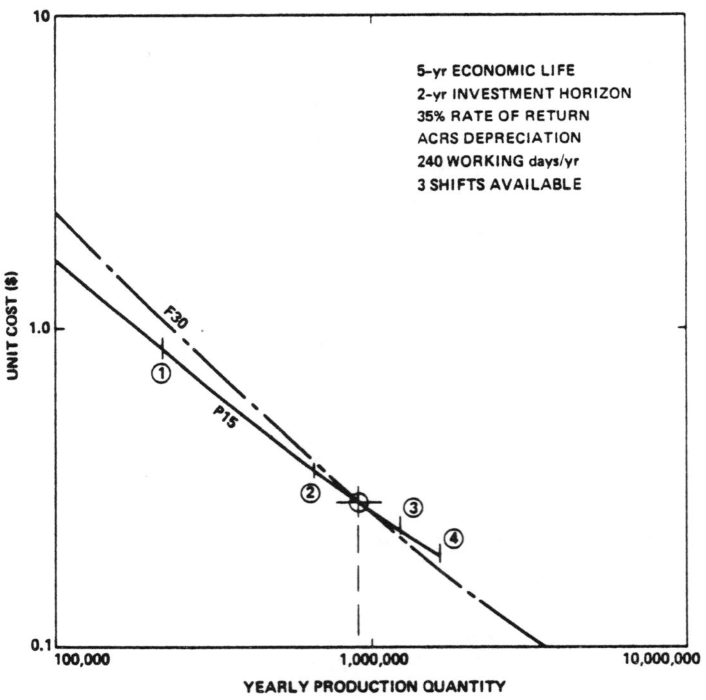

10
1.0
0.1
UNIT COST ($)
100,000
1,000,000
10,000,000
YEARLY PRODUCTION QUANTITY
5-yr ECONOMIC LIFE
2-yr INVESTMENT HORIZON
35% RATE OF RETURN
ACRS DEPRECIATION
240 WORKING days/yr
3 SHIFTS AVAILABLE
F30
P15
①
②
③
④

Figure 10

Presented at the SME 13th ISIR/Robots 7 Conference, April 1983

Robotic Assembly: Design, Analysis and Economic Evaluation

by Peter B. Scott
Tom. M. Husband
Imperial College

INTRODUCTION

Ever since the first tentative introduction of practical industrial robots over twenty years ago, there has been an ever increasing requirement for the construction of a robust yet sensitive methodology which could provide enlightenment on such areas as: the cost-effectiveness of introducing robots into a particular environment; the choice of the most suitable areas for such introductions; the design and selection of the best layouts, tools, and especially robots for a given application; as well as the suitable approach to follow to allow sufficient sensitivity to the changes which production engineers constantly face. It is salutary to consider that of all the areas of robotics work, perhaps among the most vital are those which might result in, as with the search for the methodology above, a direct increase in the numbers and acceptance of robots in industry. At the end of the day, if there are no substantial markets for a product, however worthwhile that product may be, there is unlikely to be much progress in its subsequent development.

Out of the recognition of this need has grown the current work on the Cost-Effectiveness of Assembly by Robot being carried out at the Centre for Robotics at Imperial College, London. The decision to concentrate for the moment on the assembly aspects of robotics arises from the knowledge that, in many countries' manufacturing industries, assembly accounts for more than 50% of the total manufacturing cost of a product and more than 40% of the labour force [1]. Nevertheless, despite the clear incentives for firms to automate such assembly, it is obvious that they should not embark upon installation of robots before first arriving at sensible assessments for the likely savings and possible consequences. It is also apparent that many firms do not enter into robotic assembly because they are unaware of, or unable to estimate, these savings. It therefore seems a particularly appropriate area for a university team to investigate, for few robot users are likely to make, or even be able to make, the necessary effort, and no single supplier of robots is likely to develop models which might show deficiencies in their product compared to others.

It must be stressed that the area of concern is very much wider than the strictly economic aspects of robotic assembly, for in choosing between manual, robotic or dedicated systems, it is vital to compare the optimum solutions for each. Thus, cost-effectiveness of assembly by robot must of necessity stretch right from the design of the original product to be assembled, through the choice of assembly sequence, the design of the flexible assembly system (FAS) including choice of feeders, conveyors, robot, and overall layout, right up to the eventual assessment of cycle times,

---

This paper covers the preliminary two-month investigation by Peter Scott into "The Cost-Effectiveness of Assembly by Robot" (funded by the British SERC), supervised by Professor T M Husband.

and the analysis of costs (where even the choice of an economic model for depreciation which satisfactorily takes into account the robot's inherent flexibility is far from obvious). Therefore, any methodology that succeeds in covering robotic assembly cost-effectiveness will, due to its very nature, be comprehensive and at many points fundamental to both assembly robotics in particular, and probably also many aspects of industrial robotics in general.

1 CURRENT SITUATION

Upon detailed investigation of industrial case studies, design proposals, and general robotic assembly research, a hierarchy tends to emerge of the various stages that must be passed through before a decision can be arrived at on the cost-effectiveness of robotising particular assembly tasks (Figure 1).

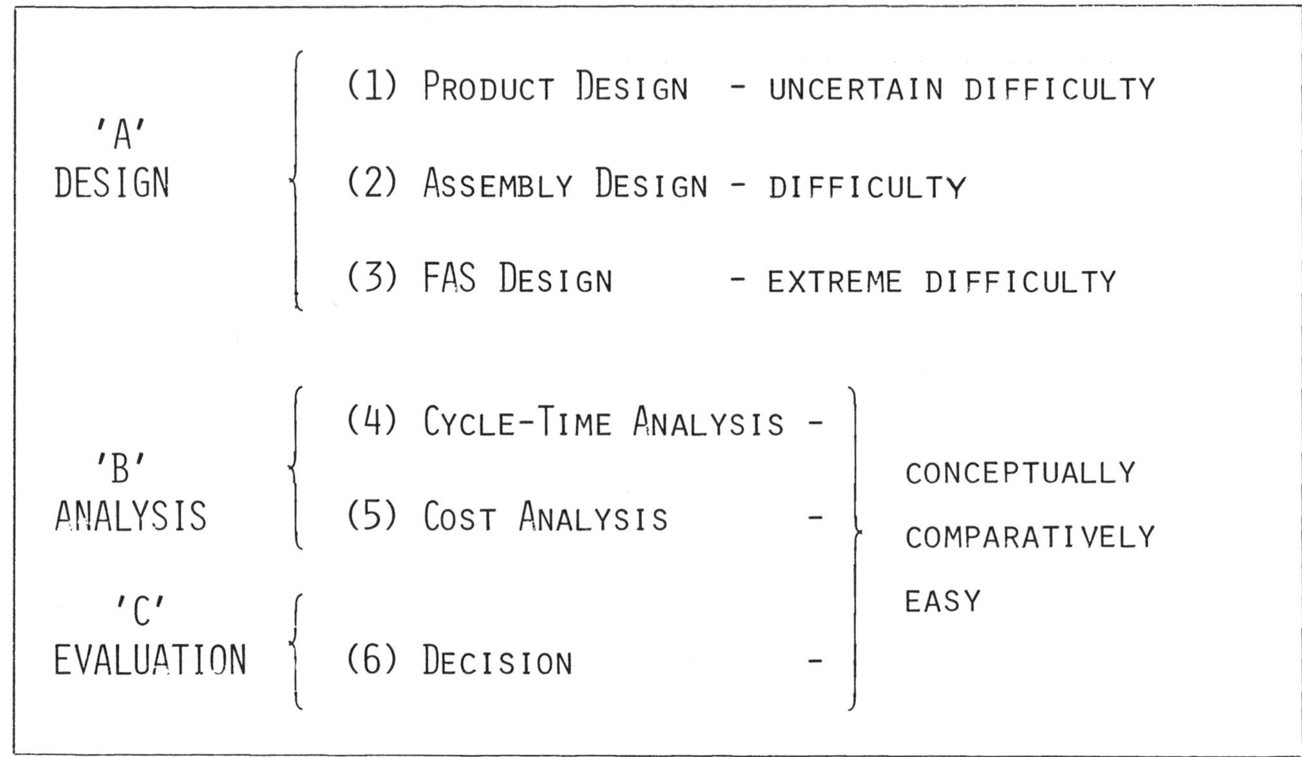

Figure 1: The "orthodox" hierarchy followed when considering a potential FAS application.

Although in specific cases details may vary from this basic sequence, on the whole it does appear to represent the "orthodox" approach, and it is indeed an obvious sequence to follow. In this section each stage will be looked at in turn to see what the current state-of -the-art is, and a note made of the "Conceptual Difficulty" of accomplishing it.

1.1 Design

There are three fundamental design stages in the orthodox

hierarchy. Firstly, starting from the desired purpose of the product the choice must be made of the product design most suitable for manual, robotic or dedicated assembly. Secondly, given the particular product, comes selection of the optimum assembly sequence (in this case for a robot). Finally, there is the design of the FAS itself, from layout to choice of feeders, conveyors and robot.

The product design stage is a comparatively new concept. In assembly, due to the extreme difficulty of designing equipment with even a small proportion of a human's inspection, selection and manipulative skills, there has traditionally been too heavy a reliance placed on the versatility of an assembly operator. Thus, unlike much other design work, in assembly a product is hardly ever designed to suit the capabilities of either operators or mechanisms and, indeed, where the assembly equipment is mechanical, it is usually referred to as a special-purpose machine, implying that it was conceived after the product design had been finalised [1]. Conceptually such 'design for assembly' is of uncertain difficulty, as the field (certainly as applied to robots) is so new. However complex it may be though, it seems necessarily to be the first stage of any hierarchy, but what impact the stage may have is itself not clear. It has been suggested that many supposed savings due to automatic assembly, may in reality have arisen not from the fact that the product was redesigned and subsequently assembled automatically, but simply because it was redesigned at all [2].

Deciding on an optimal assembly sequence (the second stage) may be complex, largely because it is not at all obvious when the optimum has been reached. Creative skill is needed on the part of a designer to come up with potential solutions which can then be compared (though it is not even clear by what criteria). Work at the Charles Stark Draper Laboratory (CSDL) has made use of a graphical representation 'Parts Tree Analysis' to assist in comparison of different sequences [3], but there are still no real guidelines for the initial design of the sequences, and as such the conceptual difficulty for this stage is high.

Deciding with any certainty on the optimum FAS for a given family of products is currently impossible even with the aid of computers. The FAS design process must somehow reflect the desired product-design flexibility, quality, and volume requirements in its choice of robot, feeders, conveyors, orientation systems, inspection devices, and overall final layout. On the one hand, there are so many possible configurations that any "brute force" attempt to compute them all is infeasible, while on the other, the results of attempts at an Operations Research (O.R.) type approach appear over-simplistic. Such an O.R. based design tool has been developed at CSDL as an Analytical Design Evaluation System (ADES) [4]. ADES can mathematically solve the static cost, annual volume, technology problem by synthesising and comparing assembly systems, given data on available work station capabilities, assembly task requirements, and cost and annual volume constraints. Unfortunately, although the results may be optimal economically (according to its simplistic criteria) they are frequently not technically even feasible. Due to the comparative enormous complexity of FAS design, even when using the most skilled designers (let alone a computer program), it is generally considered to be the single most conceptually difficult stage in the whole hierarchy.

1.2 Analysis and Evaluation

Having decided upon a particular design of FAS it is subsequently necessary to conduct some kind of cycle-time analysis (to obtain the production time for each assembly), followed by the resultant cost analysis, so leading to a decision on whether the projected robotic assembly would be cost-effective.

Time and Motion study of robots is still in its infancy, and various approaches have been suggested. Rogers [5] argues that when the dynamics and control algorithms of a robot are known, it is straightforward to calculate assembly times. He proposes three predictive models of cycle-times from a simple, approximate one to a detailed, accurate one, and has tested them using a Unimate 6000 robot to assemble a governor. Although his "Detailed" model displayed accurate results, the calculations involved are very tedious and require a detailed layout of the assembly station with the design of all tooling and other ancillary equipment plus a complete choreography of arm movements for the complete assembly cycle of each part. Despite even this complexity, Rogers' models cannot incorporate the effects of either sensory feedback (which may result in uncertain variance of assembly time), or the deviation of cycle-time caused by down-time due to the presence of faulty parts. It may well be that any great accuracy of prediction would be unattainable if they were included, so rendering his detailed model excessively complex, and suggesting that his simpler (less accurate) models may be of more practical benefit.

Both Rogers' work, and research at Purdue University [6], utilise an established methodology in human work analysis, Methods Time Measurement (MTM). Using MTM, an industrial task can be broken down to its elementary motions, and the time for each motion and total estimated time for the task can be derived [7]. Purdue have developed a methodology analogous to MTM termed Robot Time and Motion (RTM), in which they have defined new task motion elements for robots, because many MTM elements aggregate basic robot motions, while others are impossible for a robot to perform [8]. Given a task described in MTM it is possible to translate it directly into RTM, though many elements will translate into impossible tasks for a robot, indicating that the work method must be changed. Although the list of RTM elements will yield an estimate of the robot cycle-time, it is very unclear how an optimal conversion of a manual work method to a robotic one could be accomplished.

The confused area of cost analysis for robotics is in bad need of rationalisation. Robot manufacturers tend to offer Simple Payback and Production Rate Payback formulae or methods such as Return on Investment Evaluation [9], all of which ignore the cost of money; the calculation of the present value and therefore the timing of the flows of income and expenditure relative to the investment are not considered. This is unreasonable, given the long working life of robots, and their flexibility which allows redeployment when a product line changes. Benedetti [10] suggests that among the more sophisticated methods available, Present Value and Discounted Cash-Flow Rate of Return methods are best suited to the determination of the internal rate of return of robots, and also proposes a suitable analysis of the qualitative aspects and possible interdependencies of the variables which characterise their use.

Cost justification is further complicated by what a given company accountant will accept. Those used to dedicated machinery will expect early obsolescence and be unfamiliar with how best to satisfactorily take into account the robot's far greater flexibility. Likewise, straight line depreciation may sometimes be used, but the tax schemes of different countries may involve depreciation artificially weighted to early years, and special tax credits to encourage capital investment may also influence cost justification and the decision whether or not to buy.

Some basic work on the justification of robot systems in comparison to existing methods has been carried out by Lynch [11] who developed an economic model which suggests an economic region suitable for robotic assembly. The comparison in Figure 2 of unit assembly costs using manual, robotic, and dedicated methods, shows robotic assembly to be the most cost-effective choice in the medium volume region.

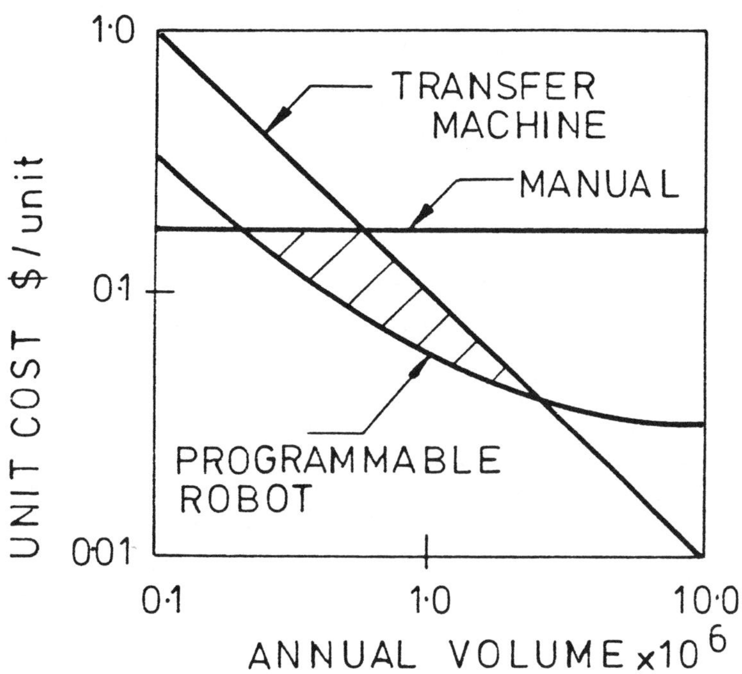

Figure 2: Regions of economic advantage for three assembly methods.

Extensions of this work by Boothroyd [12] show similar trends, largely dependent on various aspects of assembly such as levels of quality of parts, costs of tooling, and frequencies of design changes, suggesting that robot assembly systems are more sensitive to non-fixed costs associated with tooling and arm speed than to fixed costs such as those associated with arm hardware.

1.3 Problems

Due to the nature of the "Orthodox Hierarchy", it is not possible to reach an optimal solution for one stage before proceeding to the next, simply because it is only once all stages have been executed that their compound effects can be evaluated - and it is not even then possible to state how close to its potential optimum the actual system is. At best the hierarchy is used iteratively, constantly returning to its first stage and working down through it, hoping for an improvement, but of course, with no guarantee that the result will not in fact be worse. In practice, however, the early steps of the hierarchy are so unwieldy that there is every temptation to merely modify parts of the later stages without bothering to rework the whole procedure. Thus, for example, it might be decided late in the game to introduce chamfers for some minor reason onto all holes of a product, without realising that this then allowed the use of a far less repeatable robot fitted with a Remote Centre Compliance device, which in turn required different feeders, and so layout and so cost.

One major reason for the unwieldiness of the orthodox approach, and so its infeasibility, is the comparatively very large number of possible solutions to the early stages. The detailed design involved in the first three steps requires the least algorithmic yet most creative approaches of the whole hierarchy. Consequently, there is every possibility to choose the wrong track early on and so have no hope of an optimal system. If, somehow, these potentially disasterous stages could instead be preceded by the comparatively straightforward stages, then at least one would be closer to the desired goal when the "uncertain" decisions of the detailed designs were tackled.

In addition, as with the FAS-design stage alone, the prodigious number of potential solutions for the whole sequence of stages utterly prevents any "brute force" approach, even by computer, so rendering the whole procedure almost unusable. Yet the disadvantages of such an impractical hierarchy are more far reaching than just the initial evaluation of cost-effectiveness. Even assuming that a satisfactory design of FAS is reached, because it is so difficult to re-run through the whole methodology, it is far from sensitive to changes of design of product or robot. Instead, all that tends to occur is that attempts are made to "modify" the existing FAS if gross changes demand it. For a technology specifically designed for flexibility, this is a great shortcoming. The possibility of designing an FAS, truly suitable for a whole family of products, by frequent re-iteration of the hierarchy, is such a daunting task that, in practice, it is rarely attempted. Thus, overall, the "obvious" and "orthodox" approach to design, analysis, and economic justification of assembly by robot is very far from being satisfactory, or maybe even usable.

2 ALTERNATIVE APPROACH

2.1 Prediction driven hierarchy

As an alternative to the problem fraught hierarchy above, a different approach is currently being evaluated. In this new hierarchy, the global stages of:

Design
Analysis
Evaluation

become instead:

PREDICTION
Analysis
Evaluation
Directed Detailed Design

Thus, the Analysis and Evaluation is accomplished using PREDICTED data, and only after an encouraging evaluation of the products' suitability for robotic assembly is the detailed design of an FAS attempted - and even that is directed towards the calculated goal "ideal" system. Provided that the Prediction stage is relatively straightforward, and yet sufficiently accurate, then the whole process through to evaluation is easy enough to re-run to be very responsive to changes of design (unlike the previous hierarchy). Similarly, not only is little time wasted should an alternative form of assembly be suggested by the evaluation stage, but also, should the decision be made to go ahead, because a suitable framework for the FAS has already been decided, there is less chance of "taking the wrong path".

2.2 Working Model

Clearly, for accurate prediction of the framework for an "ideal" configuration of FAS, heuristics from a suitable working model are required. The fundamental economic model currently being employed is that the unit cost of assembly is equal to the product of the cycle time and the joint cost rates of the flexible and dedicated portions of the FAS:

$$C_u^R = t\ (F + D)$$

where C_u^R is the unit cost of assembly by robot
t is the number of seconds taken to assemble the product (cycle time)
F is the "cost per second of the flexible portions of the FAS"
D is the "cost per second of the dedicated portions of the FAS"

The distinction between D and F is basically the difference between that equipment which is specific to a particular product (and is not used except when assembling batches of that product), and that remaining equipment which is flexible enough to be used when assembling batches of any of the product runs for that FAS (largely the robotic equipment). It must be stressed that in practice although D might be easy enough to calculate (if the life of the dedicated equipment was relatively short), a value for F is far more difficult, as the cost of a robot should probably not be written off over less than at least five years.

This model, although straightforward, nevertheless successfully agrees with the predictions of more esoteric equations, such as Boothroyd's results for a robot assembly system, which suggested that halving the assembly time, by doubling the robot speed, had far more effect on unit cost than halving the price of the robot [12]. This is a direct conclusion from the working model, when it is realised that an FAS is always paced by the robot, not the parts-feeding equipment. This in turn ties in with work done at Salford [1] which showed that a special-purpose assembly machine could assemble at about one fifth of the cost of an assembly robot - not primarily due to the cost of the robots but to the low speed of operation which led to excessively large parts-feeding costs. A further prediction from the working model (remembering that "dedicated" equipment can only be used for assembling one type of product) is that as the numbers of different products to be assembled increases, so the amount of dedicated equipment for each product should be reduced, and the sophistication of the flexible equipment (common to all batches) should be increased to compensate; however, because it is always quicker (and so cheaper for large runs) to assemble using fixed-automation, there is always a balance to be maintained, dependent upon how frequent the design changes are to be (Figure 3).

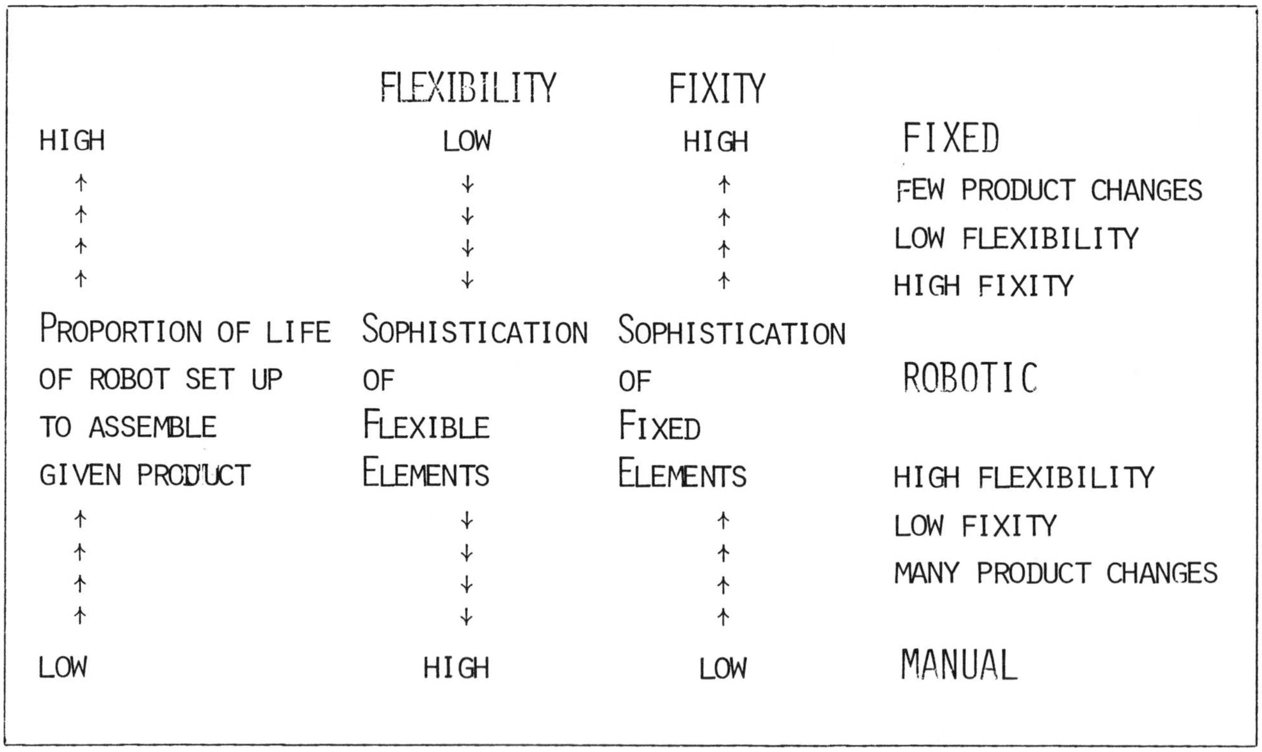

	FLEXIBILITY	FIXITY	
HIGH	LOW	HIGH	FIXED
↑	↓	↑	FEW PRODUCT CHANGES
↑	↓	↑	LOW FLEXIBILITY
↑	↓	↑	HIGH FIXITY
PROPORTION OF LIFE OF ROBOT SET UP TO ASSEMBLE GIVEN PRODUCT	SOPHISTICATION OF FLEXIBLE ELEMENTS	SOPHISTICATION OF FIXED ELEMENTS	ROBOTIC HIGH FLEXIBILITY
↑	↓	↑	LOW FIXITY
↑	↓	↑	MANY PRODUCT CHANGES
↑	↓	↑	
↑	↓	↑	
LOW	HIGH	LOW	MANUAL

Figure 3: The altering balance of the fixed and flexible elements of an "ideal" system depending on the frequency of design changes of product assembled.

The transition from Fixed, to Robotic, to Manual as the most cost effective methods of assembly agrees with Boothroyd and with Schraft, and, of course with common practice. The merge from Fixed to Flexible is a gradual one (much so-called Dedicated Assembly can these days assemble a variety of parts [13]). In contrast, the supplantation of robotic assembly by manual is quite sudden because although the required flexibility increases steadily, there is a marked point at which a human, although vastly more sophisticated than necessary, is nevertheless cheaper than would be a robot of the required extra sophistication (so conforming with Lynch's model in Figure 2).

2.3 Principle of Fixity

From consideration of the underlying workings of the above transitions, there has arisen a suggested "Principle of Fixity" which states:

> "For optimal robotic assembly, the proportion of the total sophistication of the FAS which is due to dedicated equipment is directly related to the total time spent throughout the life of the system assembling a given type of product."

Algebraically:

$$T \propto \frac{S_D}{S_F + S_D}$$

where T is the sum of all the periods the FAS was set up to assemble a given product.

S_D is the "sophistication" of all the equipment dedicated to the assembly of only one product.

S_F is the "sophistication" of all the remaining equipment, flexible enough to be used for all product assemblies.

It should be noted that it is the degree of "Fixity" not "Flexibility" which is paramount (though they are, of course, both closely related): it is the "fixed" automation which must be reduced if there are to be many product changes. Also, it should be recognised that the total sophistication of the FAS (i.e. $S_F + S_D$) must be kept to the minimum capable of performing the desired assembly.

Clearly, "sophistication" is such a nebulous concept that for the Principle of Fixity to be of any practical benefit some alternative substitution is necessary. Given the close similarities of the technologies and markets for both flexible and dedicated equipment, it seems valid to adopt the philosophy that "you get what you pay for" and so to approximate the ratio of "sophistications" by a ratio of costs. These costs would, of course, be the "list prices" of the equipment; the fact that in reality some form of discount might have been obtained on some of the equipment is irrelevant here. We are not interested in the true economics, only in the apparent "sophistications".

Given this substitution, the Principle of Fixity approximates to:

$$T \propto \frac{C_D}{C_F + C_D}$$

where C_D is the list price of the dedicated equipment
$C_F + C_D$ is the total list price of the system.

In this form, the Principle agrees with common sense: as T gets shorter (more frequent design changes) more of the money that would have been spent on equipment dedicated to assembling just one product should instead be channelled into buying flexible equipment (such as robots) which can be used for assembling all the products. For example, in designing an FAS destined for assembling products with four times the frequency of design change compared with an otherwise similar FAS (i.e. $\frac{1}{4}$ T), the new FAS should have only $\frac{1}{4}$ C_D spent on its fixed equipment, and 3/4 C_D channelled into buying robot power greater than the old FAS. Thus, the initial total system cost will be the same for both FAS's. However, although both systems will on average make obsolescent the same cost of dedicated equipment over a given timespan, the new FAS with the shorter T initially had more spent on its robotic systems. Thus overall, as T gets shorter, the total cost of assembly rises, until eventually it becomes cost-effective to assemble manually instead.

It is, of course, easy to derive from the Fixity Ratio the proportion of total cost due to flexible equipment (or Flexibility Ratio):

$$\frac{C_F}{C_F + C_D} = (1 - kT) \qquad (1)$$

where k is an empirically determined constant of proportionality.
In practice, C should always be chosen to be the least expensive robot capable of achieving the maximum sophistication required of it (i.e. for minimum T).
As, by definition,

$$t \equiv \frac{T\, n_s}{3N} \implies T = \frac{3t\, N}{n_s} \qquad (2)$$

where t is the average cycle time (in same units as T)
n_s is the number of shifts per day the FAS operates
N is the total number of a given product assembled by the FAS

then it is possible to rewrite the Flexibility Ratio in terms of cycle time, production volume, and shifts worked:

$$\frac{C_F}{C_F + C_D} = 1 - \frac{ctN}{n_s}$$

where c is a constant of proportionality,

It is this relationship that allows the Principle to be employed in the subsequently suggested Methodology.

Upon investigation of the Flexibility Ratio, it becomes clear that one of the consequences of adopting the Principle of Fixity is the implication that a given FAS is best suited for products all with approximately the same value for T. If an FAS, once built, has to adapt to a shorter T than originally designed for, the robot may be hard pressed to compensate for the necessary loss of dedicated equipment; if T is made longer, maintenance of the correct Flexibility Ratio will result in an over-sophisticated system. Thus, an FAS should be designed with a given value of T in mind for all products to be assembled on it. Clearly, within that value, the cycle times, batch sizes, and total production volumes for different products may vary considerably.

3 RESULTANT METHODOLOGY

Employing the approach detailed in the last section, it is possible to construct a tentative methodology designed to act as a replacement for the problematic "orthodox" hierarchy (Figure 4).

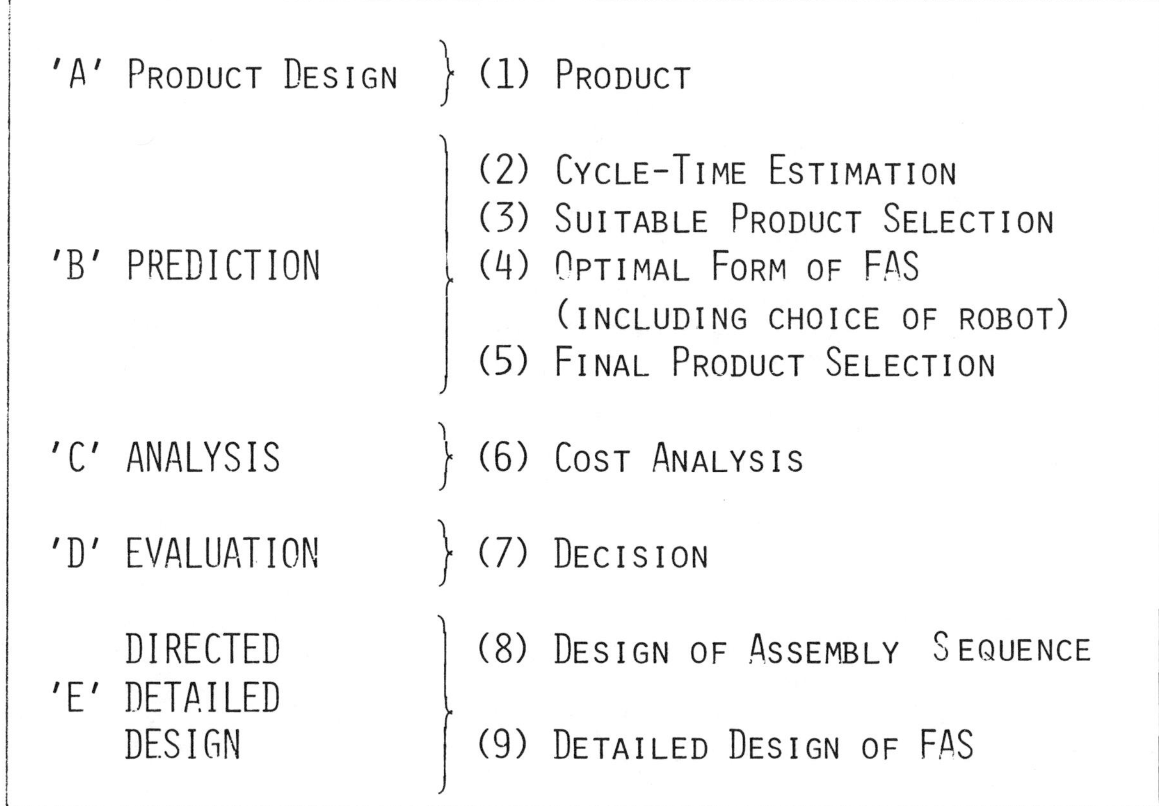

Figure 4: An alternative hierarchy to follow when considering a potential FAS application.

Such a construction highlights those research areas of particular relevance to this approach. In this section, an explanation will be given of the detailed sequence of actions (as far as they are known) that would be followed by management in developing an FAS according to the suggested Methodology.

3.1 Product Design

Of necessity, the initial design (or re-design) of products suitable for assembly by robot is the first stage. [Such "design for assembly" should also be conducted for the alternative Manual and Fixed methods of assembly (unless it is very clear that these alternatives are unsuitable), for the eventual comparison between assembly by each method should be made using optimal designs for each.] Although research in this area is still at an early stage, it is already possible to suggest some of the factors which can affect the ease with which a product can be assembled [2]:

(i) reduction of handling times of parts - handling represents a considerable portion of assembly costs, and may be affected by: size, weight, symmetry, roughness of surface, fragility and rigidity, sharp corners, tangling and nesting, and orientation;
(ii) reduction in the number of parts;
(iii) provision of a suitable base component;
(iv) assembly in layer fashion from above along a vertical axis;
(v) elimination of lengthy fastening techniques;
(vi) design for ease of insertion by the provision of chamfers.

3.2 Prediction

Knowing the numbers and forms of the parts to be assembled for each product, the next stages involve the prediction of a reasonable assembly cycle-time for each product, grouping together of products suitable for assembly by the same FAS, and prediction of the optimal broad form of that FAS (including selection of an appropriate robot) together with final product selection.

Prediction of a reasonable cycle-time (t) for a given product is at present largely a matter of experience. However, it seems likely that the solution to such a problem could be based purely upon a knowledge of the parts to be assembled without any reference to potential FAS designs, and as such would be susceptible to a largely algorithmic solution along the lines of:

$$t \propto \text{"complexity for a robot of the assembly"}$$

The "complexity" would be likely to take account of such factors as the numbers and types of each part, the frequency of each different method of part-mating, and measures of many of the factors covered in Section 3.1. The "reasonable" cycle-time so obtained should be the time the assembly would take on an FAS ideally balanced between high speed and low system-cost to achieve the lowest unit-assembly-cost.

At this point, it is necessary to decide upon the desired total production volumes (throughout the life of the FAS) for each product being evaluated (N) and also to make a decision on the number of shifts per day for which the FAS is to operate (n_s). Using the equality $T = 3tN/n_s$ (2), the various products should be grouped according to T. Only products all with a similar required value for T are suitable for assembly by the same FAS. If N (or even n_s) is not too inflexible, it may be possible to group products that otherwise would not be compatible.

Having now decided on a value of T for the required FAS, it is possible to substitute it into the Flexibility Ratio (1) which relates T to the ratio of cost of flexible equipment to cost of the whole FAS. The required constant of proportionality may well vary for different types of assembly: assembly of PCB's may need a different constant to assembly of pumps; assembly in layers a different constant to assembly from many angles. The constant corresponding to the most appropriate assembly description in a published list of empirically determined values would therefore be chosen. By the end of this stage, the broad balance between flexible and dedicated equipment in the FAS is known.

Choice of the most suitable robot for this particular FAS involves looking for the least sophisticated robot (or robots) capable of coping with all the various assembly tasks, given the proportion of its cost that can be spent on dedicated equipment to help it. It is envisaged that an estimation of the "sophistication" of robot required for the various assemblies could be calculated from a suitable formula based upon similar lines to that for estimating cycle-times above, though it would, of course, have to include the flexibility ratio. In contrast to the cycle-time estimation, this formula would be more concerned with such considerations as the required number of degrees of freedom, the necessary sensory feedbacks, the types and sizes of movements needed, and the possibilities for multiple arms. After estimating the flexible-sophistication requirements for each product, additional selection may be required, as only products requiring similar robotic power to assemble them should be included in the same FAS. It is unlikely that assembly of those products for which theFAS would otherwise be oversophisticated, could ever be cost-effective. After this final sifting, a suitable robot/s could be selected from a published "league table" of robot sophistications.

3.3 Analysis and Evaluation

Having selected the appropriate robotic components for the FAS, and therefore knowing the actual costs of the flexible and so (from the Flexibility Ratio) the dedicated portions of the system, it should now be possible to insert these costs into whatever the company accountants consider to be the most suitable unit-cost formula. The form of such a unit-cost analysis will, as mentioned in Section 1.2, very much depend upon the methods of depreciation considered most suitable for the robot (with a life of 10+? years) and for the dedicated equipment (which may rapidly become obsolescent). Whatever its form, it would seem unreasonable if it did not take into account rates of interest and of inflation, tax schemes and incentives, and the flexibility of the robotic components over their long working lives.

Once the unit cost-of-assembly by robot is known for each product, this can be compared with the equivalent costs of manual and fixed-automation assembly and an evaluation made for each product on whether it really is cost-effective to assemble it by FAS. In certain cases, it may even become apparent that because so few of the given products are suitable for robotic assembly, it is overall more cost-effective not to build an FAS at all - even though, as a result, certain products will cost more to assemble (using manual or fixed methods) than they would with an FAS. Similarly, considerations of required product quality may also influence the choice between manual, flexible or fixed assembly techniques. Finally, the decision on whether to build an FAS may involve factors which, on the surface, do not appear purely commercial. For example, worker-dissatisfaction caused by splitting up the manual workforce might, in the medium term, prove counter-productive. Such considerations can only be adequately evaluated by those with an intimate knowledge of the particular environment for which the FAS would be destined.

3.4 Directed detailed design

It is only after a decision has been made to definitely go ahead with installing an FAS that any thought has to be given to the design of the optimal assembly sequence and to the detailed design of the FAS layout itself. Even then, the designer is not working completely in the dark - he is aiming for a goal total cycle-time for a given robot and with a given broad functional breakdown of the FAS layout. This is very much a "Top Down/Bottom Up" approach, and, as such, is much preferable to a purely "Bottom Up" design.

An approach to this "directed" detailed design that at present appears potentially rewarding is the use of Metalanguages similar to MTM and RTM mentioned in Section 1.2. Such an approach would involve the description by one metalanguage of a manual assembly task (very easily conceptualised by a human even when the task has never actually been performed), the translation of that metalanguage to a second metalanguage which describes the task as performed by a robot (not so easily conceptualised by a human), and finally the reduction of the second metalanguage into an actual program to run on the FAS (Figure 5).

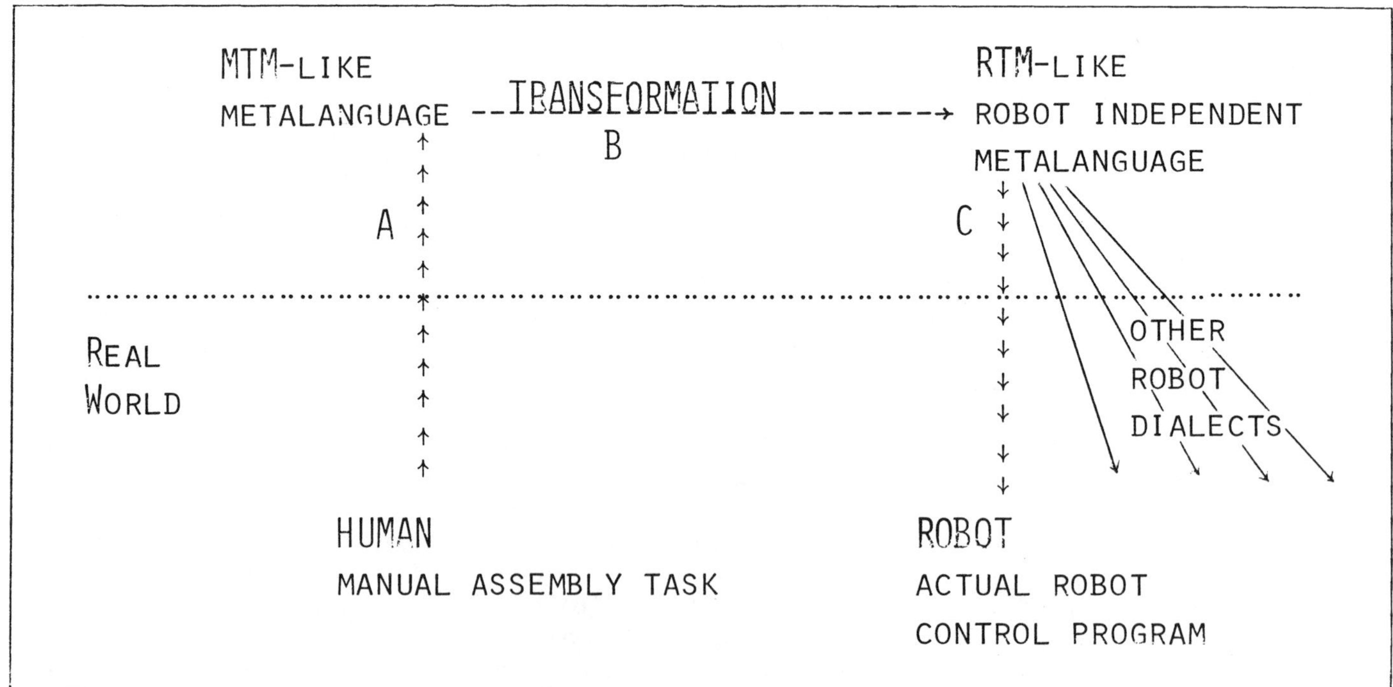

Figure 5: A diagrammatic representation of metalanguage transformation.

The description of human tasks by dialects of MTM has been common since the late '40's [7] and so step 'A' in Figure 5 does not really hold any problems. Similarly, although nowhere near so perfected, the production of useable robot programs from a robot task description in RTM, step 'C', is highly feasible. Already, Purdue has developed an RTM Analyser to compute performance measures from an RTM work-method specification [14], and at a later stage, it is expected to interface the RTM specification directly with a robot control program generator [15]. Thus, the real challenge now lies in the development of "mechanical translation" between metalanguages such as MTM and RTM, step 'B', which in essence is very similar to the application of program transformation to computer program synthesis [16,17]. Clearly, such a mapping of one metalanguage onto another cannot be accomplished at a syntactic, word to word level (many actions within a human's repertoire are quite impossible for a robot) but must be conducted more at a semantic level. It may be possible to draw up some powerful heuristics for this process; alternatively, restrictions on the vocabulary and range of MTM-expressable tasks may indeed allow a far lower level translation to be sufficient.

CONCLUSION

There are clearly very many areas in the proposed methodology requiring a significant research effort to test and develop them. It is this work which is programmed for the remainder of the three year research project on robotic assembly cost-effectiveness now running at the Imperial College Centre for Robotics, and it will, of necessity, make heavy use of both in-house robot systems and of strong links with industry. Some of the principal areas listed for detailed investigation include:

- testing the validity of the Principle of Fixity
- searching for empirically determined constants of proportionality for the Flexibility Ratio equation
- developing heuristics for product design suitable for robotic assembly
- development of a formula for estimating optimal cycle-times
- evolution of a formula for the robotic sophistication required for a given assembly
- evaluation of existing robots with respect to the last formula
- investigation into suitable unit-cost formulae for FAS analysis
- examination of suitable metalanguages for tasks representation
- investigation of possible program transformations between metalanguages.

It is hoped that even if only some of these goals are fully realised, the results will nevertheless both offer substantial assistance to those who want to use robots, and so, through increased demand, help those who want to make them as well.

BIBLIOGRAPHY

[1] "Handbook of Design for Manual and Automatic Assembly" University of Massachusetts and University of Salford, 1980.

[2] G.Boothroyd, "Design for Economic Manufacture", Computer Aided Manufacturing - Reference Book II. MIT, 1978.

[3] (see for example) J.L. Nevins, D.E. Whitney and S.C. Graves, "Programmable Assembly System Research and its Applications - a Status Report", CSDL, 1980.

[4] S.C. Graves and D.E. Whitney, "A Mathematical Programming Procedure for Equipment Selection and System Evaluation in Programmable Assembly", Proceedings IEEE Decision and Control Conference, 1979.

[5] P.F.Rogers, "A Time and Motion Method for Industrial Robots" The Industrial Robot, December, 1978.

[6] (see for example) S.Y. Nof, J.L. Knight, G. Salvendy, "Effective Utilisation of Industrial Robots - A Job and Skills Analysis Approach", AIIE Trans, Vol 12, No 3, 1980.

[7] H.B. Maynard, G.J. Stegemerten and J.L. Schwab "Methods-Time Measurement", McGraw-Hill, 1948.

[8] R.L. Paul and S.Y. Nof, "Human and Robot Task Performance", Symposium on Computer Vision and Sensor-Based Robots, GM Research Labs, 1979.

[9] J.F. Engelberger, "Robotics in Practice - Management and Applications of Industrial Robots", Kogan Page, 1980.

[10] M. Benedetti, "The Economics of Robots in Industrial Applications", The Industrial Robot, September, 1977.

[11] P.M. Lynch, "Economic-Technological Modelling and Design Criteria for Programmable Assembly Machines", PhD Thesis, MIT Mech Eng Dept, June 1976.

[12] G. Boothroyd and C. Ho, "Performance and Economics of Programmable Assembly Systems", SME Paper No AD77-720, presented at Autofact 1, November 1977.

[13] "Grouping Parts into Families for Automatic Assembly", p. 22, Assembly Automation, February 1982.

[14] S.Y. Nof and H. Lechtman, "Robot Time and Motion System Provides Means of Evaluating Alternate Robot Work Methods", Industrial Engineering, April 1982.

[15] S.Y. Nof and H. Lechtman, "Now It's Time for Rate-Fixing for Robots", The Industrial Robot, June 1982.

[16] R. Balzer, N. Goodman and D. Wile, "On the Transformational Implementation Approach to Programming", USC/151 Marina del Rey, Los Angeles, 1975.

[17] J. Darlington, "Application of Program Transformation to Program Synthesis", Proceedings of the International Symposium on Proving and Improving Programs, 1975.

Reprinted from the *Journal of Manufacturing Systems*, Volume 1, Number 1

Economics of Assembly Systems

G. Boothroyd, Dept. of Mechanical Engineering
University of Massachusetts, Amherst, Massachusetts

Abstract

This paper presents a description of typical assembly systems including manual, special-purpose automatic and programmable assembly systems. For each case, mathematical models are developed to describe economic performance. Comparisons of the economics of the various systems are also made and the future role for programmable assembly robots discussed.

Keywords: *Assembly, Robots, Assembly Economics, Automatic Assembly, Assembly Costs.*

In this paper, the economics of common assembly systems used in industry will be considered. In each case the effects of the following variables will be analyzed:

1. Parts Quality (*PQ*).
2. Number of Parts in the Assembly (*NA*).
3. Annual Production Volume per Shift (*VS*).
4. Product Style Variations (*SV*).
5. Design Changes (*ND*).
6. Number of Products to be Assembled (*NP*).
7. Economic Climate (*RI*).

Parts Quality (*PQ*)—A major problem in applying automation to assembly processes is the loss in production resulting from stoppages of automatic workheads when defective component parts are fed to the machine. With manual workstations on an assembly line, the operators are able to discard defective parts quickly and little loss of production occurs. However, a defective part fed to an automatic workhead can, on an indexing machine, cause a stoppage of the whole machine and production will cease until the fault is cleared. The resulting downtime can be very high with assembly machines having several automatic workheads. This can result in a serious loss in production and a consequent increase in the cost of assembly. The quality of the parts to be used in automatic assembly must, therefore, be considered when an assembly system is being analyzed.

In the following analyses, the effects of the quality levels of parts on the performance and economics of assembly systems are included. These quality levels will be represented by a factor *PQ* which is the average ratio of defective to acceptable parts. A typical value for *PQ* would be 0.01 for mechanical parts. This would mean that roughly 1% of parts supplied to an assembly system would be defective.

Number of Parts in the Assembly (*NA*)—A major factor in the choice of an assembly system is the number of parts (*NA*) in the assembly. It will be shown, for example, that for assemblies containing more than about 10 parts, special-purpose automatic machines using indexing transfer mechanisms become increasingly uneconomic as the number of assembly stations increases.

Annual Production Volume per Shift (*VS*)—Clearly one of the most important factors affecting the choice of an assembly system is the required annual production volume. It has been found that the most meaningful way to express this factor is by the annual production volume per shift (*VS*). When the annual production volume per shift is very small, only manual assembly can be considered; when it is very large, special-purpose automation is worthwhile. Somewhere between these extremes, programmable assembly might be economical.

Product Style Variations (*SV*)—Often assembly systems must be designed to assemble a variety of styles of the same product. To allow for this situation in the present work, it will be assumed that different product styles can be obtained by assembling alternative parts at the various stations on the machine or assembly line. This arrangement can be suitably described by a factor SV which is the ratio of the total number of parts available (NT) to the number NA actually used in the product. Then, on a special-purpose indexing assembly machine for example, the number of workheads and feeders is equal to NT the total number of parts available. On a programmable system, the workhead can make the necessary selection of parts so that the number of stations is equal to NA, the number of parts used; but the number of part feeders or magazines in equal to NT, the number of parts available.

Design Changes (ND)—It is generally considered that automatic assembly machines would only be economical for products whose design is not likely to change for several years. Of course this limitation should not exist with programmable assembly systems using robots. To allow for this, a factor ND is used which is the number of parts which are changed (necessitating new feeders and workheads, etc.) during the life of the machine.

Number of Products to be Assembled (NP)—Clearly, special-purpose assembly equipment is likely to be uneconomical if more than one product is to be assembled. To consider this effect, a factor NP—the number of different but similar products—is employed. For the same system (even a robot system) to assemble various products, the products must be of approximately the same size, of the same basic type and with a similar number of parts. This means that in a system using a transfer device, the transfer device can be re-used but the workheads, feeders, grippers, etc., must be replaced unless these are programmable.

Economic Climate (RI)—In analyses of the economics of automation equipment, it is advantageous to use a factor to convert the capital cost of the equipment to an equivalent operator rate. For this purpose a factor QE will be defined as the cost of the capital equipment that can economically be used to do the work of one operator on one shift. This figure, which must be determined for a particular company, is the basis for all the economic comparisons in the present paper. For example, if a simple piece of automation equipment were being considered and it can do the job of one operator, then the economic cost of the equipment would be QE for one shift working, $2*QE$ for two shifts working, etc. The annual rate for a piece of equipment initially costing CE is then given by $CE*WA/(SH*QE)$ where WA is the annual cost of employing one assembly operator and SH is the number of shifts. In fact, many companies will determine QE by multiplying WA by a factor which usually lies between 1 and 3. In other words, the ratio QE/WA is a constant for a particular company and is not likely to vary with wage rates. For the purpose of analysis, the economic climate for investment in automation can be represented by the ratio RI which is given by:

$$RI = SH * QE / WA \qquad (1)$$

When this ratio is low, investment in automation is less likely to be profitable.

In addition to the seven variable factors discussed above, it should be remembered that manual assembly systems will be necessary if the fitting or adaption of parts is required or if there are wide fluctuations in demand for the product.

Basic Cost Equations

The cost of assembly CA for a complete assembly is given by:

$$CA = TP * (WT + CE * WA / (SH * QE)) \qquad (2)$$

where:

- TP = the average time between delivery of complete assemblies for a fully utilized system.
- WT = the total rate for the machine operators.
- CE = the total capital cost for all equipment including engineering setup and debugging cost.

For the purposes of comparing the economics of assembly systems, the cost of assembly per part will be used and will be nondimensionalized by dividing this cost per part by the rate for one assembly operator, WA and the average manual assembly time per part TA. Thus, the dimensionless assembly cost per part CD is given by:

$$CD = CA / (NA * WA * TA) \qquad (3)$$

Substitution of Eq. (2) into Eq. (3) gives:

$$CD = (TP / TA) * (WR + (CE / NA) / (SH * QE)) \qquad (4)$$

where:

$$WR = WT / (WA * NA) \qquad (5)$$

and is the ratio of the cost of all operators compared with the cost of one manual assembly operator and expressed per part in the assembly. Thus, the dimensionless assembly cost per part for an assembly operator working without any equipment will be unity, which forms a useful basis for comparison purposes.

Finally, it should be pointed out that for a particular assembly machine, Eq. (4) only holds true if the required average production time (TQ) for one assembly is greater than or equal to the minimum production time (TP) obtainable for the machine. In other words, if $TP <= TQ$, then TQ must be substituted for TP in Eq. (4) because the machine is not fully utilized. If $TP > TQ$, then more than one machine will be needed to meet the required production. Now, for example, two machines producing X assemblies per hour will give the same assembly costs per assembly as one machine producing $X/2$ assemblies per hour. Hence, it will be assumed that as the average required production time for one asembly (TQ) is reduced below the production time (TP) obtainable from one machine, then the assembly costs become constant and are given by Eq. (4). To obtain TQ, it is necessary to know the required annual production volume per shift (VS), and the plant efficiency (PE); the latter being defined as the time actually worked in the plant divided by the time available. Thus:

$$TQ = 0.072 * PE / VS \qquad (6)$$

In this formula, TQ is given in seconds, VS is in millions, PE is expressed as a percentage and the factor 0.072 arises from the assumption that 7.2 Ms is the maximum time available in one shift year.

These are the basic cost equations necessary to compare the economics of assembly systems.

Assembly Systems

Figure 1 shows the typical assembly systems which will be studied in this paper. It can be seen that all systems are categorized as either manual or automatic assembly systems. However, it should be realized that various degrees of automation can be applied to manual assembly systems and that some degree of manual assembly is usually incorporated into so-called automatic assembly machines.

Also, the types of assemblies considered here are basically limited to those where the various parts are individually manufactured and must be handled and assembled individually. For example, this study does not include the production of a plastic molding having several metal inserts. It would, however, include the assembly of the molding (treated as a separate part) into a larger subassembly.

Each basic assembly system will now be described and analyzed in terms of the variables enumerated above. The results of the analyses will finally be presented in the form of a two-digit classification system which allows the most economic assembly system to be determined quickly for any given product.

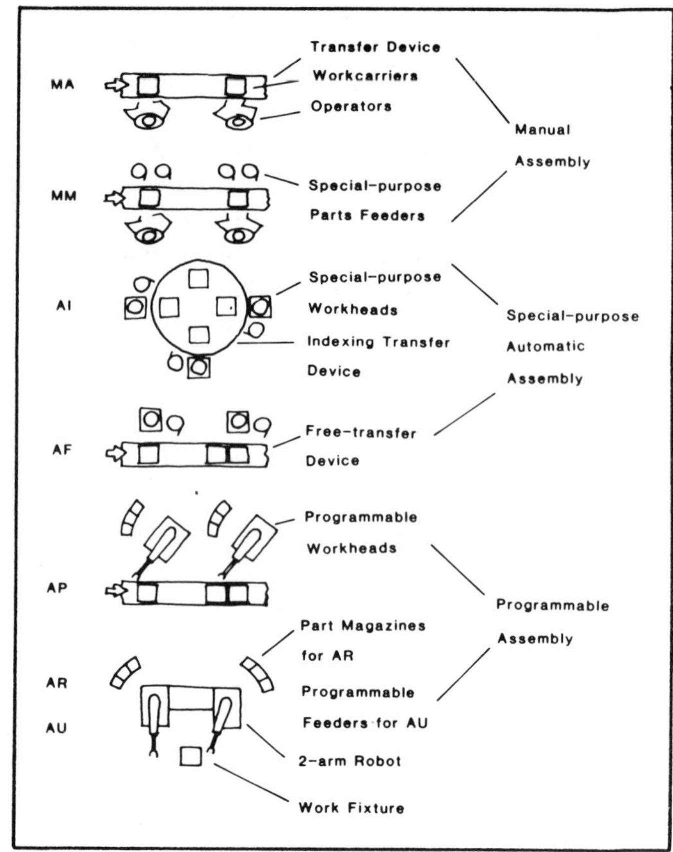

Figure 1
Typical Assembly Systems

Automatic Assembly Machines (*AI* and *AF*)— In assembly systems using special-purpose machinery, the various individual assembly operations are generally carried out at separate workstations. For this type of assembly, the partly completed assemblies are transferred automatically from work-station to workstation, and a means is provided of ensuring that no relative motion exists between the assembly and the workhead while the operation is being carried out. As the assembly passes from station to station, it is necessary that it be maintained in the required attitude. For this purpose, the assembly is usually built up on a base or "work carrier" and the machine is designed to transfer the work carrier from station to station.

Assembly machines are usually classified according to the system adopted for transferring the work carriers. Thus an "in-line" assembly machine is one where the work carriers are transferred along a straight slideway and a "rotary" machine is one where the work carriers move in a circular path. Also, transfer all the work carriers are known as "indexing" those assembly machines which simultaneously machines and on these machines a stoppage of any individual workhead causes the whole machine to stop.

In the other group of machines, which are known as "free-transfer" assembly machines, the workheads are separated by buffer stocks of assemblies, and transfer to and from these buffer stocks occurs when the particular workhead has completed its cycle of operations. Thus, with a free-transfer machine, a fault or stoppage of a workhead will not necessarily prevent another workhead from operating because a limited supply of assemblies will usually be available in the adjacent buffer stocks.

With the rotary indexing machine *(Figure 2)*, indexing of the rotary table brings the work carriers under the various workheads in turn and assembly of the product is completed during one revolution of the table. Thus, at the appropriate station, a completed product may be taken from the machine after each index.

The in-line indexing machine *(Figure 3)* works on a similar principle but in this case a completed

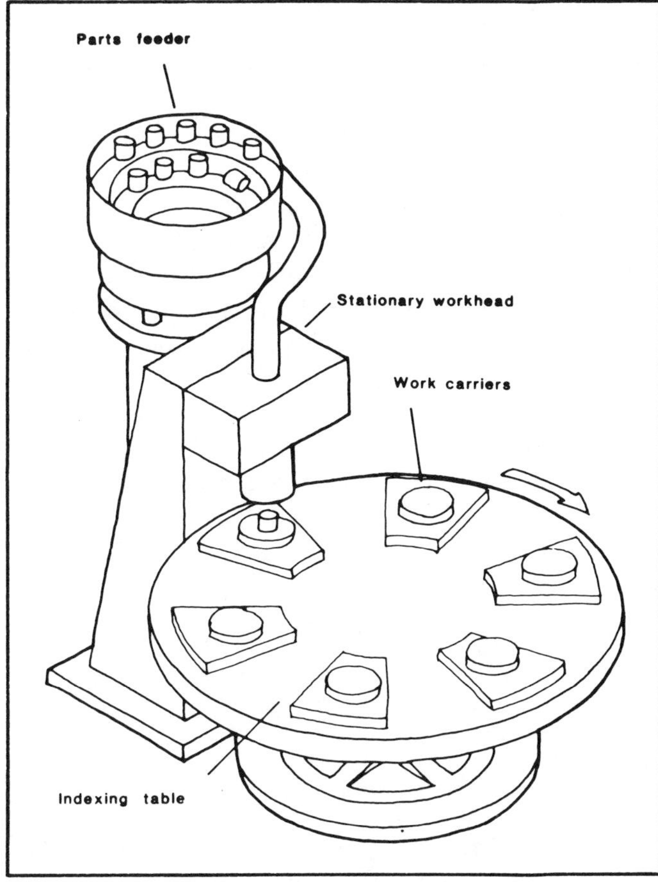

Figure 2
Rotary Indexing Machine

product is removed from the end of the line after each index. With in-line machines, provision must be made for returning the empty work carriers to the beginning of the line.

In the in-line machine, known as a free-transfer machine *(Figure 4)*, the spacing of the workstations is such that buffer stocks of assemblies can accumulate between adjacent stations. Each workhead or operator works independently and the assembly process is initiated by the arrival of a work carrier at the station. The first operation is to lift the work carrier clear of the continuously moving conveyor and clamp it in position. After the assembly operation has been completed, the work carrier is released and transferred to the next station by the conveyor, provided a vacant space is available. Thus, on a free-transfer machine, a fault at any one station will not necessarily prevent the other stations from working. It will be shown that this is an important factor when considering the relative economics of indexing and free-transfer machines.

Special-Purpose Indexing Assembly Machine (*AI*)—Remembering that NA is the number of parts

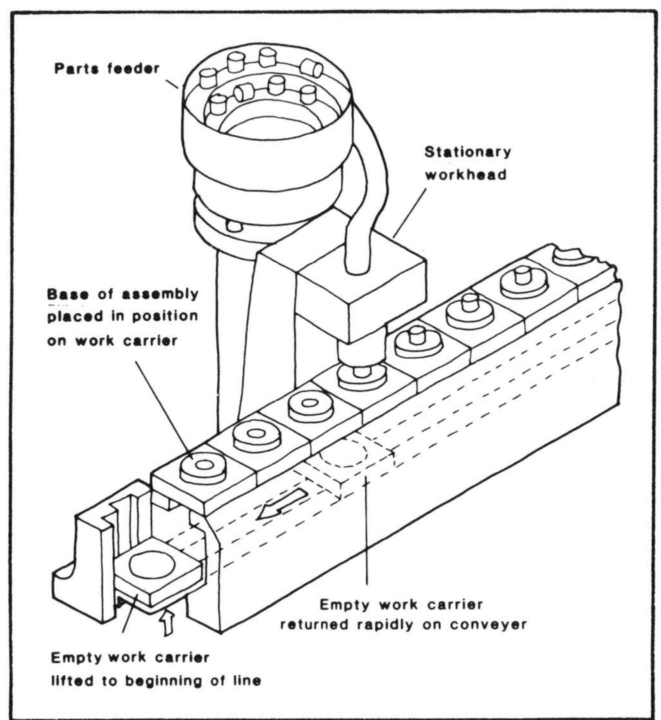

Figure 3
In-line Indexing Machine

in an assembly, then for N assemblies to be produced on an indexing machine, $N*NA$ assembly operations will be needed using this same total number of acceptable parts. If PQ is the average ratio of defective parts to acceptable parts, then the number of machine stoppages will be $N*NA*PQ$. If each stoppage causes a downtime of TD, then the total downtime will be $N*NA*PD*TD$. The machine working time to produce N assemblies is $N*TW$, where TW is the operation time for the workheads (the cycle time for the machine). Finally, adding the working time and the downtime and dividing by N gives the average production time obtainable for the system TP. Hence:

$$TP = TW + NA * PQ * TD \qquad (7)$$

It is considered that each isolated assembly machine will require one full-time machine supervisor together with perhaps one assembly operator to keep the machine feeders loaded with parts, and to load the basc part at the first station. On an in-line machine, even a second assembly operator may be needed to unload complete assemblies from the end of the machine, whereas on a rotary machine the same assembly operator can load base parts and unload complete assemblies since these operations occur at adjacent workstations. Experience also dictates that rotary indexing machines are not practicable for the assembly of more than six parts. Thus, if WA is the annual cost of an assembly operator and WS the annual cost of a machine supervisor, the total cost of personnel for a rotary machine is:

$$WT = NR * WA + WS \tag{8}$$

and for an in-line machine:

$$WT = NI * WA + WS \tag{9}$$

In this case, NR and NI are the numbers of assembly operators additional to the machine supervisor on rotary and in-line machines respectively. Thus, combining Eqs. (5) and (8), and (5) and (9) to give the dimensionless personnel cost per part, WR gives:

when:

$$NA <= 6$$

then:

$$WR = (NR + WS / WA) / NA \tag{10}$$

when:

$$NA > 6$$

then:

$$WR = (NI + WS / WA) / NA$$

It should be noted that Eq. (10) is a general expressions for personnel costs and that usually NI will be $NR + 1$, and that frequently, NR will be zero

The following equipment costs are defined for the individual items making up an indexing machine:

CC = the cost of a work carrier.
CF = the cost of an automatic feeding device and delivery track.
CT = the cost of a transfer device and its associated controls per station.
CW = the cost of a special-purpose workhead.

The total equipment cost (CE) for a machine to assemble NA parts is therefore given by:

$$CE = NA * (CC + CF + CT + CW) \tag{11}$$

However, if there are several product styles or the effects of other factors enumerated earlier are considered, this equation must be modified. To allow for various product styles built from alterna-

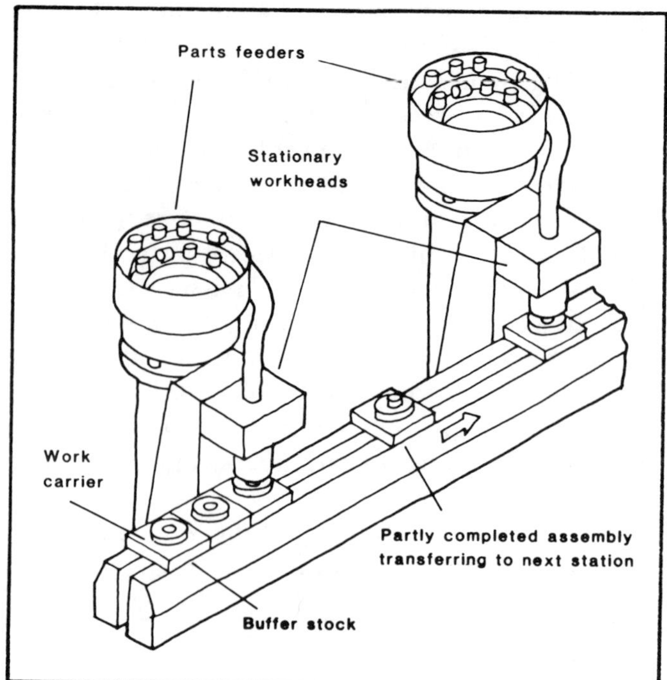

Figure 4
Free-transfer Machine

tive parts, all the terms in Eq. (11) must be multiplied by SV, which is the ratio of the number of parts available to the number of parts used. To allow for ND part design changes, each one necessitating a new feeder and workhead, then $ND*CW$ must be added in Eq. (11). Finally, for each new product (having the same number of parts) to be assembled, all items will be replaced except, perhaps, the transfer device. Hence, all terms except CT will be multiplied by NP which is defined as the number of products to be assembled during the life of the machine.

Applying all of these factors to Eq. (11) gives:

$$CE = NA * (SV * CT + NP * (SV * CC + (SV + RD)(CF + CW))) \tag{12}$$

where:

$$RD = ND / NA \tag{13}$$

Eqs. (7), (10) and (13) can now be used in Eq. (4) to give the dimensionless assembly cost per part (CD) for system AI—a special-purpose indexing machine.

Before proceeding with the analysis of a free-transfer machine, it is interesting to consider the effects of parts quality on the performance of an indexing assembly machine and the effects of production volume on assembly costs.

Figure 5 first shows how the machine downtime (DM) for the machine is affected by the parts quality. These curves are obtained from:

$$DM = \text{downtime} / \text{production time} = NA * PQ / (TW / TD + NA * PQ) \tag{14}$$

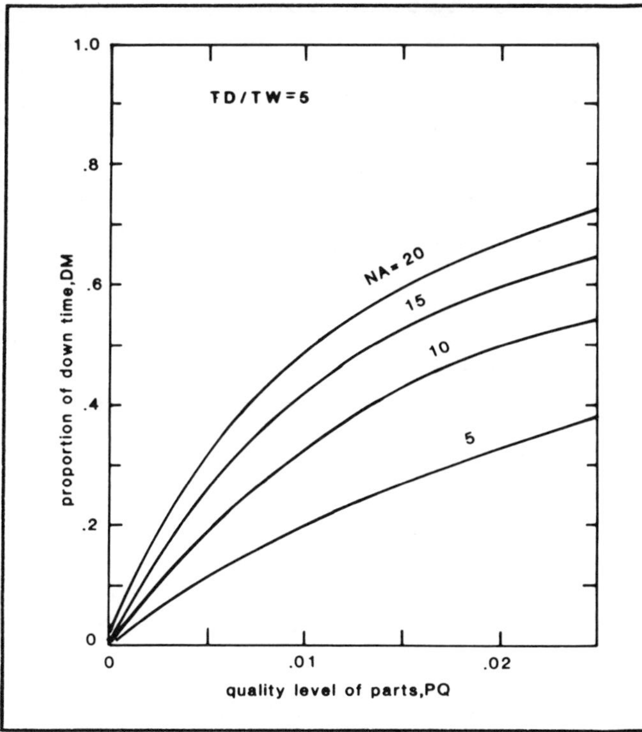

Figure 5
Downtime on Indexing Machine

It can be seen from the curves that for a given quality level and ratio of downtime to cycle time *TD/TW*, the number of parts in the assembly (*NA*) (i.e., the number of stations on the machine) has a marked effect on the downtime. If it is considered that a machine downtime due to faulty parts greater than 50% would be unacceptable, then an indexing machine could be termed inappropriate if:

$$NA * PQ * TD > TW \tag{15}$$

This condition will be used to exclude the system *AI* from economic comparisons when the number of parts *NA* is too large.

Turning now to the effects of production volume, it should first be realized that a special-purpose assembly machine is designed to work at a particular rate. This means that under normal circumstances, the output can be obtained from a knowledge of *TP*, the mean production time obtainable. Thus, the obtainable annual production volume per shift (*VS*) in millions is given by:

$$VS = 0.072 * PE / TP \tag{16}$$

In this formula, *PE* is the plant efficiency (percent) and is defined as the time actually worked in the plant divided by the time available, and where it is assumed that 7.2 million seconds are available in one shift year. Thus, on a graph of assembly cost versus annual production volume such as that shown in *Figure 6*, a special-purpose assembly machine is represented by a single point corresponding to the production volume obtainable from the machine working at maximum capacity. However, for the purposes of comparison, it is desirable to be able to show the relation between assembly cost and the annual production volume per shift required.

If the volume required is less than that obtainable from the machine, then as explained earlier, the assembly cost will be a function of the average production time required (*TQ*), given by Eq. (6). In this case, when the assembly costs given by Eq. (4) and the annual production volume per shift are plotted on logarithmic scales, a linear relationship results as shown in *Figure 6*. This relationship simply shows how costs increase because the machinery is not being fully utilized.

If the required production volume is greater than that obtainable from the machine, then it is assumed that backup will be provided in the form of manual assembly stations. Then, when the required volume approaches twice that obtainable, a second machine will be employed and so on. Since the assembly cost is independent of the number of assembly machines, the relation between dimensionless assembly cost and volume will be a horizontal line (*Figure 6*) because of the smoothing provided by the use of manual assembly where required.

Special-Purpose Free-Transfer Machine (*AF*)—*Figure 7* shows the layout of a portion of a free-transfer machine where *BS* is the size of the buffer storage between workstations.

This buffer storage helps to insulate the work-heads from the effects of stoppages on neighboring

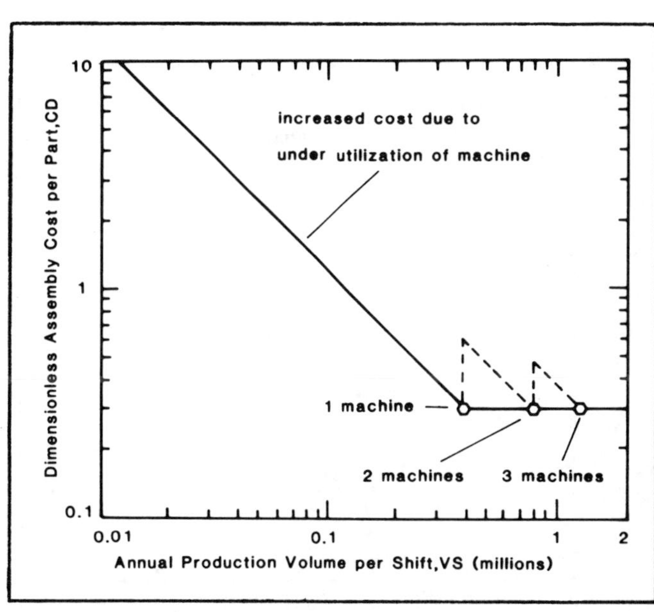

Figure 6
Performance of Indexing Machine
(Relationship obtained if conditions not smoothed by making up production volume with manual assembly)

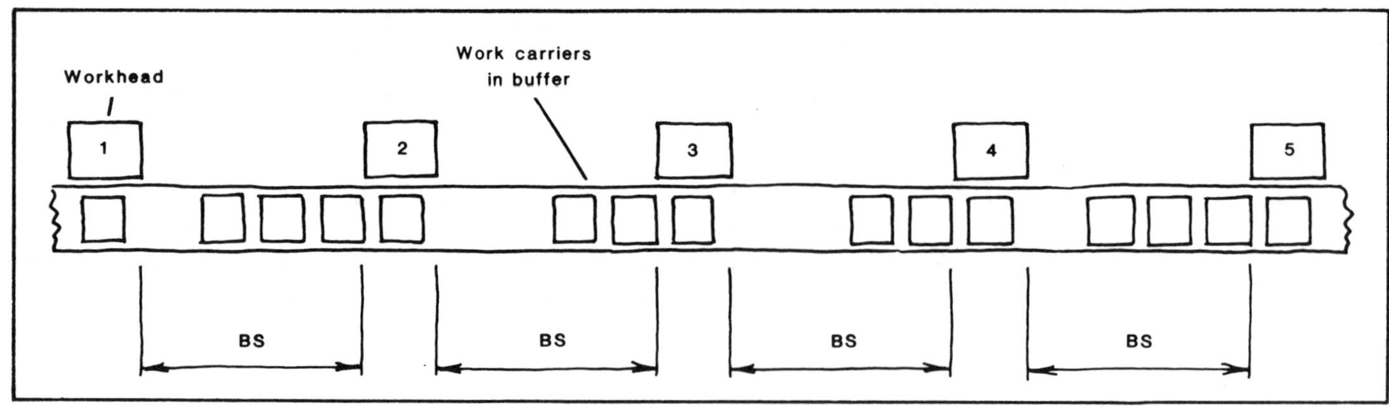

Figure 7
Buffers in Free-transfer Machine

workheads so that as BS is increased, the overall downtime on the machine reduces as shown in *Figure 8*. However, if BS is too large, then the machine will be unnecessarily expensive resulting in high assembly costs. If BS is too small, then assembly costs will again be high because the downtime will approach the high values obtained with indexing machines. Analysis of the economics of individual machines[1] shows that a value of BS equal to TD/TW results in assembly costs closely approaching the minimum and that the resulting downtime for the whole machine approaches that for an individual workstation.

Hence, for a free-transfer machine of any size, the production time obtainable can be estimated from:

$$TP = TW + PQ * TD \tag{17}$$

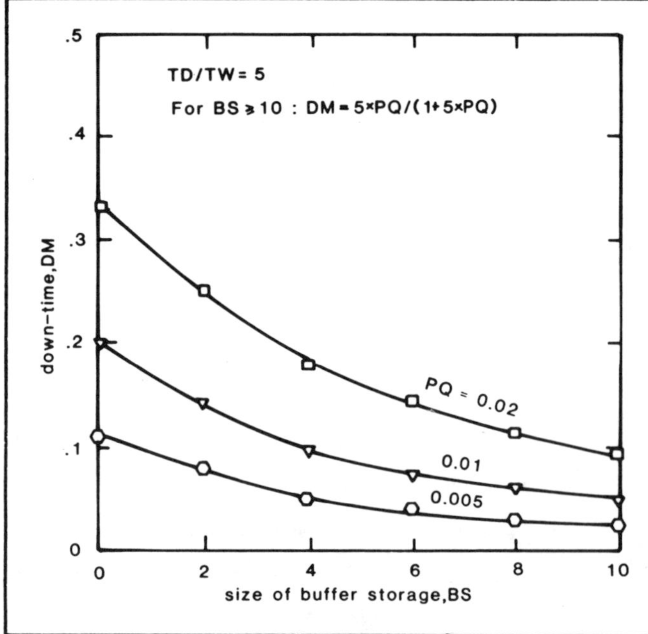

Figure 8
Downtime on Free-transfer Machine

It can be assumed that the number of operators would be the same as that for an in-line indexing machine and their cost would, therefore, be given from Eq. (10), by:

$$WR = (NI + WS / WA) / NA \tag{18}$$

If CB is the cost of the transfer device per workstation or buffer space on a free-transfer machine and the remaining cost (feeders, workheads, etc.) are identical to those for an indexing machine, then the total equipment cost for a machine to assemble NA parts is:

$$CE = NA * (CF + CW + (1 + TD / TW) * (CB + CC / 2)) \tag{19}$$

It will be noted that the number of work carriers is assumed to be one-half of the number of spaces on the machine. Clearly, if the machine were filled with work carriers, it would operate as an indexing machine and if there were too few work carriers, the advantages of buffer storage would be lost. Analysis has shown[1] that the optimum number of work carriers would be one-half the maximum capacity of the machine.

Adding the effects of product style variations (SV), number of design changes (ND) and number of products to be assembled (NP) as before, gives:

$$CE = NA * (SV * (1 + TD / TW) * (CB + NP * CC / 2) + NP * (SV + RD) * (CF + CW)) \tag{20}$$

again, where:

$$RD = ND / NA \tag{21}$$

Eqs. (17), (18) and (20) are used in Eq. (4) to give the dimensionless assembly cost per part (CD).

It is now possible to compare the economics of the two machines selected to represent special-purpose assembly equipment. *Figure 9* shows how assembly costs vary with the number of parts in the assembly, for typical values of the equipment costs given in the ***Appendix*** and for a situation where special-purpose equipment is likely to be economically applied. For these conditions, and as the number of parts increases, the assembly costs for the free-transfer machine are continuously reduced. For the indexing machine, the assembly costs initially decrease but then increase sharply due to the increase in operator cost at the changeover from a rotary to an in-line machine.

It can be seen from *Figure 9* that indexing machines would be preferred for small assemblies of 10 parts or less; otherwise a free-transfer machine would be employed. There is evidence that systems based on small indexing units connected by buffer storage—which can be called hybrid systems—can sometimes be economically applied since they exhibit the best features for both systems described here. However, the economics of such systems are

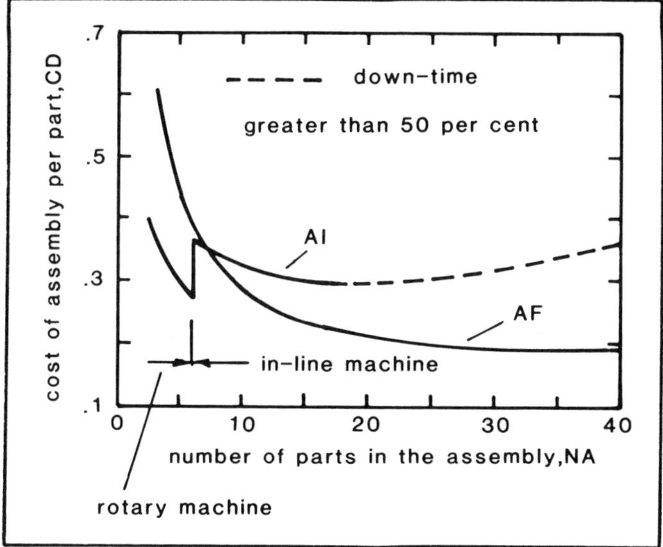

Figure 9
Performance of Special-purpose Machines
Effect of Number of Parts

found[1] to closely approach those for a free-transfer machine and hence the latter can be chosen as a representative system.

As with an indexing machine, a free-transfer machine is designed to operate at a constant production rate. This means that the effects of changes in required annual production volume will be similar and give a similar shaped relationship to that illustrated in *Figure 6*.

The multistation assembly machines considered above are representative of special-purpose machines and are 'dedicated' to one product. Special-purpose machines are virtually the only types of machines to have been successfully applied in industry as yet. However, it should be realized that the output from one machine is on the order of one-half million assemblies per year per shift, and that application of these systems is limited to mass production situations when the product design is stable for a few years. It should also be realized that mass production only comprises about 15-20% of all production in the U.S., and for this reason much attention is presently being given to the possibilities of automating batch assembly processes. It is clear that for these applications it will be necessary to employ 'programmable' or 'flexible' automation-in fact, assembly robots.

Programmable Assembly Machines (*AP* and *AR*)—The application of programmable equipment to automatic assembly can be divided into three areas:

1. Applications of computer control to dedicated multistation machines to allow for various product styles.
2. Applications of programmable workheads (or robots) to multistation machines so that more than one assembly operation can be carried out at each station.
3. Completely programmable assembly centers to accommodate product changes.

When a product is required in large quantities but is subject to style variations, then multistation machines can be employed where the individual stations, although dedicated to one task, can be commanded by a computer to carry out that task if required. For example, suppose a product has either a yellow or a red cap. In this case, two workstations can be provided, each with its own workhead and feeding device—one for yellow caps and one for red caps—the computer can trigger the operation desired. If this kind of option is repeated for other components, then with only a few alternatives, a large number of product styles can be produced.

Such a system has already been allowed for in the analysis of special-purpose machines. The principle disadvantage remains that production volumes on the order of one-half million per year per shift are necessary, and since this is still a dedicated machine, the product must be of stable design for several years.

Free-Transfer Machine with Programmable Workheads (*AP*)—These machines are basically special-purpose machines, where the single task workheads are replaced by programmable workheads capable of performing more than one assembly task. This scheme provides flexibility in that the machine can be designed to match more closely the annual production volume required. In addition, product style changes can be accommodated since the workheads are under computer control and can be commanded to select parts among alternatives available at a particular station.

Programmable workheads can be applied to

indexing machines, free-transfer machines, or hybrid machines but analysis shows[1] that free-transfer machines can be taken as representative of all these systems. It will be assumed in the analysis that the cost of the workheads (or robots) is given by:

$$CR = C1 + DF * C2 \qquad (22)$$

In this formula, DF is the number of degrees of freedom available and $C1$, $C2$ are constants for particular robots. Also, it will be assumed that the cost of the gripper for each workhead is CG per part to be handled, that if the workhead is to handle only one or two parts, only two degrees of freedom are required, for from three to six parts, the number of degrees of freedom will be equal to the number of parts, and for more than six parts, six degrees of freedom will be required.

The output from a free-transfer machine with programmable workheads can be varied by selecting an appropriate number of parts (NS) to be assembled at each station. Thus, the number of stations on the machine will be given by NA / NS and the cycle time for the machine will be given by $NS*TR$ where TR is the average assembly time per part. Allowing for downtime due to defective parts as before, we get:

$$TP = NS * (TR + PQ * TD) \qquad (23)$$

and when the required production time is TQ, given by Eq. (6), then the appropriate value of NS will from Eq. (23) be given by:

$$NS = TQ / (TR + PQ * TD) \qquad (24)$$

rounded to the nearest integer.

The number of operators will be assumed the same as that for a free-transfer machine with special-purpose workheads. Thus:

$$WR = (NI + WS / WA) / NA \qquad (25)$$

Previous studies[1] have shown that the use of special-purpose feeders and delivery tracks will be uneconomical when applied to programmable assembly equipment and that parts should be presented in hand-loaded magazines. If the cost of each magazine is CM, then the basic total equipment costs will be:

$$CE = (NA / NS) * (CR + (1 + TD / TR) * (CB + CC / 2)) + NA * (CG + CM) \qquad (26)$$

Each additional part made available to generate different product styles and each design change will require a new magazine. Also, each new product to be assembled will require new magazines and grippers and new work carriers. Incorporating these effects into Eq. (25) gives:

$$CE = (CR + (1 + TD / TR) * CB) / NS + NP * ((SV + RD) * CM + CG + (1 + TD / TR) * CC / 2 / NS) \qquad (27)$$

Finally, Eqs. (22) through (26) can be used in Eq. (4) to give the dimensionless assembly cost per part (CM).

Clearly, in practice, the number of operations at each station (NS) given by Eq. (24) must be an integer greater than or equal to unity and the different values of NS will each give a fixed production rate. Again it will be assumed that the relation between assembly cost and required production will be smoothed (*Figure 10*) by making up any deficiencies with manual assembly.

It can be seen from *Figure 10* that free-transfer machines with programmable workheads provide much greater flexibility than those with special-purpose workheads because of the possibility of assembling multiple parts at each station. However, it should be realized that if NS becomes greater than $NA / 2$, the need for a transfer device is eliminated and a single robot working at one assembly station is more appropriate.

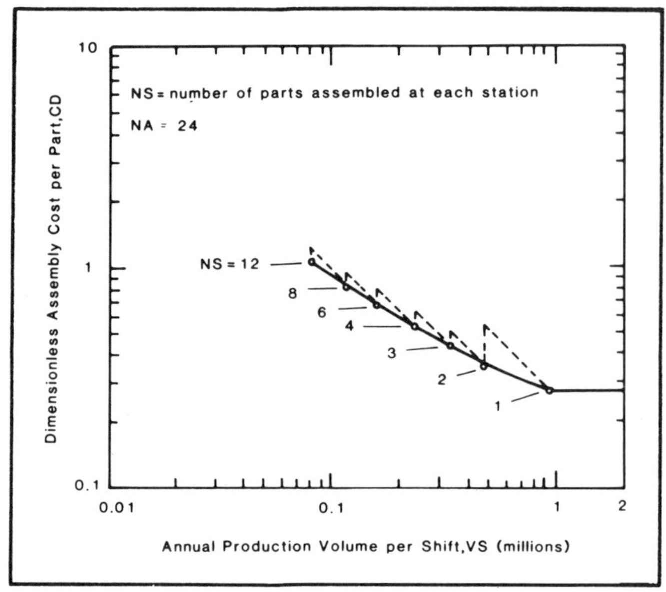

Figure 10
Performance of Free-transfer Machines with Programmable Workheads (AP)

Robot Assembly (*AR*)—For low production volumes, the single-station assembly system with one or two sophisticated robot arms, can be employed to build an assembly from a complete set of parts. This system provides perhaps the greatest potential for batch assembly work, where batches of differing products are to be assembled or where large numbers of product styles are required. However, the following important questions require consideration:

1. What approach should be used in gripper design?
2. What approach should be used for parts presentation?

Although sophisticated robots are available which can be quickly programmed to perform a complicated series of manipulations, the gripper fixed to the end of the robot arm is usually very limited in its capabilities and often, must be tailor-made for a particular part. The economics of the various solutions to this problem have been considered elsewhere[1].

One possible solution is to provide individual grippers arranged in racks within reach of the robot arm. Before performing an assembly operation, the robot must replace the previously used gripper in the rack and select the new gripper. Clearly the time for this manipulation can easily be on the order of the time taken for the assembly operation. This will not only increase the total production time but will also increase the cost of assembly due to the nonproductive time on the machine owing to gripper changes. It is this factor that presents the biggest economic problem in systems where gripper changes are necessary.

The use of a programmable or universal gripper should be considered in these circumstances. However, since such grippers have not yet been developed, it will be assumed that to minimize the nonproductive time due to gripper changes, two robot arms are used where one arm is performing an assembly operation while the other is changing grippers. Under these circumstances, it will be assumed that gripper costs are CG per part to be handled.

Parts Presentation Techniques—It is likely that in any single-station assembly system for batch production the means of part presentation for a given product will form a combination of (1) operator loading of large parts and some medium parts, (2) pallets of pre-positioned large or medium parts, (3) special magazines containing medium parts and some small parts, (4) special feeders for some medium and small parts, and (5) adjustable or programmable feeders for some small parts.

As programmable feeders have not yet been developed and since, in general, special-purpose feeders are not economical, it will be assumed that special magazines or pallets for each part (each costing CM) that are hand loaded on-line will be used for the robot assembly system.

Part of the production time taken to perform an assembly operation is used in selecting a part from a magazine or feeder and transporting it to the assembly area. By utilizing two arms on a robot system, a reduction in the mean assembly time might be achieved whereby one arm is selecting the next part (and changing grippers if necessary), while the other arm is performing the assembly operation with the previous part. If it can be assumed that the operations of one arm never interfere with the time of operation of the other, then the mean assembly time obtainable will be given by:

$$TP = NA * (TR / 2 + PQ * TD) \qquad (28)$$

The time $TR/2$ (where TR is the average assembly time per part) is used since it has been found that lengths of time.

It can reasonably be assumed that one machine supervisor will tend the robot assembly system—correcting faults when necessary and providing parts for the magazines or pallets. In this case:

$$WR = WS / WA / NA \qquad (29)$$

If hand-loaded magazines or pallets (cost CM) and special-purpose grippers (cost CG per part) are used then the total equipment costs are:

$$CE = 2 * CR + CC + NA * (CM + CG) \qquad (30)$$

In this formula, CC is the cost of one work carrier or assembly fixture and CR is given by Eq. (22) when DF is 6 and is the cost of a single arm.

For every new product, a new assembly fixture, new magazines and new grippers will be required and thus the terms representing the cost of these items should be multipled by NP, the number of products to be assembled. Additional product styles or design changes should only affect the special-purpose magazines and so the equipment costs become:

$$CE = 2 * \mathrm{CR} + NP * (CC + NA * CG + NA * (SV + RD) * CM) \qquad (31)$$

Eqs. (28) through (31) can now be used in Eq. (4) to give the dimensionless assembly cost per part (CD).

Like most other systems the single-station robot assembly system is designed to operate at a fixed rate. Thus, the relation between assembly costs and required production volume (*Figure 11*) follows the same pattern as these other systems.

Manual Assembly Systems

Manual Assembly (MA)—To represent manual assembly systems it will be assumed that the assembly process is broken down into individual tasks and performed in sequence by a series of operators arranged in 'assembly line' fashion. An individual operator will continually repeat the same operation or limited series of operations and the rate of output from the line will be dependent on the time taken by the slowest operator.

One advantage of this scheme compared with a multistation assembly machine is that by providing

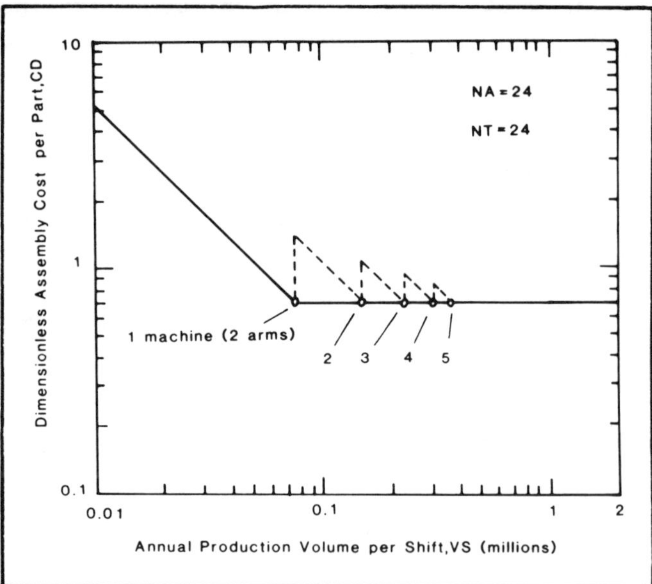

Figure 11
Performance of Assembly Robot (AR)

each operator with more than one assembly task (several parts assembled at each station), the output from the assembly line can be matched more closely to the production rate required. Another advantage is that defective parts do not create the severe problems encountered on assembly machines. An operator can quickly recognize a defective part and discard it with little loss in production.

If it is assumed that each defective part involves a simple repetition of the assembly task and that the average time taken by the slowest operator is *TA* seconds per task, then the mean production time obtainable is given by:

$$TP = NS * TA * (1 + PQ) \qquad (32)$$

and if *TQ* is the required production time, then:

$$NS = TQ \;/\; (TA * (1 + PQ)) \qquad (33)$$

where *NS* is the number of parts assembled by each operator.

Since there is one operator at each station and there are *NA*/*NS* stations, the total dimensionless cost of operators per part is:

$$WR = 1 \;/\; NS \qquad (34)$$

Assuming that a free-transfer device with work carriers is provided and that each operator has two spaces on the transfer device to allow for minor delays in assembly time, then the cost of equipment is:

$$CE = (NA \;/\; NS) * (2 * CB + CC) \qquad (35)$$

For *NP* products to be assembled this becomes:

$$CE = (NA \;/\; NS) * (2 * CB + NP * CC) \qquad (36)$$

Manual Assembly with Mechanical Assistance (*MM*)—In some situations, it is found that assembly times can be reduced by providing the operators with mechanical assistance; for example, by providing oriented parts from automatic feeding and orienting devices. In this case, if the operator assembly time is reduced to *TM,* the equations for production time and cost of equipment become:

$$TP = NS * TM\ (1 + PQ) \qquad (37)$$

and:

$$NS = TQ \;/\; (TM * (1 + PQ)) \qquad (38)$$

and the cost of equipment becomes:

$$CE = (NA \;/\; NS) * (2 * CB + NP * CC + NS * (NP * (SV + RD) * CF)) \qquad (39)$$

Combining Eqs. (32) through (36) or Eqs. (34), (37), (38), and (39) with Eq. (4) gives the dimensionless assembly cost per part for manual assembly systems *MA* or *MM* respectively.

Figure 12 shows how the assembly costs vary with required production volume and it can be seen that, when the production volume is low the manual assembly system without mechanical assistance gives the lowest costs, but when the production volume is high, mechanical assistance in the form of feeders, for example, becomes worthwhile.

Assembly Centers

One final system to be included in the present analysis is an "assembly center" meant to represent the kind of batch automatic assembly machine that might become a reality in the future.

In considering the application of robots to assembly, it is often not realized that the robot itself

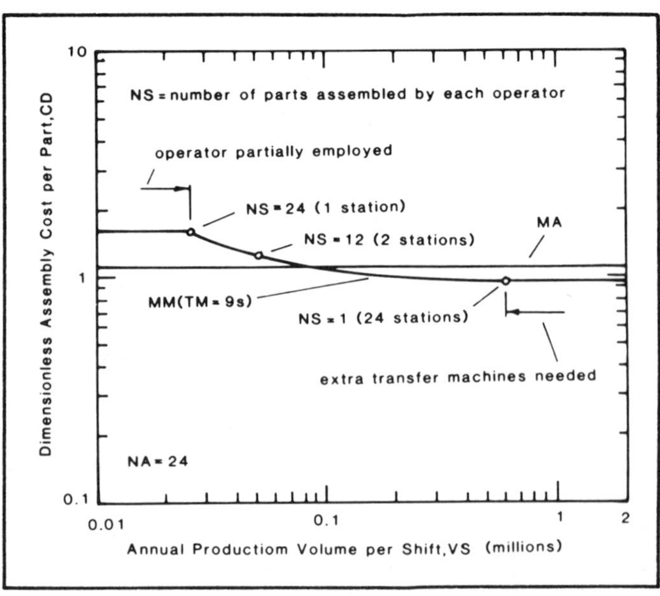

Figure 12
Manual Assembly

forms only a portion of the assembly system. In fact, the greatest difficulties arise not with the robot, or its programming, but in three specific areas:

1. Grippers.
2. Parts presentation devices.
3. Product design for ease of assembly.

At present, the grippers and the parts presentation devices used in a specific robot assembly system will cost at least as much as the robot itself and will be dedicated to one task only. Typically, each part to be assembled will require a special feeder and, in many cases a special gripper (the robot changes grippers during the assembly cycle as required) as previously discussed. Thus, when it becomes necessary to change the system over to a different product, only the robot can be re-used. In fact, most items which make contact with the parts to be assembled will generally have to be replaced and this situation usually results in robot assembly systems being economically unjustifiable.

Another problem encountered by those interested in applying automation to batch assembly is that the product is often not suitable for this kind of assembly. The individual parts are difficult to feed and the manipulations during assembly are difficult to carry out automatically. As a result of a recent research program carried out at the University of Massachusetts[3], it is now possible to evaluate designs for manual assembly and automatic assembly of the mass production type—it is not possible to provide the same information for robot assembly. This is because the assembly systems necessary for robot assembly have not yet been developed.

Universal Assembly Center (*AU*)—This system is meant to describe the economic performance of two robot arms working at a single work station and supported by inexpensive "programmable" feeders and using "universal" grippers.

Clearly, feeders that can be programmed to feed any part are not likely to become available. A universal gripper that will grip any part would be prohibitively expensive. Thus, one requirement for the universal assembly center is that the products have been designed for this type of assembly. This means that each part has been designed so that it can be feed in one of the programmable feeders and designed so that it can be gripped by the "universal" gripper. Finally, the product must have been designed so that it can be easily assembled by the robot. Of course, the rules for design cannot be developed until the gripper and feeders have been developed.

To make universal assembly centers as flexible as possible, it is proposed that they should be designed so that they could be arranged in-line and able to hand the partly completed assemblies to the next station in the line. With this arrangement it is assumed that one machine supervisor will 'tend' the whole line.

If *NS* parts are assembled at each station then the production time obtainable is given by:

$$TP = NS * (TR \ / \ 2 + PQ * TD) \qquad (40)$$

and if TQ is the required production time then:

$$NS = TQ \ / \ (\mathrm{TR} \ / \ 2 + PQ * TD) \qquad (41)$$

For one operator to tend the line of machines gives:

$$WR = WS \ / \ WA \ / \ NA \qquad (42)$$

Since there are *NA*/*NS* stations, each one assembling *NS* parts, the total cost of equipment will be given by:

$$CE = (NS \ / \ NS) * (2 * CR + 2 * CU + CC) + NA * CP \qquad (43)$$

In this formula, *CP* is the cost of a programmable feeder and *CU* is the cost of a "universal" gripper.

Assuming that different product styles only affect the number of programmable feeders and that new products only affect the design of the work fixtures, the equipment costs become:

$$CE = (NA \ / \ NS) * (2 * CR + 2 * CU + NP * CC) + NA * SV * CP \qquad (44)$$

CR is given in this formula by Eq. (22) when *DF* is 3.

Combining Eqs. (40), (41), (42) and (44) with Eq. (4) will give the dimensionless assembly cost per part *CD*. *Figure 13* shows how this is affected by the production volume per shift.

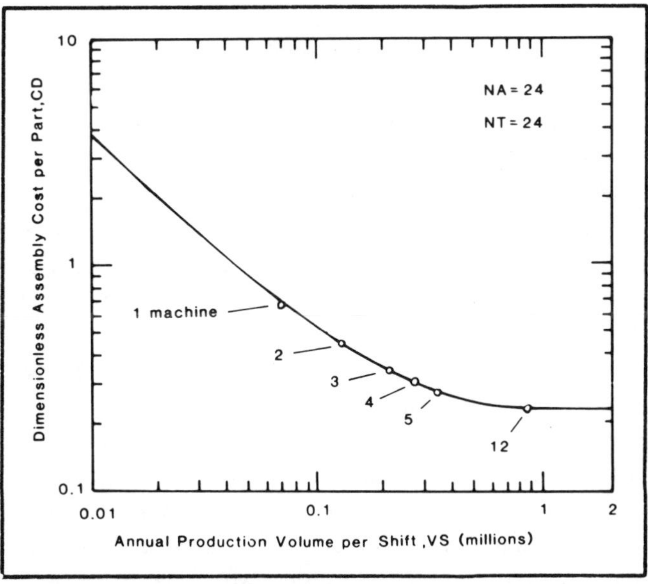

Figure 13
Universal Assembly Center (AU)

Comparison of Assembly Systems

For the same conditions as with the previous figures, *Figure 14* shows how the dimensionless assembly cost per part varies with annual production volume per shift for the following four situations.

Special-purpose Automatic Assembly (*A*)—For this curve either system *AI* or *AF* is used—whichever gives the least cost. For the present conditions, *AF*, a free-transfer machine with special-purpose workheads, is the best because of the large number of work stations. In fact, *AI* is inappropriate because of excessive downtime.

Robot or Programmable Assembly (*R*)—This curve represents system *AP*—a free-transfer machine with programmable workheads except for low production volumes where *AP* is inappropriate and single-station robot assembly *AR* is used.

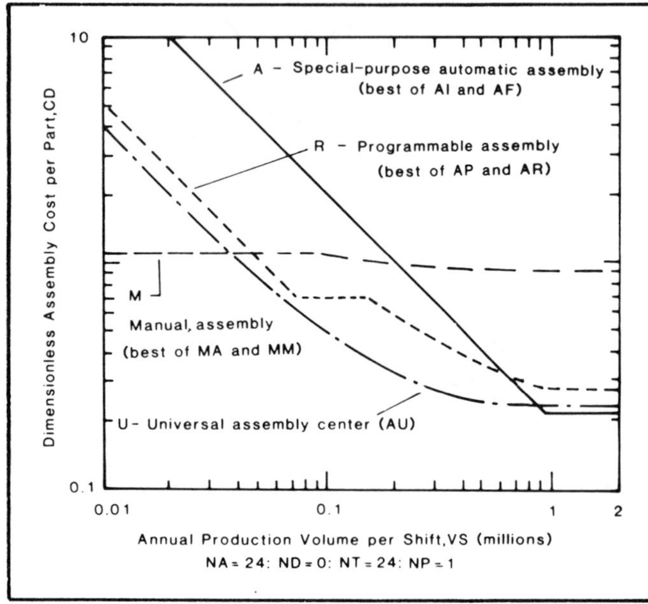

Figure 14
Comparison of Assembly Systems

Manual Assembly (*M*)—For low production volumes, *MA* gives the minimum cost, otherwise *MM* is used for this curve.

Universal Assembly Center (*U*)—This curve represents system *AU* and, therefore, represents the economics of an ideal system that may be approached in the future.

The conditions employed for *Figure 14* included one product style (*NT*=*NA*), no design changes (*ND*=0) and one product (*NP*=1) having a long market life with no significant fluctuations in demand. It can be seen from the figure that, as expected, when the production volume is high the special-purpose automatic assembly machines are economical. When the production volume is low, manual assembly must be used. However, there exists a range of medium production volumes where the free-transfer machines with programmable workheads can, because of their flexibility, be economically employed. On the other hand, the situation considered is where programmable assembly would not be expected to demonstrate its greatest potential.

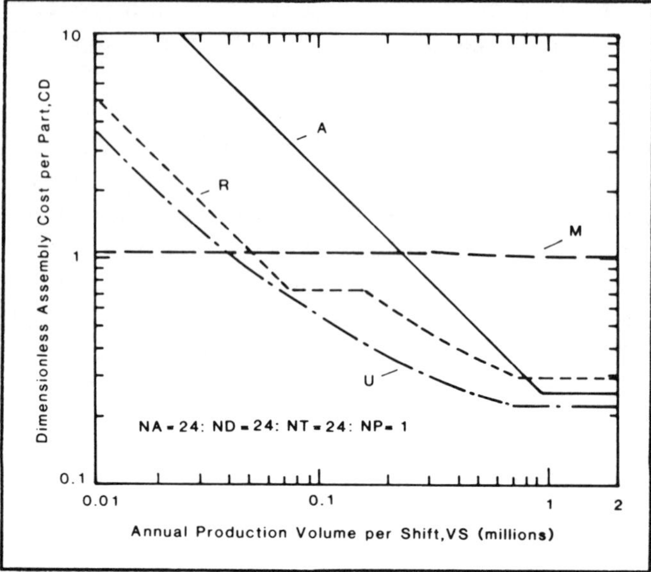

Figure 15
Effects of Design Changes

Figures 15 and *16* show the effects of design changes (*ND*=24) and product style variations (*NT*=48), respectively. These parameters give similar effects and indicate that for high production volumes, the programmable equipment (*R*) using present day technology becomes competitive. For both medium and low production volumes, however, manual assembly (*M*) is the most economical. The hypothetical universal assembly center (*U*) would be

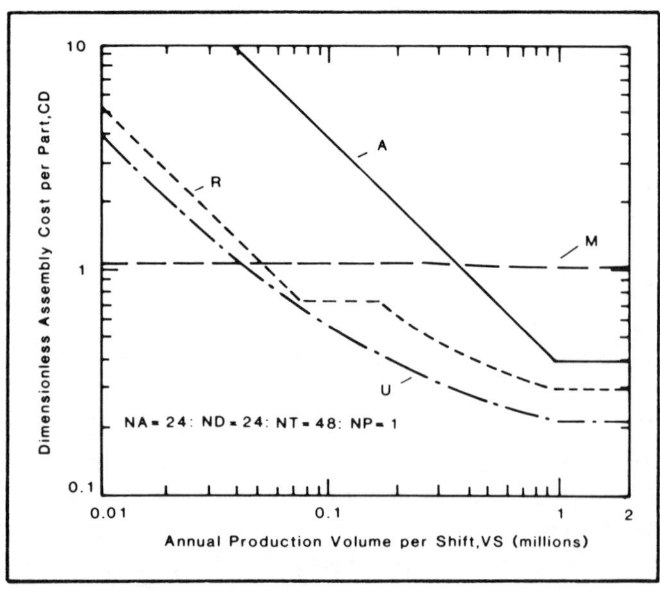

Figure 16
Effects of Product Style Variations

economical for both medium and high production volumes (above 500,000 per shift per year).

Figure 17, where the effect of assembling 20 different products (*NP*=20) is illustrated, shows more startling results. It can be seen that the special-purpose equipment (*A*), as would be expected, becomes completely uneconomical. However, the free-transfer machine with programmable workheads has also become uneconomical, whereas the hypothetical universal assembly center (*U*) shows considerable savings compared with manual assembly (*M*) at medium and high production volumes.

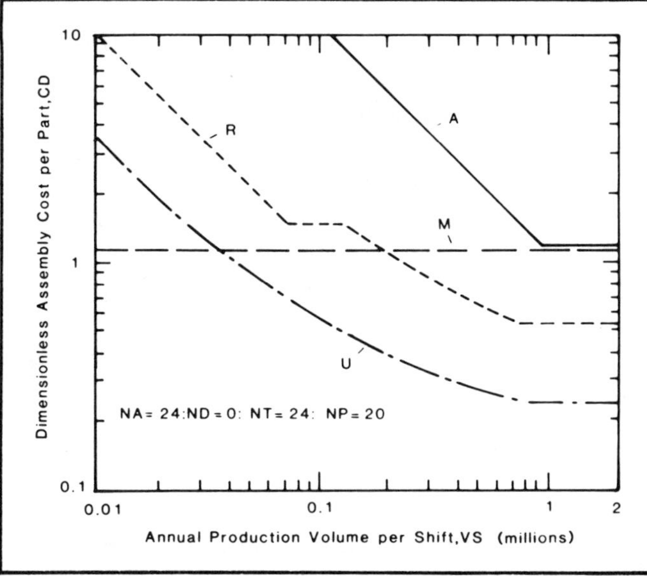

Figure 17
Effects of Several Products

Concluding Remarks

From these comparisons, the following conclusions can be drawn:

1. Special-purpose assembly machines should be considered when the production volume is high and when one product with few styles is to be assembled. However, for high production volumes and even where design changes and product style variations are anticipated, the special-purpose machines might still be the most economical.
2. Free-transfer machines with programmable workheads can be economical with medium and high production volumes and when product design changes and/or various product styles are involved. These systems should also be considered for high production volumes when different products are to be assembled.
3. The hypothetical universal assembly centers could change the picture radically. These systems might be competitive for all but very low production volumes (i.e., when at least the full output from one assembly center is required) and would be the most advantageous when different products are to be assembled.

It is clear that the development of programmable feeders and universal grippers and the design of products for ease of robot assembly could considerably extend the areas of application of programmable assembly automation.

References

1. Boothroyd, G., Poli, C. and Murch, L. E., *Automatic Assembly*, Marcel Dekker Inc., New York, 1981.
2. Lynch, P. M., "Economic Technological Modeling and Design Criteria for Programmable Assembly Machines", Report T-625, The Charles Stark Draper Laboratory, Inc., Cambridge, Massachusetts, June 1976.
3. Boothroyd, G., *Design for Assembly,* Department of Mechanical Engineering, University of Massachusetts, Amherst, Massachusetts.

Appendix

This Appendix includes definitions of constant factors and variable parameters together with the base values employed.

I. Constant Factors

A. Equipment Costs

C1 = Basic cost of robot or programmable workhead = 25 k$.

C2 = Additional cost of robot or programmable workhead per degree of freedom = 8 k$.

CB = Cost of transfer device per work station or buffer space on free-transfer machine = 5 k$.

CC = Cost of work carrier = 1 k$.

CF = Cost of automatic feeding device and delivery track = 5 k$.

CG = Cost of gripper devided by *NA* (number of parts in the assembly) = 0.5 k$.

CM = Cost of manually loaded magazine = 0.5 k$.

CP = Cost of one programmable feeder—only used for system *AU* = 1 k$.

CT = Cost of transfer device per work station for an indexing machine = 10 k$.

CU = Total cost of universal gripper—only used for sytem AU = 10 k$.

CW = Cost of special-purpose workhead = 10 k$.

B. Times

TA = Manual assembly time per part = 10 s.

TD = Machine downtime due to each defective part = 30 s.

TM = Manual assembly time per part when mechanical assistance is provided = 9 s.

TR = Assembly time per part with robot or programmable workhead = 5 s.

TW = Assembly time per part with special-purpose workhead = 5 s.

C. Other

PE = Plant efficiency—average time worked divided by time available = 69%.

PQ = Ratio of faulty parts to acceptable parts = 1% (0.01).

NI = Number of operators additional to machine supervisor on in-line assembly machine = 1.

NR = Number of operators additional to machine supervisor on rotary indexing machine = 0.

II. Variable Parameters

NA = Number of parts in the assembly.

VS = Annual production volume per shift.

QE = Capital expenditure to replace one assembly operator on one shift.

SH = Number of shifts.

NT = Total number of parts available for building different product styles.

ND = Number of part design changes expected during first three years.

NP = Number of different products to be assembled during first three years.

WA = Annual cost of one assembly operator = 36 k$.

WS = Annual cost of one machine supervisor = 1 56 k$.

III. Derived Parameters

NS = Number of operations performed at each work station.

$SV = NT/NA$; measure of style variations.

$RI = SH*QE/WA$; measure of investment potential.

$RD = ND/NA$; measure of design stability.

Note: Although the numerical values assigned for equipment costs may seem high, it should be realized that these values include the basic equipment costs; the assembly machine design; engineering and debugging costs; and the cost of controls, etc. It has been estimated[2] that basic equipment costs often form only 40% of their total costs.

Author Biography

Geoffrey Boothroyd is Professor of Mechanical Engineering at the University of Massachusetts, Amherst, Massachusetts. Since joining the University of Massachusetts faculty in 1967, Professor Boothroyd has developed a laboratory for research and teaching in metal machining, machine tools, and mechanized assembly. He was the first director of both the full- and part-time graduate programs leading to the M.S. degree in manufacturing engineering. Professor Boothroyd has also been principal investigator of numerous grants, and has acted as a consultant to various machine tool, cutting tool, and automation equipment user industries. He has published many books and articles, and was the recipient of the Western Electric Distinguished Teacher Award of the ASEE (1974) and the SME Education Award (1979). He received his Ph.D. degree (1962) and his D.Sc. degree (1974) in mechanical engineering from the University of London.

Reprinted by permission of IFS (Publications) Ltd. from *Assembly Automation*, February 1984

Automated assembly in the electrical industry

Bruno Lotter, Director, E.G.O. Elektrogerätebau GmbH, Sulzfeld, W. Germany.

In the electrical industry 50 to 75% of the total production costs of a product are in assembly. This shows that the main focus for rationalisation should be in assembly. This paper illustrates planning methods for rationalisation, describes a completed assembly line and outlines the economies obtained.

A THERMAL switch, as shown in Figs 1 and 2, consisting of 37 component parts is to be assembled at a net output of 1,000 units per hour. In order to attain economic efficiency the operating efficiency of the assembly line laid down must be at least 80%. Based on these assumptions the individual cycle time for the assembly line is calculated as follows:

$$^{t}T = \frac{t \times \eta}{P} = \frac{3600 \times 0.8}{1000} = 2.88 \text{ sec.}$$

where P is the production rate.

To determine optimum production and the most economic method a 4-stage planning method was used.

- Stage 1: is an **ABC-analysis extended to assembly.** This analysis determines the cost of a component until it reaches its operational purpose, i.e. assembly into the product. From this analysis the requirements of the quality of the component parts emerges both in their delivered condition, and joining capability. In the thermal switch the result of this analysis was that the base (part 1, see Fig. 4), produced in a ceramic material, must be handled manually. Also the microswitch element (part 5) and connector strips (part 11) cannot be fed automatically as they are either too sensitive or they cannot be sorted owing to their geometry. This means that these parts must be handled manually or, especially for the strips (part 11), a different production method must be found in order to achieve automatic feeding. This requires processing operations to be integrated in the assembly processes. The ABC-Analysis also finds that the standard tolerances of common parts, such as screws and nuts, are not good enough to provide high availability at the automatic stations. All the other parts are capable of being sorted and joined.
- Stage 2: considers the **product construction and the direction of joining.** The thermal switch can be divided into three assembly products:

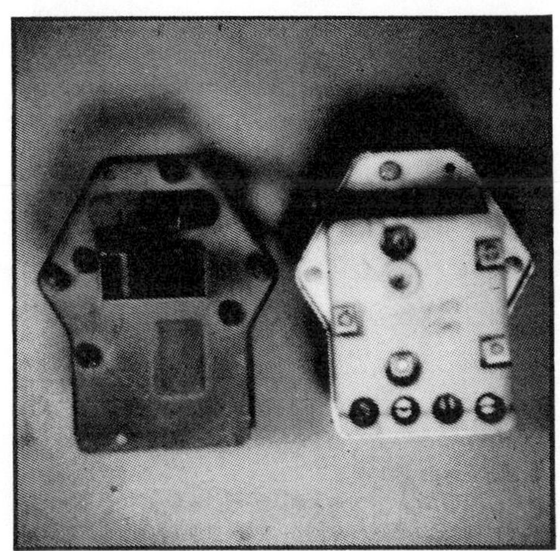

Fig. 1. Thermal switch.

- ○ the base sub-assembly;
- ○ the cover sub-assembly;
- ○ the complete assembly of base, lever and a few additional parts.

The construction of the product and the resulting direction of joining can be seen in Fig. 3 – the base sub-assembly, Fig. 4 – the cover sub-assembly, and Fig. 5 – the complete assembly. The analysis regarding product construction and direction of joining shows that *pre-assembly operations* are required. For example the screw (part 9 – see Fig. 4) must be fitted to the terminal (part 8), the terminal with the screw must be pre-assembled with part 7 so that the complete sub-assembly can be joined to the base (part 1). The same applies to the connector strips (part 11) with the terminal (part 8) and screw (part 9).

- Stage 3: examines the **order of the**

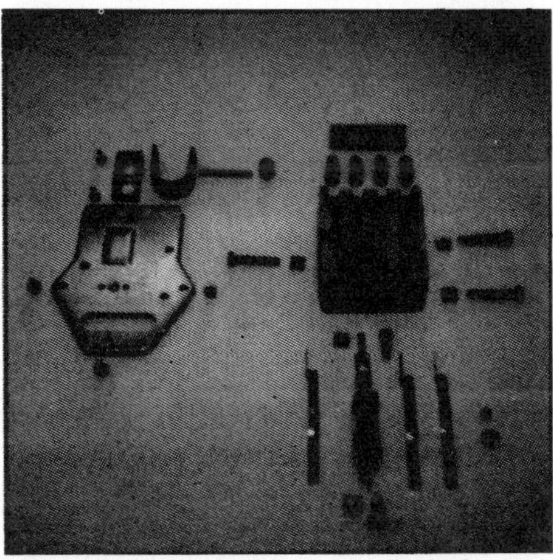

Fig. 2. Thermal switch (exploded view).

assembly processes. The order of the assembly and processing operations is based on the construction of the product and the directions in which joining occurs as shown in Fig. 6.

- Stage 4: is a **function analysis.** Starting from the specified output and the resulting cycle time of 2.88s the function analysis examines to what extent this given cycle time can be attained by the individual handling, joining, and machining operations. An analysis must be carried out for each component part. Some parts can be grouped together. Fig. 7 shows the functional analysis for the thermal switch. By using the symbols for handling, joining, and machining processes in accordance with DIN 3239, the individual functions of the switching and stopping times of the transfer device being used are allocated and the individual times determined by calculation. As can be seen from the functional analysis on Fig. 7 it is possible to carry out the handling and joining processes within the calculated cycle time of 2.88s. But, the necessary machining operations to be integrated, such as setting the contact distance, running-in the microswitch element and measuring and adjusting the working pressure, cannot be accommodated.

The result of these four planning stages and analyses lead to the following guidelines for the construction of the automated assembly line for the thermal switch.

□ The base part of the thermal switch (part 1) must be handled manually due to the material, so a manual workplace must be provided for this. However, owing to its geometry it is possible to transport the base part without a carrier and not lose its orientation.

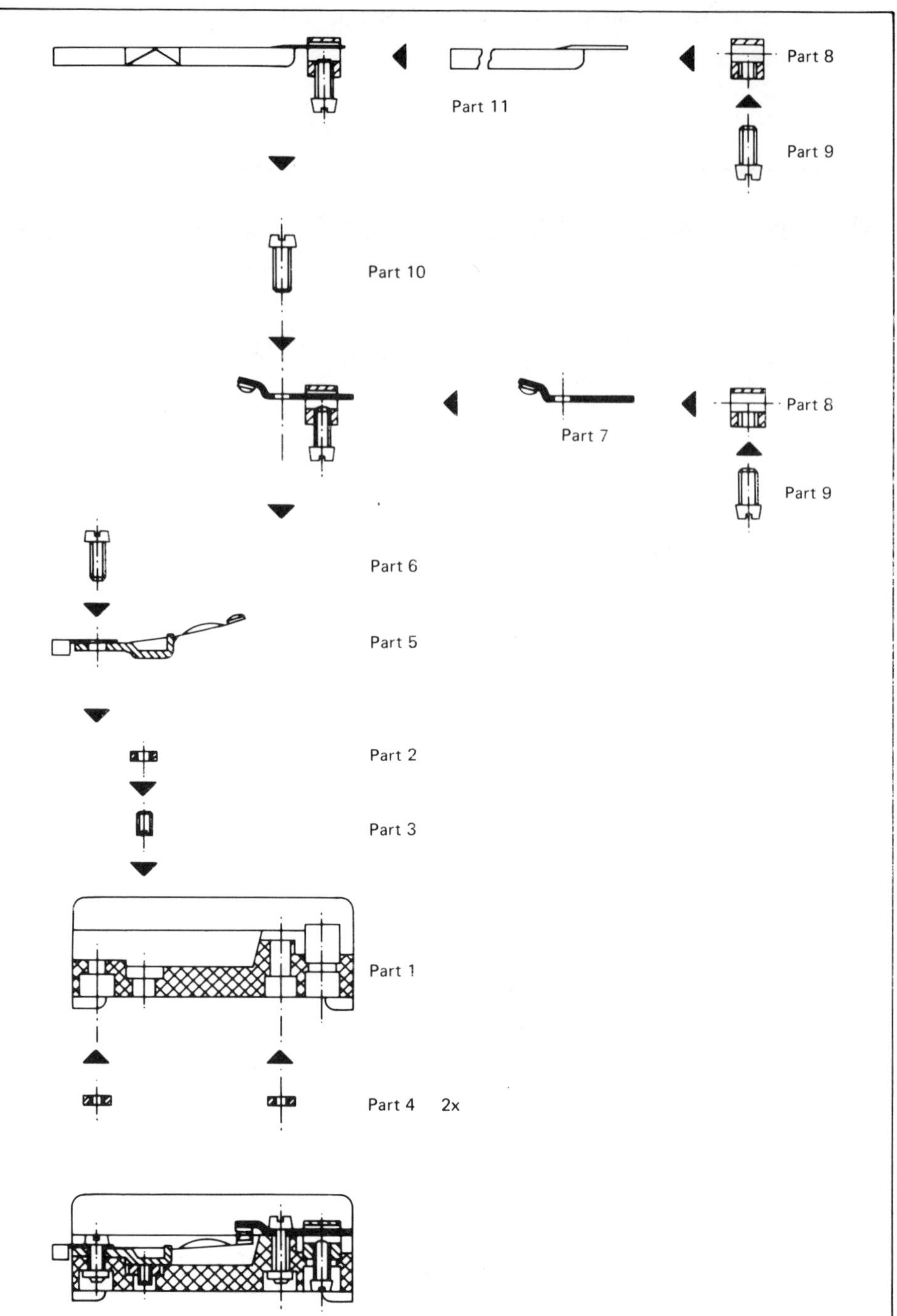

Fig. 3. Base sub-assembly.

□ The microswitch element (part 5) is delicate and must be handled manually so another manual workplace is necessary.

□ For standard components such as screws and nuts, special discussions must be held with the suppliers, regarding reduced tolerances and increased quality.

□ In order to avoid handling, the three connector strips per switch (part 11), they are transferred to the assembly line magazined in blank form. The necessary forming process is integrated into the assembly line.

□ The product construction and the order of the assembly processes indicate that assembly can be largely sub-divided, so that it is possible to use rotary indexing machines loosely linked to a line.

□ The operating processes allocated to individual assembly machines must be kept as low as possible, in order to obtain the highest overall availability of the individual machine.

□ In order to increase and ensure the efficiency of each individual machine, each individual supply station is equipped with lead length control, shown schematically in Fig. 8. Experience shows that most faults occur during sorting of components in the vibratory conveyor. Similarly, a large number of the faults occur in the transfer between the latter and the electromagnetically driven outflow rails. Lead length control achieves a buffer effect after the conveyor by using suitable lengths of outflow rails. Faults within the vibratory conveyor and the transfer position to the outflow rails no longer directly affect the efficiency of the assembly machine since enough good parts are buffered in the lead length to allow removal of faults arising in the conveyor, in sufficiently good time that idling does

Table 1 – Costings

Operating costs

Planned quantity = 16,000/16h/day = 1,000 pieces/h.

Calculable depreciation $(K_A) = \frac{\text{replacement value}}{\text{machine life } T_N \text{ (4 years)}}$

Calculable interest

$(K_Z) = \frac{\text{Replacement value}}{2} \times$ interest rate (10%)

space cost	(K_R)	= $m^2 \times$ DM per annum (120.00).
energy cost	(K_E)	= KWh $\times$ (DM 0.20) $\times T_N$ compressed air $m^3 \times$ (DM 0.08) $\times T_N$
maintenance cost	(K_I)	= 10% of replacement value
machine life	(T_N)	= 220 days $\times$ 16h = 3,520h
machine hour rate (K_{MH})		= $K_{MH} = \frac{K_A+K_Z+K_R+K_E+K_I}{T_N}$

Personnel expenses per hour:

production personnel	= number of personnel × wage + 110% of incidental wages costs
non-productive personnel (charge-hand, setter)	= number of personnel × wage + 110% of incidental wage costs
allowances (shift)	= $\frac{\text{shift allowances per day}}{16h}$ = average per hour
supervision (foreman)	= $\frac{\text{wage + incidental wage cost}}{172.33h}$

The replacement value with a useful life of four years and an assumed inflation rate of 5% per annum is calculated as purchasing price × factor of 1.215.

not occur in the lead length.

☐ Pre-assembly operations must be carried out at 'satellite' machines separate from the assembly machines or integrated in the workpiece carrier of the individual assembly machines.

☐ The function analysis shows that the specified cycle time of 2.88s is insufficient for a number of the working and testing processes. For those processes that take longer than 2.88s duplexing must be arranged.

☐ Interlinking of the individual assembly machines is achieved by conveyor belts which serve as intermediate buffers between individual machines in the sense of lead length control, or lead length control of individual insertion stations.

☐ The cycle time of 2.88s dictates rotary indexing unit controlled by cams or Geneva-motions.

☐ Each assembly or work station is followed by a test station with an immediate go/no go decision. The individual workpiece carriers are fitted with coding pins which indicate the result of the test. A single *no* signal does not shutdown the machine, only three consecutive *no* signals will cause a stoppage. The cause of the breakdown is signalled simultaneously to the operator. Faulty or incomplete assemblies must be sorted out automatically at each individual machine so that only acceptable sub-assemblies are offered to the following machine.

The assembly line for the thermal switch was constructed on the basis of the above findings and guidelines. It consists of nine loosely linked assembly machines of the circular indexing type as shown in the layout of Fig. 9.

The assembly line operates (with reference to Fig. 3) as follows:

Machine 1 assembles the lower microswitch element into the base. The base is loaded by the operator onto the input conveyor at a rate independent of the machine cycle. A stud and nut (parts 3 and 2), fastened together on a satellite rotary indexing machine are automatically placed into the base. A second operator places the lower microswitch element (part 5) into the base. A nut and screw (parts 4 and 6) fasten the element in place.

Machine 2 fits the upper microswitch element. First, a second nut (part 4) is automatically loaded into the carrier shown in Fig. 10b and then the base placed over the nut locating on a fixed pin and an axially moveable pin (see Fig. 10c) which also locates the nut. Another section of the carrier is designed to pre-assemble the upper microswitch element (part 7) with a terminal and screw (parts 8 and 9) which themselves are fastened together on a satellite machine – a special vertical rotary indexing machine shown in Fig. 11. The terminal plus screws are placed into a locator which is mounted on a slide – BB in Fig. 10b (see also Fig. 12). The lower element (7) is loaded adjacent to the terminal on the carrier. As the table of machine 2 indexes the slide is forced by a cam-roller inwards and the element is pushed into the terminal hole. This sub-assembly (parts 7, 8 and 9) is then placed automatically into the base sub-assembly, the retaining hole in the element locating with the moveable pin of carrier. In order for the upper element to be placed without interference from the lower element, the latter is held during assembly of the former by a finger, which is rotated down by a rack and pinion actuated by the slide (see Figs. 10a and 12). Finally, the lower element is automatically fastened by a screw (part 10), which as it is screwed down depresses the moveable pin. The completed base

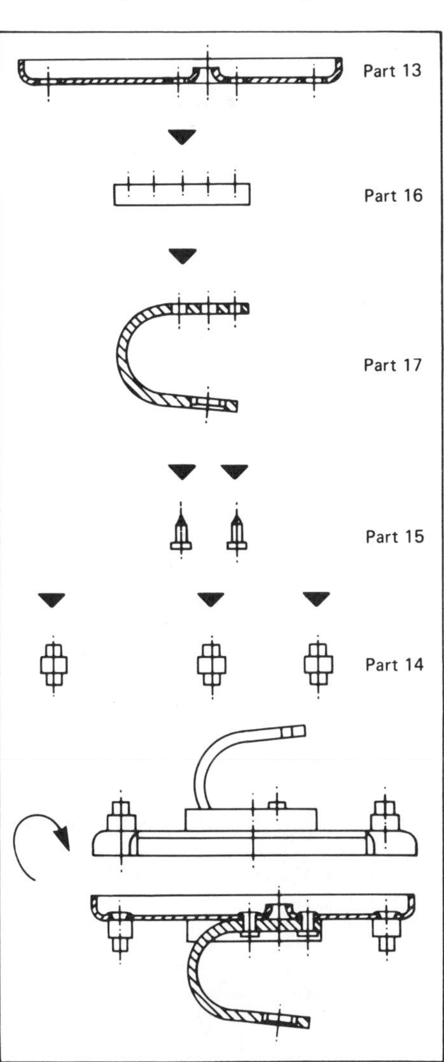

Fig. 4. Cover sub-assembly.

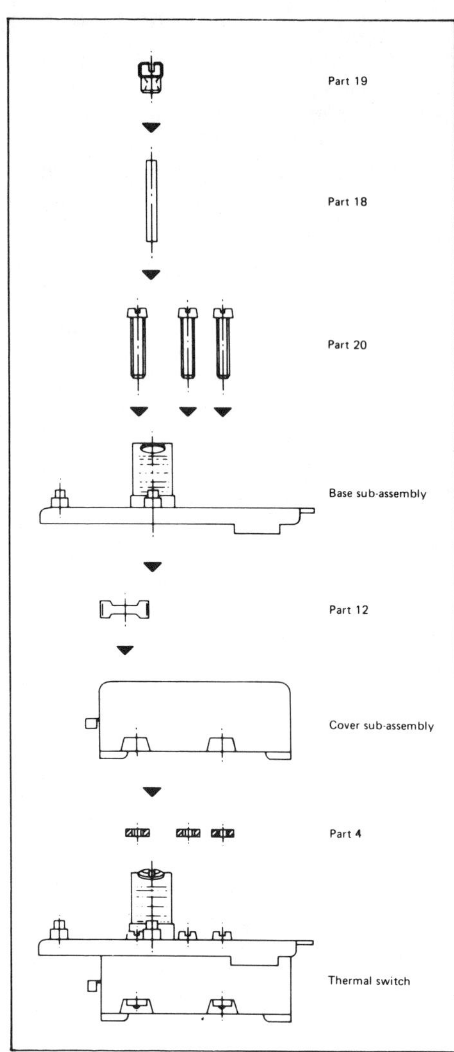

Fig. 5. Final assembly.

sub-assembly is fed to the following machine on a twin track conveyor (see Fig. 13).

Machine 3 performs the operations of setting the contact gap, running-in the microswitch element and measuring the microswitch spring pressure. Because the cycle for these operations is longer than 2.88s a duplex arrangement is employed and two sub-assemblies are processed in each cycle. The tested sub-assemblies are fed out on a single belt conveyor to the next machine.

Machine 4 assembles two connector strips (part 11) each with a terminal and screw (parts 8 and 9) into the base sub-assembly. The two terminals plus screws are fastened together on two satellite machines similar to those of machine 2 (see Fig. 10). The connect strips are fed as blanks to the indexing plates of the two satellite machines, finish formed and pre-assembled with the two terminals. The strips plus terminals are then loaded into the base sub-assembly two-at-a-time.

Machine 5 assembles a third connector strip and terminal and fits this into the base in a similar way to that of machine 4. In theory it would be possible to assemble all three connector strips and terminals into the base part on one machine but this would require a large number of work stations. Reliability of the assembly line would be reduced by such a large machine so this second machine is used.

Machine 6 commences the assembly of the cover shown in Fig 4. On this machine the cover lid (part 13) is fitted with three rivets (part 14) which are fastened using the orbital rivetting process.

Machine 7 completes the assembly of the cover. The bi-metallic strip (part 17) and cap (part 16) are fastened to the cover lid with two rivets (part 15) using orbital rivetting. On completion the cover sub-assembly is lifted out, turned through 180° and placed on the conveyor for transfer to the next machine.

Machine 8 assembles the base sub-assembly to the cover sub-assembly (see Fig. 5). The two sub-assemblies together with a terminal insulating cover (part 12) are fastened together by three screws and nuts (parts 20 and 4).

Machine 9 fits the thermostat actuating pin (part 18) and adjusting screw (part 19) to the sub-assembly and carries out adjustment. As can be seen from the function analysis this adjustment process takes 3.5s compared to the required cycle time of 2.88s. Therefore, this operation is duplicated. When the adjustment has been completed the adjusting screw (part 19) is secured by applying a resin and the finished assembly is automatically unloaded.

In order to assess an investment the

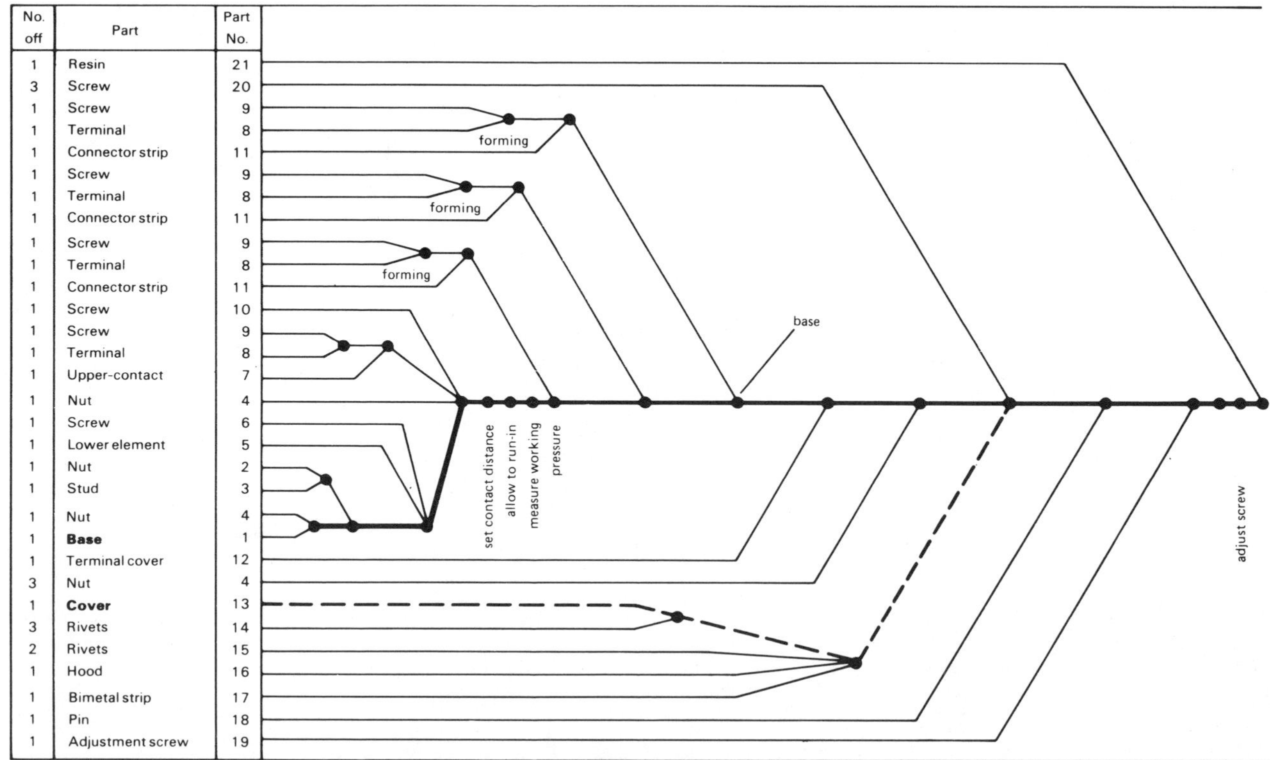

Fig. 6. Order of assembly processes – sequence of operations.

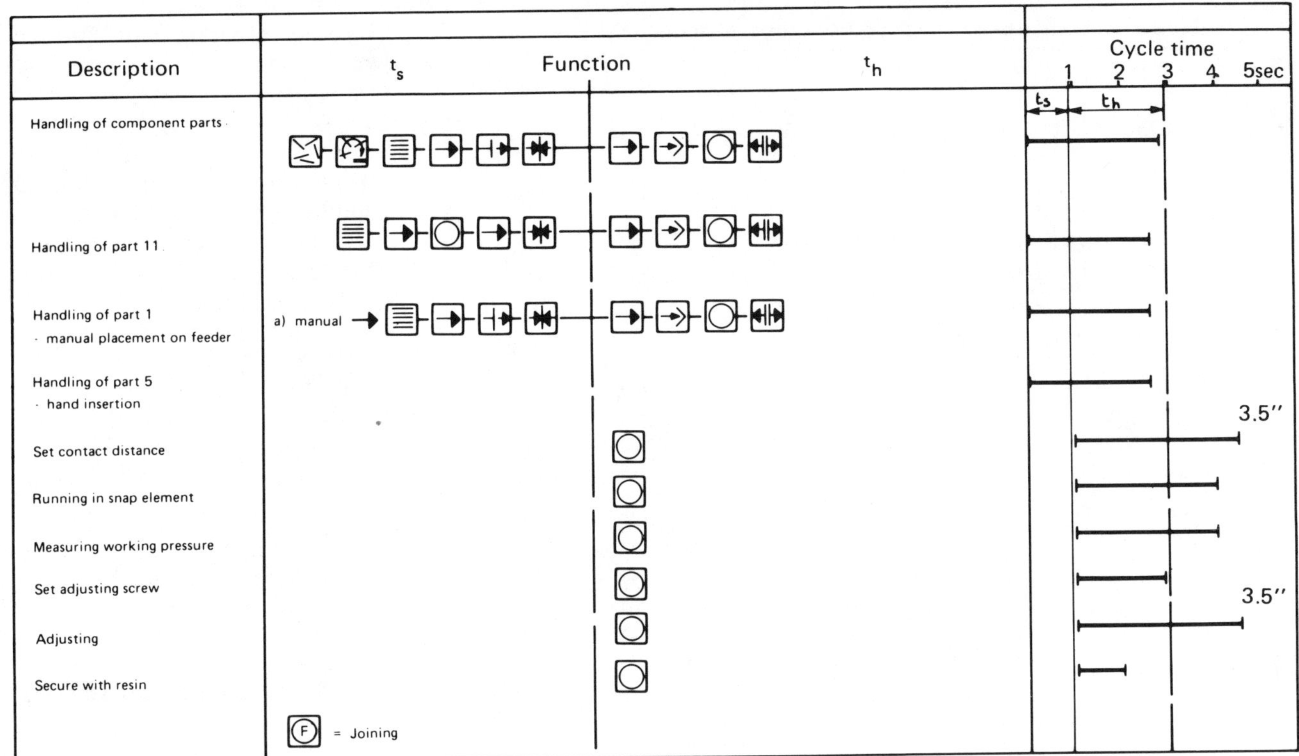

Fig. 7. Function analysis.

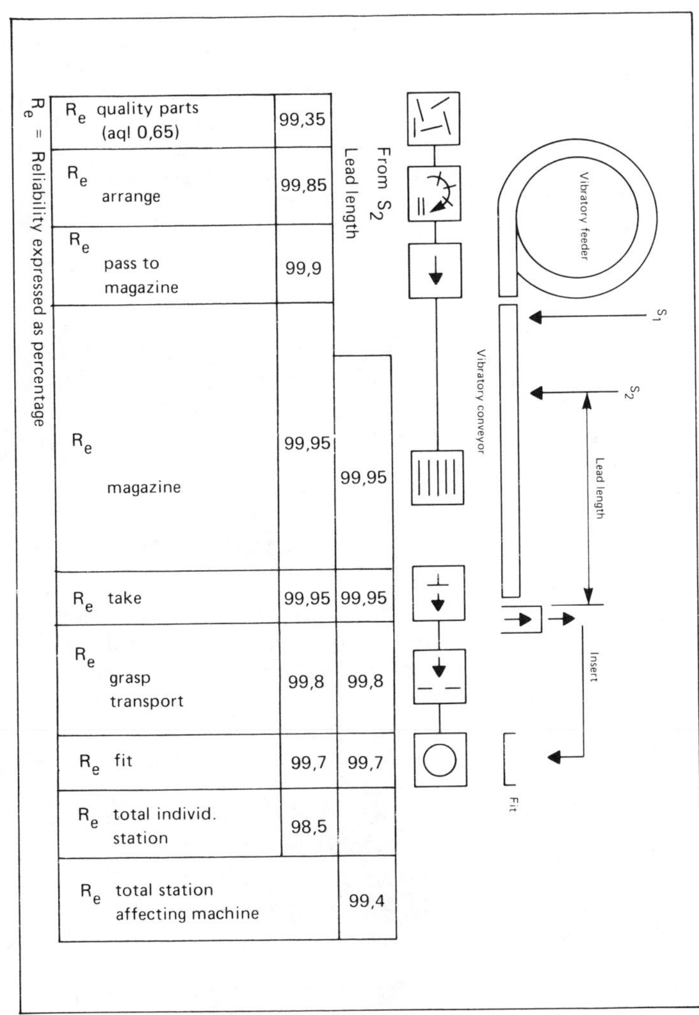

Fig. 8. Illustration showing reliability of a feed-fit station with lead length (S_1–S_2 = signal generator).

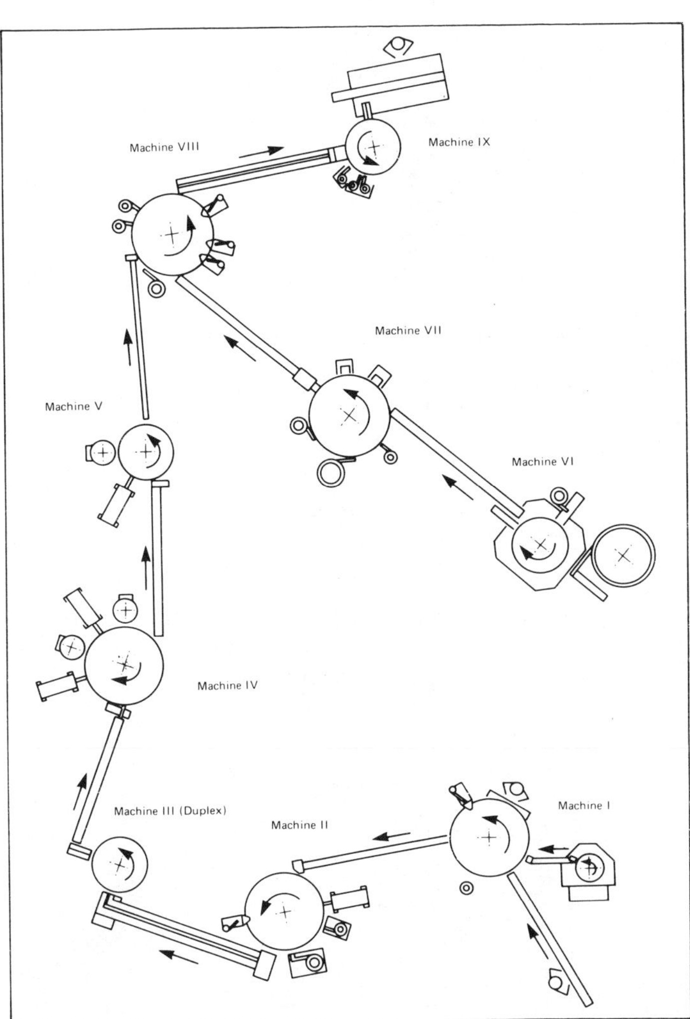

Fig. 9. Layout of assembly line for thermal switch.

Table 2. Calculation of assembly costs for manual method (MTM)

1. Machine – hour rate	$K_A = \frac{\text{DM } 704{,}700.00}{4 \text{ years}}$	DM 176,175.00
	$K_Z = \frac{\text{DM } 704{,}700.00}{2} \times 10\%/\text{year}$	DM 35,235.00
	$K_R = 280\text{m}^2 \times \text{DM } 120.00/\text{year}$	DM 33,600.00
	$K_E = 30 \text{ KWh} \times \text{DM } 0.20 \times 3{,}520$	DM 21,120.00
	$= 60 \text{ m}^3 \times \text{DM } 0.08 \times 3{,}520$	DM 16,896.00
	$K_I = \text{DM } 704{,}700.00 \times 10\%$	DM 70,470.00
		DM 353,496.00
	$K_{MH} = \frac{\text{DM } 353{,}496.00}{3{,}520^h}$	DM 100.42
2. Personnel expenses/hr.	49 persons (women) × DM 11.42 + 110%	DM 1,151.14
	2 Foremen × DM 13.90 + 110%	DM 58.38
	Shift allowance 48 × DM 0.53	DM 25.44
	2 × DM 0.65	DM 1.30
	Supervisor $\frac{\text{DM } 3{,}600.00 + 110\%}{172.33^h}$	DM 43.86
	Expenses per hour	DM 1,280.12
3. Assembly costs	1. Machine-hour rate	DM 100.42
	2. Personnel expenses	DM 1,280.12
	Assembly costs	DM 1,380.54
	Production rate 1,000 pieces/hour	
	Assembly costs per piece $\frac{\text{DM } 1{,}380.54}{1{,}000}$	DM 1.38

Table 3. Calculation of assembly costs for automated assembly method

1. Machine – hour rate	$K_A = \frac{\text{DM } 2{,}369{,}250.00}{4 \text{ years}}$	DM 592,312.50
	$K_Z = \frac{\text{DM } 2{,}369{,}250.00}{2} \times 10\%/\text{year}$	DM 118,462.50
	$K_R = 160\text{m}^2 \times \text{DM } 120.0/\text{year}$	DM 19,200.00
	$K_E = 40 \text{ KW}^h \times \text{DM } 0.20 \times 3{,}520$	DM 28,160.00
	$= 130 \text{ m}^3 \times \text{DM } 0.08 \times 3{,}520$	DM 36,608.00
	$K_I = \text{DM } 2{,}369{,}250.00 \times 10\%$	DM 236,925.00
		DM 1,031,668.00
	$K_{MH} = \frac{\text{DM } 1{,}031{,}668.00}{3{,}520^h}$	DM 293.08
2. Personnel expenses/hr.	11 persons (women) × DM 11.42 + 110%	DM 263.80
	2 Set-up Men × DM 15.20 + 110%	DM 63.84
	Shift allowance 11 × DM 0.53	DM 5.83
	2 × DM 0.71	DM 1.42
	Supervisor $\frac{\text{DM } 3{,}600 + 110\%}{172.33^h}$	DM 43.86
	Expenses per hour	DM 378.75
3. Assembly costs	1. Machine-hour rate	DM 293.08
	2. Personnel expenses	DM 378.75
	Assembly costs	DM 671.83
	Production rate 1,000 pieces/hour by 80%	
	Assembly costs per piece $\frac{\text{DM } 671.83}{1{,}000}$	DM 0.672

economic efficiency of the available production methods has to be investigated. It is necessary to assign to each of the alternative methods the same relevant costs assuming the same output and the same methods of calculation. A method using machine-hour rates, or alternatively a place cost investigation is suitable for this task.

The investigation of the described assembly line for the thermal switch was based on the comparison of two production methods:

- Method 1 – manual assembly in accordance with MTM on individual assembly station with partial loose interlinking.
- Method 2 – mechanised assembly.

The costings for the comparison are as itemised in Table 1.

Individual assembly

For **manual assembly** (method 1), planning in accordance with MTM on individual assembly places with partial loose interlinking gave a standard time of 3.9 min. per unit. Assuming an average efficiency of 135% this gives an effective working time of 2.88 min. per unit. With a planned outyput of 1,000 units per hour 2,880 min. of working time must be available which requires $\frac{2880}{60} = 48$ work places or productive personnel.

The calculated investment required to create 48 work places in accordance with the production process in DM 580,000.00. Its replacement value is DM 580,000.00 × 1.215 = DM 704,700.00. The calculation of the assembly costs are given in Table 2.

Procurement

The **mechanised assembly** method as described (method 2) requires the following production personnel:

Machine 1 – 1 operator to load the base,
– 1 operator to place microswitch element.
Machines 2-8 – 1 person each = 7 operators (supervision).
Machine 9 – 1 person (supervision),
– 1 person (removal and packing).

This gives a total of 11 operators for method 2.

The procurement cost for the automatic system is DM 1,950,000.00, which has a replacement value of DM 1,950,000.00 × 1.215 = DM

2,369,250.00. The calculation of the assembly costs for this method are shown in Table 3.

A comparison of the two manufacturing methods gives:

- assembly costs with manual production = DM 1.380 per unit
- assembly costs with mechanised production = DM 0.672 per unit
- Thus the saving with the mechanised method =

DM 0.708 per unit

This means an annual saving at 16,000 units per day × 220 days × DM 0.708 of **DM 2,492,160.**

The amortisation period of the assembly line for the thermal switch is calculated as follows:

$$\frac{\text{capital investment}}{\text{saving} + \text{annual depreciation}} = \frac{\text{DM } 1{,}950{,}000.00}{\text{DM } 2{,}592{,}160.00 + \text{DM } 592{,}312.00} = \mathbf{0.63\ years}$$

The short amortisation period is due to two-shift operations. With single shift operation the K_{MH} is increased from DM 293.08 to DM 586.16, so that the saving in assembly costs is reduced from DM 0.708 to DM 0.422 per unit. So the annual saving is 16,000 × 220 × 0.422 = **DM 1,485,440.**

Thus the amortisation period for single shift operation is:

$$\frac{\text{capital investment}}{\text{saving} + \text{depreciation per ann.}} = \frac{\text{DM } 1{,}950{,}000.00}{\text{DM } 1{,}485{,}440.00 + \text{DM } 592{,}312.00} = \mathbf{0.94\ years}$$

These calculations show that with a high capital investment it is necessary to operate more than one shift in order to reach a short amortisation period.

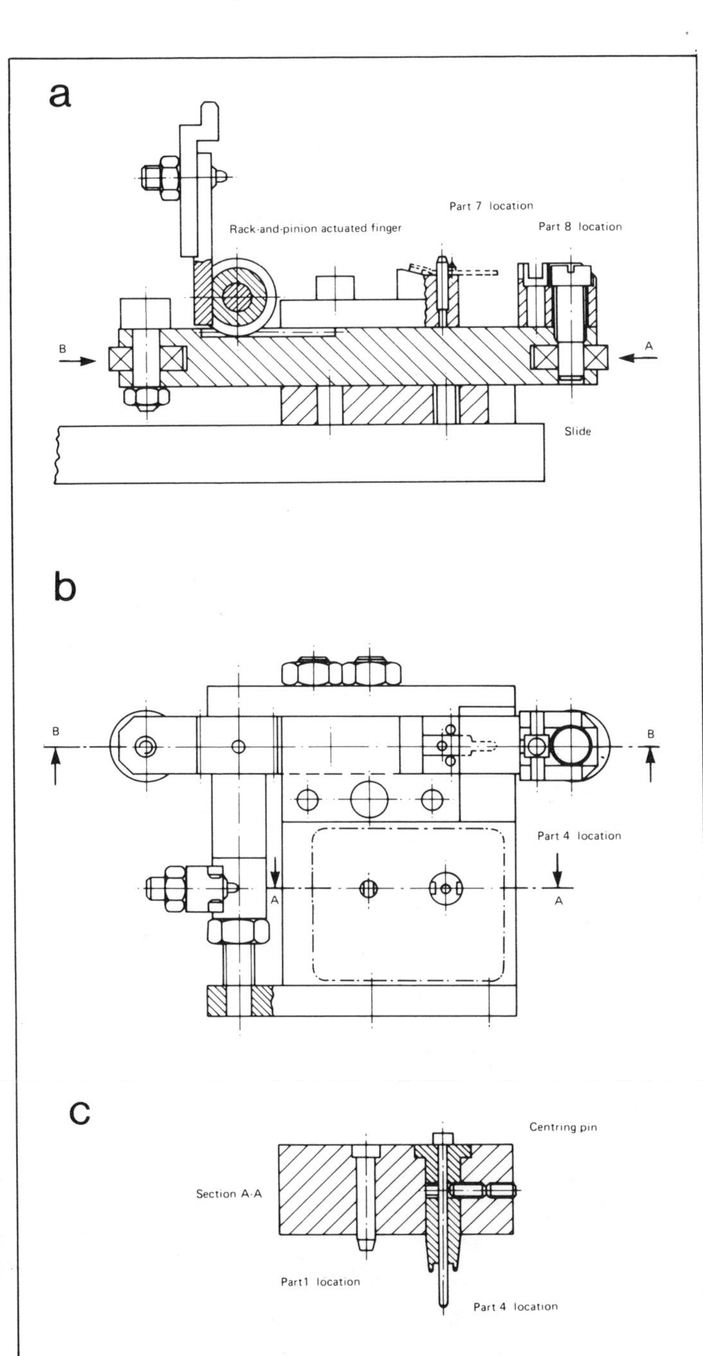

Fig. 10. Work carrier design for machine 2.
a. Section BB. b. Plan. c. Section AA.

Bibliography

Bruno Lotter, Workbook for assembly techniques (Arbeitsbuch der Montagetechnik). Vereinigte Fachverlage Krausskopf-Ingenieur Digest, Mainz 1982.

Bruno Lotter, Assembly primer for management (Montagefibel für das Management). Vereinigte Fachverlage Krausskopf-Ingenieur Digest, Mainz 1982.

Fig. 11. Satellite assembly machine for fastening terminal and screw (parts 8 and 9).

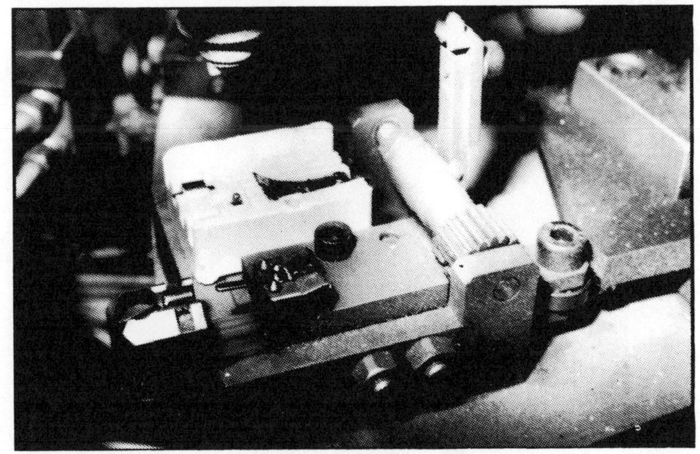

Fig. 12. Work carrier for machine 2.

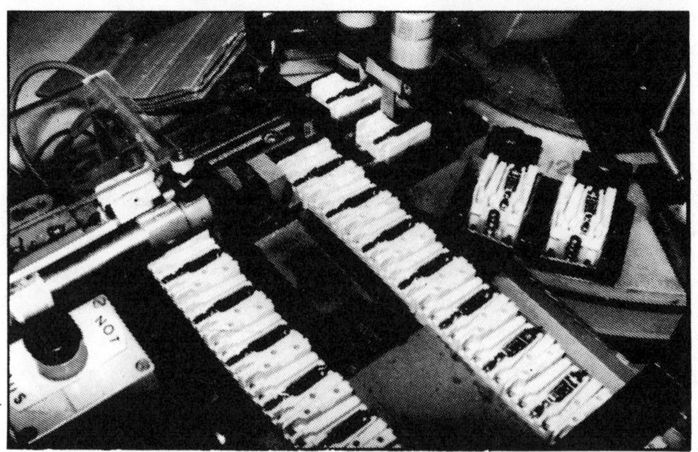

Fig. 13. Duplex loading station on machine 3.

Reprinted courtesy of Robert L. Douglas,
Gilman Engineering & Manufacturing Company

Cost Effectiveness of Automatic Assembly Systems

by Robert L. Douglas
Gilman Engineering & Manufacturing Company

Introduction

Low-cost computing power available in personal computers, coupled with studies published by G. Boothroyd,[1.,2.] can provide the tools for analyzing the cost effectiveness of automatic assembly systems. In the following examples, the personal computer is used to:

1. Simulate an assembly system to determine the mean time to produce each completed assembly;
2. Calculate the cost to produce each good product.

The cost calculation is based on modifications to Boothroyd's equations which have the form:

$$C = M*t + M*K*T$$

Where M is the cost of operating the machine per unit time if only good parts are produced;

t is the machine cycle time when producing only good parts;

K is a downtime factor dependent on incoming parts quality, number of loading stations, the up-time of each station and the characteristics of this machine;

and T is the time to repair the machine and repair the faulty product.

Other variables also considered in these terms include: machine costs, manual operator costs, annual production required, number of shifts of operation, years of product life and the value of the machine at the end of the life of the product.

The following illustrates the application of Boothroyd's study to existing assembly systems, tests his conclusions and describes how the addition of computer simulation, based on personal computer graphics, significantly extends the value of this landmark work.

Simulation has long been touted as a useful tool for predicting system performance.[3.,4.] With the emphasis on flexible manufacturing systems, simulation is gaining widespread acceptance. It is, in fact, becoming a necessity in testing configurations to determine system performance prior to building the actual system.[5.]

Basic Analysis

Boothroyd develops the equations relating the cost to produce an acceptable, completed assembly for various assembly systems. The assembly cost per acceptable assembly, when plotted as a function of annual production volume, identifies the annual production domain in which a specific system is most economical.

For instance, Figure 1, generated using Boothroyd's formulae, reveals that manual assembly is the low cost means of production up to about 200,000 units per year. From 200,000 units per year to 1,000,000 units per year, programmable assembly is most cost effective. When the annual production exceeds 1,000,000 units per year, dedicated, hard automatic assembly is most cost effective.

The example shown in Figure 1 incorporates some specific assumptions including:

1. The assembly includes 50 parts to be loaded at a rate of 5 seconds per part;

2. There are no design changes during the 3-year life of the product;

3. Manual labor is calculated on a rate of $25,000/yr.

4. Automatic assembly is calculated on the basis of $40,000 per loading station for a free-transfer assembly machine.

5. One percent of all parts are defective and cause the load station to stop for 30 seconds when the defective part is loaded.

Figure 1 represents the assembly system application for a mature, stable product. Historically this has been the case in the auto industry. Programmable assembly, defined as robots in a cell configuration, or as stations on a parts transporter, is only now becoming a reality.

The cost effectiveness of programmable assembly systems remains academic, inasmuch as there are few system to use as a measure. Robot manufacturers complain that they are "handicapped by traditional economic justification guidelines".[6.] However, in most companies, robots will ultimately be judged on traditional guidelines just as hard automation is judged.

Boothroyd states a commonly held belief that, "It is clear that the design of products for assembly would extend enormously the area of application of programmable assembly automation."[2.] This

Figure 1

DEDICATED
MANUAL
PROGRAMMABLE
COST TO PRODUCE COMPLETE PRODUCT
DOLLARS($)
2.0
1.0
0
1
2
ANNUAL PRODUCTION VOLUME
(MILLIONS OF ASSEMBLIES PER YEAR)

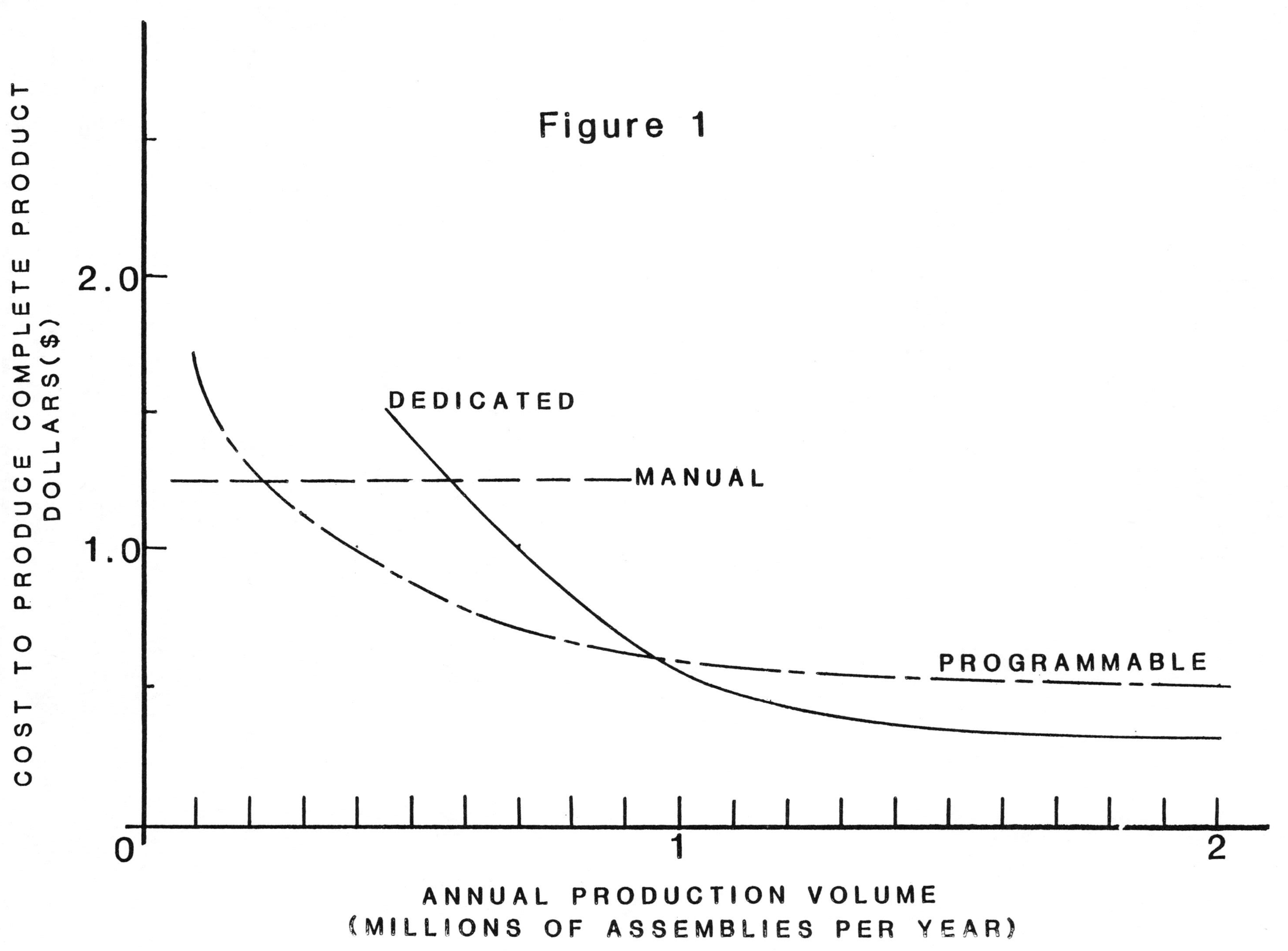

belief must be critically examined before it is accepted. There is evidence that when a product _is_ designed "for assembly", one of two scenarios follow:

1. The ease of manual assembly makes automation uneconomical at all, or

2. Hard automation costs are reduced and it becomes cost effective at lower annual production rates.

Therefore, the area of application for programmable assembly is reduced under these conditions, as it is squeezed by manual assembly and hard automation simultaneously from both sides.

Boothroyd has also developed equations for the various product conditions. Conditions considered include; short product life, multiple design changes during the product's life, several styles of the same product, and even multiple products to be assembled on a given system. These conditions require flexibility in the assembly system and programmable assembly should become cost effective over a much wider range of annual production volumes.

Using these equations a reasonably accurate assessment of the best assembly system to use, given the required production volume of the product or products to be produced, can be made. While one may wish to quarrel with specific assumptions made by Boothroyd, the analysis, as a whole, is first rate and generalized enough to be usable.

One element not covered by Boothroyd is the effect of downtime due to machinery malfunction. He covers the effect of downtime due to unacceptable part quality and it is a relatively simple addition to his equations to include the effect of machinery failure.

Another element not covered by Boothroyd is the effect of machine memory on the cost of producing each good part. Most assembly machines offered for sale today have some sort of machine memory. That is, if a reject assembly is detected, the parts continue to move through the machine with no further work done. The assembly is segregated from good assemblies at the unload area automatically to allow for manual repair.

Machine memory in the form of control logic, or mechanical pins remembers the status of every assembly on the machine and controls the subsequent stations accordingly. This is a powerful feature of modern assembly machines because the machine is allowed to run without stoppages due to poor part quality. Of course, the cost to repair the parts, off-line must be included in the analysis.

Figure 2 illustrates how the cost to produce each good, final assembly is affected by the number of parts in the assembly when assembled automatically on an indexing machine. The parts are considered to be 99% good parts. In addition, the upper curve illustrates the cost when each rejected part stops the machine and 30 seconds is required to repair the assembly. The lower curve illustrates the cost when machine memory is used and the machine does not stop on reject, but rather an operator repairs all rejects, off-line in 30 seconds.

From Figure 2, it is apparent that an indexing machine stopping when each reject is produced requires considerably more dollars to produce each good assembly. For instance, if there are 25 parts to be assembled on the machine, the cost with the machine running "stop on reject" is .665 dollars. When the same machine is running with the machine memory, the cost to produce each good, completed assembly is .414 dollars. This is approximately 25 cents difference per completed product. If the machine produces 1,000,000 parts per year, the cost saving to use the machine with memory is $250,000. The cost to provide machine memory on a 25-station machine is probably less than 1/10 of that figure. Therefore, machine memory is a powerful feature which is used to reduce automatic assembly cost.

It should be noted that if a machine is running slowly, with a long cycle time, for instance approaching that time required to repair a reject, the cost effectiveness of machine memory is greatly diminished.

Figure 3. is another way of presenting the information as shown in Figure 2. In this case, the Y axis is shown to be cost to assemble each component into the product. This is compared to Figure 2. where the Y axis is the sum total cost of all parts added to the product. It is noted in Figure 3. that there is a definite minimum cost and ideally it is at this minimum point that the optimize indexing machine would function. For instance, if the machine is running as "stop on reject", the lowest cost to produce a part is found at about 6 or 7 parts in the assembly. Many products require more parts to be loaded than that. The minimum point for an indexing machine using memory, however, is shown to be in the area of 15 to 16 parts, and the curve is relatively flat in this area. It would appear that 40 parts can be assembled on

Figure 2

STOP ON
REJECT
MACHINE
MEMORY
.665
.414
1.0
.8
.6
.4
.2
0
10
20
30
40
COST TO PRODUCE COMPLETE PRODUCT
DOLLARS($)
NUMBER OF PARTS TO BE ASSEMBLED IN EACH PRODUCT

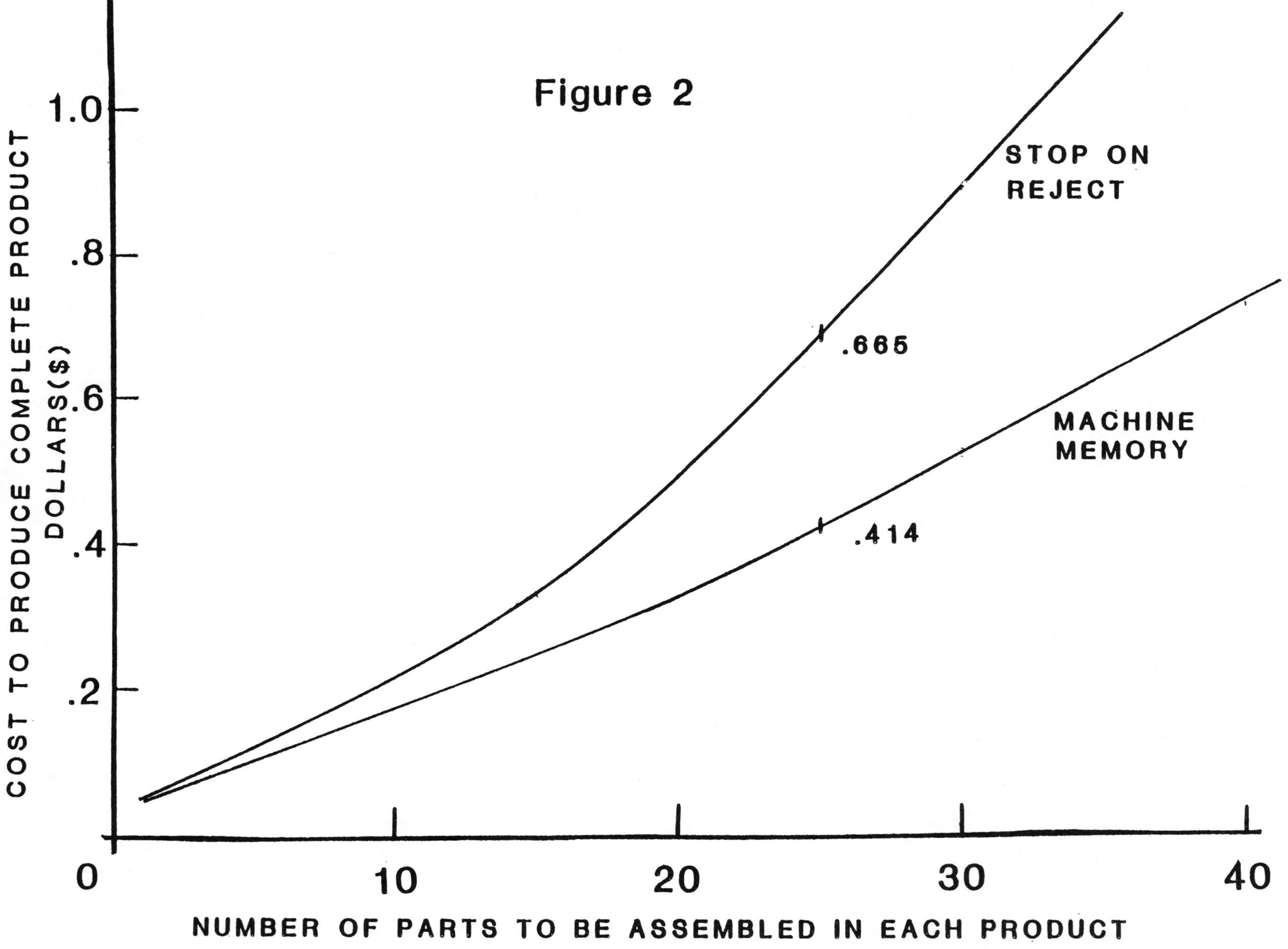

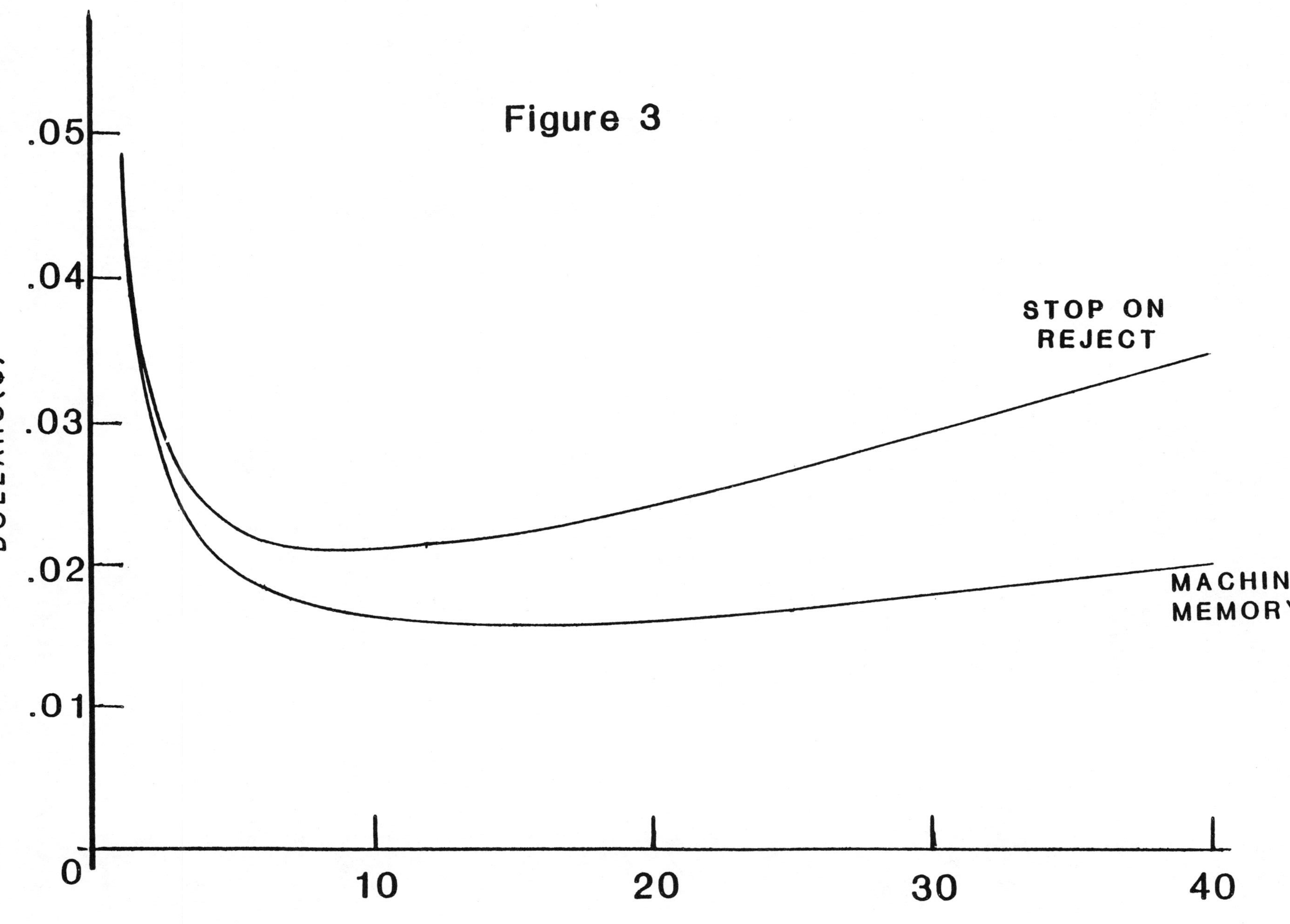
Figure 3
.05
.04
.03
.02
.01
0
10
20
30
40
COST TO ASSEMBLE EACH COMPONENT INTO PRODUCT DOLLARS($)
STOP ON REJECT
MACHINE MEMORY
NUMBER OF PARTS TO BE ASSEMBLED IN EACH PRODUCT

an index machine with memory at about the same cost as it would take to assemble a product with 5 parts. As has been known in the assembly industry, the addition of machine memory allows for many more load stations to be added to an indexing machine and maintain essentially the same cost effectiveness.

Figures 1, 2 and 3 all are developed using Boothroyd's formulae.

MACHINE SIMULATION

Boothroyd generates a series of sophisticated equations which allow for approximation of the net production rate of a free transfer or power and free assembly machine. A free transfer machine is one in which stations are positioned along a conveyor which moves the part from station to station to be worked upon. Each station works independently of every other station in the system and works on a part when the particular part arrives at the station. The system works asynchronously and storage spaces, or storage queues, are provided between stations. When a station has experienced a malfunction, the performance of a free transfer system is something of an average. This average is considerably higher than the performance of an indexing machine wherein each and every malfunction causes a machine stoppage.

Boothroyd's equations present difficulty in using. They require an itertive process whereby answers are achieved and checked. Then assumptions are made and the whole process repeats. It has been found that much more suitable answers can be achieved by computer simulation of asynchronous free transfer systems. Gilman Engineering & Manufacturing Company has developed a simulation program that runs in Basic on a personal computer and allows for system productivity data to be generated for almost any configuration assembly system. The simulation allows for different types of stations to be positioned in the system at any prescribed position required by the analyzer and will allow for stations of different failure rate, cycle time, and efficiencies to be incorporated into one system. In addition, the graphic display or pixels available on the personal computer display allow for a human understanding of blockages, and other phenomena that may cause a deterioration in the system performance.

The simulation data which as one of its outputs provides the average time to produce each good, completed assembly is coupled with the equations provided by Boothroyd to provide for a clear and valuable technique for evaluating a specific assembly system.

For instance, by simulation, it can be shown that by adding machine memory to an 8-station assembly machine the production rate, is increased by 10% for an indexing machine and 6% for a free transfer, or power and float machine. The results agree well with the purely analytical result shown in Figure 2.

Another simulation reveals that if a repair loop is added to an 8-station free transfer machine, to allow for repair of all rejects before they leave the system, the system will increase its production rate, by 3% if memory is already present.

Percentage increases of 3 and 6% may not appear to be significant, however, when assembly machines of this type are producing at a rate of 1,000,000 or more parts per year, per shift, the improvement of a couple percentage points provides huge returns to the user's equipment. For instance, for each percentage improvement in thruput, along with this corresponding 1% drop in assembly cost, the manufacturer can realize 10,000 more parts produced per year at no additional cost.

Conclusion

Machine memory and repair loops are but two of many features that can be analyzed using computer simulation and Boothroyd's formulae. Use of the tools described can provide for:

1. Cost justification for proposed assembly systems;
2. Justification for using state-of-the-art features;
3. Cost effective alternatives and options to be studied analytically rather than intuitively; and
4. Optimization of design features of any system under consideration.

REFERENCES

1. Boothroyd, G., "Economics of Assembly System", Journal of Manufacturing Systems, Vol. 1, No. 1, 1982, p. 111.
2. Boothroyd, G., Poli, C., Murch, L., Automatic Assembly, Marcel Dekker, Inc., New York, 1982, p. 201-254.

3. Metzger, R. W., "The Performance of a Series System of Production Stations Separated by Limited Inventories", PhD Dissertation 63-8069, The University of Michigan, 1963.

4. Church, Janis, "Simulation for Cost Avoidance in Assembly", SME Technical Paper AD76-671.

5. Finkel, James I., "Computer Model for Flexible Manufacturing", Computer--Aided Engineering, July, August, 1983, Vol. 2, No. 4, p. 52.

6. Van Blois, J. P., "Strategic Robot Justification: A Fresh Approach", Robotics Today, p. 44, April, 1983.

CHAPTER 2

DESIGN FOR ASSEMBLY

M.F. Cobb

Design-for-assembly is the key to increased productivity

A new software system provides the design engineer with an effective method for incorporating into the product design the requirements for efficient assembly.

In our society, manufacturing is the major contributor to the wealth-producing activities that establish our standard of living. Recently, other countries seeking to raise their standard of living have begun to effectively compete with the U.S. in a world market. Studies of these new competitors show that manufacturing productivity is a key indicator of a nation's ability to compete.

Historically, productivity has been increased through innovation, price increases, and large investments in capital equipment; however, in recent years, increased competition has limited price increases, and higher interest rates have diminished returns from capital. The most innovative product with the most up-to-date manufacturing methods will not attain its profit goals if it cannot be efficiently manufactured.

Abstract: *In a competitive market such as consumer electronics, companies can no longer afford the lost profit opportunities and low productivity resulting from redesign cycles required to correct assembly problems. Research has shown that 70 percent of the product's cost is determined during design and only 20 percent during actual production. In other words, to achieve maximum productivity, design and manufacturing must be considered as one continuous process. The task then is to blend advanced technology—and the creative efforts of the design and manufacturing engineer—into the new product, without sacrificing time to market. To assist in this effort, a system called "Design for Assembly" or "UMASS" has been developed by Dr. G. Boothroyd and his associates from the University of Massachusetts, in collaboration with the University of Salford Industrial Center in England. This paper is a brief discussion of this "Design for Assembly" system.*

Final manuscript received January 17, 1984.
Reprint RE-29-2-1

Many companies, recognizing the diminishing returns of conventional productivity-improvement remedies, have placed greater emphasis on the study of assembly. Traditionally, assembly problems have been regarded as product oriented, to be solved during the production process. Research has shown, however, that assembly costs are fundamentally established in product design.

In general, the ease with which a product is assembled is dependent upon communication between design engineers (in different disciplines) and manufacturing. Ineffective communication, absence of guidelines, and the design engineers' lack of manufacturing experience can create serious manufacturing problems. As a result, productivity suffers and profit goals are compromised.

In an attempt to eliminate these problems within the VideoDisc player, RCA Consumer Electronics' Manufacturing Technology was given the responsibility of working in concert with Design to provide information that would result in a player that would:

- Provide the necessary function at the lowest cost;
- Be compatible with the manufacturing system;
- Assemble easily; and
- Achieve high "first-pass" yields through manufacturing.

To achieve these goals, everyone involved needed to know the requirements for effective assembly early in the design cycle. However, because of the vast amount of information involved in manufacturing, it was difficult to develop *comprehensive* assembly guidelines for design. Consequently, an inordinate amount of time could be spent in meetings defining assembly requirements on a part-by-part, assembly-by-assembly basis.

The "Design for Assembly" program

During this time, we became aware of a system called "Design for Assembly," or "UMASS," developed by Dr. G. Boothroyd at the University of Massachusetts under a grant from the National Science Foundation. "Design for Assembly" is the result of approximately 15 years of research at the University of Massachusetts, in conjunction with the University of Salford in England. Recently, the UMASS system has been developed into a series of six interactive microcomputer software programs compatible with the IBM Personal Computer in a 128K configuration. Before getting into the actual details of the system, I would like to relate a recent example as an illustration of the system's potential value.

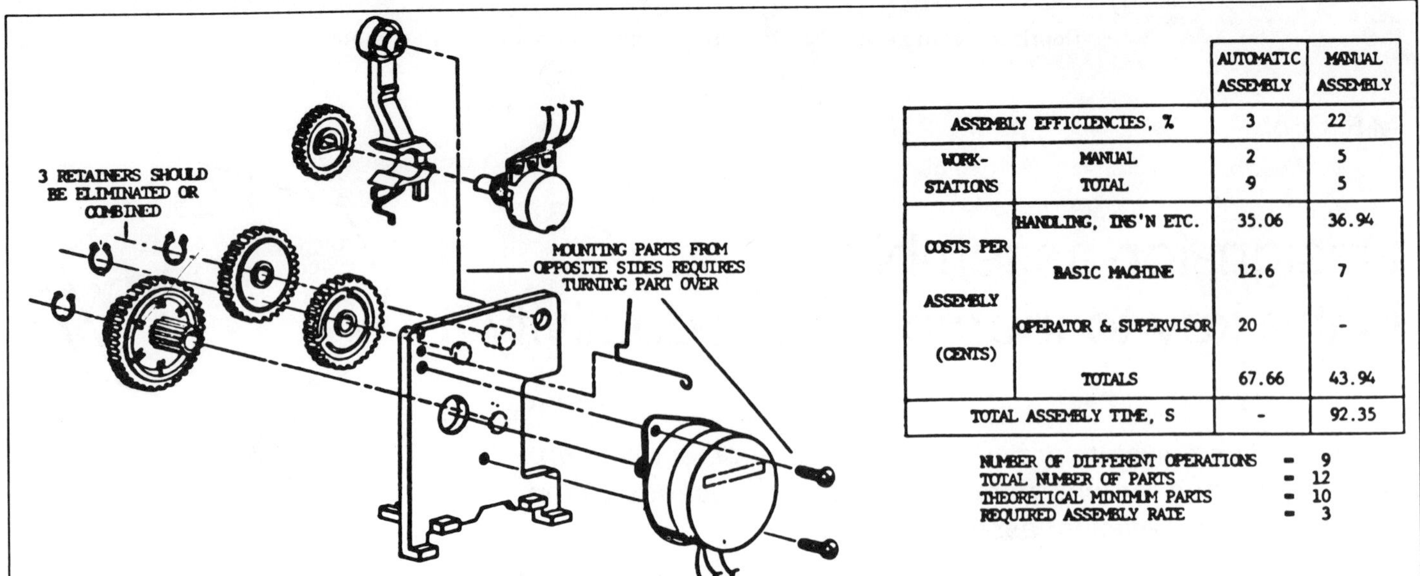

		AUTOMATIC ASSEMBLY	MANUAL ASSEMBLY
ASSEMBLY EFFICIENCIES, %		3	22
WORK-STATIONS	MANUAL	2	5
	TOTAL	9	5
COSTS PER ASSEMBLY (CENTS)	HANDLING, INS'N ETC.	35.06	36.94
	BASIC MACHINE	12.6	7
	OPERATOR & SUPERVISOR	20	-
	TOTALS	67.66	43.94
TOTAL ASSEMBLY TIME, S		-	92.35

Fig. 1. *Arm-drive gear assembly and related UMASS analysis summary. Failure to achieve the theoretical minimum number of parts is the main determinant for further analysis.*

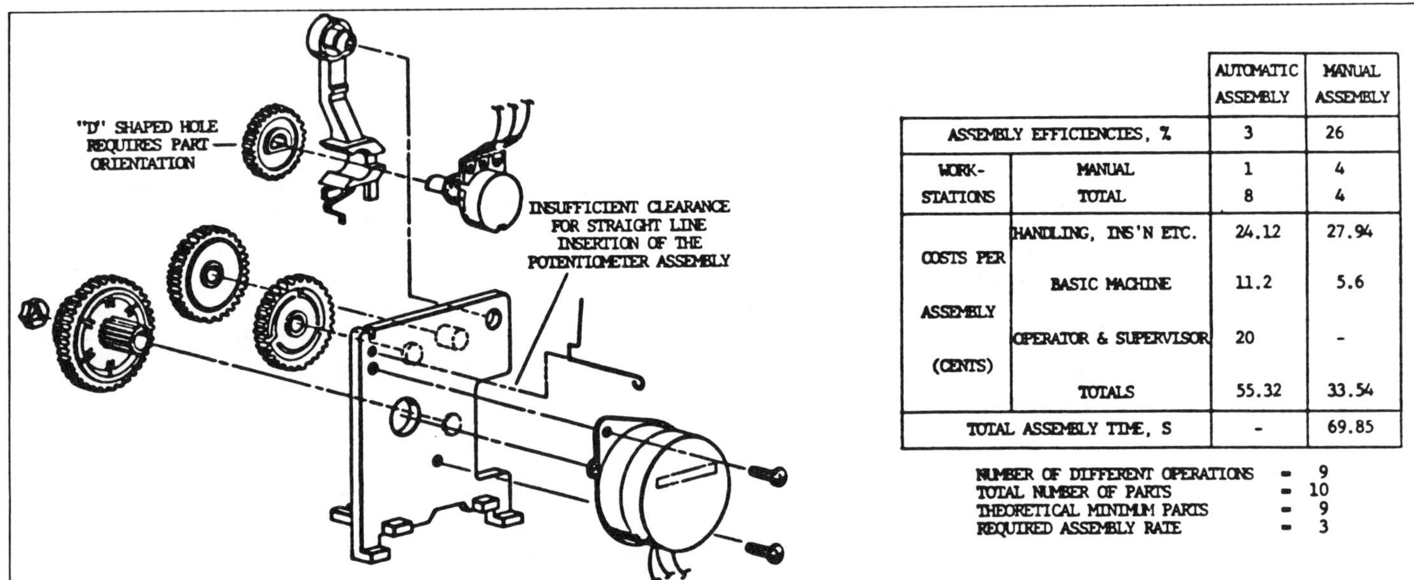

		AUTOMATIC ASSEMBLY	MANUAL ASSEMBLY
ASSEMBLY EFFICIENCIES, %		3	26
WORK-STATIONS	MANUAL	1	4
	TOTAL	8	4
COSTS PER ASSEMBLY (CENTS)	HANDLING, INS'N ETC.	24.12	27.94
	BASIC MACHINE	11.2	5.6
	OPERATOR & SUPERVISOR	20	-
	TOTALS	55.32	33.54
TOTAL ASSEMBLY TIME, S		-	69.85

Fig. 2. *Gear assembly as released to production, incorporating some UMASS recommendations that reduced assembly time by 22 seconds, but did not achieve the theoretical minimum parts count.*

Design for assembly and robotics

During the development of the 1983 VideoDisc player, the arm drive (Fig. 1) was identified as a mechanism having high potential for assembly by robotics. To enable the robot to achieve the desired output, engineers from Manufacturing Technology, Product Design, and RCA Laboratories, made several recommendations for simplifying the assembly. Figure 2 is the assembly as released to production, incorporating a few of the recommendations. The changes eliminated two retainers and substituted a push-on nut in place of the third retainer, resulting in major improvements in the ease of assembly. These studies and recommendations required approximately 300 person hours and $6,000.00 in tooling costs.

To evaluate the "Design for Assembly" system, we requested Dr. Boothroyd's analysis of the arm drive assembly (Fig. 1). The results were impressive. Dr. Boothroyd's recommendations closely paralleled our previous studies, but required only 12 hours, instead of 300 hours. Further, if either study had been performed at the design-concept stage, the $6,000.00 die cast tool, and related ECN and print change costs would not have been required.

Accompanying Dr. Boothroyd's recommendations was a software analysis and summary of results. This summary, with its assignment of relative measurements to design parameters, amplifies the value of the "Design for Assembly" system.

The analysis summaries shown with the assembly drawings can be used to compare design alternatives. They provide direction for achieving the lowest-cost assembly. Figure 3 is the gear assembly modified using the suggestions and recommendations

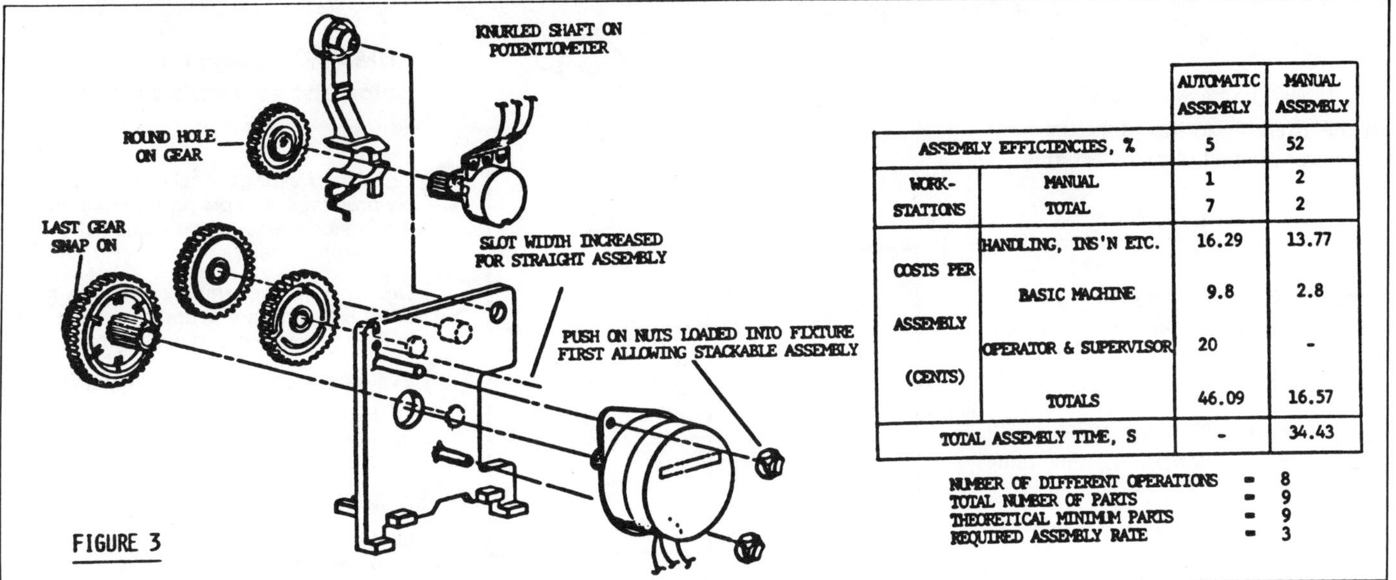

		AUTOMATIC ASSEMBLY	MANUAL ASSEMBLY
ASSEMBLY EFFICIENCIES, %		5	52
WORK-STATIONS	MANUAL	1	2
	TOTAL	7	2
COSTS PER ASSEMBLY (CENTS)	HANDLING, INS'N ETC.	16.29	13.77
	BASIC MACHINE	9.8	2.8
	OPERATOR & SUPERVISOR	20	-
	TOTALS	46.09	16.57
TOTAL ASSEMBLY TIME, S		-	34.43

NUMBER OF DIFFERENT OPERATIONS = 8
TOTAL NUMBER OF PARTS = 9
THEORETICAL MINIMUM PARTS = 9
REQUIRED ASSEMBLY RATE = 3

Fig. 3. *Recommended gear assembly achieving a stackable assembly and the theoretical minimum number of parts. The parts count was reduced by three and assembly time was reduced by 58 seconds.*

within the design-for-assembly guidelines. For all studies, the lever and potentiometer were considered as pre-assembled. Combining the three summaries in Fig. 4 illustrates the potential value that can be achieved by simplifying the assembly and eliminating parts. The savings shown are significant and achievable. These savings, if initiated in the original design, could be realized without capital investment, and without pilot runs, prototypes, or time-consuming group discussions.

Use of the system

This is not meant to imply that the software is creative. Rather, it is an easy-to-use interactive system that forces the designers to systematically review and consider each part in an assembly as to need, geometric relationship, ease of handling, and method of assembly. By assigning relative quantified values to criteria such as design efficiency, assembly time, and the theoretical minimum number of parts, the system provides direction to improve assembly and shows the relationship between a product's ease of assembly and its cost. In this sense, the software encourages the designer to be as creative as possible in examining the best options.

The six software programs are independent and can be used in any sequence; however, the procedure illustrated below is possibly the most time efficient:

Tips & practice

- Become familiar with design alternatives

Design for automatic assembly

- Achieve discrete part reduction
- Achieve layered assembly
- Simplify part mating to require the fewest degrees of part rotation
- Provide part locators to reduce assembly time

	No. parts	Manual assembly efficiency (%)	Assembly time (sec)	Relative % cost reduction	Increase pieces/ hour	Productivity (%)
Arm drive 1	12	22	92.35	BASE	40	BASE
Gears retained with 3 snap retainers						
Arm drive 2	10	26	69.85	24	52	+ 30
Gears retained with (1) locknut						
Arm drive 3	9	52	34.43	63	105	+162
3rd reduction designed for snap-on						
Pot gear redesigned to eliminate end-to-end orientation						
Mounting bracket cutout enlarged to allow perpendicular insertion of the lever & pot assembly						
Changed sequence of assembly loading pushnut into fixture first—achieving a stackable assembly						

Fig. 4. *Comparison of the three summaries, illustrating the main factors relating to reduced product cost and showing a potential gain in productivity of 162 percent.*

Assembly system economics

- Test the assembly to determine the most economical method of manufacturing

Either design for manual assembly . . .

- Optimize part geometry for manual assembly

Or design for automatic handling . . .

- Optimize part geometry for feeder design
- Estimate cost of automatic part-handling equipment
- Estimate cost of part-insertion equipment

Machine simulation

- Evaluate the effect of part quality on machine downtime
- Determine optimum buffer size
- Determine optimum number of support operators

Tips and practice

In this program, the user is provided tips to assist in assessing the features of individual parts for automatic handling. In addition, sample parts with a variety of geometric shapes for which part-handling codes have been determined are displayed. The user is then tested for selection of significant feature determination and part-size envelope. A pass/fail grade is displayed at the end of each attempt. This testing and prompt feedback assists the user in gaining familiarity and confidence with the methodology used in the "Design for Automatic Assembly" program.

Design for Automatic Assembly

This is the main program to be used for optimizing a product for manufacturing. The actual procedure for analyzing a product is relatively straightforward, as all inputs are the result of prompt statements and interactive screen displays. The flow chart (box) illustrates the procedure, and highlights the important elements in each step.

During execution, the program will determine for each part its general fastening method, geometric part relationship to insertion axis, physical size, envelope shape, amount of symmetry or asymmetry, potential handling or feeding problems, and poten-

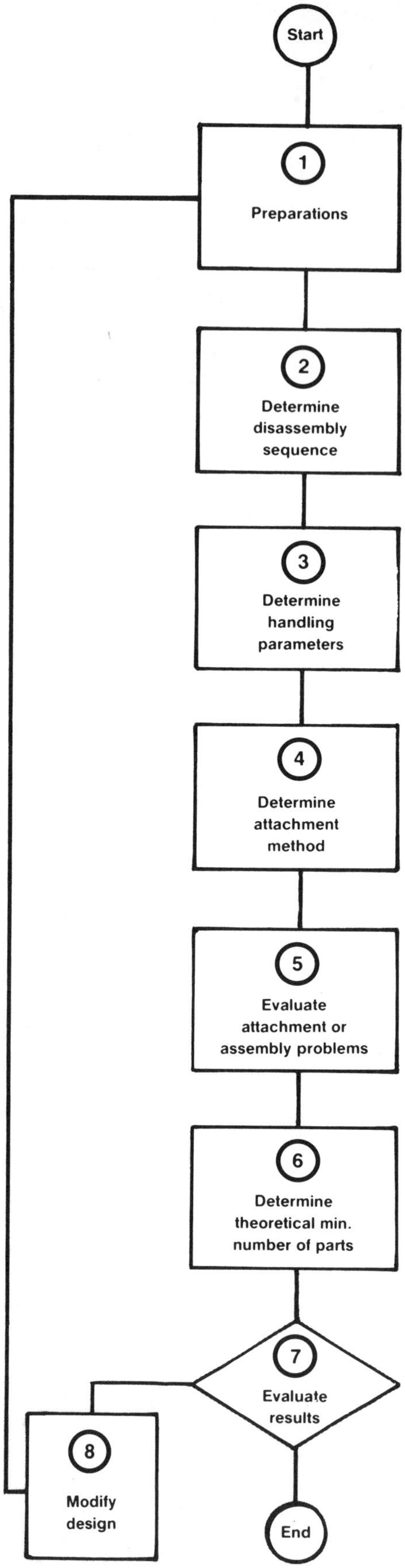

Start

1 Preparations

2 Determine disassembly sequence

3 Determine handling parameters

4 Determine attachment method

5 Evaluate attachment or assembly problems

6 Determine theoretical min. number of parts

7 Evaluate results

8 Modify design

End

"UMASS" design for automatic assembly procedure

1. Concept drawings, final drawings, assembly drawings, sample parts, exploded three-dimensional views, and if possible, prototype
2. Determine disassembly sequence, and enter the part names and number of parts in their sequence of disassembly
3. Evaluate each part geometrically for:
   (a) Handling
   (b) Feeding
   (c) Orienting
4. Determine method of attachment:
   (a) Screwing
   (b) Riveting
   (c) Snap fit
   (d) Bending or crimp
   (e) Bulk deformation
5. Evaluate each part attachment or fastening method for assembly insertion problems due to:
   (a) Alignment
   (b) Obstructions
   (c) Nonlinear motion during insertion
   (d) Resistance to insertion
6. Determine theoretical minimum number of parts. Parts should be combined unless:
   (a) They move during the assembly operation
   (b) A different material is required
   (c) They are required for assembly of other parts
7. Review and evaluate results printout for:
   (a) Cost
   (b) Minimum number of parts
   (c) Assembly efficiency
8. Use subprograms as necessary
   (a) Part design for automatic handling
   (b) Provide part locators
   (c) Quantity of parts reduction

Table I. *Basic assembly systems description.*

Code	Description
MA	Manual assembly using an in-line free-transfer machine with one buffer space between each manual work station.
MM	System MA, with mechanical assistance provided for the operators in the form, for example, of parts presentation devices.
AI	Indexing of synchronous automatic assembly machine; a rotary machine with six or fewer work stations, otherwise an in-line machine.
AF	In-line free-transfer or nonsynchronous automatic assembly machine with optimum buffer storage space for work carriers between each work station.
AP	In-line free-transfer or nonsynchronous assembly machine with robot work stations and parts presented in hand-loaded magazines; the number of work stations and the number of parts assembled at each station determined by the required assembly rate; the complexity of robots (three to six degrees of freedom) determined by the number of operations to be carried out by each robot.
AR	Two robot arms, each with six degrees of freedom sharing one work fixture; the parts presented in hand loaded magazines.

tial insertion problems due to obstructions or nonlinear assembly motions. From this information, the program determines relative values for assembly time, equipment cost for automatic and manual assembly, design efficiency, and the theoretical minimum number of parts.

The unique program capability to predict a *theoretical minimum number of parts* for a design is an important part of the UMASS system. Following considerable study, the approach adopted was that separate parts should be combined into one part, unless one of the following criteria are met:

1. Because the part *moves* during the operation of the assembly, and the movement cannot be achieved through a combination of parts in a flexible material.
2. Because a *different material* is required or because the part must be *isolated* for insulation purposes, and so on.
3. Because the part must be *disassembled* or because it must be separate to allow *assembly* of other parts.

It is important at this point in the program that the designer not be concerned with technical feasibility, but with identifying potential parts for elimination. For example, referring to Fig. 1, the three gear retainers are not justified by any of the three criteria and therefore should be eliminated.

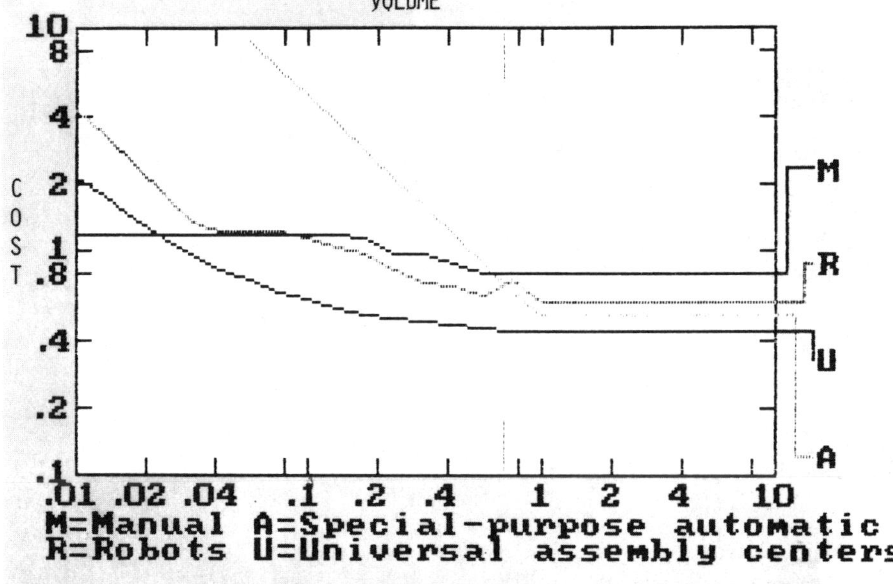

Fig. 5. *Plot, relating assembly cost to assembly systems, that considers factors of volume, parts quality, economic climate, and the number of parts, style variations, and design changes.*

Whether or not the parts are eliminated will depend on the creativity of the engineer.

The theoretical number of parts is then combined with values derived for discrete part assembly time and a comparable ideal part assembly time, to determine the design efficiency of the assembly. This efficiency is given as a relative value, reflecting how effectively an assembly has achieved the design guidelines. It also can be used to compare alternate designs. This does not mean an assembly should achieve a design efficiency of 100 percent. In fact, the typical electromechanical assembly efficiency will approximate 25 percent to 30 percent.

At the end of the Design for Automatic Assembly program, a results summary is displayed, comparable to Fig. 1, which gives a breakdown of the total cost for both manual and automatic assembly. Prior to or after the summary is displayed, the opportunity is provided to change cost and volume parameters to reflect actual costs or to evaluate future cost impact.

This feature provides the opportunity to quickly model and evaluate the cost effectiveness of different assembly concepts. In addition, the program provides a list of all parts in the assembly with their handling and insertion cost. In this way, the assembly cost associated with each part can be studied to determine how the total cost is distributed and to identify problem areas where changes would be most beneficial. Parts are listed for redesign if they are unnecessarily expensive to feed or impossible to feed and orient automatically. In these cases, a five-digit code is given for the part; the code can be used in the "Design for Automatic Handling" program to obtain suggestions for redesign.

During the analysis process, it should become obvious that maximum system efficiency and productivity will be realized only when the requirements of the manufacturing system and product design are integrated.

Assembly system economics

Assuming the Design for Assembly program has been completed and has resulted in parts reduction, has provided for mating part locators, and has led to the achievement of a layered assembly, the next step would be to determine the most economical method of manufacturing. To accomplish this, the "Assembly System Economics" program uses mathematical models to assimilate approximately thirty fixed and variable parameters of the assembly and work climate to determine the most economical assembly system from those listed in Table I.

During operation of the program, all parameters are shown on an interactive screen display. The option to change any value provides an opportunity to quickly model and evaluate the system sensitivity to a variety of conditions.

When the data base presented is satisfactory, the program will calculate and list relative assembly cost for each manufacturing system. The system also provides a graph (Fig. 5) that illustrates the assembly cost of manufacturing systems plotted against annual production volume per shift.

Design for Automatic Handling

This program is used to analyze individual parts for automatic feeding and orienting. Numerous suggestions for design changes are presented, allowing parts to be automatically handled more efficiently.

Design for Manual Assembly

This is used when automatic assembly cannot be justified. The program is similar to the automatic assembly program. Emphasis is again placed on the elimination or combination of parts, and simplification of the part-orienting required for assembly.

Assembly Machine Simulation

The effect of four parameters on the performance of an automatic assembly system are investigated in this program. These parameters are:

- Machine cycle time;
- Parts quality at each work station;
- Average time to clear a stoppage due to a faulty part; and
- Number of service operators available.

The first program is a graphical simulation of an automatic assembly system with five work stations, four buffer storage spaces between each work station, and one operator to clear malfunctions due to defective parts.

The second program is a free-form numerical simulation. In this program, buffer space between work stations, number of operators, time to clear malfunctions, and the ratio of defective parts at each station can be specified. In both systems an on-screen display shows the total system output and the percentage of downtime resulting from defective parts.

The ability to model a manufacturing system provides an efficient tool for determining the system size and studying the operating behavior in relation to parts quality. Additionally, the program provides a tool for studying the cost effectiveness of increased parts quality versus system downtime, and the resulting loss of product.

Summary

Several major companies such as IBM, DEC, Xerox, and General Electric are using "Design for Assembly" and reporting material and labor savings of 25 percent to 30 percent. A similar system in notebook form was developed by Hitachi in about 1975. The fact that they understand and apply the principles of design for assembly is clearly demonstrated in their products.

Within RCA, the "Design for Assembly" program was implemented by Player Manufacturing Technology and is being used in the VideoDisc player program by Product Design, Manufacturing Technology, and RCA Laboratories Manufacturing Research. The goal was to provide not only operating hardware, but a system that would allow interaction of ideals, and sharing of analysis between Design and Player Manufacturing Technology.

To expand the performance of the personal computer and achieve the sharing of ideals, required a network system with a common storage. The network chosen was "Corvus Omninet" with Constellation software and a Corvus Winchester disc for mass storage. The Omninet is a shared-access local network using RS422 twisted pair cable, which can be 4000 feet in length and handle 64 devices. The network uses "bus topology," which means that stations can be added anywhere along the network by tapping into the trunk line.

The network system as installed consists of:

Work station

- IBM PC 128K configuration
- IBM DOS 1.1
- Dual disc drive
- Color graphics adapter card
- RGb color monitor
- Epson 100 dot-matrix printer

Network

- 20 Mb Corvus hard disc
- Transporter card for each IBM PC
- Omninet disc server with Constellation software

With this system, the analysis results from the "Design for Assembly" programs are stored on a common sector of the hard disc. Access to the sector directory provides a list of all analysis results that can be retrieved for review or further analysis. Collision is avoided by a modification to the "UMASS" software that assigns a user prefix to the file name. This allows multiple access, and also allows additions or changes to be made to previous analyses without changing the original analyses. In this manner, all the information is shared and there is no duplication of effort when more than one analysis of the same assembly is desired.

"Design for Assembly" has been an effective tool for product analysis and improved

Malcolm Cobb joined RCA in 1979 as Senior Industrial Engineer in the VideoDisc Player Manufacturing Technology Department, where he is project engineer coordinating the efforts between Design and Manufacturing Technology to reduce player manufacturing cost. As part of this responsibility, he recently implemented the "Design for Assembly" program.

He received his B.S. Degree from the University of Tennessee in 1961. Since then he has accumulated a wide variety of Industrial Engineering experience, serving as supervisor of a wage incentive program for Eaton-Yale Inc., and as project engineer at Hamilton Beach and Magnavox. Most of his assignments in consumer electronics have involved the implementation of new products into manufacturing. Mr. Cobb is an active member of the Institute of Industrial Engineers, currently serving on the Board of Directors of the Indianapolis Chapter.

Contact him at:
RCA Consumer Electronics Division
Indianapolis, Ind.
TACNET: 426-3348

part design for assembly. Equally important has been the development of a common language between research, design, and manufacturing technology. This common language has been the catalyst for increased cooperation and understanding necessary to achieve the common goal of increased manufacturing productivity.

Acknowledgments

The author would like to acknowledge A. Sproul for the software enhancements and system maintenance that made the total system operational; J. Aceti for his continued support and use of "UMASS" in advanced player design at RCA Laboratories; and Doctors G. Boothroyd and P. Dewhurst at the University of Massachusetts for their assistance and cooperation.

References

1. G. Boothroyd, C. Poli, and L. Murch, *Automatic Assembly*, Marcel Dekker Inc., New York, N.Y. (1982).
2. G. Boothroyd, *Design for Assembly Designers Handbook*, Mechanical Engineering Department, University of Massachusetts Amherst, Amherst, MA 01003; phone: (413) 545-0054 (1983).
3. M. Anderson, S. Kahler, and T. Lund, *Design for Assembly*, Springer-Verlag., IFS (Publications) Ltd., United Kingdom (1983).
4. G. Boothroyd, "Economics of Assembly Ssytems," *Journal of Manufacturing Systems*, Vol. 1, No. 1, p. 111.
5. G. Boothroyd, "Design for Manufacturability," NSF Grant No. APR 77 10197 (September 1979).
6. G. Boothroyd, C. Poli, and L. Murch, *Handbook of Feeding and Orienting Techniques for Small Parts*, Mechanical Engineering Department, University of Massachusetts Amherst, Amherst, MA 01003 (1983).
7. B. Machrone, "Battle of the Network Stars," *PC Magazine*, Vol. 2, No. 6, pp. 93-104 (November 1983).
8. Corvus Systems, 2029 O'Toole Avenue, San Jose, CA 95131; phone: (408) 946-7700.

Reprinted from *MACHINE DESIGN*, January 26, 1984

DESIGN FOR ASSEMBLY: AUTOMATIC ASSEMBLY

Assembly cost depends crucially on product design, especially when a high-speed dedicated assembly system is to be used. The suitability of a product for automatic assembly and the cost of assembly can both be assessed by systematically classifying design features — even when details of the assembly process are not known.

PETER DEWHURST
Professor

GEOFFREY BOOTHROYD
Professor
University of Massachusetts
Amherst, MA

DESIGN of assemblies and the processes that assemble them presents a chicken-and-egg problem. The production process cannot be well defined until the assembly is designed, but the assembly cannot be well designed until the process is defined. The difference between an economical assembly process and a costly failure lies almost entirely in the hands of the product designer, not the process designer.

Production costs are minimized when design and production engineering are coordinated. That is, when the least costly production process is selected and the product is designed for that process. A method for finding the least costly process was detailed in "Design for Assembly: Selecting the Right Method," MD, Nov. 10, 1983, p. 94. A method of designing products for manual assembly was presented in "Design for Assembly: Manual Assembly," MD, Dec. 8, 1983, p. 140. A future article will cover software systems that aid in analysis and redesign.

When a product is to be assembled automatically, the design can be assessed for suitability even before the details of the assembly process are known. Design features can be classified and numerical ratings assigned to indicate areas where assembly can be made easier.

Suitability of a product for automatic assembly can be determined by an analysis that consists of three major steps for each component part in the assembly:

- Estimate the cost of handling the part automatically in bulk and delivering it in the correct orientation for insertion on an automatic-assembly machine.
- Estimate the cost of inserting the part automatically into the assembly and the cost of any extra operations.
- Decide whether the part must necessarily be separate from all other parts in the assembly.

These three pieces of information can be combined to estimate the total cost of assembly and to estimate the "design efficiency," which is a numerical rating for the ease of automatic assembly.

Handling cost

The most important and difficult consideration in automatic assembly is the efficiency with which individual parts can be handled automatically. Some parts are impossible to feed and orient automatically, even with expert redesign, and the assembly system must then include one or more manual workstations.

The handling efficiency of parts is presented in four charts in the *Design for Assembly Handbook*, and the chart for rectangular parts is reproduced in this article. The charts make it possible to estimate the relative cost of a parts feeder, C_r, and the efficiency with which the part can be oriented automatically, E_o.

With values for C_r and E_o, plus the basic cost of a standard feeding and orienting device, simple equations give the cost of automatic part handling. For example, if the cost of a basic feeder that can deliver parts at 1500 mm/min is $5,000, then the cost of feeding and orienting any part, C_f is

$$C_f = 0.03\,D_f\text{¢/part}$$

where

$$D_f = 60C_r/F_r \text{ if } F_r < F_m$$
$$D_f = 60C_r/F_m \text{ if } F_r > F_m$$
$$F_m = 1500E_o/Y$$

Insertion cost

The cost of using an automatic workhead for part insertion can be estimated from the Automatic Insertion Data Chart. This chart classifies insertion processes and gives relative workhead costs.

The basis for cost comparisons is the cost of a simple pick-and-place device that performs easy insertions from directly above

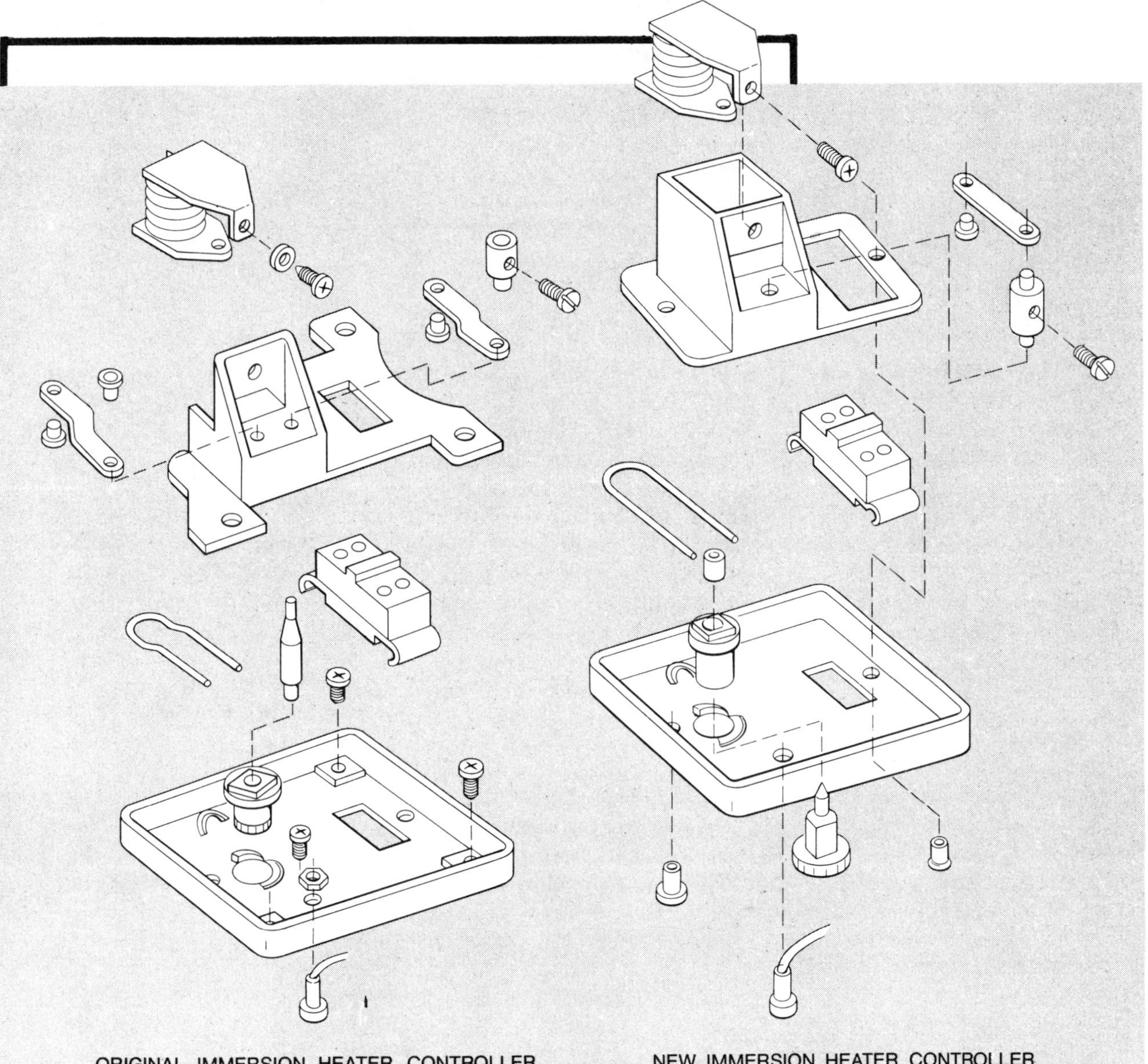

ORIGINAL IMMERSION HEATER CONTROLLER

NEW IMMERSION HEATER CONTROLLER

From manual to automatic assembly

When executives at Satchwell Sunvic Ltd. wanted to cut costs by automating production of an immersion heater controller, they consulted experts at the University of Salford Industrial Center. Initial analysis showed that of the 21 component parts in the original controller, 14 were unsuitable for automatic assembly. As a result, automation was uneconomical.

The controller was then redesigned and simplified, using information from the analysis. The new controller had 16 parts, of which only six were unsuited to automatic assembly. Now it was possible to automate the process, even though some manual assembly remained.

Such solutions are not uncommon in automatic assembly systems. One or more parts of an assembly are often impossible to handle automatically, so the assembly machine includes manual workstations. For the immersion heater controller, the mixed automatic and manual assembly system saved about $150,000 per year in assembly costs.

1	2	3	4	
Part I.D. No.	Number of times operation is carried out simultaneously	Automatic handling code	Orienting efficiency, E_o	
5	1	214	0.45	
4	1	000	0.70	
3	1	100	0.70	
2	1	645	0.10	

the assembly at a maximum rate of 60 parts/min. If such a workhead, installed and operating on an assembly machine costs $10,000, then the cost of automatically inserting a part is

$$C_i = 0.06D_i \text{ ¢/part}$$

where

$$D_i = 60W_c/F_r \text{ if } F_r < 60$$
$$D_i = W_c \qquad \text{if } F_r > 60$$

Minimum parts

Assembly costs usually increase in proportion to the number of parts in the product. Therefore, no part should escape scrutiny, no matter how low its intrinsic value. Such parts as fasteners, clips, and washers may seem insignificant in themselves, but they add enormously to assembly cost. Taken as a group, they often account for the majority of assembly cost.

The cost influence of small parts is particularly evident in automatic assembly, because each part in the product requires a feeding and orienting device, a workhead, at least one extra work carrier, a transfer device, and an increase in the size of the basic machine structure.

It is not unusual for the elimination of a single fastener to save $20,000 or more in the cost of an assembly machine. Moreover, because the resulting machine has fewer work stations, it generally operates more efficiently.

For these reasons, estimating the theoretical minimum number of parts is a particularly important step in the analysis. For a part to be judged essential, it must satisfy one of three criteria:

1. Does the part move with respect to all other parts already assembled?

2. Must the part be made of a different material or be isolated from all other parts already assembled? (Only fundamental reasons concerned with material properties may be considered here.)

3. Must the part be separate from all other parts already assembled because necessary assembly or disassembly would otherwise be impossible?

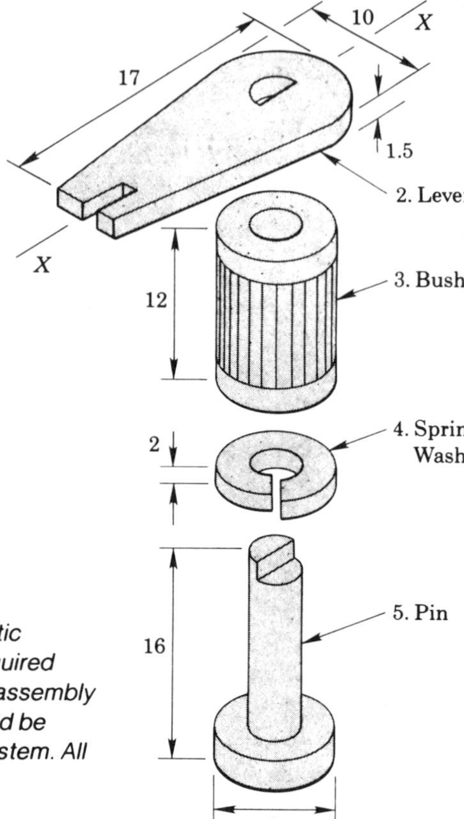

The design efficiency of this lever assembly is only 33% with automatic assembly. However, much of that low percentage stems from the required assembly rate of only 30/min. With that assembly rate, an ideal lever assembly would have an efficiency of only 50%. Efficiency of the assembly could be raised to 43% if the lever were symmetrical and were press fit on the stem. All dimensions are in mm.

5	6	7	8	9	10	11	12	13	14	required rate of assembly Fr (per minute)
Relative feeder cost, $C_r = F_c + D_c$	Maximum basic feed rate, F_m	Difficulty rating for automatic handling, D_f	Cost of automatic handling per part, $C_f = 0.03D_f$	Automatic insertion code	Relative workhead cost, W_c	Difficulty rating for automatic insertion, D_I	Cost of automatic insertion per part, $C_i = 0.06D_I$	Operation cost, cents (2) x [(8) + (12)]	Figures for estimation of theoretical minimum parts	Name of assembly Adjusting Lever
1	48.2	2	0.06	00	1	2	0.12	0.18	1	Pin
1	150	2	0.06	00	1	2	0.12	0.18	1	Spring washer
1	105	2	0.06	00	1	2	0.12	0.18	1	Bush
1.5	8.82	10.2	0.31	00	1	2	0.12	0.43	1	Lever
				91	0.9	1.8	0.11	0.11		Separate operation (rivet pin head)
								1.08	4	Design efficiency $= \frac{0.09N_m}{C_a} =$ 0.33
								C_a	N_m	

Automatic Handling — Data for non-rotational parts (first digit 6, 7 or 8)

First digit	E_o	C_r
6	0.7	1
7	0.45	1.5
8	0.3	2

		$A > 1.1B$ and $B > 1.1C$	$A \leq 1.1B$ or $B \leq 1.1C$ (Code the main feature or features which distinguish the adjacent surfaces having similar dimensions.)							
			Steps or chamfers parallel to —			Through grooves parallel to —			Holes or recesses $>0.1B$ (cannot be seen in silhouette.)	Other — including slight asymmetry, features too small etc.
			X axis and $> 0.1C$	Y axis and $> 0.1C$	Z axis and $> 0.1B$	X axis and $> 0.1C$	Y axis and $> 0.1C$	Z axis and $> 0.1B$		
		0	1	2	3	4	5	6	7	8
Part has 180° symmetry about all three axes	0	0.8 1 0.9 1 0.6 1	0.8 1 0.9 1 0.5 1	0.2 1 0.5 2 0.15 2	0.5 1 0.5 1.5 0.15 1.5	0.75 1 0.5 1 0.5 1	0.25 1 0.5 1.5 0.15 1	0.5 1.5 0.6 1 0.15 1.5	0.25 2 0.5 1 0.15 2	MANUAL HANDLING REQUIRED

			Code the main feature, or if orientation is defined by more than one feature, then code the feature that gives the largest third digit							
			Steps or chamfers parallel to —			Through grooves parallel to —			Holes or recesses $> 0.1B$ (cannot be seen in silhouette.)	Other — including slight asymmetry, features too small etc.
			X axis and $>0.1C$	Y axis and $>0.1C$	Z axis and $> 0.1B$	X axis and $> 0.1C$	Y axis and $> 0.1C$	Z axis and $> 0.1B$		
			0	1	2	3	4	5	6	7
Part has 180° symmetry about one axis only.	About X axis	1	0.4 1 0.5 1 0.4 1	0.6 1 0.15 1 0.6 1	0.4 1.5 0.25 2 0.4 2	0.4 1 0.5 1 0.2 1	0.3 1 0.25 1 0.3 1	0.7 1 0.25 1.5 0.15 1	0.4 2 0.25 3 0.1 2	MANUAL HANDLING REQUIRED
	About Y axis	2	0.4 1 0.4 1 0.5 1	0.3 1 0.2 1 0.15 1	0.4 1.5 0.25 2 0.5 2	0.5 1 0.4 1 0.2 1	0.3 1 0.25 1 0.15 1	0.4 1 0.25 1 0.15 2	0.4 2 0.25 2 0.15 2	MANUAL HANDLING REQUIRED
	About Z axis	3	0.4 1 0.3 1 0.4 1	0.3 1 0.2 1 0.2 1	0.4 1.5 0.25 2 0.4 2	0.4 1 0.3 1 0.2 1	0.3 1 0.25 1 0.15 1	0.4 1.5 0.25 2 0.15 2	0.4 2 0.25 2 0.15 2	MANUAL HANDLING REQUIRED
Part has no symmetry (code the main feature(s) that define the orientation.)	Orientation defined by one main feature	4	0.25 1 0.25 1 0.15 1	0.15 1 0.1 1.5 0.14 1	0.15 1.5 0.24 2 0.15 1	0.1 1 0.2 1 0.1 1	0.15 1 0.1 1.5 0.05 1	0.1 1.5 0.15 2 0.1 1.5	0.1 2 0.15 3 0.08 2	MANUAL HANDLING REQUIRED
	Orientation defined by two main features and one is a step, chamfer or groove	6	0.2 2 0.1 3 0.05 2	0.15 2 0.1 3.5 0.05 2	0.1 2.5 0.1 4 0.05 2.5	0.1 2 0.1 3 0.05 2	0.15 2 0.1 3.5 0.05 2	0.1 2.5 0.1 4 0.05 2.5	0.1 3 0.1 5 0.05 3	MANUAL HANDLING REQUIRED
	Other — including slight asymmetry etc.	9	MANUAL HANDLING REQUIRED							

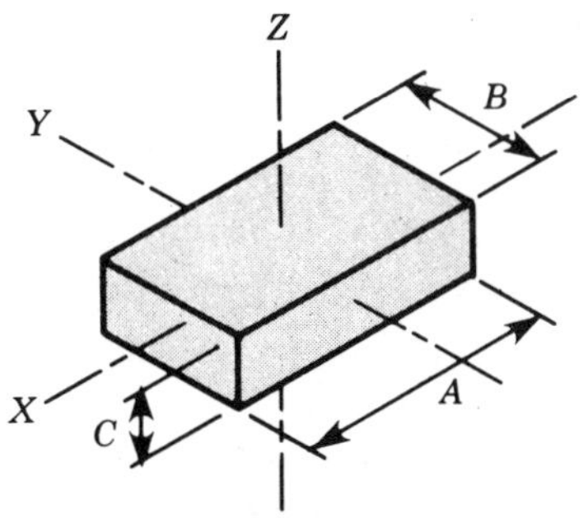

Automatic insertion — relative workhead cost

		After assembly no holding down required to maintain orientation and location.				Holding down required during subsequent process(es) to maintain orientation and location.			
		Easy to align and position		Not easy to align or position (no features provided for the purpose)		Easy to align and position		Not easy to align or position (no features provided for the purpose)	
		No resistance to insertion	Resistance to insertion	No resistance to insertion	Resistance to insertion	No resistance to insertion	Resistance to insertion	No resistance to insertion	Resistance to insertion
		0	1	2	3	6	7	8	9
Addition of any part where no final securing is taking place: Straight line insertion — From vertically above	0	1	1.5	1.5	2.3	1.3	2	2	3
Addition of any part where no final securing is taking place: Straight line insertion — Not from vertically above	1	1.2	1.6	1.6	2.5	1.6	2.1	2.1	3.3
Addition of any part where no final securing is taking place: Insertion not straight line motion	2	2	3	3	4.6	2.7	4	4	6.1

Key: shaded box = Part added but not secured

When these criteria have been applied to all parts. the sum of the essential parts is the theoretical minimum number for the assembly. The criteria should be applied without regard to the apparent feasibility of eliminating parts or combining them with others. Feasibility and practicality are matters to be addressed by the designer after the analysis. The analysis itself indicates the possible directions for simplification and the cost benefits that result.

Generating data

These principles are carried out in practice with the aid of the Automatic Assembly Worksheet. To illustrate the procedure, consider an adjusting-lever assembly.

The first step is to obtain the best information about the assembly. Here, an exploded three-dimensional view is used, but useful information might also be found in engineering drawings, existing versions, or prototypes.

The second step is to take the assembly apart, or imagine how it might be done. Assign an identification number to each item; the complete assembly is numbered 1, and individual parts are numbered in the order of disassembly.

Next, begin to reassemble the product, starting with the part with the highest identification number. Complete one row of the worksheet for each part as it is added to the assembly, or for each separate assembly operation. One row of the completed worksheet for the lever assembly serves as an example.

Enter the identification number of the part in Column 1. For the lever, this number is 2.

The lever is only added once, so enter a 1 in Column 2.

The part feeding and orienting code is entered in Column 3. The lever is basically a flat rectangular (nonrotational) piece, with its length less than three times its width and more than four times its thickness, so the first digit of the code is 6. The second and third digits come from the Automatic Handling Chart. The lever is not symmetrical about any of its axes, so the second digit must be 4, 6, or 9. Its proper orientation is defined by the semicircular hole, so the second digit is 4. The through groove is parallel to the Z axis and is longer than 0.1 times the width, so the third digit is 5.

Code 645 indicates that the orientation efficiency $E_o = 0.1$ and that the relative cost of a parts feeder $C_r = 1.5$. These values are entered in Columns 4 and 5 of the worksheet.

The lever is 17 mm long, and the standard operating rate for parts feeders is 1500 mm/min, and $E_o = 0.1$, so a standard feeder delivers 8.8 parts/min. Therefore, 8.8 is the value of F_m, and is entered in Column 6.

The required assembly rate $F_r = 30$ parts/min, so the difficulty rating for automatic handling is given by

$$D_f = 60C_r/F_m = 60(1.5)/8.8 = 10.2$$

This value is entered in Column 7.

The cost of feeding and orienting each part is entered in Column 8. For the lever,

$$C_f = 0.03D_f = 0.03(10.2) = 0.31¢$$

The lever is inserted onto the pin, which has enough clearance to allow easy alignment and positioning from directly above the pin. The appropriate two-digit code from the Automatic In-

Redesigning the lever assembly

Design efficiency for automatic assembly depends both on the features of the design and the required assembly rate. Assembly takes place using equipment that typically operates at speeds up to one cycle per second. Thus, if the required assembly rate is only 30 per minute, as it is for the lever assembly, then even an ideal design has an efficiency of only 50%.

The lever assembly has a design efficiency of 33%, so it is already fairly good. However, it is not ideal for two reasons. First, the assembly involves a separate securing operation that requires an extra workstation on the assembly machine. Second, Column 7 of the worksheet shows that the lever is difficult to feed and orient automatically.

The Automatic Handling Data Chart shows that the lever would be easier to handle if it were symmetrical about its *X* axis. In the existing design the lever is asymmetrical because of the offset slot and the semicircular hole, neither of which is essential to the product. A redesigned lever with a centralized slot and circular hole would have an automatic handling code of 615, with an orienting efficiency $E_o = 0.7$ and a relative feeder cost $C_r = 1$. The estimated cost of automatic handling would then be 0.06¢ per lever instead of 0.31¢ per lever for the original design.

If the lever were also designed to be press fit onto the pin, the estimated total assembly cost would become 0.83¢, a 23% reduction from the original assembly cost. The efficiency rating for this new design would be 43%, quite close to 50%.

sertion Data Chart is thus 00. This number is entered in Column 9.

The relative workhead cost from the Automatic Insertion Data Chart is $W_c = 1$, which is entered in Column 10.

The difficulty rating for automatic insertion is entered in Column 11. For the lever,

$$D_i = 60W_c/F_r = 60(1)/30 = 2$$

The cost of insertion is entered in Column 12. For the lever,

$$C_i = 0.06D_i = 0.06(2) = 0.12\ ¢$$

The total cost of feeding, orienting, and inserting the lever is the sum of the separate costs per part for these operations. In other words, it is the sum of Columns 8 and 12, multiplied by Column 2. For the lever this amounts to 0.43¢, which is entered in Column 13.

If the part must be separate according to the three criteria for the minimum number of parts, a 1 is entered in Column 14. Otherwise, a 0 is entered. The lever must be separate from the pin to allow the bush and spring washer to be assembled, so Criterion 3 is satisfied, and the lever must be a separate part.

When data for each part has been entered on the worksheet, the total cost of automatic handling and insertion can be found by adding the numbers in Column 13. The theoretical minimum number of parts is found by adding the numbers in Column 14. In this example, $C_a = 1.08$ and $N_m = 4$.

Design efficiency is calculated from the equation on the worksheet. For the adjusting-lever assembly, it is 33%.

Assessing data

The data generated in filling out the worksheets cover two main areas: potential for eliminating parts and potential for improving automatic handling and assembly. Areas where parts may be eliminated or combined with other parts are easy to locate from Column 14 of the worksheet. However, parts identified with a 0 in that column often cannot be eliminated because of other constraints. Manufacturing combined parts may be impossible with available equipment, or just too costly. However, a reduction in parts is usually the most effective way to reduce assembly cost.

The second major way to cut cost is to simplify the handling and assembly of the parts that remain. Start by checking Column 6 on the worksheet. If this maximum basic feed rate is less than the required feed rate, then check the feeding and orienting efficiency in Column 4. A low value for E_o indicates considerable scope for improvement.

The Automatic Handling Data Chart serves as a guide to the part features that create problems. Consider changes to the part that make the second and third digits of the feeding and handling code as close as possible to 00. The same charts can point out helpful changes if the relative feeder cost in Column 5 of the worksheet is greater than 1. Use the Automatic Insertion Chart as a guide to possible improvements in the insertion or fastening process if the relative workhead cost W_c is greater than 1. MD

Nomenclature

A = Longest orthogonal dimension of part, mm
B = Middle orthogonal dimension of part, mm.
C = Shortest orthogonal dimension of part, mm.
C_a = Cost of automatic assembly, ¢/assembly
C_i = Cost of automatically inserting a part, ¢/part
C_f = Cost of feeding and orienting a part, ¢/part
C_r = Relative cost of parts feeder
D_i = Relative difficulty of inserting a part
D_f = Relative of difficult of feeding a part
E_o = Efficiency of orienting a part automatically
F_m = Maximum basic feed rate, part/min
F_r = Required feed rate, parts/min
N_m = Theoretical minimum number of parts
W_c = Relative workhead cost
Y = Maximum dimension of the part, mm

This article is largely based on the *Design for Assembly Handbook*, which has been developed by Boothroyd and Dewhurst at the University of Massachusetts. The handbook is available from Professor Boothroyd, Automatic Assembly Program, Department of Mechanical Engineering, Univ. of Mass., Amherst, MA 01003, (413) 545-0054.

Presented at the SME Assemblex '83 Conference, May 1983

Product Design for Automatic Assembly

by Frank J. Riley
Bodine Corporation

INTRODUCTION

With the introduction of the vibratory feeder in the 1950's, there was at last available the tool to feed a wide variety of parts from bulk condition. Automatic assembly has grown with pauses to mark each economic recession in its ability to handle a wide variety of assemblies. Until the advent of the vibratory feeder, there was no good way to feed highly complex part pieces presented to the assembly line in bulk condition.

In a most important SME paper, Arthur LaRue of IBM quickly highlighted the basic problem with efficient mechanized assembly, that of variation in the part pieces themselves as they came to the machine.

In recent years, there has been increasing focus on the correlation of efficient mechanized assembly and product design and now many people are aware of the essential nature of designing parts to facilitate mechanized assembly.

As builder's of moderate to large assembly systems, we have often been asked to assist in the design of products for mechanized assembly. In other instances, we are asked for principles of designing for automatic assembly that may be incorporated into the design procedure.

While these opportunities to participate in the design are indeed welcome, there are certain basics that cannot be ignored. Any product intended for high volume production (and hopefully high volume sales) must be designed first of all for its functional requirements; secondly, for its aesthetic consideration; thirdly, for the cost of fabrication and material. Only when there has been a true effort to respond to these three needs can the question of assembly capability be considered.

There is a second major consideration; that is that the vast majority of all assembly work probably will continue to be done by hand. Assembly costs still remain the majority of all direct labor costs in manufacturing and probably will do so for the foreseeable future, the claims of robotic enthusiasts not withstanding. Once a product has been conceived and designed for its functional, aesthetic and fabrication cost considerations, one has to first of all insure that the design leads to efficient manual assembly. With the exception of very miniature parts that cause problems in finger manipulation, one could almost place as his first principle, that it is difficult to automatically assemble any product that cannot be easily assembled by hand. The correlary, however, is not true. Even though a product be designed for efficient hand assembly it does not mean that it can be easily automatically assembled.

One does not have to be a truly religious person to recognize the incredible dexterity, sensory ability and judgment value of even the most illiterate and ill-motivated human assembler. The entire thrust of intelligent robotic assembly has been to duplicate in tactile and visual sensors and in control sophistication the crudest imitations of some human assembler who has not the slightest degree of formal education.

In order to make certain, however, that a part that has been designed for good manual assembly can also be assembled automatically there are further considerations.

Before going into these considerations, however, it is imperative to stress that automatic assembly is but a tool for insuring that the best quality production is performed for the least unit cost and that it must be subservient to the marketability of the product. There will be many occasions where the functional or aesthetic requirements demanded by the marketplace will not allow one to reach the ultimate design for the most efficient mechanized assembly. There are a series of slides accompanying this paper to illustrate this point.

When looking at a product from its suitability for mechanized assembly we must look at the overall product design, the design of the individual components, the provision for automatic joining and the quality characteristics viewpoints

PRODUCT DESIGN FOR MECHANIZED ASSEMBLIES

It is important to recognize that there are two different types of assemblies. In one type of assembly the ability to efficiently add a new component is dependent upon the success of the installation of the previous components. This type of assembly might be called an additive assembly. One part is piled on top of one another and failure to insert one part properly precludes the ability to continue the assembly successfully.

A second type of assembly involves multiple insertion of common or similar components on a common base (such as circuit board assembly). Failure to insert one component correctly will not really affect the successfulness of subsequent insertion components. It should be noted that these two basic types of assemblies, and many assemblies are hybrids, will greatly affect the decision as to the type of machine control system used for mechanized assembly. It also will have a direct correlation to the probability of salvage of incomplete or defective assemblies. When looking for efficient automatic assembly, one should consider the design and the viewpoint of the planes in which part pieces must be inserted into the basic assembly. The ideal is to insert all components from a single direction. Each deviation from the single plane of insertion will add significantly to fixture cost and station efficiency.

Modular design may be significant in that it may allow a family of parts, each of relatively low volume, to be assembled on a common machine system. A basic chassis or base is capable of accepting a variety of options that may or may not be necessary on any given model.

The ability of the assembly to grow up with increasing physical integrity is particularly important were the total number of components exceeds 10 or 12 parts. For efficiency's sake it may be necessary to process the assembly which has a large number of components across a series of machines, with float or buffer storage between each machine. It becomes essential, therefore, that these growing subassemblies have physical integrity that allows for transport or storage between each machine operation. In this regard the use of shop aids is a most useful one. Very few people would consider introduction of a totally foreign component into a subassembly to maintain its physical integrity until the assembly continues to its final configuration. These shop aids may remain in the product or may be removed in a subsequent operation.

COMPONENT DESIGN FOR AUTOMATIC ASSEMBLY

While ideally it would be wonderful to retain the orientation established in the original fabricating operations, pipe line costs, the necessity of secondary operations and the incompatibility time-wise of fabricating operations to assembly usually will prohibit retention of orientation from fabrication to assembly. The whole question of how to reestablish orientation from bulk condition is currently being examined by many research groups under the somewhat nebulous term bin-picking. For most of us, however, and for the foreseeable future, the use of vibratory and other mechanical part feeders will be the basic means by which parts are returned from a disoriented bulk condition into a useable attitude. Fortunately, there is a great deal of literature available on this topic. Where parts will not, however, be susceptible to mechanical part feeding, the product designer should be reluctant to release his component design for manufacturing without consultation with productivity engineers, advanced manufacturing engineers or manufacturing engineers to determine whether or not the part can be partially fabricated off the machine and transported to the machine efficiently in bandoleer reels or some other type of carrier device.

DESIGNING FOR AUTOMATIC JOINING

Often times excellent component design work is visciated by neglecting the real problems of joining in high volume assembly. Fortunately, the widespread use of plastic parts and ultrasonic welding has greatly reduced this problem in mechanized assembly. Threaded fasteners and rivets remain a problem because of their relatively poor quality levels when coming to the automtic assembly machine. Perhaps the worst problem of all is where product designers feel compelled to use some form of adhesives in order to insure bonding or sealing of the assembly. The real problem comes where it is necessary to hold two parts joined under pressure for prolonged periods of time in order to insure that sufficient cure strength is accomplished. This is very difficult and costly to do on most automatic assembly systems. A great deal of care and experimentation should be done to determine whether or not rapid curing or accelerated adhesives could be used, in which half of the cure strenth is sufficient for final assembly purposes. Very often half strength can be achieved very rapidly when contrasted to the total cure time to achieve ultimate strength.

Where it is necessary to use mechanical fasteners such as screws or rivets, and the quantity of screws or rivets is relatively high (in excess of two) every effort should be made to determine whether or not these components can be placed in the assembly in one station and fastened in the next. It is always more efficient to break down each increment of mechanical assembly into the simplest components and thus isolate the feeding and the securing of mechanical fasteners. This often will involve significant consideration in the design of the product.

QUALITY REQUIREMENTS FOR EFFICIENT ASSEMBLY

In past years as potential users of mechanized assembly became more aware of the requirements for quality parts for efficient automatic assembly, there was a tendency to begin to tighten tolerances to levels which are unrealistic considering the variables in manufacturing quality which in turn result from the caliber of production machines utilized to produce the components.

Wherever possible component, products should be designed so that their orientation in parts feeders be determined by relatively gross physical configuration rather than the holding of some specific tolerance. This is particularly true when the components come from multiple sources such as different vendors or different molds. Another area that must be looked at when doing matched assemblies such as fuel injectors or bearings is the fact that the sigma curves of relevant dimensional characteristics do not necessarily overlap and may affect the tolerancing of the matching part pieces.

In my personal viewpoint, the most overlooked quality characteristic is the provision in product design for monitoring through inspection devices such as probes and transducers and even optical scanners of the ongoing quality of the product as it develops. Very often the cumulative tolerances of the stacked up parts makes it impossible to determine the presence, position or relative attitude of a new component when gauging from the machine slide or the fixture surface. There must be reference surfaces within the parts previously fed that allow the gauge to determine the correct placement or insertion of new components.

It is my belief that, in fact, designing a product for mechanized assembly will essentially be a reevaluation of the product that has been previously designed for functional, aesthetic and cost considerations in order to make minor and sometimes major modifications to facilitate mechanized assembly. This is probably best done by a series of check lists utilizing the approach developed by Gerald Hock and others at General Electric's Corporate Consulting activity or those such as were produced by Honeywell's producibility engineers in many of their ordnance programs. Mr. Walter Firm of Honeywell produced such a list some years ago for an ordnance seminar. After this excellent paper, there was some negative reaction among the audiance who first heard his talk. They were looking for guide lines or principles when in effect what was really needed was to evaluate existing design in the light of assembly requirements. This may be difficult for some to accept but I believe from a functional standpoint to be probably the most efficient way to insure that product design does make proper provision for mechanized assembly.

Presented at the SME Assemblex VIII Conference, March 1982

Computer-Aided Design for Automatic or Manual Assembly

by G. Boothroyd
P. Dewhurst
University of Massachusetts

INTRODUCTION

A substantial research program into automatic assembly has been underway at the University of Massachusetts for about twelve years. This program resulted first in a handbook of feeding and orienting techniques for small parts(1) and then in a handbook dealing with design for automatic and manual assembly(2) and which incorporates much of the experience and data contained in the first handbook. Experience in applying the design for assembly handbook in industry has shown that considerable improvements in assembly productivity and reductions in assembly cost can be achieved. Taking one example(3) it was found that manual assembly time could be reduced to one third of the previous time with corresponding reductions in assembly costs. These reductions were obtained first by reducing the number of separate parts in the assembly and then by ensuring that the remaining parts were easy to handle and assemble. These reductions and improvements were brought about by a systematic application of simple procedures described in the "Design for Assembly" handbook(2). Surprisingly it was also found that the reduction in part cost was about equal to the reduction in assembly costs.

However, the present authors were aware that the design analysis methods developed in the handbook could be expanded and applied more powerfully in the form of a software system. This software is now available for use on an Apple II microcomputer and a case study involving the application of the system is presented in the present paper. The system uses an extensive data base of manual assembly times and automatic handling and assembly costs in order to determine the total automatic and manual assembly costs for any product and the total manual assembly time. In addition, parts which are either impossible or uneconomic to feed and orient automatically are identified, and the number of manual workstations that would be needed on an automatic assembly machine is included in the results. Total costs determined for automatic assembly include total machine and operator costs. The cost for manual assembly assumes the cost of a free transfer line with one buffer station between each workstation. Alternatively if bench assembly is preferred then the number of bench operators needed to achieve the desired assembly rate is given. Finally a printout gives a breakdown of costs associated with each component and suggestions are made as to which parts should be considered for elimination or be combined and which parts would benefit from redesign for economic automation.

SYSTEM OPERATION

The software system consists of four main modules and a main menu as shown in Fig. 1. The menu program allows the user to discontinue an analysis at the end of modules #1, #2, or #3. Subsequent typing of the next module number in the menu will retrieve the data from disk storage and continue with the next part of the analysis. Option number 4 from the menu can also be used to read the results of any previous analysis from a data disk in order to make any changes to the assembly procedure or simply review the results. The system is written for an Apple II plus microcomputer with 48k RAM and a 3.3 disk operating system. Full use is made of color, so that colored and enlarged portions of text and color graphics menus will prompt the user, with only little experience, without the need for reading smaller displayed text. Detailed results of an assembly analysis can be output to any printer by module #4.

The operation of the system will be described by showing steps used in the analysis of the product shown in Fig. 2. The exploded view shows the injection molded parts and rubber seals of a domestic water filter presently being marketed in large numbers in the U.S. The first task in applying the software system is to obtain an explode view of this type or the actual assembly if it exists and select option 1 from the main menu to analyze a new product. Module #1 will then be loaded and will prompt the user to disassemble the assembly (mentally if it is a new

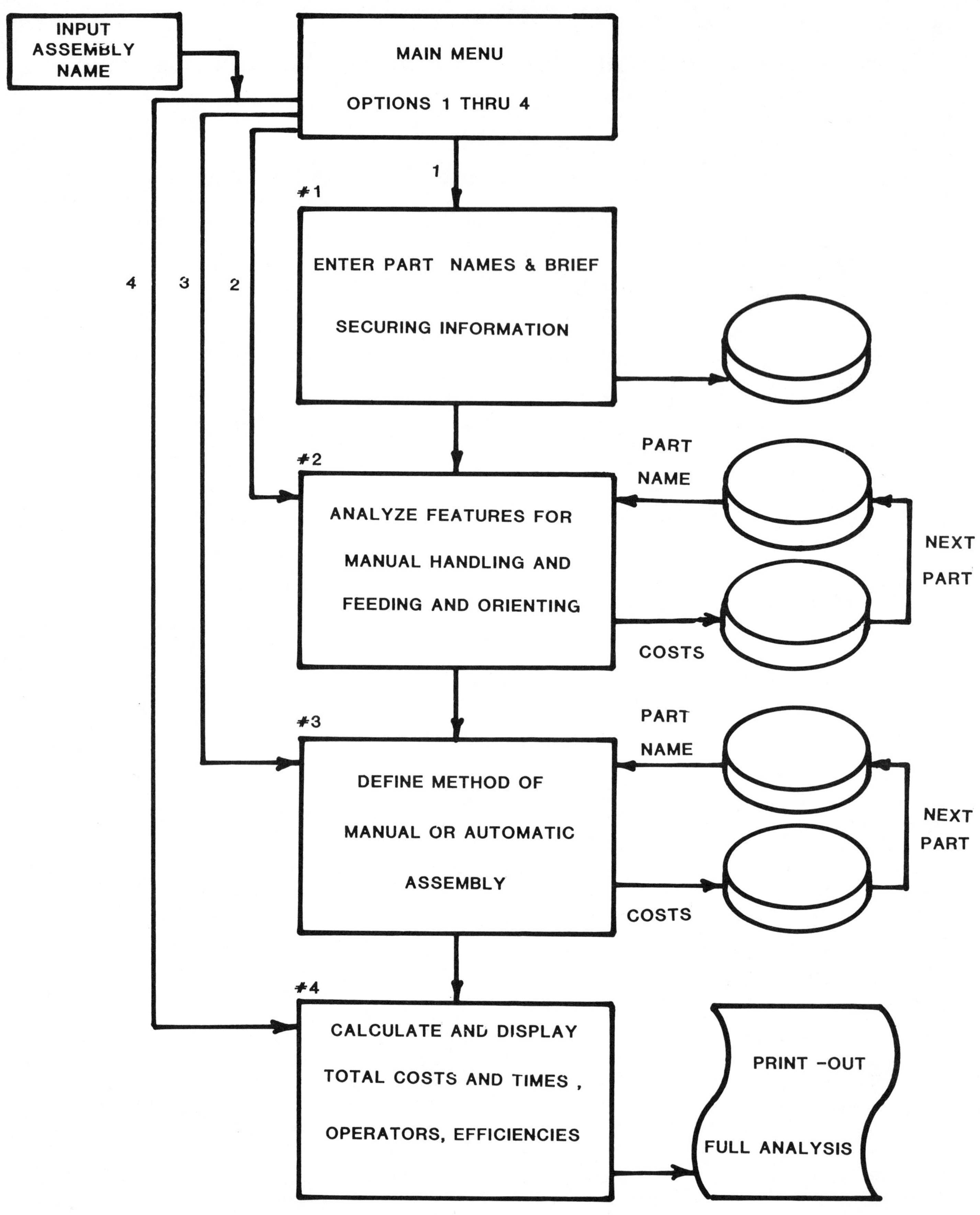

Figure 1 Block Flow Diagram of Software System

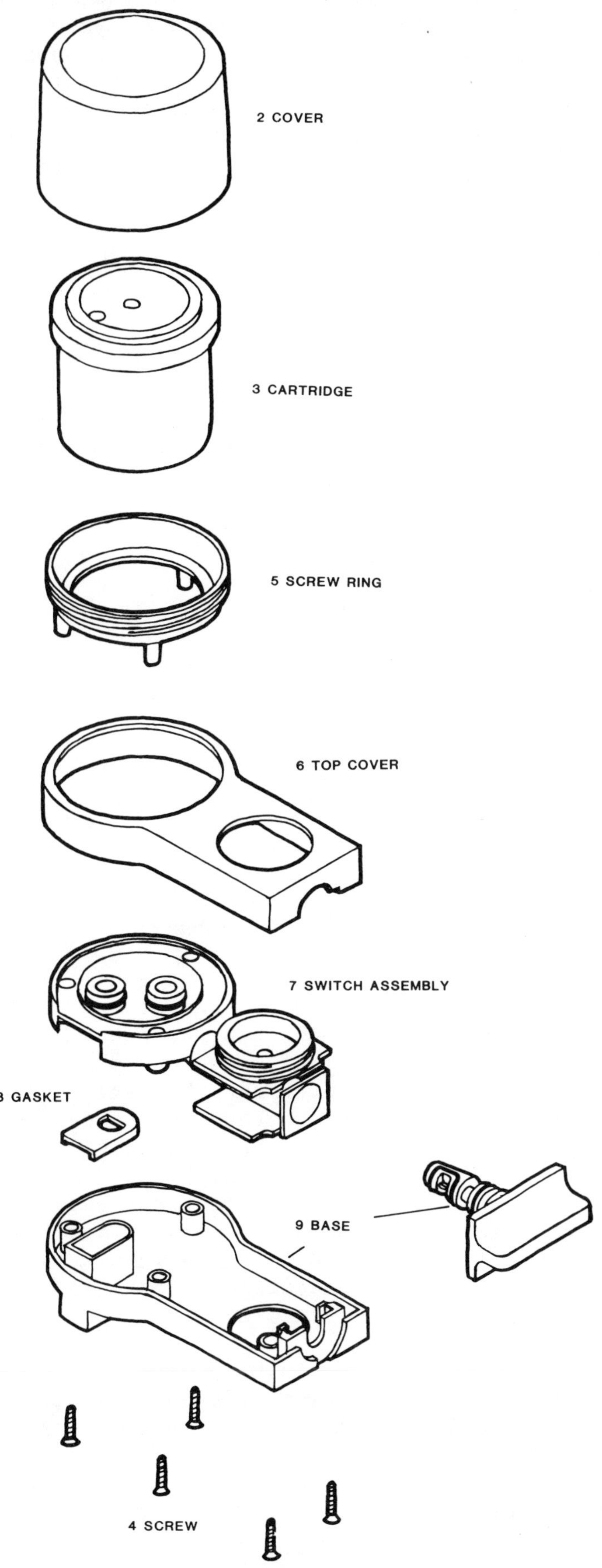

Figure 2 Water Filter - Present Design

design), and enter the names of the parts or subassemblies in turn. The number 1 is reserved for the name of the complete assembly and the system increments this number after each entry. Thus the number 2 is the first part or subassembly removed. After typing each name, and the number of times the part is repeated when appropriate, the system requests whether any reorienting is needed to remove the part, whether it is secured, and if so what type of general securing method is used. Depending upon the replies made at this stage, additional information will be requested in module #3 to determine the precise nature of the assembly operation. Figure 3 shows the replies

```
The SCREW RING

N can be removed without
  reorienting other parts

N is secured

N is secured mechanically
  (e.g. screw, twist tab, crimp, rivet,
  press fit, snap fit, etc.)

N is secured by resistance
  or friction welding

N is secured by other welding
  or brazing

N is secured by soldering

N is secured by adhesive

N is secured by a mechanical fastening
  element separate from
  the part to be removed          OK?
```

Figure 3 Computer Display - Module 1

made for the screw ring (part no. 5). Response 'N' to the first statement was needed because the assembly had to be turned over at this point after removal of the 5 screws from the base. Mutually exclusive statements are monitored throughout the system, so that response 'N' to the second statement (unsecured), immediately produced 'N's for all the remaining statements and the prompt 'OK?' at the bottom right. Response 'Y' or 'N' at this point allows the user to continue to the next part or alter the responses made for the screw ring. Figure 4 shows the partially completed parts list. The list is terminated by entering 'END' and the system then goes into browse mode so that it is possible to browse through and check a long list extending over several pages, and correct any mistakes or omissions. Continuing with the program immediately or after checking, produces a system request for the required assembly rate, F_r. This is a vital piece of data for the following analyses, since it has a major influence on automatic assembly costs. The effect of changing the feed rate on the final results can be determined in module #4, but the choice of F_r at this stage will influence some of the decisions to be made in module #2. For the present assembly an assembly rate of 10 per min. was given.

Module #2 prompts the user to give mainly geometric information about each part in turn, starting with part number 2. Figure 5 shows the first page to appear in this module. Selections are made by pressing 'N' which moves the curser to each picture in turn. Key 'Y' is then pressed at the appropriate prismatic shape. In the present case the cover is clearly cylindrical and pressing 'Y' immediately will give the second page shown in Fig. 6 requesting the dimensions of the envelope.

```
NO.   PART NAME
 1    $WTR FILTER
 2     COVER
 3     CARTRIDGE
 4     SCREW                    *5
 5     SCREW RING
 6    █ - - - - - - - - -
```

```
        After entering part name
type * then no. of parts else 'RETURN'
  Begin name with '$' for subassembly
    Type 'END' when list is complete
     Use back arrow key for editing
```

Figure 4 Computer Display - Module 1

```
Neglecting all features such as holes,
 grooves, projections, etc., enclose
          Part #2 COVER
    with a prism having one of these
          cross-sections:
```

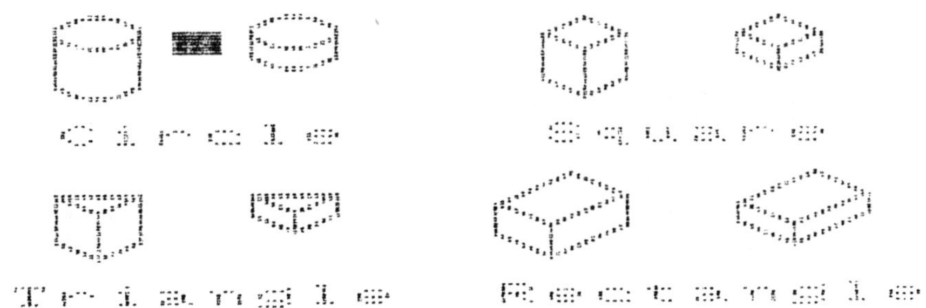


```
Press 'N' for NO, then press 'Y' at the
  most appropriate prismatic shape
```

Figure 5 Computer Display - Module 2

```
     Consider the smallest cylinder
  that can completely enclose the part
       -including projections-

     Give dimensions in millimeters
```

```
          Diameter, d=66

              Length, l=54

OK? █
```

Figure 6 Computer Display - Module 2

If part is not to be
oriented with respect to
some features- ignore them

Does the part require:

[illegible] orientation?   YES

orientation about the [illegible]? NO

Is the [illegible]

the axis of insertion? Y/N

Figure 7 Computer Display - Module 2

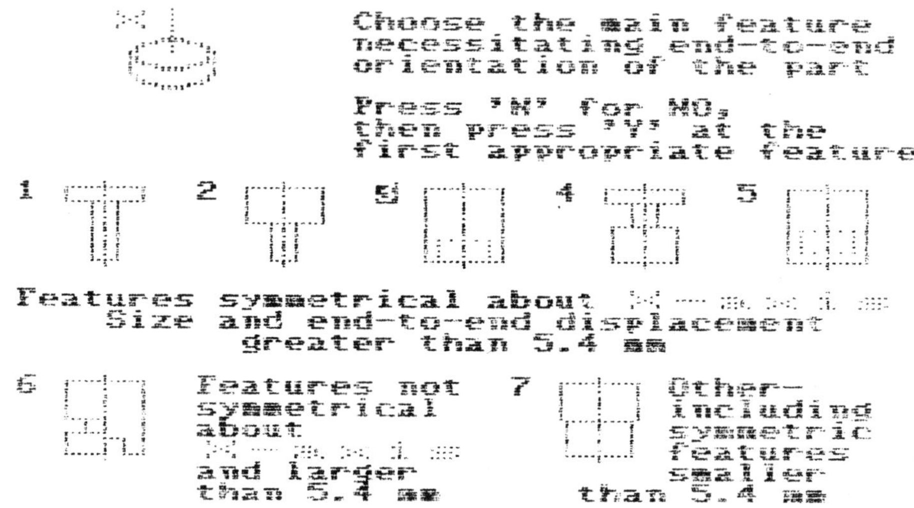

Figure 8 Computer Display - Module 2

Figures 7 and 8 show the other menu options made for the cover. From this information the system estimates the manual handling difficulty level for the part and determines the suitability of the part features for automatic feeding and orienting. From a data base of empirical time standards and costs stored on disk, estimated handling costs for the part are determined for both automatic and manual handling as shown for the filter cover in Fig. 9. In this way the user is prompted to analyze each part in turn by name. At the end of each analysis information is presented regarding the suitability of the part for automatic handling and assembly. The reasons why automatic handling may not be appropriate include: low assembly rate, large component size, or unsuitable component features. If the cost of automatic feeding and orienting is determined to be greater than the cost of manual handling then the user is given the choice of proceeding with automatic assembly of the part or deciding upon a manual workstation for the automatic assembly machine.

Module #3 in the system is used to investigate the costs of building up the assembly with correctly oriented and positioned parts, either with automatic workheads or at manual workstations. This module also assesses whether each part in turn is necessarily a separate one or if it could be combined with other parts of

```
            Part #2 COVER

    -Automatic Feeding and Orienting-

Codes      150        985

Maximum rate from one feeder    4.5
(parts per minute)

Cost per part                   3.96
at required rate (cents)

          -Manual Handling-

           Code        10

         Cost per part    .4

OK? ▮
```

Figure 9 Computer Display - Module 2

```
            Part #6 GASKET

Initial insertion of the part involves:

Y straight line motion
   from vertically
   above

N straight line motion
   other than
   vertically above


If the part is assembled manually:

Y operator can view
   the operation clearly

Y has unrestricted
   access to perform
   the necessary manipulations

OK?▮
```

Figure 10 Computer Display - Module 3

```
             Part #6 GASKET

           must be a separate part
  from all those parts already assembled

 N because it moves during
    the operation of the assembly

 N because it must be of a
    different material
    or must be isolated
    for insulation purposes etc.

 N for reasons of assembly
    or disassembly

 OK?▮
```

Figure 11 Computer Display - Module 3

the assembly. In this way a theoretical minimum number of parts is obtained for the product. See Figs. 10 and 11 for typical pages from this module.

The final module displays the total costs involved in assembling each part of the final product or subassembly, as shown for the complete water filter assembly in Fig. 12. The user can interrogate the system to find the handling and insertion costs, both manual and automatic, for each part in the assembly; see Fig. 13. In this way problem components or securing techniques which will make assembly costly or automation impossible can be identified. A quantiative assessment of the design's assemblability is given by efficiency figures (Fig. 12) which compare the actual assembly cost with the cost of assembling the theoretical minimum number of parts each having an ideal design.

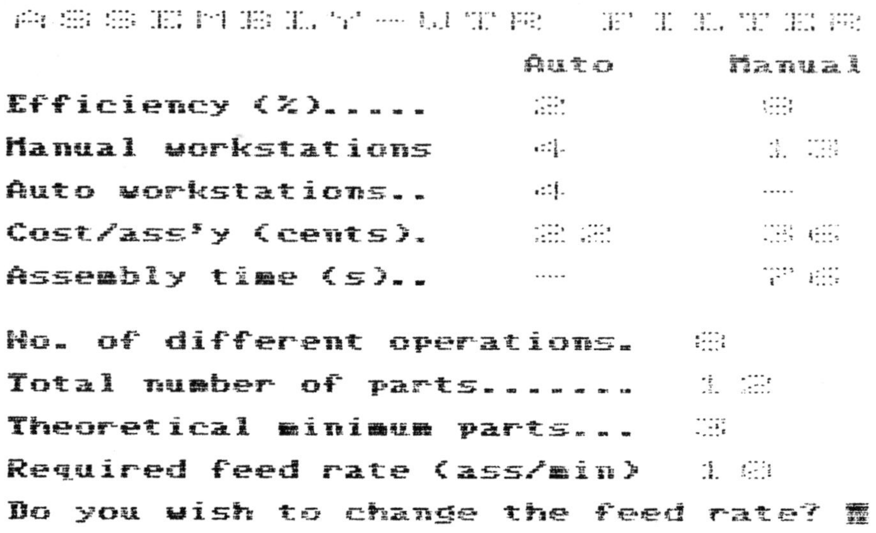

ASSEMBLY-WTR FILTER

	Auto	Manual
Efficiency (%).....	[illegible]	[illegible]
Manual workstations	4	1[illegible]
Auto workstations..	4	-
Cost/ass'y (cents).	[illegible]	[illegible]
Assembly time (s)..	-	[illegible]

No. of different operations. [illegible]

Total number of parts....... 1[illegible]

Theoretical minimum parts... [illegible]

Required feed rate (ass/min) 1[illegible]

Do you wish to change the feed rate?

Figure 12 Computer Display - Module 4

Part #2 COVER

Automatic assembly
costs/part (cents)

Feeding and orienting	[illegible]
Insertion or assembly	[illegible]
Extra operation(s)	
Total	4.1[illegible]

Manual assembly
costs/part (cents)

Manual handling	[illegible].4
Insertion or assembly	[illegible].4
Extra operation(s)	
Total	[illegible]

Manual assembly
assumed

Press:
P=Print N=Next L=Last C=Change
F=Find I=Insert D=Delete S=Save

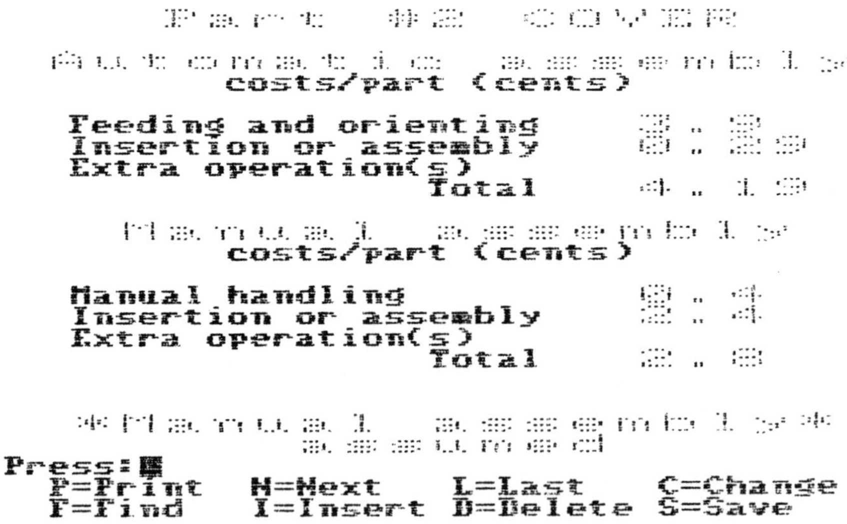

Figure 13 Computer Display - Module 4

Changes in securing methods or redesign of components which present difficulties in automatic handling can be entered subsequently into the system. The effects of individual changes on the cost of assembling the complete product can then be readily determined.

The system offers considerable flexibility in use so that errors can be corrected, parts changed or deleted and additional parts inserted into the assembly if required. The complete results of an analysis can be saved to a data disk. This can be simply for future review, but it will also allow the effect of any subsequent design changes to the product to be quickly assessed.

The print-out given by the system includes details of each component and a summary sheet. The summary sheet for the water filter and the two pages of component details are given in Figs. 14 and 15. It can be seen from Fig. 15 that part #6, the top cover for example is a suggested candidate for elimination. However if this were considered technically not feasible or undesirable for other than economic reasons, then the user is informed that the part is also a candidate for redesign. This means that the estimated costs for automatic assembly could be reduced if design changes were made to facilitate automatic orienting of the part. It is intended that the part code, 84000 in this case, will be used to obtain suggestions for redesign, from a 'Suggestion for Redesign for Automation' module presently being developed by the authors.

For the water filter assembly the analysis suggested elimination of parts #4, #5, #6, #7 and #8 (or their combination with other parts) in order to reduce the total number of twelve parts to the theoretical minimum of three. Figure 16 shows an alternative design with this minimum number of parts. (It should be noted that the switch is now numbered separately since there is no longer any reason to first produce a subassembly or the switching mechanism). This new design could be achieved by making the switch (part #4 in the new design) a snap fit into the body. In the original design the switch spigot is grooved and retained between the base and the top cover.

Interestingly, this new design is now more efficient to assemble manually than automatically, taking only 17 seconds at a cost per assembly of 8¢, a figure which is approximately 25% of the original manual assembly cost.

RESULTS AND CONCLUSIONS

An interactive software system has been developed which will allow analyses to be made of a product design for ease of automatic or ease of manual assembly. Experience with the use of this system has high-lighted two main advantages:

1. All of the tedious work of looking-up data from classification sheets is eliminated.

2. Errors made during analysis for automatic assembly are reduced considerably and in most cases eliminated.

ACKNOWLEDGEMENTS

The authors wish to thank Mr. Shiow-Chyuan Yang for his help with the analysis of the water filter.

Figure 14 WTR FILTER

SUMMARY OF ANALYSIS

		Automatic Assembly	Manual Assembly
Assembly efficiencies, %		2	8
Work-stations	Manual	4	13
	Total	8	13
Costs per assembly (cents)	handling, ins'n etc.	12.49	30.4
	basic machine	3.36	5.46
	operator & supervisor	6	-
	Totals	21.85	35.86
Total assembly time, s		-	76

Number of different operations = 8
Total number of parts = 12
Theoretical minimum parts = 3
Required assembly rate (ass/min) = 10

NOTES: 1. Indexing machine assumed in estimation of automatic assembly costs
2. The figure for the number of manual work-stations represents one or more free-transfer lines with one buffer space between each operator.This number also represents the minimum number of bench assembly operators necessary to achieve the required assembly rate
3. Part code to be used in 'Suggestions for Redesign' software module
4. These results obtained from microcomputer software package 'Design for Automatic Assembly' -Copyright 1981 G.Boothroyd & P.Dewhurst

Figure 15a DESIGN FOR ASSEMBLY ANALYSIS

WTR FILTER

Part number and name	Assembly method	Costs (measured in cents per part) Handling	Insertion	Extra op.	Total
2	Auto	3.9	.29	-	4.19
COVER	Manual	.4	2.4	-	2.8
Because this part would be expensive to feed and orient automatically, manual handling and insertion are assumed This part is a candidate for redesign (part code 15085)					
3	Auto	-	.43	-	-
CARTRIDGE	Manual	.4	1.2	-	1.6
Because it would not be feasible to handle this part automatically, manual handling and insertion are required This part is a candidate for redesign (part code 17885)					
4	Auto	.18	.29	.11	2.9
SCREW	Manual	.4	2.4	.72	17.6
There are 5 of these and 5 are candidates for elimination					
5	Auto	.29	.36	.54	1.19
SCREW RING	Manual	.4	.4	3.6	4.4
This part is a candidate for elimination This part is a candidate for redesign (part code 02300)					
6	Auto	.93	.36	-	1.29
TOP COVER	Manual	.4	.4	-	.8
This part is a candidate for elimination Because this part would be expensive to feed and orient automatically, manual handling and insertion are assumed This part is a candidate for redesign (part code 84000)					
7	Auto	.63	.54	-	1.17
SWITCH ASS	Manual	.4	1.2	-	1.6
This part is a candidate for elimination Because this part would be expensive to feed and orient automatically, manual handling and insertion are assumed					

Figure 15b WTR FILTER

Part number and name	Assembly method	Costs (measured in cents per part) Handling	Insertion	Extra op.	Total
8	Auto	.83	.36	-	1.19
GASKET	Manual	.4	.4	-	.8
This part is a candidate for elimination Because this part would be expensive to feed and orient automatically, manual handling and insertion are assumed This part is a candidate for redesign (part code 66600)					
9	Auto	.56	.36	-	.92
BASE	Manual	.4	.4	-	.8
Because this part would be expensive to feed and orient automatically, manual handling and insertion are assumed					

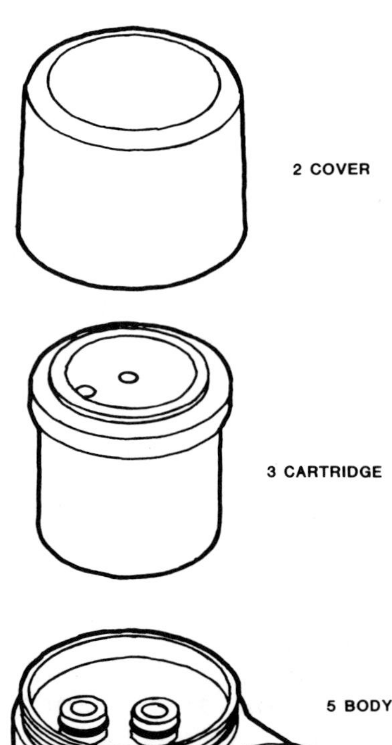

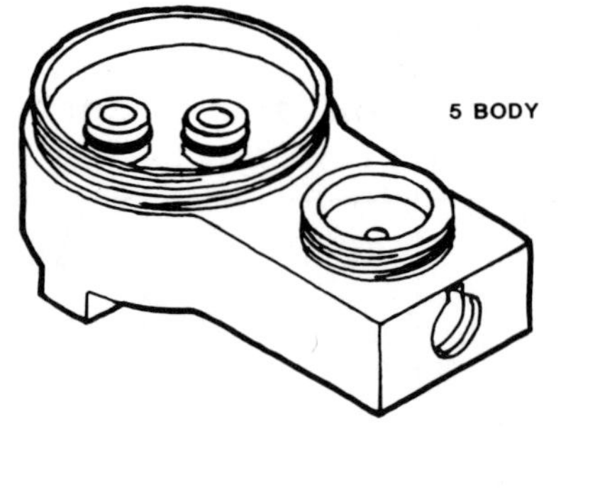

Figure 16 Water Filter - Alternative

REFERENCES

1. Boothroyd, G., Poli, C., and Murch, L.E., "Feeding and Orienting Techniques for Small Parts," Handbook, Department of Mechanical Engineering, University of Massachusetts, Amherst, Massachusetts 01002.

2. Boothroyd, G, "Design for Assembly," Handbook, Department of Mechanical Engineering, University of Massachusetts, Amherst, Massachusetts 01002.

3. Boothroyd, G., "Design for Producibility - The Road to Higher Productivity," Assembly Engineering, March 1982.

Presented at the SME 13th ISIR/Robots 7 Conference, April 1983

Product Design for Robotic Assembly

by J. Ronald Bailey
IBM Corporation

Introduction

In the first wave of industrial deployment, robots have been used primarily for stand-alone applications such as material handling, spray painting, and spot welding. As we move into the next phase of deployment, intelligent robots are now being used for assembly which offers the greatest potential but is also the most challenging because it requires integration of the robot with special end-of-arm tools, fixtures, feeders, material handling systems, and various pieces of ancillary process equipment. Thus, programming the motion of the robot is only one part of the total problem. Sensing, logic, data processing, and communications with equipment and people are essential elements in a robotic assembly system. Furthermore, products must be designed for robotic assembly if maximum utilization of this emerging technology is to be obtained.

The purpose of this paper is to focus attention on product design for robotic assembly. The intent is to provide a set of general guidelines which will enhance the design of products for robotic assembly.

Design Philosophy

Increasing international competition is a primary driving force in today's manufacturing environment. As industry strives to meet this competition, tremendous pressure is now placed on the designer to deliver innovative products which can be manufactured in a most efficient manner. In addition, the design/development cycle has been compressed in many industries because of the accelerated rate at which new products are introduced. This accelerated pace puts great emphasis on "doing things right the first time" since product life tends to be short and production decisions tend to be based on delivery schedules. There is a temptation in this environment to relax design requirements for producibility, particularly when sourcing decisions are made prematurely. For example, design for assembly may be given a low priority if it has been decided that a product will be manufactured off shore. However, this relaxation may prove to be strategically foolish because of the exposure to value engineered copies by the competition with the accompanying pressure on profit margins and market share.

A better philosophy is to design new products for robotic assembly. This approach will often lead to basic improvements in overall product design because of the discipline associated with robotic assembly. Furthermore, it allows maximum flexibility in the actual assembly process, since design for robotic assembly will lead to products which can be assembled manually or with hard automation equipment. Thus, all manufacturing options are preserved beyond the design stage.

Design Strategy

A successful strategy for robotic assembly design obviously requires designers to be aware of robot capabilities. In addition, a general awareness of product cost is essential in that changes in design must be justified from a product cost or out-of-pocket standpoint. Furthermore, successful implementation of a design for robotic assembly philosophy requires early involvement. This is essential because the net savings tend to decrease as the product progresses from the conceptual stage through validation and development into production. In fact, beyond a certain point, the cost to implement a design change for robotic assembly may exceed the savings resulting in a net loss. Thus, the greatest potential for success is at the conceptual stage, where a design change will not require revisions in paperwork, documentation, testing, or tooling.

A final strategic requirement is the development of specific design ground rules related to mechanical assembly, testing, and overall product philosophy. The remainder of this paper will be devoted to outlining some of these guidelines.

Design Guidelines

The most important general design guideline for robotic assembly is to minimize the number of parts. Fewer parts to procure, fabricate, transport, orient, feed, and assemble will almost always lead to lower product costs. Value Engineering(1) offers a scientific approach for minimizing part numbers. The essence of Value Engineering is to evaluate function and cost. This is accomplished in a systematic manner by asking key questions such as the following:

- What is the item or service?
- What does it do?
- What does it cost?
- What else would do the job?
- What would the alternative cost?
- What is it worth?

Application of Value Engineering often leads to a 40% reduction in part numbers. This translates into a shorter assembly cycle time, less expensive tooling, reduced need for error recovery, and shorter robotic system lead time for tooling and programming.

From an overall product design standpoint, several concepts are worthy of consideration. For example, as shown in Figure (1), the number of assembly directions can be minimized.

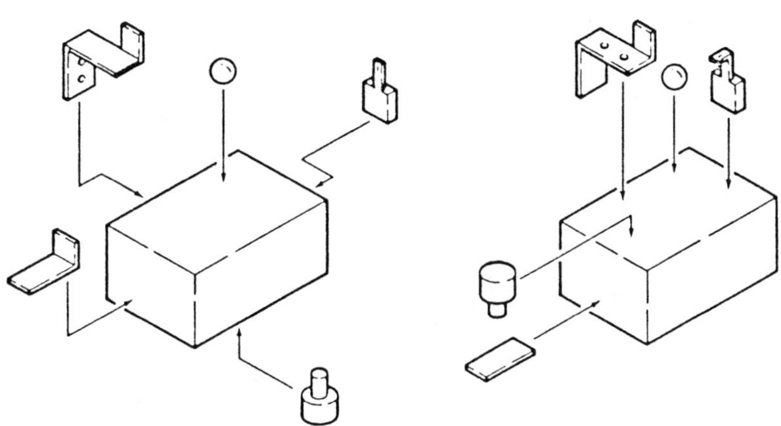

Figure (1)

Providing unobstructed access is a second important product design concept as illustrated in Figure (2).

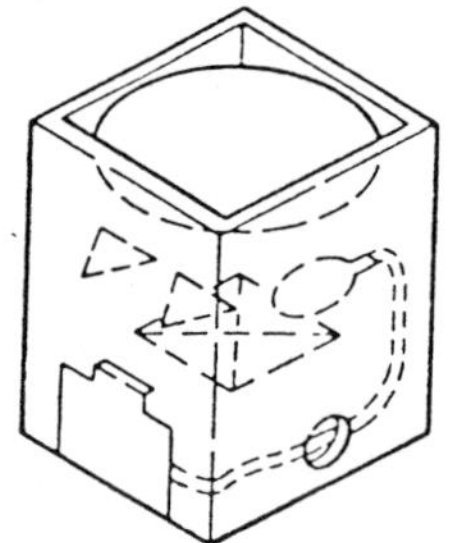

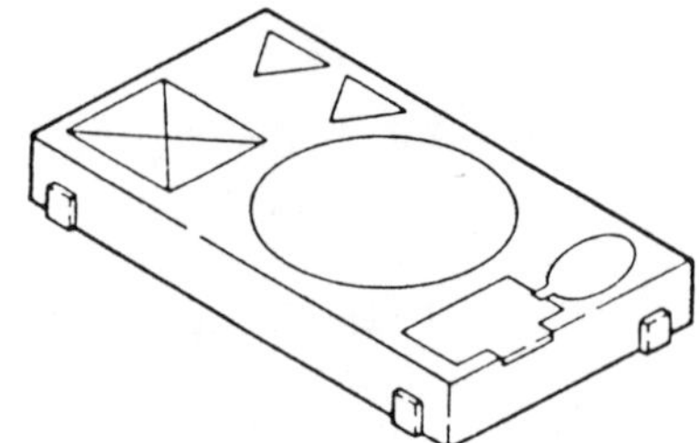

Figure (2)

Individual component part design should be examined from a part handling viewpoint as illustrated in Figure (3).

difficult to grip

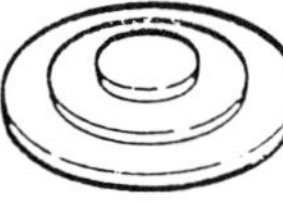

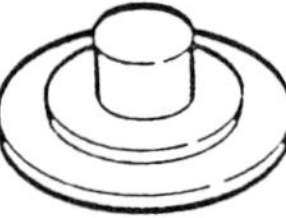

large, flat, smooth top surface for vacuum

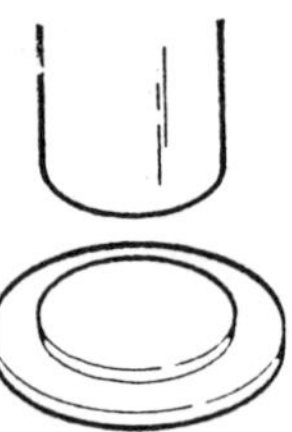

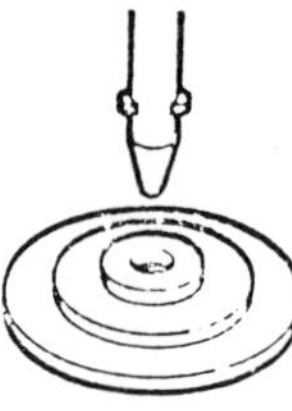

Figure (3)

Self-aligning and self-fixturing parts, as shown in Figure (4) will aid robotic assembly.

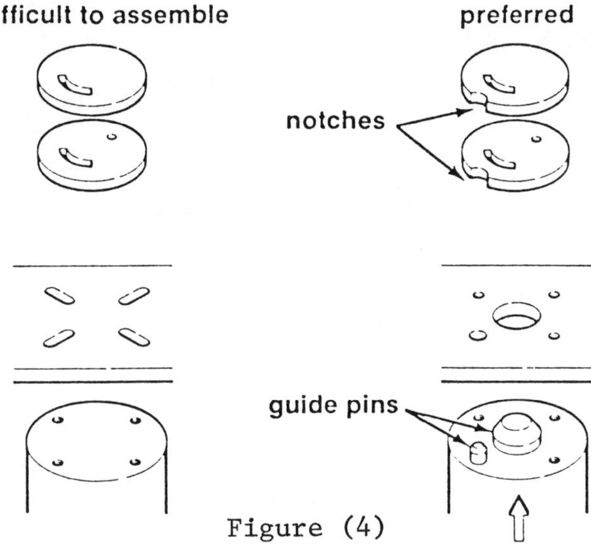

Figure (4)

Small parts can be designed for automatic feeding (2). Vibratory bowls, as shown in Figure (5) can be used to feed and orient a wide variety of small parts.

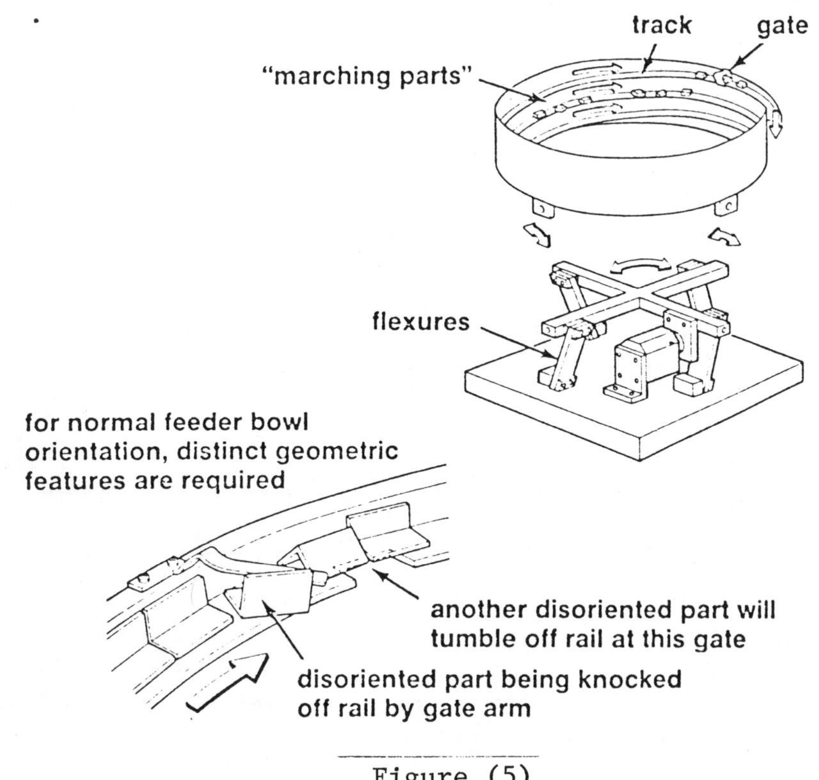

Figure (5)

The use of symmetry as shown in Figure (6) will help with part orientation. External features may be needed, as shown in Figure (7), to provide orientation of hidden features.

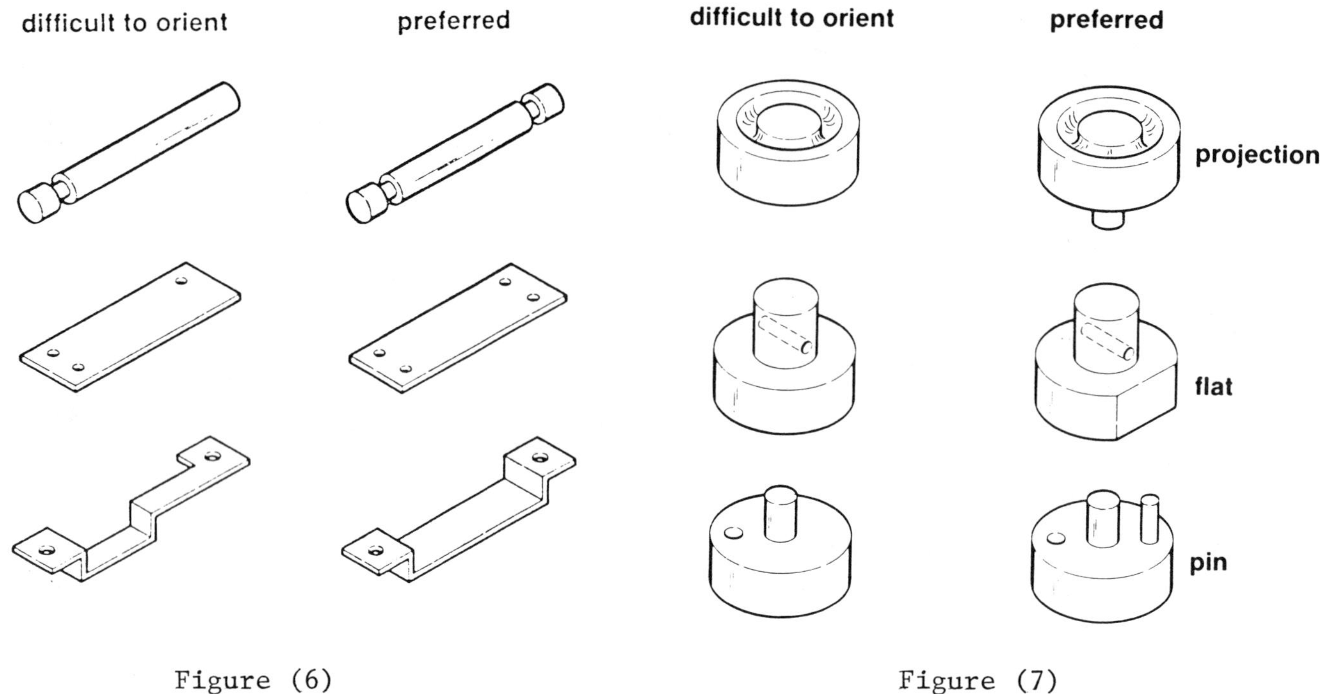

Figure (6)

Figure (7)

Figures (8) through (10) illustrate guidelines to facilitate feeding of small parts.

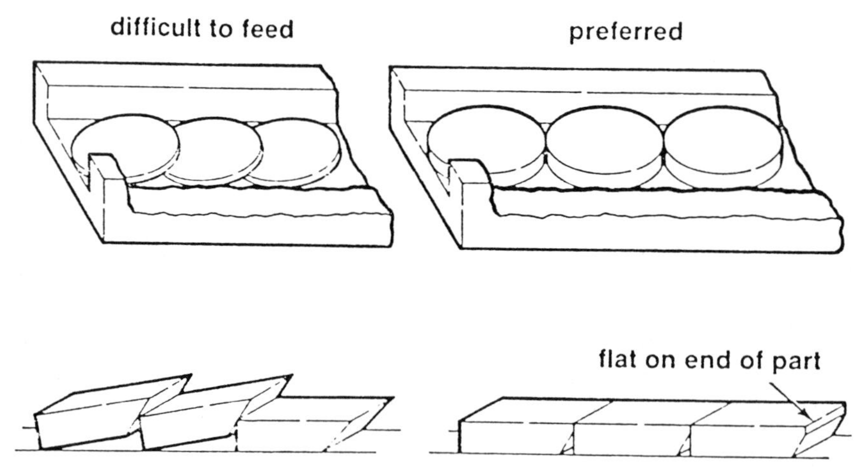

Figure (8)

Avoid parts which shingle.

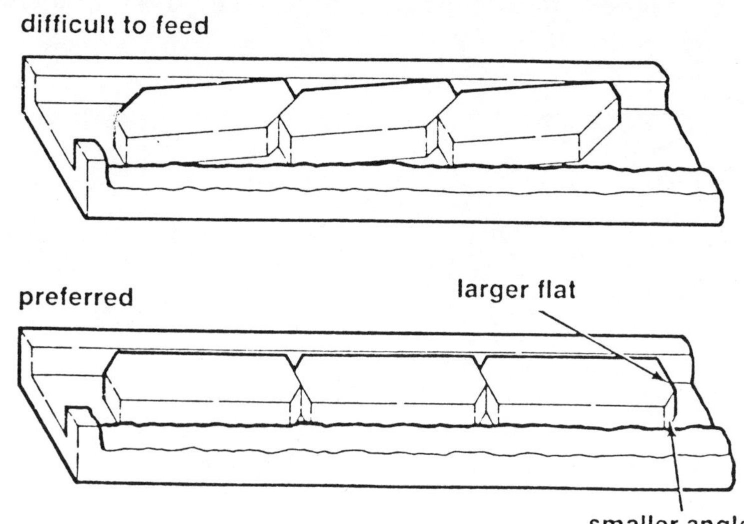

Figure (9)

Avoid parts which wedge

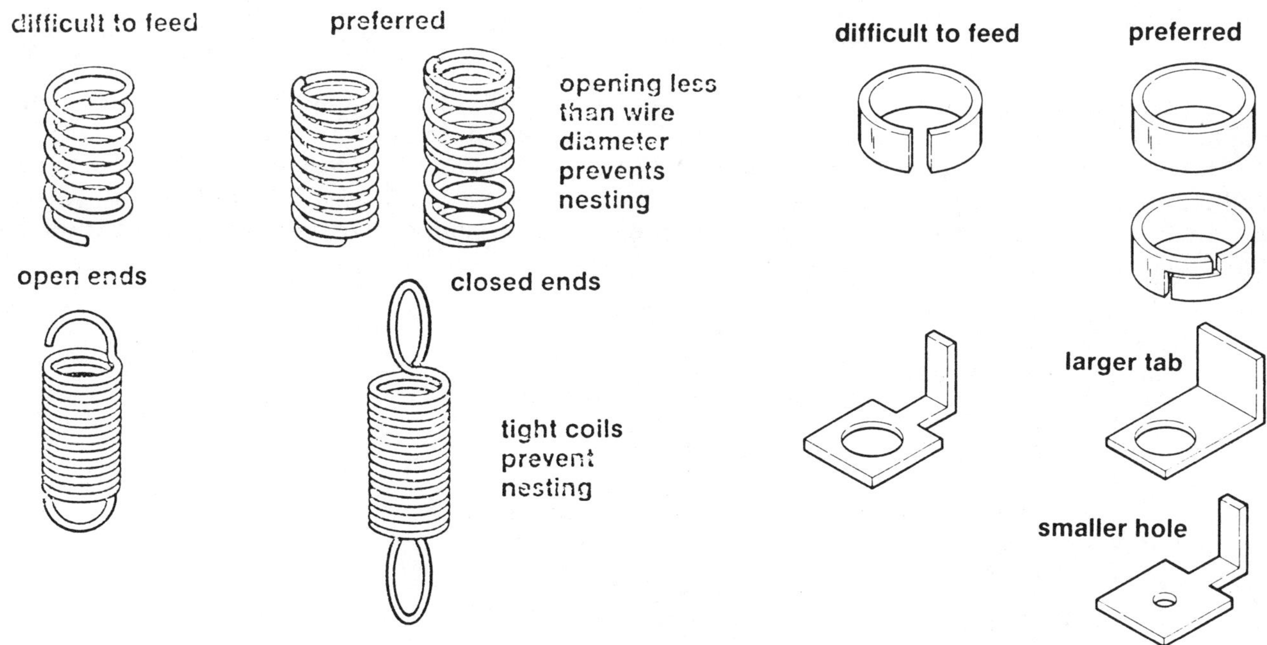

Figure (10)

Avoid parts which tangle or daisy chain.

Alternatives to feeder bowl handling of small parts are shown in Figure (11)

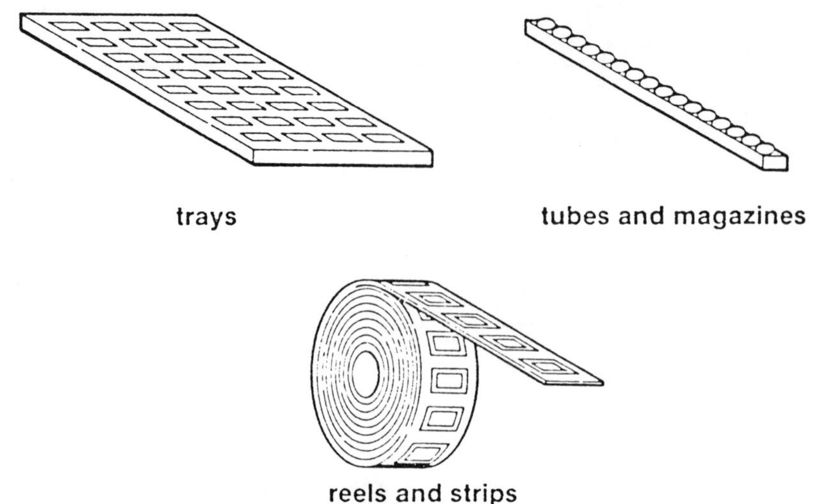

Figure (11)

For mechanical assembly, a high proportion of all tasks can be represented as peg-in-hole (3). This task can be greatly facilitated by a combination of compliance in the robot arm and/or compliance in the holding fixture. Chamfers and corner breaks as shown in Figure (12) can increase significantly the probably of part mating, especially when combined with compliance. Force sensing and error recovery routines can be utilized when passive measures fail.

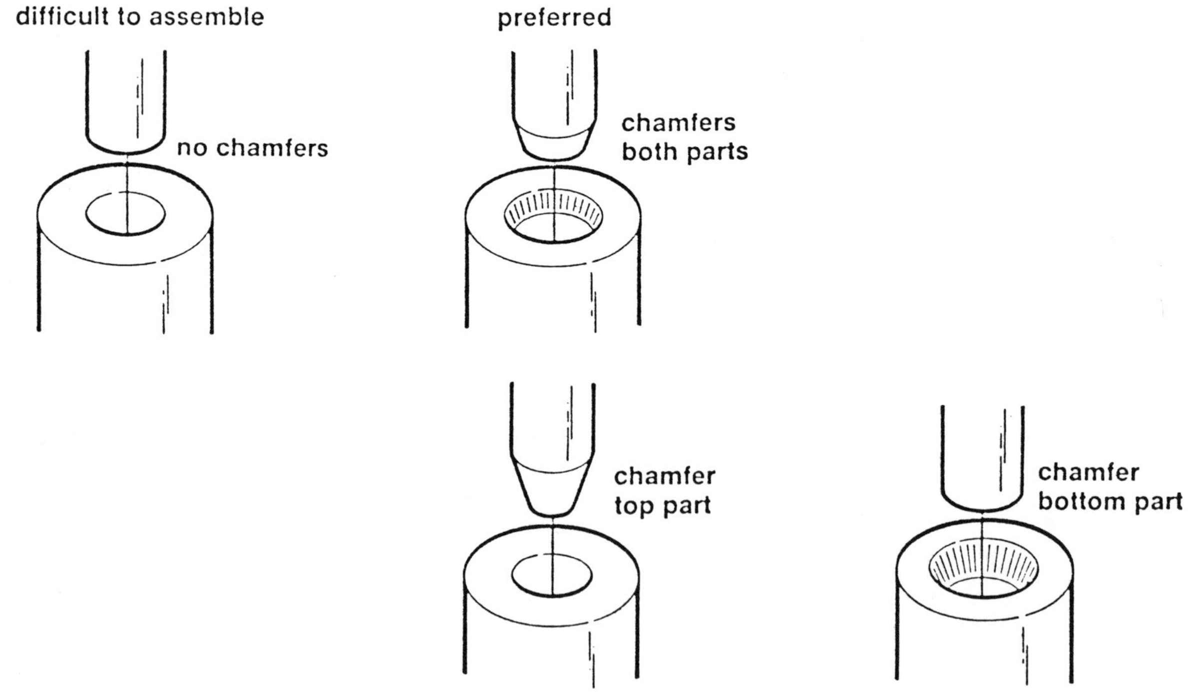

Figure (12)

Snap-together parts, as shown in Figure (13) are ideally suited to robotic assembly. However, when disassembly is required, traditional screw fastening techniques may be required. When using screws, minimize the number, type, and sizes, provide tool clearance, consider combining screw and washer, and use screws which can be automatically fed. Figures (14) and (15) illustrate two of these points.

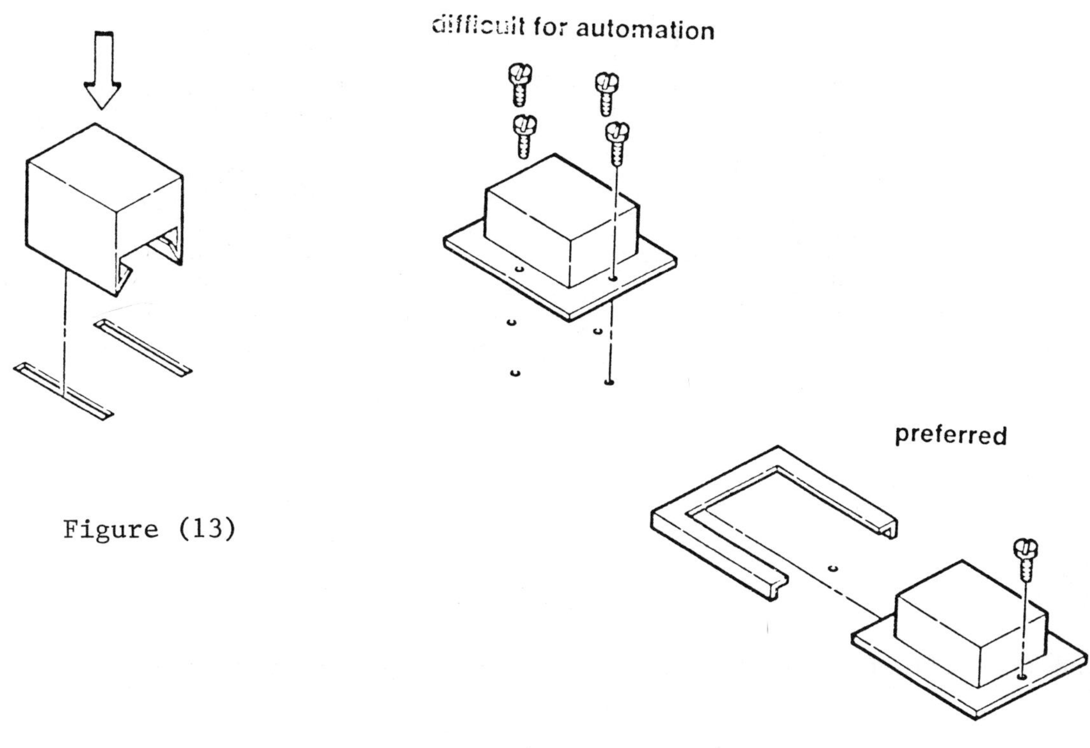

Figure (13)

Figure (14)

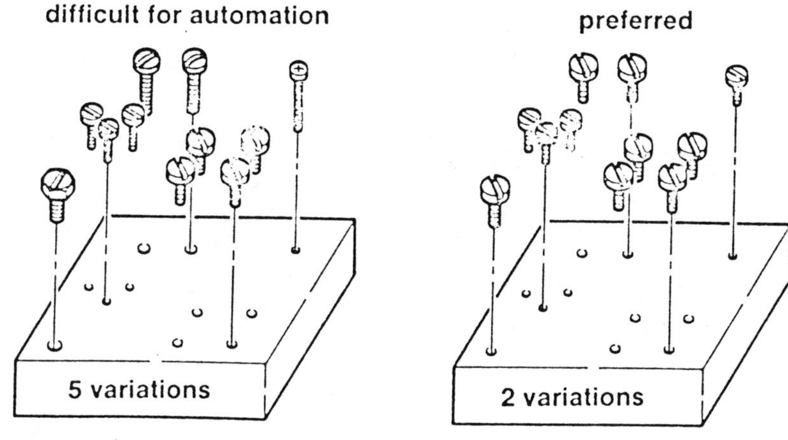

Figure (15)

Figure (16) illustrates the concept of one-handed adjustment which is highly desirable for robotic assembly.

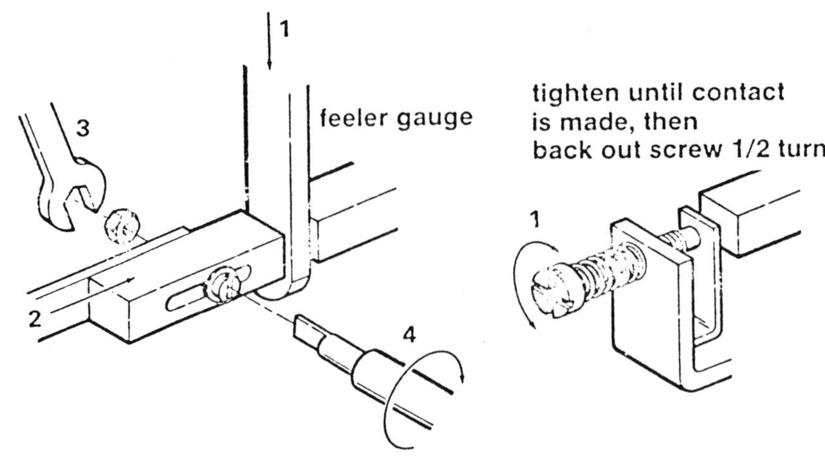

Figure (16)

Figure (17) illustrates the concept of providing common mechanical and electrical interface.

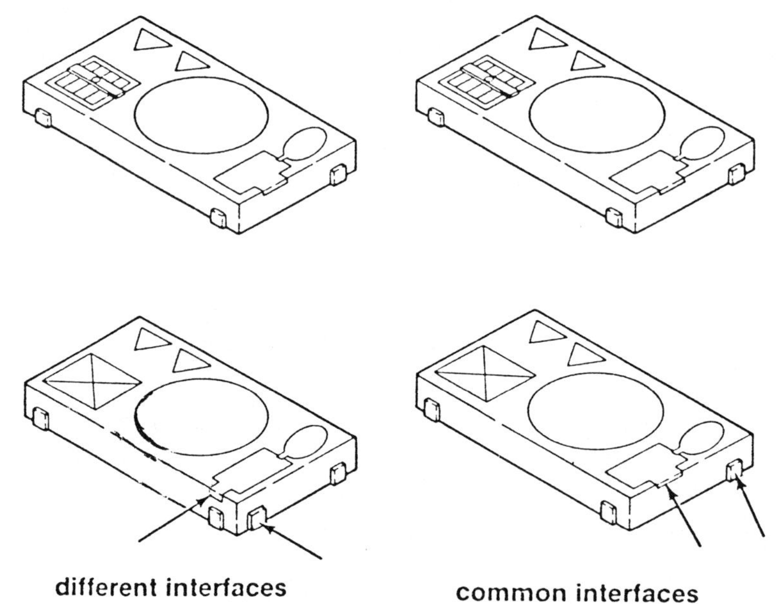

Figure (17)

Case Study

Shown in Figure (18) is the original design of a mechanism used to feed single sheets of paper into a printer. It can be seen that several component parts cannot be fed automatically, several assembly directions are required, extension springs which are difficult to attach are used, and numerous adjustments are required.

A redesigned paper feed mechanism is shown in Figure (19). The number of parts has been reduced from 27 to 14. A single compression spring has replaced the original extension springs. The motor can now be installed with a push-and-twist motion. Access has been provided for the spindle and a "handle" has been provided to facilitate grasping of the spindle. Thirteen of the fourteen parts can now be assembled robotically with the IBM RS1. (The only exception is the lift mechanism which would require further work to eliminate a two-handed motion.)

The redesign mechanism is now in production in the IBM plant in Austin, Texas. The original reason for redesign of this mechanism was so that it could be assembled robotically. However, it can be seen that the resulting effort has produced an improved design for assembly manually, robotically, or with fixed automation. In fact, the remaining manual assembly time is now too short to justify the use of a robot for assembly, thus illustrating the main point of this paper: namely, that a philosophy of design for robotic assembly will result in improved value, reduced cost, and better designs regardless of the final assembly method.

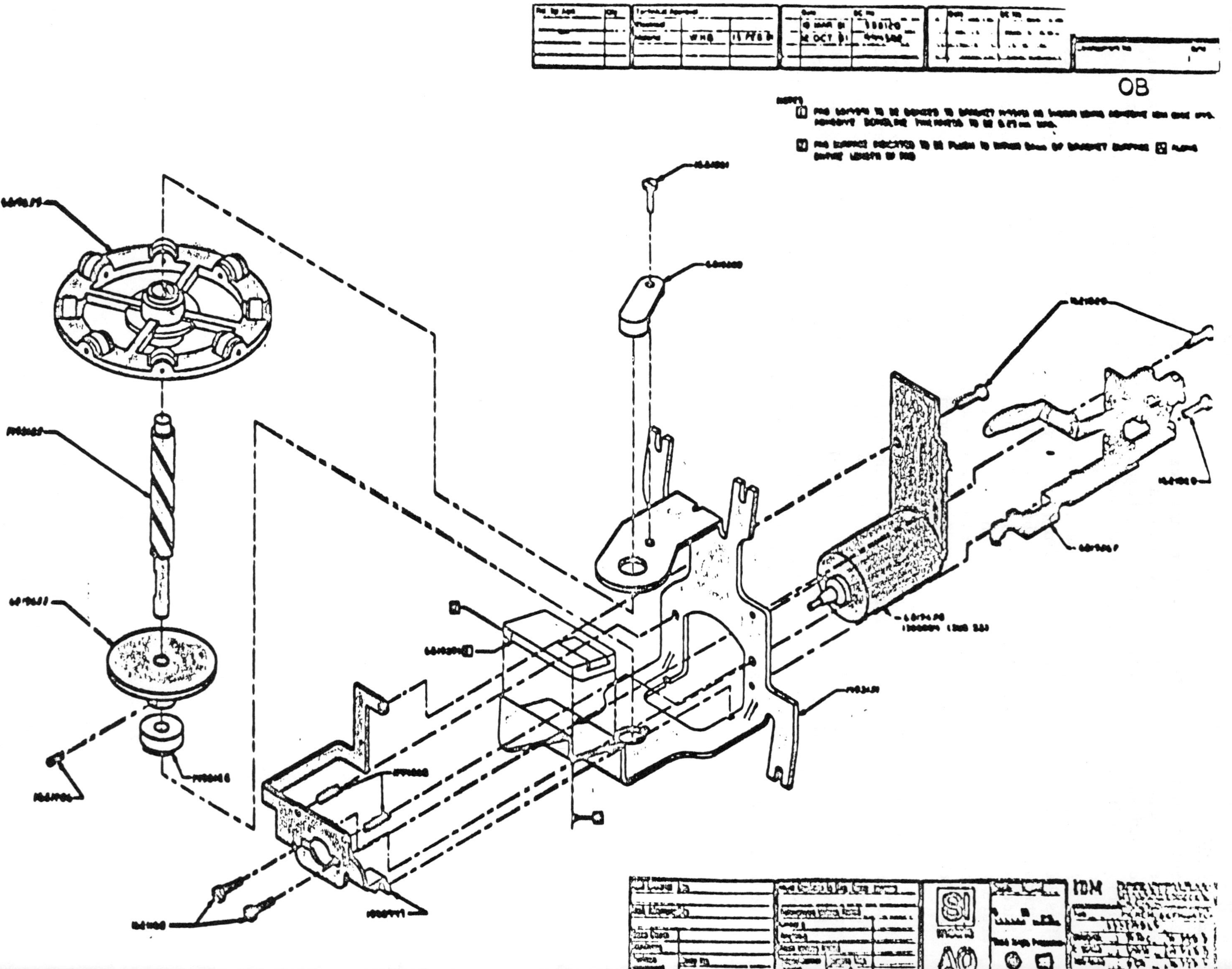
OB

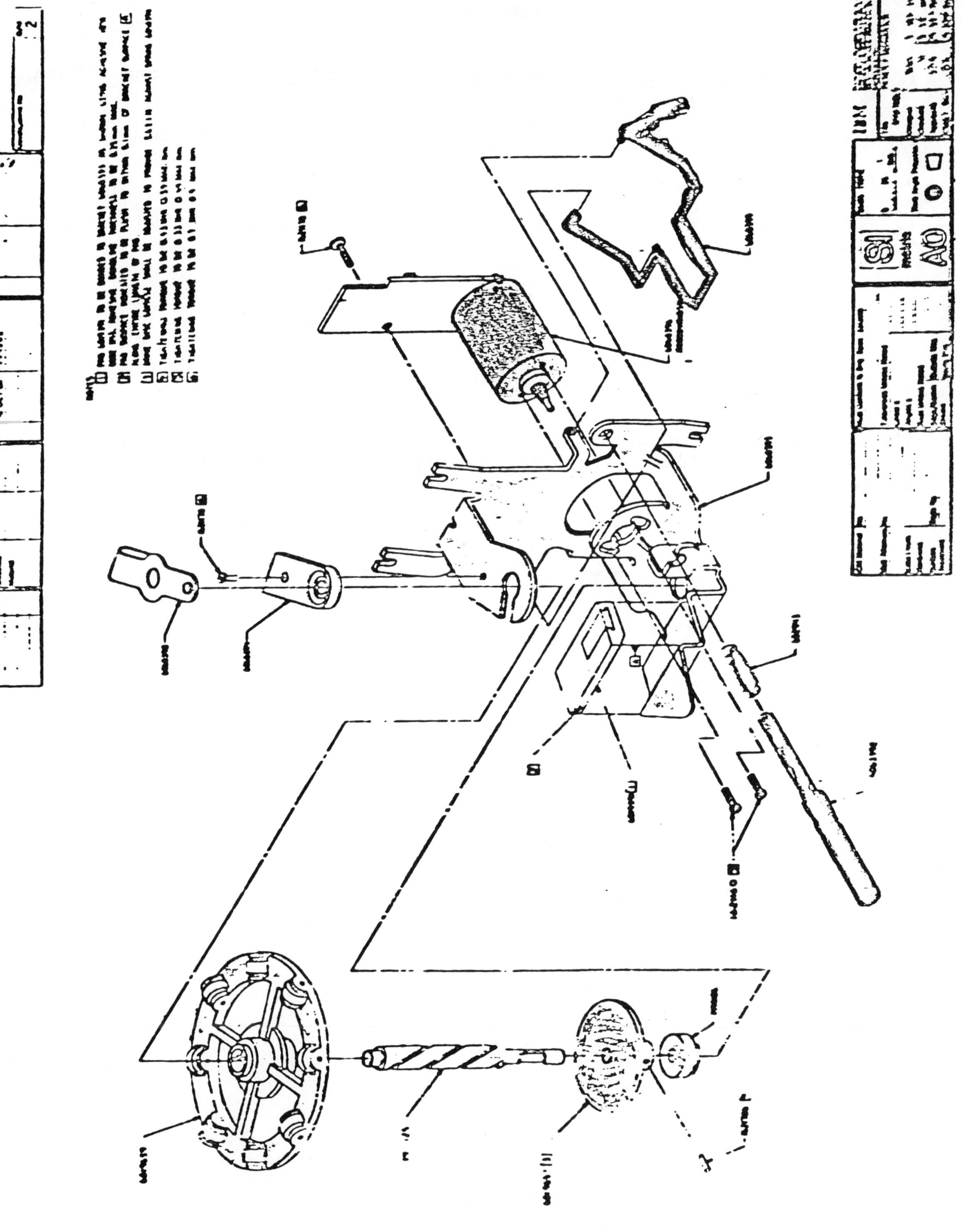

BIBLIOGRAPHY

1. Miles, Lawrence D., Techniques of Value Analysis and Engineering, Second Edition, McGraw-Hill, 1972.

2. Boothroyd, G., Design for Assembly Handbook, University of Massachusetts.

3. Kondoleon, A., M.I.T. Thesis, 1976.

4. Harvey, John J. et. al., Design Guide for Assembly and Automation, IBM Corp., 1982.

ACKNOWLEDGEMENT

The contribution of Bill Adair and his collegues at the IBM plant in Austin, Texas to the redesign of the paper feed mechanism is gratefully acknowledged.

Presented at the SME Ultratech '86, Robots West Conference, September 1986

Automation: Teacher and Test for Good Product Design

by Henry W. Stoll
Industrial Technology Institute

INTRODUCTION

In the late 1970's, American manufacturers suddenly realized that their preeminent position in the global marketplace was being challenged by declining U. S. productivity, a threat from world-class manufacturers, and the emergence of new automation processes. This precipitated a quest for short-term solutions that quickly focused on fledgling technologies such as robotics, vision and optical processing, and flexible manufacturing systems. Failure to achieve promised productivity improvements using these and other advanced automation technologies in a variety of industrial settings has taught much about the product design/automation interface.

Perhaps the most important lesson learned is that the design, function, and implementation of advanced automation technology is directly related to the product being manufactured. Hence, when implementing these technologies, product design and automation design can no longer be treated as separate entities, but rather, must be integrated into one common activity. It is especially essential that interactions between product and process enter into consideration very early in the product design process. Product concept and automation decisions must be made in parallel in order to obtain an integrated manufacturing system optimally configured to satisfy both product and process needs.

Design for automation, or "automation friendly" design, is concerned with defining product design alternatives which facilitate optimization of the product/process as a whole. This paper reports on an approach to "automation friendly" product design which involves the application of concisely stated design principles and guidelines to each decision-making step. The approach has three objectives:

1. Identify product concepts which are inherently easy to automate.

2. Focus on component design for ease of automated processing.

3. Integrate automation system design with product design to achieve optimal matching of needs and requirements.

In developing this approach, the problem of designing for automation is first briefly reviewed. A set of codified design principles and guidelines, distilled from automation experience, are then concisely stated and discussed to show how they can be used to both teach and test for automation friendly design. Using these principles and guidelines as a basis, a design for automation methodology is then proposed and its use discussed.

THE PROBLEM

A manufacturing system comprises a large number of distinct processes or stages which, individually and collectively, affect product cost, product quality, and productivity of the overall system. The interactions between these various facets of a manufacturing system are complex, and decisions made concerning one aspect have ramifications which extend to the others. This is essentially the problem in design for automation. Many of the product decisions made during the course of product development impact the automation concept in a variety of direct and indirect ways. For example, if the product family has been differentiated into a series of different sizes, these sizes, if not carefully considered, could make production of one model incompatible with another. This incompatibility results in added automation complexity, tooling changes, and other adjustments. Numerous other examples of interactions between product and process can be cited and the cost, quality, and productivity problems caused because these interactions were not properly considered early in the product design process are legion.

The result of not considering the interaction between product and process is suboptimal design of both facets of the manufacturing system. Compounding the productivity loss due to this suboptimality is the needless time, effort, and money spent solving automation problems which could have been avoided in the first place through proper product design. Even after the problems are ironed out and the engineering changes made, the automation still falls short of providing hoped for productivity gains because the product and process are improperly matched. Most devastating of all, the extra manufacturing cost and loss of productivity incurred continues throughout the life of the product.

To correct this problem, it is clear that the multitude of possible product/process interactions must be anticipated and properly accounted for early in the product/process development cycle, before concept and equipment decisions have been made. This requires a new, more disciplined, approach to product development. But the method for providing this discipline is the other side of the problem. At present, the knowledge base which completely and accurately describes the complex interactions between product and process is simply not available. Hence, current product/process development practice consists of a non-systematic, somewhat random method in which design engineers, manufacturing engineers, and others get together, exchange views, and formulate a set of decisions which ultimately result in a product.

DESIGN PRINCIPLES AND GUIDELINES

To compensate for the lack of a complete product/process interaction knowledge base, a wealth of product design approaches, techniques, tricks, and design tips have evolved out of automation experience which help show the way to good design for automation. Knowledge of this information and the ability to correctly apply it has always been one of the hallmarks of the expert design and manufacturing engineer. The central premise of this paper is that proper codification of the empirical knowledge which automation experience has taught into a design for automation methodology can form a highly effective tool for ensuring that product/process interactions are systematically anticipated and dealt with early in the product design process. The great advantage of using concisely stated design for automation principles and guidelines is that they explicitly state what is intuitively obvious to experienced designers and commonly used by good designers. Such a system of principles and guidelines therefore provide a firm basis upon which design knowledge can be expanded in a systematic way.

Study of successful automation applications clearly shows that success depends on satisfying a few, very basic, underlying design objectives. These objectives are stated as the following design for automation principles. The methodology used in developing the principles is discussed in references [1]. Because they are intended for use in the product/process design, the principles are stated as directives rather than observations.

1. **Separate, or "decouple", product requirements from automation requirements.**

2. **Minimize product/process information content of the decoupled design.**

3. **Integrate product and process design.**

Information is contained in both hardware and software. In "hard" automation, all information is designed into hardware. This accounts for the disadvantage of hard automation which is its intolerance to change. To offset this, the tendency in flexible automation is to transfer as much information as possible to software. The second design principle requires that total information content be minimized by judicious distribution of information between hardware and software. In many cases, the information content can best be minimized by clever use of both hardware, software, and perhaps "humanware". For example, designing-in guiding and self-locating features into the product components and using a simple SCARA type robot for component insertion involves less information content than using a multi-degree-of-freedom robot combined with tactile sensing and vision. Similarly, if a component or operation is particularly difficult to automate for one reason or another, the low information solution may be to do it manually. Because hardware includes both product and automation equipment, it is essential that both be considered during design. Principle 3 is therefore an imperative which must be satisfied to assure correct application of principles 1 and 2.

In order to assist and facilitate proper applications of the design for automation principles, a set of corollaries to the principles can be stated to serve as guidelines for product design. Like the principles, the guidelines are stated as directives which act to both stimulate creativity and show the way to automation friendly design. If correctly followed, they should result in a product which is inherently easier to automate.

1. Minimize Total Number of Parts

2. Develop a Modular Design

3. Use Standard Components

4. Design Parts to be Multi-Functional

5. Design Parts for Multi-Use

6. Design Parts for Ease of Fabrication

7. Avoid Separate Fasteners

8. Minimize Assembly Directions

9. Maximize Compliance

10. Minimize Handling

Standardization is a way of separating or "decoupling" product and marketing needs from manufacture. Standardization also reduces the amount of information required for product manufacture. This can be illustrated by considering modular design. A module is a self-contained component with standardized interfaces to other product modules and to the production equipment and tooling used in the product manufacture. Individual modules can be varied to provide functional and styling diversity. Similar diversity can be provided by using different combinations of standard modules. All of this has no effect on the production line as long as the module/process equipment and tooling interfaces are standardized. Modular design also reduces final assembly information content because there are less parts to assemble and each module can be fully checked prior to final assembly.

Multi-functional and multi-use parts are another way of decoupling the product from manufacture and reducing manufacturing information content. For example, the same mounting plate can be designed to mount several different components. By having robot grippers only touch the mounting plate, the particular component being installed is decoupled from the installation process. Information is reduced because the same mounting plate and installation process can be used in a variety of applications.

Reducing or minimizing the number of individual parts in a product is perhaps the most effective way to reduce information content. Put another way, a part that is eliminated costs nothing to design, make, assemble, move, handle, orient, store, purchase, clean, inspect, rework, service. It never jams or interferes with automation. It never fails, malfunctions or needs adjustment. It requires no drawing or part number and never needs to be changed.

When viewed in this way, it is seen that all of the remaining guidelines listed above are really ways for reducing the manufacturing information content. Reduced information content explains the emphasis on near net shape processes and the need to avoid secondary processing. Often, higher material and/or processing cost can be accepted because the reduced information content leads to lower overall production cost. In automation applications, separate fasteners are difficult to feed, tend to jam, require monitoring for presence and torque, and require costly fixturing, parts feeders, and extra stations. Avoiding separate fasteners eliminates all this extra automation information. The same thing is true for assembly directions and material handling. By designing for "Z-axis" insertion, many needed degrees of freedom and the information that goes with them are eliminated. Similarly, the use of symmetry to assist in achieving proper part orientation and the use of parts magazines, tube feeders, part strips, as well as palletized trays and kitting techniques for preserving orientation greatly reduces the information required to manipulate and handle parts. Finally, use of generous tapers and other guiding, orienting, and locating features in part design greatly simplifies component insertion and assembly because it reduces the amount of information needed to perform these tasks.

APPLYING THE GUIDELINES

Application of the design for automation guidelines is not always easy or straightforward. They show the way, but do not replace the talent, innovation, and experience of the product development team. They must also be applied in a manner which maintains and, if possible, enhances product performance and marketing goals. In applying the design guidelines, they should be thought of as "optimal suggestions", which, if successfully followed, will result in a more optimal, automation friendly design. If a product performance or marketing requirement prevents full compliance with a particular guideline, then the next best alternative consistent with the design for automation principles should be selected. Use of the guidelines in this way helps to both assure a product design optimized for automated production and to delineate problem areas requiring special attention. Hints for creative application of the guidelines as well as insights to stimulate innovative design are provided in the brief discussion of each guideline which follows. Background and basis for these hints is fully discussed in reference [2].

1. Minimize Total Number of Parts.

A part is a good candidate for elimination if there is 1) no need for relative motion, 2) no need for subsequent adjustment between parts, 3) no need for service or repairability, and 4) no need for materials to be different. However, part reduction should not exceed the point of diminishing return where further part elimination adds cost and complexity because the remaining parts are to heavy, or to complicated to make and assemble, or are to unmanageable in other ways.

Perhaps the best way to eliminate parts is to identify a design concept which requires few parts. Integral design, or the combining of two or more parts into one, is another approach. Besides the advantages given above, integral design reduces the amount of interfacing information required, and decreases weight and complexity. One piece structures have no fasteners, no joints, and fewer points of stress concentration. Conversely, structural continuity leads to high strength and lightweight.

Plastic is a major key to integral design. Plastic is available for making springs, bearings, cam and gears, fasteners, hinges, and optical elements. Powder metallurgy (P/M) is a good alternative if plastic parts do not have adequate strength, heat resistance, or cannot be held to the tolerance needed. Brazed, welded, or staked assemblies of stampings and/or machined parts can often be made as one-piece P/M parts. Extrusions and precision castings as well as impacts are also good ways to eliminate subassemblies. Although switching to a different manufacturing process may lead to a more costly part, experience with part integration has shown that a more costly part often turns out to be more economical when assembly costs are considered.

2. Develop a Modular Design

Designing for modularity requires careful consideration of a variety of needs. To decouple the product from the automation, it is essential that stable groups be identified and that the interfacing information be specified in a way which facilitates the desired decoupling. Often, standardization of just a few dimensions is all that is needed to decouple product from process. An example of this is the standardization of the base module of the IBM typewriter. Because all the location and automation interface information is concentrated in the typewriter base, changes to other modules do not impact the automation system (provided the module handling interfaces are not changed).

In seeking to identify the minimum number of stable groups, it is useful to look at the distribution of functional requirements among different product models and lines to see if a particular function could be satisfied everywhere by one module. Looking for common problems within different models and product lines is another approach. Will the same solution work everywhere? Other questions to ask include:

- Can customization (diversity) requirements be satisfied using "add-on" modules?

- How will future product(s) differ from the current design? What commonalities will exist?

- What modular configuration will simplify material handling? Decouple quality from production? Decouple variation in vendor supplied components from production? Decouple style from production?

Experience has shown that products consisting of 4 to 8 modules with 4 to 12 parts per module are most automation friendly. A good design strategy is to keep the product generic for as long as possible during assembly by saving the specialized modules for last. If possible, the modules should be designed to add up to the final product thereby eliminating the need for a housing or other integrating structure. Also, information content is reduced if all modules (except perhaps the base) are approximately the same size.

3. Use Standard Components

A stock item is always less expensive than a custom-made item. Standard components require little or no lead time and are more reliable because characteristics and weaknesses are well known. They can be ordered in any quantity at any time. They are usually easier to repair and replacements are easier to find. Use of standardized components puts the burden on the supplier and makes the supplier do more.

4. Design Parts to be Multi-Functional

Combine function wherever possible. For example, design a part to act both as a spring and a structural member, or to act both as an electrical conductor and structural member. An electronic chassis can be made to act as an electrical ground, a heat sink, and a structural member. Less obvious combinations of function might involve adding guiding, aligning, and/or self-fixturing features to a part to aid in assembly. Or providing a reflective surface or recognizable feature to facilitate vision inspection. These latter examples illustrate inclusion of functions which are only needed during manufacture. Such function combinations are often the result of design for automation awareness.

5. Design Parts for Multi-Use

Many parts can be designed for multi-use. For example, the same mounting plate can be designed to mount a variety of components. Similarly, a spacer can also serve as an axle, lever, standoff, etc. Key to multi-use part design is identification of part candidates. One approach involves sorting all parts (or a statistical sample) manufactured or purchased by the company into two groups consisting of 1) parts which are unique to a particular product or model (i.e., crankshafts, housings, etc.) and 2) parts which are generally needed in all products and/or models (shafts, flanges, bushings, spacers, gears, levers, etc.). Each group is then divided into categories of similar parts (part families). Multi-use parts are then created by standardizing similar parts. In standardizing, the designer should sequentially seek to 1) minimize the number of part categories, 2) minimize the number of variations within each category, and 3) minimize the number of design features within each variation. Once developed, the family of standard parts should be used wherever possible in existing products and used exclusively in new product designs. Also, manufacturing processes and tooling based on a composite part containing all design features found in a particular part family should be developed. Individual parts can then be obtained by skipping some steps and features in the manufacturing process.

6. Design Parts for Ease of Fabrication

This guideline requires that individual parts be designed using the least costly material that just satisfies functional requirements (including style and appearance) and such that both material waste and cycle time are minimized. This in turn requires that the most suitable fabrication process available be used to make each part and that the part be properly designed for the chosen process. Use of near net shape processes are preferred whenever possible. Likewise, secondary processing (finish machining, painting, etc.) should be avoided whenever possible. Secondary processing can be avoided by specifying tolerances and surface finish carefully and then selecting primary processes (precision casting, P/M, etc.) which meet requirements. Also, material alternatives which avoid painting, plating, buffing, etc. should be considered. This guideline is based upon the recognition that higher material and/or unit process cost can be accepted if it leads to lower overall production cost, i.e., adding information content to a particular part is acceptable as long as total information content of the product/process is reduced.

7. Avoid Separate Fasteners

Separate fasteners involve large amounts of information. Even in manual assembly, the cost of driving a screw can be six to ten times the cost of the screw. One of the easiest things to do is eliminate fasteners in assembly by using snap-fits. If fasteners must be used, cost as well as quality risks can be significantly reduced by minimizing the number, size, and variations used and by using standard fasteners whenever possible. Screws that are to short or to long, separate washers, tapped holes, and round and flat heads (not good for vacuum pickup) should be avoided. Conversely, captured washers should be used for reduced part placement risk and improved blow feeding. Self tapping/forming/locking fasteners are preferred as are screws with dog or cone (chamfered) point for improved placement success. Also, screw heads designed to reduce "cam-out" problems, bit wear, and fastener damage should be used. For vacuum pickup, use screw heads having flat vertical sides.

8. Minimize Assembly Directions

All parts should be assembled from one direction. Extras directions mean wasted time and motion as well as more transfer stations, inspection stations, and fixture nests. This in turn leads to increased cost, increased wear and tear on equipment due to added weight and inertia load, and increased reliability and quality risks. The best possible assembly is when all parts are added in a top down fashion to creat a z-axis stack. Multi-motion insertion should be avoided. Ideally, the product should resemble a z-axis "club sandwich" with all parts positively located as they are added.

9. Maximize Compliance

Because parts are not always identical and perfectly made, misalignment and tolerance stack-up can produce excessive assembly force leading to sporadic automation failures and/or product unreliability. Major factors affecting rigid part mating include part geometry (accuracy, consistency), stiffness of assembly tool, stiffness of jigs and fixtures holding the parts, and friction between parts. To guard against this, compliance must be built into both the product and production process. Methods for providing compliance include highly accurate (consistent) parts, use of "worn-in" production equipment, remote center compliance, selective compliance in assembly tool (SCARA Robot), tactile sensing, vision systems, designed-in compliance features, and external effects. Although a variety of combinations of these approaches are commonly used, experience has shown that the simplest solution consists of a combination of acceptable (consistent) quality parts, designed-in compliance features, accurate (rigid) base components, and selective compliance in the assembly tool (SCARA Robot).

Designed-in compliance features include the use of generous tapers or chamfers for easy insertion, use of leads and other guiding features, and use of generous radii where possible. A clever trick, if possible, is to design one of the product components, perhaps the largest, to act both as the part base (part to which other parts are added) and as the assembly fixture (avoid need for a special fixture to hold the assembly). In any case, the part base should be made as stable and rigid as possible to improve insertion accuracy and simplify handling. If fixturing is required, "fixture friendly" features such as accurate location points, generous tapers and other guiding features which provide easy compliance between base part and fixture should be provided. Gravity is an extremely useful external effect which assists compliance and costs nothing. In addition to assisting with insertion, gravity is useful for feeding parts and for ejecting finished and defective product.

10. Minimize Handling

Position is the sum of location (x,y,z) and orientation (α,β,γ). Position costs money. Therefore, parts should be designed to make position easy to achieve and the production process should maintain position once it is achieved. The number of orientations required during production equates with increased equipment expense, greater quality risk, slower feed rates, and slower cycle times. To assist in orientation, parts should be made as symmetrical as possible. If polarity is important, then an existing asymmetry should be accentuated, or a very obvious asymmetry should be designed in, or a clear identifying mark provided. Orientation can also be assisted by designing in features which help guide and locate parts in the proper position. Parts should also be designed to avoid tangling, nesting, and shingling in vibratory part feeders.

To facilitate robotic part handling, provide a large, flat, smooth top surface for vacuum pickup, or provide an inner hole for spearing, or provide a cylindrical surface or other feature of sufficient length for gripper pickup. Since parts usually come off the production line properly oriented, this orientation should be preserved by using magazines, tube feeders, part strips, etc. Palletized trays and kitting are methods for supplying properly oriented parts to the assembly line. For ease of handling, avoid flexible components. Use rigid gaskets where possible; use a connector to eliminate lead wires; use a circuit board in place of a cable; etc.

Design in features which facilitate product and component packaging. Use standard outer package dimensions for machine feeding and storing, design packaging to adequately protect and insure quality at all stages of handling, and design packaging for easy handling. Consider material flows within the production facility including product flow, workspace flow, supply flow, hardware flow, trash or scrap flow, bulk material flow, container flow, and fixture flow. For each flow, consider how the product, subassembly, component, or part can be designed to simplify or eliminate the flow.

PROPOSED DESIGN FOR AUTOMATION METHODOLOGY

Application of the design guidelines will help the product development team design a product which is inherently easy to manufacture but this does not necessarily guarantee that the product will be correctly matched to the automation concept. Proper matching requires integrated design of the product and the process (Principle 3). To integrate product design with process design early in the product design process before concept decisions have been made is difficult because so much of the design is fluid and unspecified. To maximize the probability of achieving the best possible design/automation solution under these uncertain conditions requires a systematic approach. Such an approach must minimize the chance of forgetting or over-looking an important consideration, be flexible enough to meet the needs of widely differing situations, and yet be simple enough to become habitual in its use. To meet this need, a systematic approach, based on the design for automation principles, is proposed.

In the proposed approach, merging of the product and process designs is implemented using iterative analysis-redesign cycles (Figure 1) in which a complete conceptual design (product and process) exists at each iteration. The process starts with initial product/process concepts and converges toward an integrated solution which, when fully optimized, is in complete agreement with the design for automation principles. The initial process concept should be a target toward which to work. The initial product concept can be either an existing design or a new concept depending on the particular circumstances of the problem. The integrated product/process design is first analyzed for possible couplings and excessive information content. Based on the analysis, an evaluation is made of its acceptability. If the design is acceptable, that is, if it agrees with the principles within acceptable limits, then the solution is complete. If not, a redesign is performed using the principles and guidelines to find innovative "fixes" to the identified problems and the process returns iteratively to the analysis phase.

In the analysis phase, the product/process development team should concentrate on anticipating and identifying all possible product/process interactions. To do this, all aspects of the production process such as material handling, inspection and quality control, functional testing, tooling, scheduling, etc. must be looked at critically. Product packaging and shipping as well as component packaging and orientation should also be questioned. In this phase, the guidelines act as a checklist for identifying possible problem areas and opportunities for improvement. Hence, when used in the analysis phase, the guidelines act as a test for automation friendly product design.

In the evaluation phase, the design for automation principles are used to judge the acceptability of the product/process integration. Product, marketing, and production requirements and constraints which impact the product/process interface are also considered. Requirements define the problem to be solved, constraints establish boundaries on acceptable solutions. Product and model mix and production sequence; range of sizes and number of variations; range of lot order sizes; amount, type, and location of data collection points and assembly test points; etc., are all examples of product/process interface requirements. Quality needed for part feeding, alignment, tolerance stack-up, etc. are examples of product/process interface constraints. Other constraints might include acceptable cost, OSHA requirements, adaptability to existing systems, etc.

In the redesign phase, innovative solutions which decouple the interactions identified in the analysis stage and minimize information content are sought. In all cases, the feasible solution space is constrained to maintain product function and marketing needs. Emphasis is on solutions which decouple product performance and marketing needs as well as product diversity and component variability requirements from the automation requirements. Design to minimize information content focuses on component design for ease of manufacture and assembly as well as minimizing number of parts, avoiding designs that require fasteners, and standardizing whenever possible. In this phase, the design guidelines are used to show the

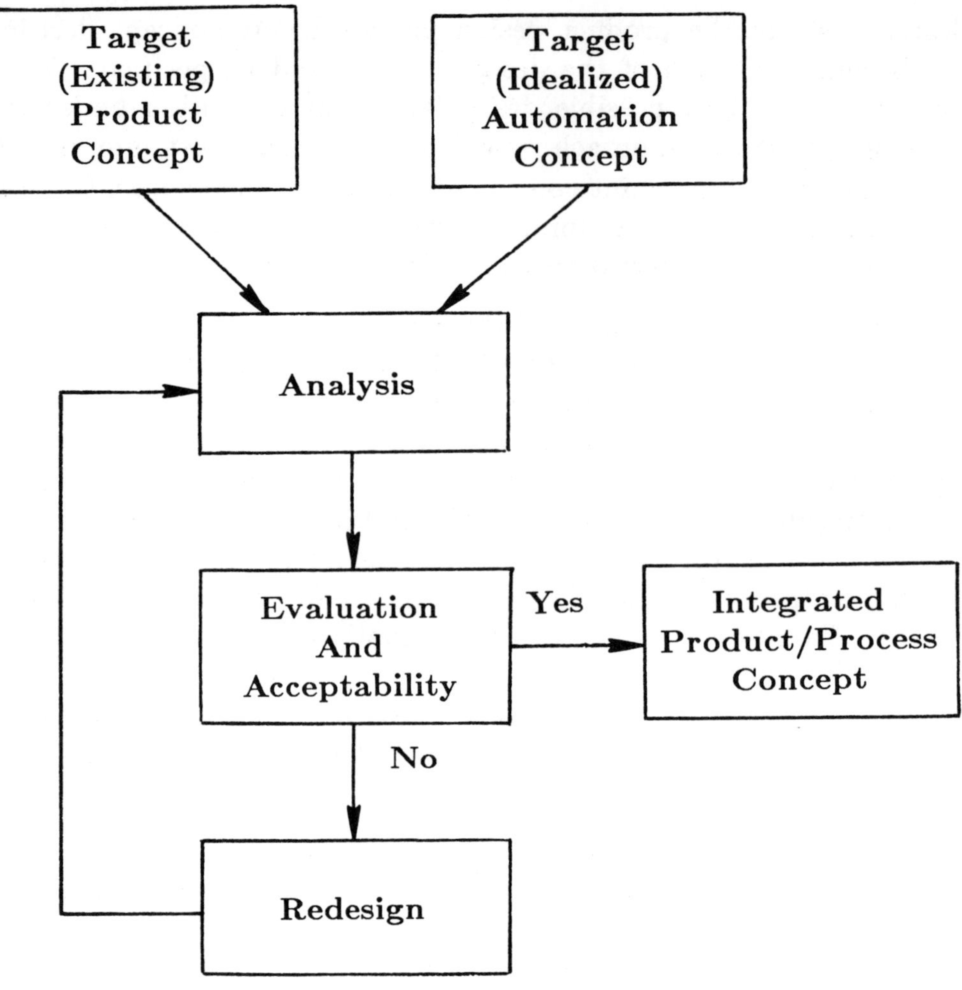

Figure 1: Design for Automation Analysis-Redesign Model

way to proceed. They teach good design by providing insight and stimulating creative design approaches.

The initial or starting process concept is a particularly important consideration in the methodology. Design for automation experience has shown that it is often best for the product to be *process driven.* For this reason, it is essential that the most desirable automation approach be targeted from the start. Ideally, the initial automation process should be predicated on the assumption that all product related obstacles can be overcome through clever design of the product/process as an integrated system.

Experience with the proposed methodology has also shown that it is best to start with a product design which is "ideal" from a performance and marketing perspective and which assumes that all automation difficulties can be overcome. By starting with the ideal, the product/process development team has a clear idea of where they want to be as the product and process concepts are merged. The ideal product and process represent two poles or extremes which bracket the globally optimal solution and can be used as a measure of "optimality" of the final integrated solution. By starting with the ideal or most desirable product concept and automation concept, the product/process development team also has a constant appreciation for how information should best be distributed between product, automation hardware, and computer software as the design proceeds toward minimization of information.

In summary, the proposed design for automation method can be considered as a four step iterative procedure (Figure 2):

1. Propose a target (starting) product design concept and automation (process) concept. To maximize the probability of developing the *best* integrated product/process solution, the target automation (process) concept should be "ideal" from a manufacturing perspective. Likewise, the target product concept should be "ideal" from a product design and engineering point of view.

2. Analyze all possible interactions between the product and process. Use the design for automation guidelines as a checklist to guide the analysis and to identify opportunities for design for automation improvement.

3. Use the guidelines to seek product/process concepts which "decouple" the identified interactions. Redesign the resulting decoupled product/process concept to minimize information content. In redesign, use the guidelines to both show the way and to stimulate innovative solutions.

4. Repeat steps 2 and 3 iteratively until an acceptable integrated product/process solution is achieved. Use the design for automation principles in conjunction with product/process requirements and constraints to evaluate acceptability.

DISCUSSION

Use of the proposed design for automation method requires a belief in the design for automation principles and a willingness to reason on a "global" level. We have found in using the method that it has invariably given us insight into the problem at hand and that it has enhanced and augmented the creativity which we were able to bring to bear on the problem. We have also found that the design for automation principles form the basis for a common language between product design and manufacturing. When viewed on the level of the principles, it is easy for all to visualize and grasp the common, system wide problems which must be solved. Ultimately, the principles are optimal suggestions which can be used to guide and judge design decisions. If the decisions are in agreement with the principles, then ease of automation is improved and system productivity enhanced.

The proposed design for automation method has been used in a variety of situations. One of the more interesting applications involved the redesign of a fairly complex consumer product. This particular product had been produced for years as a primarily sheet metal unit manually assembled on an extensive, continuous motion line. A new product design involving significant material substitution and a well thought out automated assembly process was proposed by the product manufacturer. The new design was modular in nature and more than 80 parts had been eliminated compared to the old design. Each module was to be independently assembled and tested on its own automated loop. Interfaces between each loop and a final assembly line facilitated automated assembly of each module to a base module to complete the product. The particular problem of concern was the automated transfer and installation of one of the modules which was heavy, flexible and generally difficult to handle.

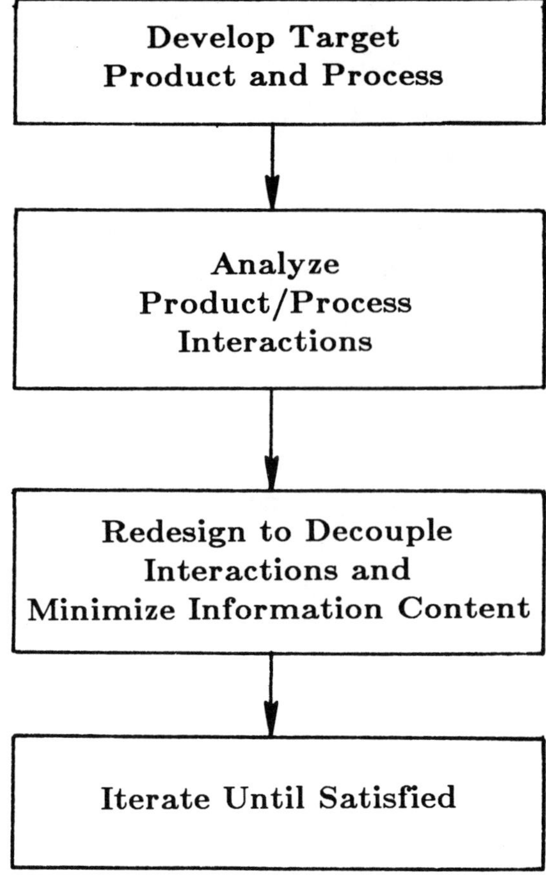

Figure 2: Proposed Design for Automation Process

This situation formed a perfect opportunity to use the proposed methodology. An entirely new (target) product concept had been proposed along with an "ideal", highly desirable automated assembly concept. All that remained was to integrate the two together into a globally optimal solution. This was done by first avoiding the original problem altogether by combining the assembly of the difficult to handle module with assembly of the base module on the final assembly line. Further iterations of the resulting product/process concept produced the following improvements:

- Elimination of an additional 35 parts and 6 tools.
- Use of a unique mounting plate which "decoupled" the shape and mounting dimensions of a major purchased component from the automation used to install it in the product. This enabled the component to be purchased from a variety of suppliers based on a cost and availability basis, without effecting tooling or assembly process.
- Elimination of all mechanical fasteners in final assembly.
- Elimination of all need for camming of molds (all straight pull molds).
- Increased use of standardized parts.
- Incorporation of labels into molded parts.

- Added features on a variety of parts, components, and modules to simplify automated handling and insertion.

- Significant improvement in product performance resulting in the use of smaller, less costly components and the addition of new marketing features.

Besides significant cost reduction, an important result of this project was the fact that product performance and perceived quality were improved along with productivity. It can also be said that many of the opportunities for improvements in both product and process were identified simply by thinking and questioning in terms of "decoupled" design and reduced information content. Once a problem area or opportunity for improved design for automation was identified, the guidelines proved extremely useful in showing the way to innovative improvements.

CONCLUDING REMARKS

The proposed design for automation method evolved out of, and is based on, very general truths distilled from automation and manufacturing experience. It is simple to use, easy to remember, and is applicable to all conceivable situations. It includes automation considerations in all phases of the design process and insists on integration of the process with the product. The proposed method is predicated on the recognition that [2]:

- Product design is the first step in manufacture.

- Every design decision, if not carefully considered, can cost extra manufacturing effort and productivity loss.

- The product design must be carefully matched to advanced flexible manufacturing, assembly, quality control, and material handling technologies in order to fully realize the productivity improvements promised by these technologies.

Automation imposes new product requirements and constraints which requires a more "disciplined" approach to product design. Experience with the proposed design for automation method shows that it is a useful tool in helping to bring this new dimension of discipline to product design. A product designed using the method should be an efficient design to manufacture, either manually or automatically, and ultimately, the extent to which this is achieved is the final test.

BIBLIOGRAPHY

1. Suh, N. P., Bell, A. C., and Gossard, D. C., "On Axiomatic Approach to Manufacturing and Manufacturing Systems," *ASME Journal of Engineering for Industry*, Vol. 100, No. 2, May 1977.

2. Stoll, H. W., "Design for Manufacture: An Overview," to be published by ASME Applied Mechanics Review.

CHAPTER 3

FEEDING FOR ASSEMBLY

Reprinted from *Industrial Engineering* magazine, May 1983

Programmable Parts Feeders

by Jerry L. Goodrich
Pennsylvania State University
and
Gary P. Maul
Ohio State University

The increase in the productivity growth rate of other nations has created serious competition in the marketplace for U.S. industry. The problem becomes one of meeting the challenge of that competition as labor wages and benefits continue their annual upward spiral. The solution appears to be to reduce the labor intensity of manufacturing operations, consequently increasing productivity per man hour. An obvious means of achieving this objective is by utilizing more of the techniques of automation. Almost all manufacturing operations require oriented workpieces. In mass production industries, workpiece orientation is usually accomplished via parts feeders or by tooling, such as transfer devices, which maintain part orientation. However, in batch production where quantities of 50 - 100,000/year are produced, workpiece orientation is usually accomplished manually. The continued use of human labor to orient workpieces is not conducive to improved productivity and it is becoming increasingly costly. The magnitude of this problem becomes apparent when it is realized that an estimated 75% of all production in this country is batch production.

Current methods of automated workpiece orientation require the use of dedicated feeders tooled specifically for one particular part. This concept is feasible when the volume of that part is high. In batch production, however, the volume of any one part is not sufficient to warrant the time and expense of a dedicated parts feeder. To make automated workpiece orientation feasible in batch manufacturing requires that the feeders be flexible and able to accommodate a wide range of parts.

Numerical control machines have, in some cases, reduced the part orientation problem because they can do multiple operations on the workpiece without the need for human re-orientation of the part. However, NC machine tools account for less than three percent of the operating stock. Researchers [1,2] have suggested some reasons for the slow introduction of NC: (1) utilization of expensive machines is low, and in-process inventory is high, resulting in decreased productivity and economic losses; (2) human involvement in the actual production process is the primary factor in this slow-down.

A logical extension of NC is the Computerized Manufacturing System (CMS). Central to a CMS is the transfer system which carries oriented parts from one machine to the next. As with an NC machine, the CMS reduces the part orientation through multiple operations. The major drawback to a CMS is the high initial cost coupled with the uncertainties of the economic pay-offs.

Although considerable effort has been directed at the problem of maintaining part orienation in batch manufacturing, little has been done in orienting parts for flexible manufacturing. Several attempts at flexible assembly systems such as G.M.'s Programmable Universal Machine for Assembly (PUMA), Westinghouse's Adaptable Programmable Assembly System (APAS), and Olivetti's Programmable Assembly System (SIGMA) have emphasized the need for flexible feeding devices. In a review of Olivetti's experiences with the SIGMA system, Salmon and d'Auria [3] state that, "The primary conclusion drawn is that the lack of flexible feeding devices causes the main limits of programmable assembly." Likewise, Beecher [4] in a report on G.M.'s PUMA system states, "Whether or not programmable assembly devices succeed could very well be determined by the success or failure of parts feeding and orienting mechanisms." From experiences such as those cited above, it becomes apparent that developments in the area of flexible parts feeding systems are essential if the benefits of flexible assembly systems and flexible manufacturing systems are to be realized.

Robots are frequently used to handle parts in flexible manufacturing and flexible assembly systems. Unfortunately, robots have the same requirement for workpieces in known orientations as do other manufacturing machines. Current robots have little or no sensing capabilities. Thus, if a robot is to pick up a workpiece, the workpiece must either be in a predetermined location in a predetermined orientation or the robot must be equipped with sensors which enable the robot to locate the workpiece and determine its orientation.

Some research is in progress which involves equipping a robot with a vision system as a means of determining workpiece orientation. In these systems, the camera can be mounted on the robot arm or in a fixed location which provides an overview of the workspace. With the camera mounted on the arm, the visual information provided is relative to the current position of the arm. When a fixed camera is used, visual information can be used to provide absolute positional information when the robot arm is not in the field-of-view or relative information if the arm is in the field-of-view. Most of these systems are based solely on the use of vision [5,6,7] while others are utilizing a combination of vision and tactile sensing [8,9]. The competence of present vision systems appears to limit their use to applications requiring only 2 dimensional feature recognition [10]. These systems require special conditions on the workpieces to be oriented. For example, the parts must be randomly placed on a moving conveyor belt, but cannot be overlapping or touching another part [5,6]. These systems also require special lighting sources or the use of special back lit stages or tables to achieve the proper lighting contrasts. Research using vision in systems for non-overlapping parts has been more successful to date than in systems which utilize a bin of parts. The bin of parts problem which deals with jumbled and randomly overlapping workpieces requires the system to have extensive interpretive capabilities [11]. These systems must cope with problems in lighting contrasts and the infinite numbers of possible workpiece orientations [12]. The complexity of the bin of parts problem can be reduced by having the robot separate few parts from the bin and place them on a flat surface [13]. All of the above research efforts have reported some success to date; however, they tend to be unreliable or slower than desirable. Correspondingly, these systems are fairly expensive to implement. Reductions in the amount and complexity of hardware have mitigated costs while introducing additional problems or new restrictions on the system. By utilizing less costly cameras which limit pixels to black or white, data compaction can be achieved, thus reducing the amount of computational support [14]. These cameras, however, have encountered special lighting problems such as blooming which occur with excessive light intensity. A more serious problem generated by video hardware is the computational requirements on the computer architecture. Present real-time black and white video systems generate 10 million pixels of data every second. This means that a present computer system must handle 60 to 80 million bits per second. Other problems arise from shadows, highlights and occlusions which may elminate expected features or cause uncertainties for the computer. The estimate of computational power required for image interpretation ranges from 1 to 100 billion instructions per second. Thus, vision appears hard to achieve from the standpoint of hardware and the uncertainties involved in present systems [15].

Another type of sensing which researchers are currently investigating is tactile sensing. This type of sensing has been incorporated into the construction of automated container-handling devices [16]. It has also been built directly into robot hands and arms. Researchers at the Naval Research Laboratory have utilized a simple micro-switch as a touch sensor on an intelligent underwater robot to collect detailed information for shape recognition [17]. Specifically designed sensors based on strain gauges were developed for another application in which the sensors supply geometric feedback about the characteristics, the path and the smoothness of a welding bead. The problem with this type of sensing and this approach is the requirement for a high degree of processor performance due to the multitude of calculations which must be performed in real time [18]. Attempts at reducing the computations result in less flexible systems where workpieces must be separated from each other, have a certain geometrical shape, and be placed within a given limited area at the input to the system.

The approach the authors are investigating to the problem of workpiece orientation is to apply present concepts of mass production automation to batch production. This had not been economically feasible. To justify the automation of workpiece orientation in batch production requires a new concept of automation called "programmable feeder automation." In this style of automation, the tooling and equipment would be designed to accommodate a wide range of parts. The volume of the parts collectively represented by this range can economically justify the applications of automation technologies. A programmable parts feeding system would have the flexibility to orient all the parts of one or more part families with down time to changeover from one part to the next not requiring more than a few minutes. Such a system could greatly enhance productivity by further reducing dependence on human labor for part orientation.

This type of feeding system could even be used as an automatic material handling element in conjunction with robots or conventional batch production machinery. There even exists the possibility that mass production industries utilizing feeders would benefit from this concepts of flexibility. For example, in mass production, a programmable feeding system could very much shorten yearly model changeover downtimes by eliminating the time and the cost to retool. Likewise, any product revisions would not require costly tooling modifications in such a system. Finally, variances in tolerances in products from vendors would not create feeding problems as in dedicated feeding equipment.

In developing a programmable feeder, the first step is to gain a thorough understanding of the parts to be fed, and the different types of conventional feeders and feeding technologies presently being used. The authors have studied these topics as a part of a research project. Parts were categorized into different part families from a feeding point of view. Different types of feeders were investigated to determine the range of part families that they could accommodate. Those feeders that were able to orient the widest ranges of parts were studied in detail to further determine the tooling strategies employed to orient parts.

From this initial research and some mock-ups, it is now apparent that the concept is highly feasible. The tooling being utilized in high volume mass production feeders can, in fact, be modified to be programmable, thus making the use of these feeding systems economical for batch manufacturing. These feeding systems with programmable tooling appear to be flexible enough to accommodate a wide range of parts from one or more part families.

The programmability of a feeding system could range from completely manual to completely automatic. In a completely manual system, an operator or technician would place a supply of parts in the feeder. The technician would operate the feeder and would manually program its tooling by trial and error. In a completely automatic system an operator would place a supply of parts in a feeder and the feeder would proceed to program its own tooling.

In either system, the tool settings, which comprise the program, could be recorded by workpiece part number. This implies programming the part in advance and then later retrieving that program to reproduce the tool settings during changeover from one part to another. This procedure will minimize the feeder downtime in the production system. Thus, the concept of programmable feeders appears to be that there could be a non-production feeder used strictly to determine the programs for workpieces, and production feeders which would be programmed.

One important consideration in this "off-line" programming concept is the amount of workpieces for which programs must be established. A considerable amount of technical support may be required if large number of parts are to be programmed manually. To reduce that support would mean designing a more automatic programming method. A solution could be to develop an adaptively controlled feeding system. In the adaptively controlled system, the feeder would automatically program its tools and record the tool-setting information. That information would be supplied to the production feeders as in the manually programmed system.

The adaptive unit would provide more efficient programs than manual programming. An index of performances could be established which would help ensure that the programming would meet certain minimum performance criteria. This index could be as simplistic as meeting a minimum feed rate. To accomplish this objective, sensors located in the feeder would monitor feeder performance. This information would serve as input to a microprocessor programmed with a search strategy. The search strategy would give the system the capability to act properly on the information supplied by the sensors, and control the positions of the tools in accordance with the index of performance. The individual tool positions could be controlled by activators such as stepper motors. When the feeder has been successfully programmed, the program of tools settings would be stored by workpiece part number.

The "programmable feeder" concept may require that part families be established for which certain groups of tools are required. The limits on the capacity of the feeder may dictate that not all part families can be accommodated by one feeder. As an alternative, both production feeders and programming feeders could be designed to handle a specific part family. If the same tools are required by a similar part family, however, it might be feasible to accommodate both part families in the same feeder.

The potential ultimate implementation of the concept could be either to automatically program the feeders with a central manufacturing computer employed by the production system or by smaller stand-alone units. This would fit the concept of computer aided manufacturing (CAM). The limits of the feeder technology could also be employed by a computer aided design system (CAD) to ensure the concept of "buildability" and design for production.

This concept certainly provides the missing link to the problems presented by the use of robots for assembly tasks. Whatever the requirement of the production system, it is consistently required that the part be in a certain orientation for the production machinery. That orientation can be supplied by flexible parts feeders. A vast knowledge presently exists about conventional feeder applications from which these flexible feeders could be developed. The concept of flexiblity could also enhance and increase the use of feeders in mass production by increasing their versatility and their useful life in the event of changes in workpiece design. Aside from the potential increases in manufacturing productivity, this concept may even offer possibilities for new ideas and philosophies of automation not imaginable, even now. The utilization of programmable feeders may spur new innovations in product design, inspection and sorting. The concepts could have a wide potential appeal for manufacturing as the need to increase productivity grows.

BIOGRAPHICAL SKETCH OF DR. GARY P. MAUL

Dr. Maul received a B.E. in Industrial Engineering from Youngstown State University in 1970, an MSIE from Purdue in 1976, and a Ph.D. from Penn State in March of 1982. His experience includes serving in various capacities with the Packard Electric Division of the General Motors Corporation, serving as a consultant to small industry, and teaching at both Youngstown State and Penn State Universities. Presently, he is an assistant professor at the Ohio State University, where his research activities are in the area of parts feeding and orientation. Dr. Maul is a licensed professional engineer in both Ohio and Pennsylvania.

REFERENCES

1. Gershwin, S.B., M. Athans, and J.E. Ward, "Progress in Flexible Automation and Material Handline Research, 1978," _Proceedings of the Sixth NSF Grantees' Conference on Production Research and Technology_, Sept. 27-29, 1978.

2. Barash, M.M., et al, "Optimal Planning of Computerized Manufacturing Systems (CMS)," _Proceedings of the Sixth NSF Grantees' Conference on Production Research and Technology_, Sept. 27-29.

3. Salmon, M., A. d'Auria, "Programmable Assembly System," _Computer Vision and Sensor-based Robots_, Ed. by George G. Dodd and Lothar Rossol, Plenum Press, New York and London, 1979.

4. Beecher, R.C., "PUMA: Programmable Universal Machine for Assembly," _Computer Vision and Sensor-Based Robots_, Ed. by George G. Dodd and Lothar Rossol, Plenum Press, New York and London, 1979.

5. Ward, M.R., Rossol, L., Holland, S.W. "A Practical Vision-Based Robot Guidance System," _Ninth International Symposium on Industrial Robots_, March 13-15, 1979, Washington, D.C., pp. 195-211.

6. Holland, Steven W., "An Approach to Programmable Computer Vision For Robotics," SME Technical Paper, MS 77-747, 1977.

7. Vanderburg, G.J., Albus, J.s., Barkmeyer, E., "A Vision System for Real Time Control of Robots," Ninth International Symposium on Industrial Robots, March 13-15, 1979, Washington, D.C., pp. 213-231.

8. Cassinis, Riccardo, "Sensing System in SuperSigma Robot," Ninth International Symposium on Industrial Robots, March 13-15, 1979, Washington, D.C., pp. 437-448.

9. Takeyasu, Kiyoo, et al., "An Approach to the Intelligent Robot With Multiple Sensory Feedback: Construction and Control Functions," Proceedings of the 7th International Symposium on Industrial Robots, October 19-21, 1977.

10. Tennenbaum, J.M., Barrow, H.G., Bolles, R.C., "Prospects for Industrial Vision," Computer Vision and Sensor-Based Robots, Ed. George G. Dodd and Lothar Rossol, Plenum Press, New York and London, 1979.

11. Rosen, C.A., "Machine Vision and Robotics: Industrial Requirements," Computer Vision and Sensor-Based Robots, Ed. George G. Dodd and Lothar Rossol, Plenum Press, New York and London, 1979.

12. Dessimoz, J.D., Kunt, M, Zurcher, J.M., "Recognition and Handling of Overlapping Industrial Parts," Ninth Symposium on Industrial Robots, March 13-15, 1979, Washington, D.C., pp. 357-366.

13. Gleason, Gerald J., and Agin, Gerald J., "A Modular Vision System for Sensor-Controlled Manipulation and Inspection," 9th International Symposium on Industrial Robots, Washington, D.C., March 13-15, 1979, 57-70.

14. Reddy, D.R., and Hon, R.W., "Computer Architectures for Vision," Computer Vision and Sensor-Based Robots, Ed. George G. Dodd and Lothar Rossol, Plenum Press, New York and London, 1979.

15. McGhee, R.B., "Future Prospects for Sensor-Based Robots," Computer Vision and Sensor-Based Robots, Ed. George G. Dodd and Lothar Rossol, Plenum Press, New York and London, 1979.

16. Sugiyama, K., et al, "Automatic Container Handling Device with Tactile Sensor," Computer Vision and Sensor-Based Robots, Ed. George G. Dodd and Lothar Rossol, Plenum Press, New York and London, 1979.

17. Dixon, J.K., Salazar, S. and Slagle, J.R., "Research on Tactile Sensors for an Intelligent Naval Robot," Ninth International Symposium on Industrial Robots, March 13-15, Washington, D.C., pp. 507-518.

18. Stute, G., Erne, H., "The Control Design of an Industrial Robot With Advanced Tactile Sensitivity," Ninth International Symposium on Industrial Robots, March 13-15, 1979, Washington, D.C., pp. 519-528.

Reprinted courtesy of Gary P. Maul,
Ohio State University

The Adaptively Controlled Parts Feeder Concept

by Gary P. Maul
Ohio State University

Programmable Feeders

The proper orientation of workpieces is a requisite for most all manufacturing operations. Parts feeders handle a substantial portion of this effort in mass production industries. These feeders are, however, typically dedicated to feeding one particular part. The cost of the dedicated feeder and its tooling is justified by the volume of the particular part which is fed.

In batch production manufacturing, yearly volumes of 50 to 100,000 units of a particular part are usually not sufficient to justify the cost of a dedicated parts feeder. Researchers (1), however, have investigated a methodology for developing parts feeders with programmable tooling. The result of this investigation has led to development of a programmable vibratory bowl feeder capable of feeding a wide range of parts from within a given part family. The collective volumes of all the parts from the family which can be fed by this unit can then be used to justify its cost.

The tools in the programmable feeder used to orient the parts must be programmed manually. This manual programming effort is one of trial and error. Programming would also have to be done "off line" away from the production floor. This programming, however, might also be accomplished automatically. The automatic programming of the tools leads to the new concept of a Programmable/Adaptively Controlled Parts Feeder. One potential application for this concept could be for production systems with extremely short runs of only a few hours and for systems which require several changeovers per day. Here the down time requirement for changeover to a new part must be extremely short to maximize system through put. The necessity to make frequent and rapid changeovers requires the full integration of the automatically programmable feeder into the scheme of the computer-aided manufacturing system. In this system, a supervisory computer could instruct the adaptively-controlled feeder to determine a program for a new part prior to changeover. The supervisory computer could then instruct the production feeders to automatically program their tools during the changeover downtime.

The production units could be equipped with only the tools required to feed a certain select part family, or a limited number of parts families, in order to minimize tooling costs.

The advantages of such a potential system are enormous. Aside from the minimal technical programming support required, the concept would be extremely responsive to changes in production schedules and permits an enormous amount of manufacturing flexibility.

Adaptive Control

The key to the Programmable/Adaptive feeder concept is the adaptive control of programming capabilities of the programmable feeder. Adaptive control of industrial processes, such as welding (2) and machining operations (3), have achieved some reasonable amount of success. This may, in part, account for the increased widespread interest in and use of adaptive control technology (4). In order to apply this technology effectively, it must first be understood. Not everyone agrees on just what is meant by adaptive control (5). There are, however, some general points of consideration to which all definitions of adaptive control comply.

Adaptive control is far more sophisticated than conventional control and can be applied to processes where changes in the input-output relationship vary over time (6). It is generally agreed that the system must control or modify its own action to accommodate changes which occur in operating conditions. The overall objective of this self-correcting ability is to improve the system's performance (7). Increasing performance increases productivity, thus ultimately reducing manufacturing costs (8).

To achieve the improved performance, the adaptive control system must perform three functions. These are the identification function, the decision function, and the modification function. The first function involves the determination of a system index of performance. This performance index is often a variable which describes how well the system is operating overall. Data which is measured or collected from the process is used in determining this index. Since the optimum value of the performance index is almost always unknown, the system must be equipped with logic to search for it. This is accomplished by the decision function. The objective of this function is that an algorithm determines how to manipulate the system to maintain or improve its operation under changing conditions (9). The system must retain information in memory to enhance the decision-making capabilities of its algorithm (10). Finally, the modification function implements the course of action prescribed by the decision function. It achieves this through activators which perform the physical manipulations necessary to alter the process variables which determine the value of the index of performance.

For the programmable vibratory bowl feeder, a suitable measure of the system's performance could be the number of correctly oriented parts that the system can feed in a given time for a minimum amount of input power. Thus, the index of performance is the feed rate. There must be an effective way to measure this index and an effective way to measure process variables which determine its value.

The index of performance could be measured by counting the parts per unit of time. In this system the process variables are the positions of the tools and the amount of power supplied to the driving unit of the feeder. The position for any given tool in the programmable feeder will be different depending on the part being fed. Also, the tool position will affect the feed rate. If the tool is improperly positioned it may reduce or impede the flow of parts. The performance of the overall system, and hence the index of performance, is dependent on how effectively each tool performs.

This effectiveness can be measured by sensors located along the track and after each tool. A sensor located at the bowl opening can be used to count the flow of parts leaving the feeder.

To make a proper measurement requires the selection of the right type of sensor. Basically, there are two major classes of sensors, contacting and noncontacting types (11). Contacting sensors, such as limit switches, are impractical for small sized parts in a vibratory feeder because of the tenuous position of the parts on the bowl track. This limits the use of potential sensors to the noncontacting type. The two most common types of noncontacting sensors are proximity sensors and photoelectric sensors. Proximity switches only function in the presence of metal parts. Parts feeders, however, are usually not limited to feeding only metal parts.

Of the noncontacting variety of sensors, photoelectric types are among the most widely used (12). This class of noncontacting sensor can be readily adapted to determining part presence. There are three basic components to a photoelectric switch. These are the emitter, the lens, and the sensor (13). Photo switches are solid state devices which can emit and detect electromagnetic radiation in the infrared, visible and ultra-violet frequency ranges (14). These sensors can be implemented in one of several different ways. In the first method, radiation from a light-emitting diode can be reflected from the surface of the part back onto the sensor. This mode is less than desirable because of variations in the surfaces of workpieces fed by the system. Another mode is to radiate a beam from a reflector back into a sensor. This is difficult to implement because it requires the accurate positioning of the extra reflecting hardware.

The most promising mode of implementation of photoelectric sensors for this application is the beam-breaking method (15). In this mode, the sensor is placed a specified distance directly opposite the emitter. A part passing between breaks a radiated beam sent from the emitter to the sensor. This creates an on or off condition in the sensor as parts travel by. Current photosensors are small enough to be mounted directly in the bowl wall and are durable enough to withstand both the vibration of the system and the constant battering of workpieces. A suitably matched infrared sensor and emitter will be very effective in an environment of flourescent lighting. An infrared emitter with a reasonably small angular response such as shown in Figure 1 is very desirable.

With a circuit design such as the one in Figure 2, the sensitivity of the photo sensor can be fine-tuned. This adjustment is important because the relative distance between the sensor and emitter is slightly different for each pair in the system. To achieve a high level of sensitivity, the photo transistor is placed in series with a resistance, R_2.

To implement this technology in a programmable vibratory feeder a sensor could be mounted in a small hole drilled through the bowl wall. Directly across the track from the sensor the emitter could be mounted in a small housing. A sensor and emitter pair would then be located before and after each tool. The information obtained from the sensors would be evaluated by the control algorithm.

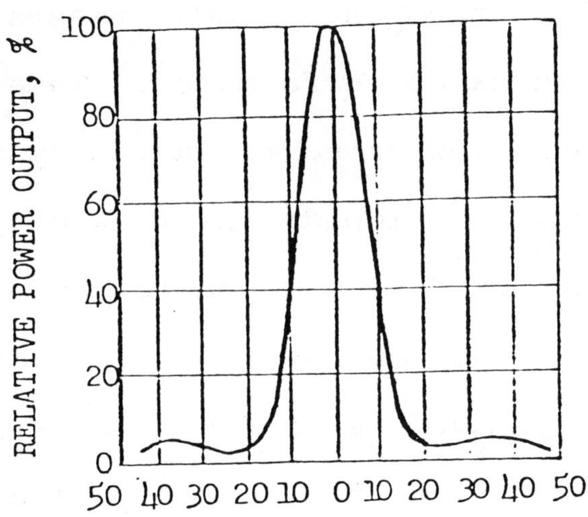

Fig. 1. Desirable Radiation Pattern for the Emitter.

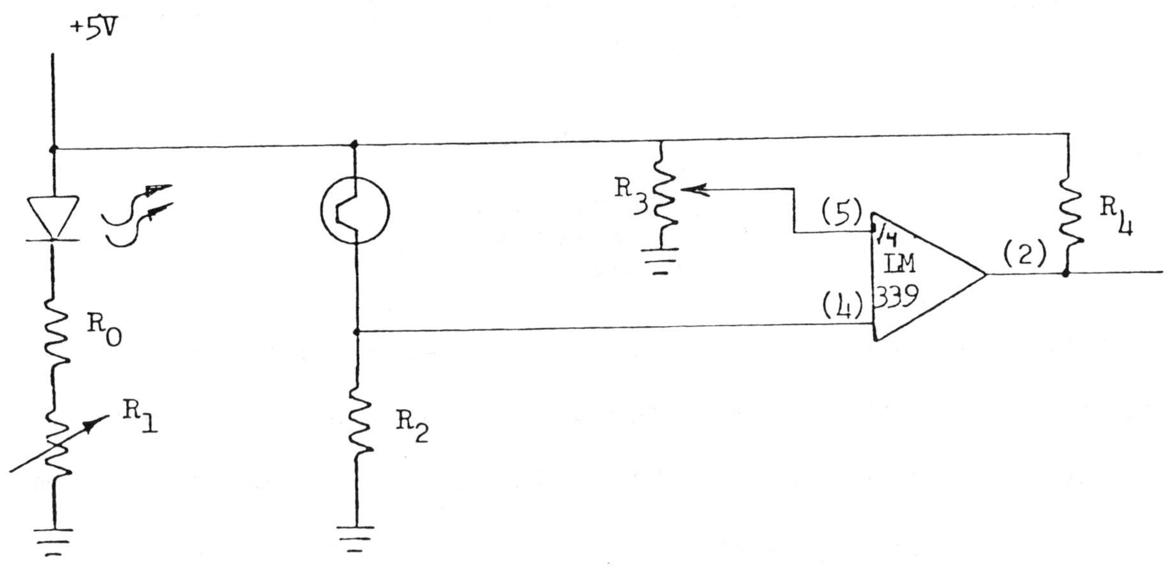

Fig. 2. Schematic of an Adjustable Circuit for Sensor and Emitter used in a Beam-Breaking Mode.

The main objective of the control algorithm would be to evaluate the sensor inputs and make corresponding adjustments in the tool settings to optimize the index of performance. To accomplish this, the algorithm must ensure a flow of parts in the system, respond to any conditions which restrict or impede the flow of parts during the execution of the algorithm, perform some basic computations, optimize tool settings, and retain those tool settings in memory. The sequence in which the tools are programmed is an important aspect in the development of the algorithm. Also, the operation and design of each tool must be considered in developing the algorithm.

The programmable tools proposed to be adaptively controlled by the algorithm are listed in Table 1. These tools are shown in the order that they would be encountered by a part traveling up the bowl track. These tools enable the feeder to accommodate various sizes of one class of parts known as headed parts.

TABLE 1. Programmable Tools Required for Orienting Headed Parts

1. Wiper
2. Pressure Break
3. Narrow Track
4. Retaining Rail
5. Slotted Track
6. Hold Down

Under the control of the algorithm the tools must operate by not permitting parts to pass until the tool is programmed. The basic operational constraint would be to have the tool move from the closed to the open position during programming to avoid damaging a tool by closing it on a part.

Initially each tool was to be individually programmed by stepping the tool to the programmed position. This scheme tended to be slow because it failed to take advantage of information learned about the part by an individual tool. Programming time could be greatly reduced if information learned about the part by a tool could be used to preset other tools. This would reduce the overall programming time by reducing the number of iterations through the algorithm for each tool to reach its programmed position. Certain tools yield very accurate estimates of critical part dimensions. For example, the narrow track tool gives a very accurate estimate of one-half of the part's head diameter. The wiper gives the tooled diameter dimension, T_d, from which an estimate of the minimum possible major diameter, D can be made (Figure 3). Programming the wiper first gives the dimension of T_d. From T_d, other tools, such as the narrow track, can be preset and the overall programming time will then be reduced.

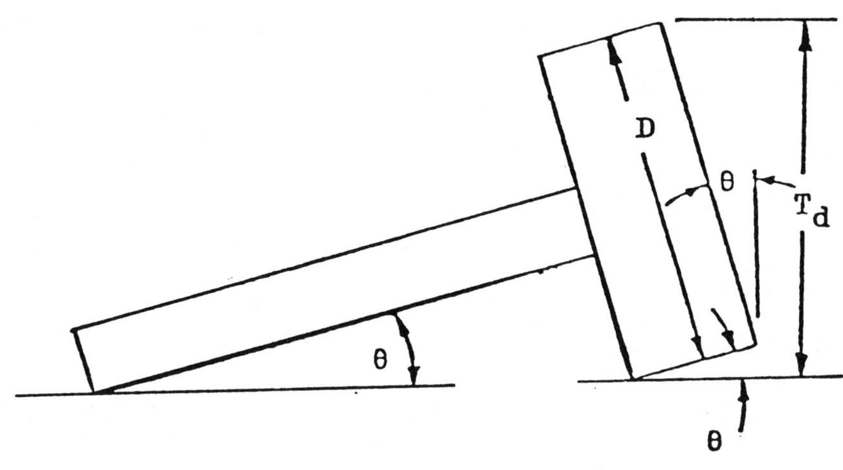

Fig. 3. Dimensions (D-d)/2, S, D, T and t for a Type II Headed Part

Overview of Algorithm

The major steps of the programming sequence finally selected for the algorithm are described by the flow chart in Figure 4. This figure represents an overview of the algorithm. The first step in this sequence performs all of the activities required to initialize the system and prepare the tools to be programmed. Each tool is then sequentially programmed.

First, the algorithm programs the wiper (Figure 5). The sensors located before the wiper inform the algorithm that parts have reached the tool and the search for the programmed position begins. The algorithm steps the tool and then polls the sensor located after the tool to see if any parts have passed through the wiper. The iterative process is repeated over and over as long as parts are detected by the first sensor. Finally, when parts do pass the wiper successfully, the optimization routine is called on to fine tune the programmed setting (Figure 6). The programming scheme for all other tools is similar to that of the wiper. With information now learned about the part by the wiper, other tools are preset.

Next, the narrow track is programmed. Additional information about the part learned by the narrow track is used to preset the pressure break. Then the programming of the pressure break and retaining rail follow.

The slotted track is then placed in the preset position and programmed to its final position by the algorithm. Now all tools but the hold-down have been programmed. The part flow must be shut off before the hold-down can be preset. This is achieved by closing the narrow track to shut off the flow of parts temporarily. With no flow of parts to the hold-down, the tool is preset. The narrow track is then reset to its optimal position and the hold-down

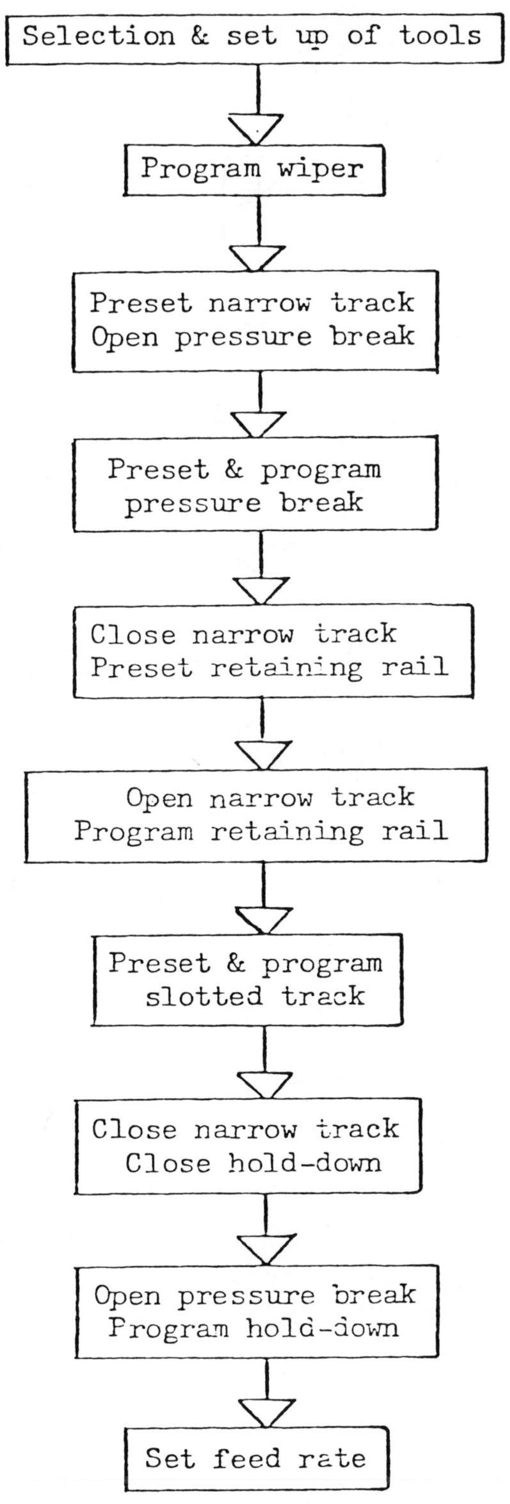

Fig. 4. Major Steps of the Control Algorithm

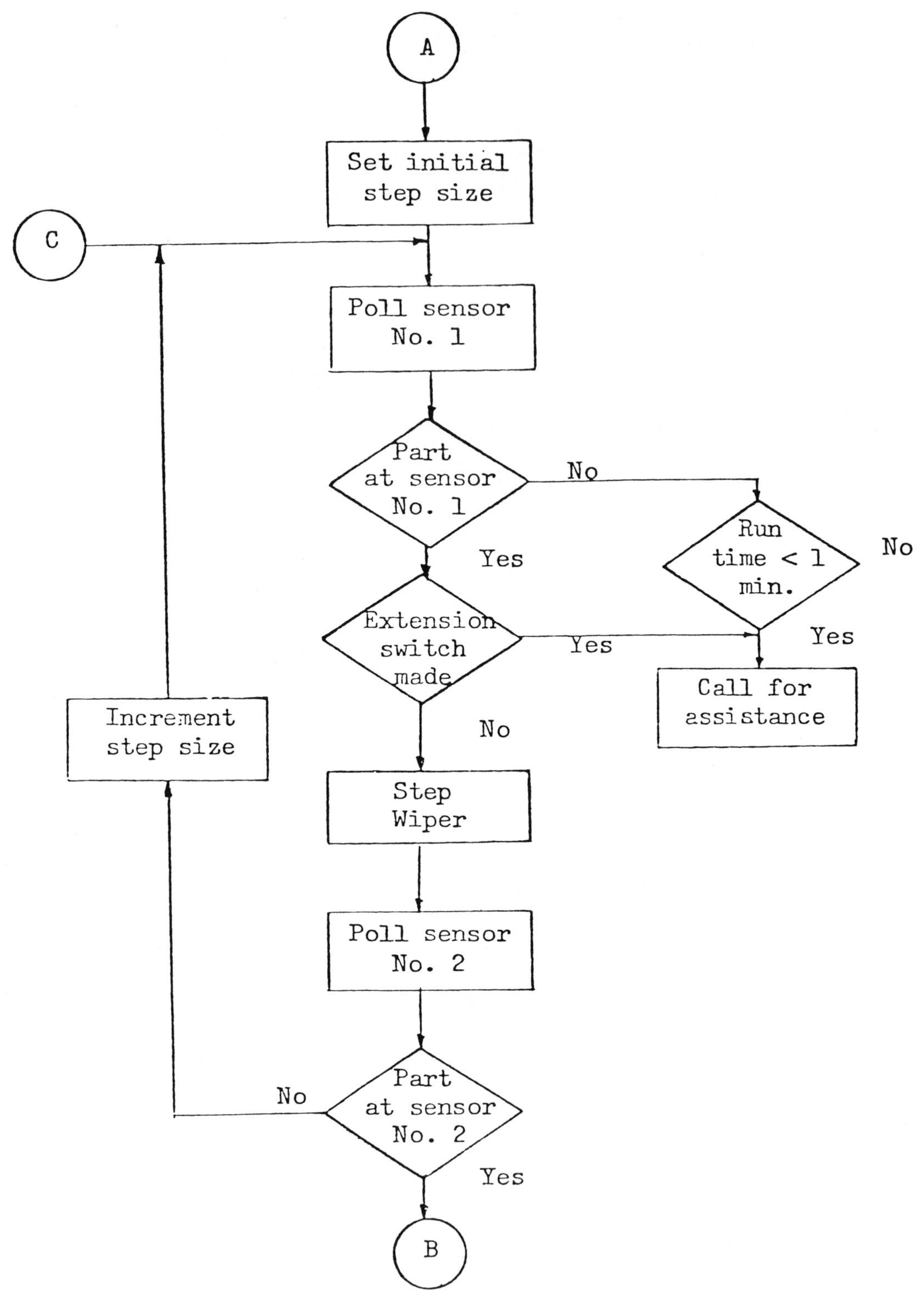

Figure 5. Programming the Wiper

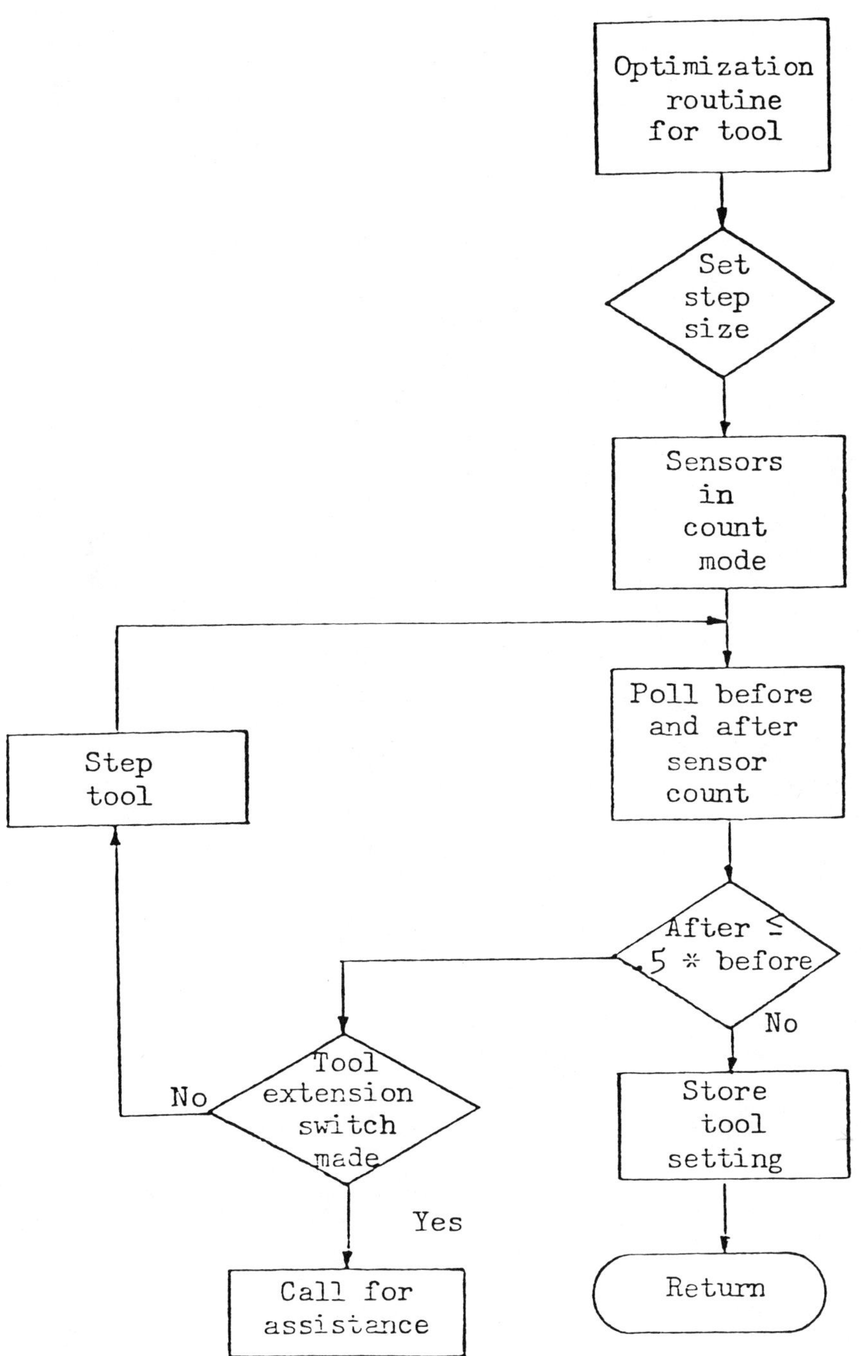

Figure 6. Optimization Routine for the Tools

is programmed. Finally, the feed rate is set (Figure 7). If no specific rate is input to the algorithm, the feed rate is set to the maximum capability of the system.

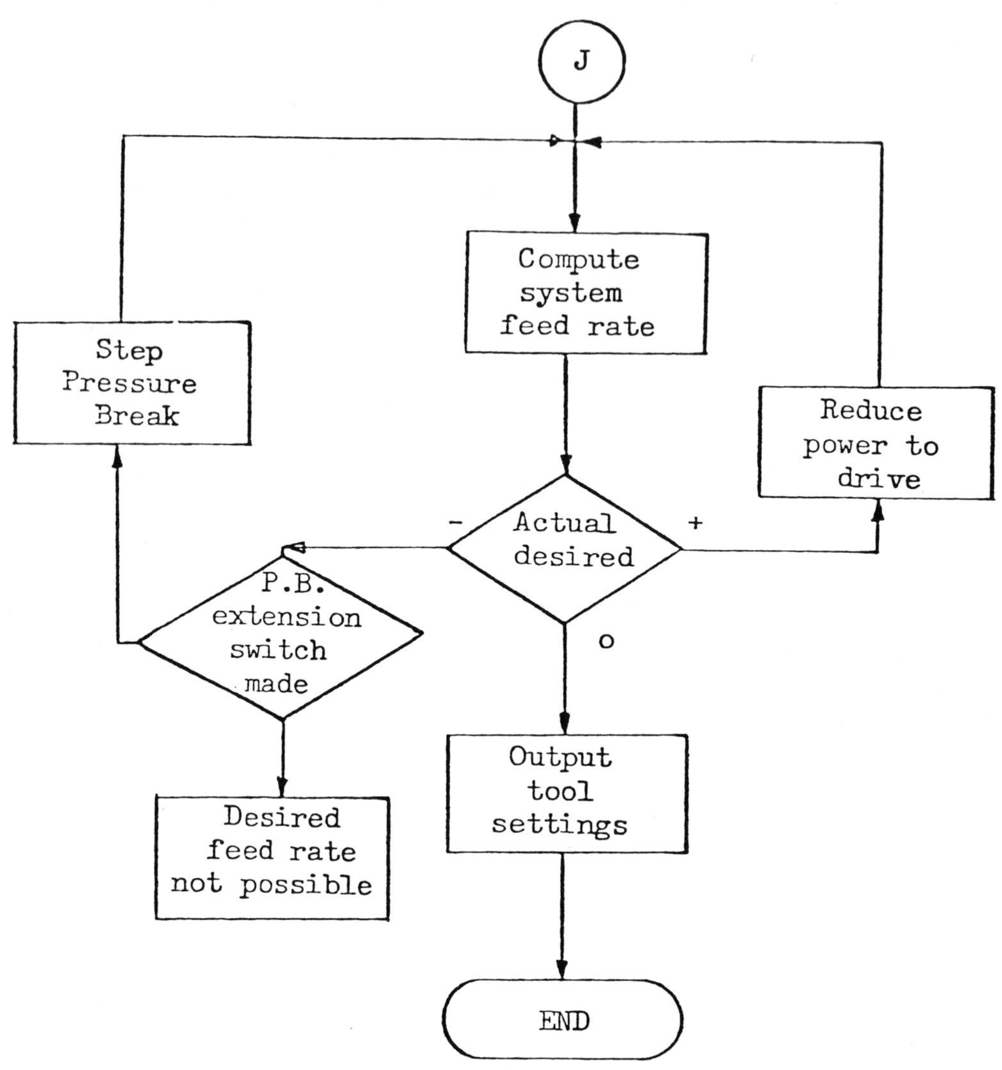

Fig. 7. Final Steps of the Algorithm Which Sets the Feed Rate

Testing

The testing of the adaptive algorithm was simulated by manually executing the steps of the control algorithm on a programmable feeder to determine the programmed tool settings for a variety of parts. The result of this simulation for a sheet metal screw was a programming time of 3.07 minutes. This part is fairly representative of typical small headed parts. The programming time for a much larger part, such as a spark plug body, was only 4.53 minutes. Thus, the time for a large part is not significantly greater than that for a smaller one. The difference in programming times is mainly attributable to the additional increments the wiper and hold-down must execute. The small difference in programming times for the two parts is a result of using an educated search routine, as opposed to an algorithm, which does not take advantage of information learned about the part.

Summary

The Programmable/Adaptive feeder concept has potential applications in mass production manufacturing, as well as batch production manufacturing. An adaptive algorithm was developed utilizing information from sensors to determine the programmed settings of tools in a programmable vibratory bowl feeder. The algorithm was simulated manually and found to yield rapid programming times for a variety of headed type parts.

References

1. Maul, G.P., A Methodology for the Adaptive Control of Workpiece Orientation. Unpublished doctoral dissertation, The Pennsylvania State University, 1982.

2. G.A. Spynu, et. al., "Adaptive Control of Welding Industrial Robot", Proceedings of the 7th Intl. Symposium on Industrial Robots, October 19-21, 1977.

3. D.A. Milner, "An Introduction to Adaptive Control", The Production Engineer, 54, March 1975, 175-180.

4. A.K. Kochhar, "Use of Computers in Manufacturing Systems - 5", Machinery and Production Engineering, 129, December 1976, 666-673.

5. Ronald Khol, "Adaptive Control Toward the Thinking Machine", Machine Design, 41, May 1, 1969.

6. Mikell P. Groover, "A New Look at Adaptive Control", Automation, 20, April 1973, 60-63.

7. James Adams, "Understanding Adaptive Control", Automation, 17, March 1970, 108-113.

8. J.F. Volk and R.P. Chase, "Adaptive Controls", Machinery and Production Engineering, 122, June 20, 1973, 784-789.

9. Mikell P. Groover, Automation, Production Systems, and Computer-Aided Manufacturing. Englewood Cliffs, N.J.: Prentice-Hall, Inc., 1980.

10. E.H. Bristol, "Adaptive Process Control: Versatile On-Line Tool", Control Engineering, 20, 4, April 1973, 41-44.

11. Robert B. Ritland, "Sensing Parts in Production", Automation, 20, 7, July 1973, 39-43.

12. Wayne Filichowski, "Photoelectric Scanners Yield Variety of Production Data", Instruments & Control Systems, 50, April 1977, 39-42.

13. Henry M. Morris, "Learn the Jargon Before Specifying Photo-electric Switches", Control Engineering, 25, 11, Nov. 1978, 47-49.

14. Fred W. Kear, "Instrumentation with Optoelectronic Sensors", Instruments and Control Systems, 47, Jan. 1974, 55-56.

15. Floyd E. Smith, "Applying Vibratory Bowl-type Feeders; Part 3: Out-of-Bowl Tooling Considerations", Automation, 10, Jan. 1963, 78-85.

Presented at the SME Assemblex VIII Conference, March 1982

Updated Art—Parts Feeding

by Richard D. Zimmerman
Spectrum Automation Company

INTRODUCTION

The purpose of this presentation is to alert applicators and users of parts feeders to features and conditions, in the industry, that can effect better decisions.

This update on the art of parts feeding is drawn from the personal experience of the author and his associate staff.

Both in-house and field experiences with the sales, design, build, debug and service of parts feeders are the basis of this information.

In-the-field experiences provide update information on those feeder features which are beyond the scope of the writer's company activity.

Update subjects are principally with new and different uses of the already established art.

ORIENTATION

Rotary, orbital and vibratory bowl feeders have been available for many years. Straightline vibratory and belt conveyor orientors are more recent developments. All have been refined to nearly triple production rates that were thought to be at peak twenty years ago.

Rotary bowls use agitator slots to lift and tumble parts onto a gravity orienting track. Simple parts, at reasonable production rates, are most frequently fed with these units. Noise, parts abuse and tooling wear are good points for critical consideration. Several feeder firms offer rotary bowls for users that prefer them to vibratory units.

Orbital bowls operate with action from a vertical axis, rotating disc within a fixed perimeter wall. Parts are moved to and along the wall by the disc's friction surface. Tooling, similar to that of a vibratory bowl, gets parts from the disc action. Certain part shapes have been fed at near 4,000 pieces, per minute. Thin flanged parts, thin discs, complex parts and tramp materials do not lend themselves to reliable performance in orbital bowls. Tramp material or a thin part edge can jam in the gap between fixed orientor tooling and the rotating disc. For the many parts that can be fed in this unit, a backup bulk storage unit may be needed to keep up with the high feed rate. A half dozen firms are experienced in this feeder concept, which was developed originally for ordnance assembly.

Vibratory bowls have added tooling features to inspect, sort and even assemble certain parts. Tooling is usually built by artisans, to suit the function, with work hardening stainless steel. Complex helical or spiral tooling cannot be heat treated without shape warpage.

Straightline vibratory orientors (Fig. 1) use the same principal of operation as bowls with some possible extra benefits. In-line orientor tooling permits simplified, true design records for preorient, orient, sort, inspect, distribute to multiple paths and live feed functions. The straightline concept permits quickchange, bolt-in tooling plates (Fig. 2) for families of parts. It also permits fully hardened part contact surfaces for long term accuracy and wear life.

Many bowl feeder manufacturers are now limiting maximum bowl size at 18-24 inches. Larger bowls are normally applied for larger parts and/or more storage; apparently, a combination that costs too much for noise control.

Large bowl production rates and related storage volumes usually call for a floor bin and elevator or overhead bulk supply unit.

It is seldom cost effective to invest in a new bowl feeder that must have a floor bin-elevator or other bulk backup units. Fully considered, a feeder package consisting of a floor bin-elevator and straightline orientor will cost less than a bowl feeder that needs a bulk supply hopper.

Gravity powered orientation is the least cost method of feeder application as long as the parts are suitable. Some experienced feeder builders have produced units that feed whole families of parts with ease and reliability at competitive prices.

Simple rolling and sliding parts can be gravity oriented off the side of the floor bin-elevator using slanted lift cleats. For certain parts, it is beneficial to carry them over the top of the elevating conveyor, into a gravity trough with suitable tooling.

Only experienced feeder applications people can safely choose gravity orientation as a recommended solution to parts handling problems. Too many poor choices by novices has hurt the reputation of gravity orienting procedures.

As part shapes become more complex and/or production rates go higher, the choice of low cost gravity orientation becomes less safe. Consider recirculation, noise and parts abuse as you evaluate a gravity orienting choice for feeding your parts.

Use experienced feeder makers for functional and economic comparisons of each feeder application.

BRUSHLON (R)

Recall how a throw rug tends to creep on top of a carpet with normal walking traffic?

Parts are moved by vibratory action, on a carpet-like surface of directionally oriented bristles. Variable angles and lengths of bristles are available to suit part weight and desired feed action.

A small amount of drive vibration is amplified by the bristles' angle/length action to gain remarkable parts feed rates.

Most vibratory feeder makers have used the bristle carpet material to solve special problems. At least one firm uses the material as a matter of course.

Advantages and/or disadvantages of the bristle carpet feed system are:

1. Quiet
2. Gentle
3. Fast feed rate
4. Light duty drive
5. Limited part shapes, size and weight
6. Material cost
7. Installation cost
8. Dirt build up
9. Cleaning cost
10. Bristle matting
11. Wear life
12. Effect of oil/chemicals

Consider the above list and discuss possible applications with experienced feeder people. Proper application success is great but with limitations that must be considered.

SOUND CONTROL

Noise control became a new demand upon feeder makers in the late 60's. Nearly all feeder concepts were hard pressed to limit sound within OSHA specifications (90 dBa, maximum employee exposure for eight hours). Users with high ambient plant noise required still lower feeder limits to avoid noise accumulation totals exceeding OSHA.

Most feeders were shrouded to contain noise (Fig. 3). Orienting units of high efficiency were developed to reduce excess recirculation. Special surfaces were applied to part contact areas to control noise production.

Stamped metal, engine rocker arms, are sleigh bell-like in noise production. Typical price to limit their sound at 80 dBa was $3,500.00 in 1980; about twenty five percent of the whole feeder price.

Fear of Federal regulators caused some feeder users to overkill noise control at great cost. Overkill was probably also caused by political pressure, in 1974, to lower the employee noise exposure limit to 85 dBa. Some feeders were purchased with 75 dBa sound limits, costing more than fifty percent of whole feeder price.

Limited cash, practical environmental attitudes and cost effective common sense have, lately, done away with much noise control overkill. Eighty five to ninety dBa is becoming, generally, acceptable as standard for a parts feeder noise limit.

Keep in mind that a cost conscious evaluation will allow a feeder to run at higher noise levels for short periods of time. Consider a feeder that runs only a few seconds, per minute, or several minutes, per hour. Should money be spent to control sound at 85 dBa or let it run at higher levels for short periods? This approach could save noise control dollars on low production or high efficiency feeders.

Discuss the aspects of production rate, orientor efficiency, storage track capacity, etc., with your feeder maker for least cost, practical sound pressure control.

FLEXIBLE FEED TRACK

Close wound spring wire tube has been used for years as flexible feed track.

More recently, we have seen high density plastic tube used for the transfer of small parts, air blown, from remote orientors.

Close wound spring wire is now made in rectangular sections for feed track. When coated with adhesive plastic (see Fig. 4), the track may be slotted for visual access, dirt removal and level sensors. Because the part carrying wire is hardened, the wear life is very good while retaining flexibility. Custom made, the product is expensive but often worth the price, compared to other solutions.

A few plastic fabrication specialists will build a mold and extrude lengths of slotted rectangular nylon feed track. Mold cost is high but track is relatively cheap. Some firms have a catalog of available molds which may produce a suitable track section without the purchase of a new mold.

Consider flex track for getting into crowded areas or where a station moves, relative to the feeder. Plastic and plastic coated track will, usually, withstand oils and chemicals but not heat.

FLOOR BIN-ELEVATOR

It's a funnel bottom steel box with an integral elevating conveyor to get bulk parts to overhead manipulators (Fig. 5). Bins are normally available in one to seventy cubic feet capacities. Popular sizes are 3, 6, 12 and 20 cubic feet.

The floor bin must flow parts to the elevating conveyor without tunneling (cavitation) or rolling them. Bin design must allow the conveyor belt to enter its bottom and pass upward without jams; even with complex parts and tramp material.

Elevating conveyor belt designs must pick-up parts reliably without jams. Belts may be fabric, rubber, plastic or steel to suit performance conditions.

Part lift cleats are made of various materials and secured to the belt. The conveyor must be designed to resist jams as parts are moved to final discharge.

Elevating conveyors should be driven at the head shaft by a variable speed motor. Conveyor angle should be shallow enough to pick up and keep parts stable and yet get rid of them at discharge.

Floor bin-elevators provide storage of parts at a convenient service level without ladder climbing or overhead mezzanines.

They are used to flow meter parts to overhead manipulator units such as:

Flow chutes
Manual pickout bins
Weigh and Dumpers
Distribution conveyors
Orientor reservoirs

and the like.

For fragile, ferrous parts, an elevating conveyor can be fitted with a plain, non-cleated fabric belt with a magnetic backup.

Most units are built to deliver parts over the top. One exception is where slanted lift cleats coax parts to one side of the conveyor belt against a fixed side rail (Fig. 6). The elevator is very steep to permit part roll or slide to a gravity orientor when the side rail is terminated at a suitable level.

Every part shape is a special challenge to the application of these units. Consult with builders that have made plenty of them to take advantage of their experience.

Nearly a dozen parts feeder makers produce floor bin-elevators with specialty features to serve production automation needs.

RETAINED ORIENTATION AND GENTLE HANDLING

Because early feeder designs tended to damage in process parts, many handling systems avoid bulk storage and reorientation between operations. Some systems use miles of retained orientation track and storage towers with related accessories, at great cost.

Recent developments in gentle handling feeders eliminated the need for retained orientation in the first five operations of making connecting rod caps. Feeders were priced less than half the retained orientation system.

Consider potential simplification and lower equipment cost benefits that could mature from a storage feeder rather than retained orientation between operations.

Many feeder features have been developed for gentle handling delicate parts. One method uses a non-cleated, magnetic backed, floor bin-elevator belt to gently remove ferrous parts from the top of the bin load for orientation.

Qualified feeder builders are experienced in applying the best combination of gentle handling features for specific parts. Such features are so numerous and varied that the subject could make a full topic presentation by itself.

BULK DISTRIBUTION

Some manufacturing process systems require feeding parts from a single machine to another operation having more than one machine. Consider bulk storage and distribution to small individual feeders at each process machine.

Recent installations proved to cost less for bulk distribution conveyors with individual orientors (Fig. 7), compared to distribution via retained orientation tracks.

Where a family of parts must be fed or a part design change occurs, a bulk distribution system is easily adapted. Changeover costs of retained orientation tracking cannot compete with bulk handling in most cases.

Distribution conveyors, with proper gating will get parts distributed for orientation at least cost. Conveyor discharge gating design is the critical area in bulk distribution. Don't use wipers or overhead sluice gates to get parts off the conveyor. Working solutions are available.

HEAVY PARTS

Until recently, most floor bin feeders were limited to handling parts weight of 3-4 pounds. A few feeder models can now meet a ten pound limit. Super duty feeders are also available to orient and feed eighty pound steel billets to forge heaters (Fig. 8).

Heavy, large and complex part designs are now feedable by those who have developed the expertise.

A European vibratory bowl feeder maker has developed a ten feet diameter unit with a ten feet high center tower to orient and feed steel billets. Part weight limit is near twenty pounds. The bowl's part storage is limited and somewhat difficult to fill from stock gons. A simple unit costs about $45,000; without noise control. Sound pressure can be well above 100 dBa. Another $30,000 may be needed for a sound enclosure with reasonable stock service access.

Floor bin-elevator feeders, with orienting tooling, can feed up to 80 pound pieces at less than 100 dBa for approximately $60,000, per unit.

Variable features on super duty feeders can greatly affect pricing. Carefully review all project aspects in practical terms with your choice of feeder maker.

TUB DUMPERS

Stock box dumpers are frequently used to get production parts into floor bin-elevator feeders. Most dumpers are designed to get gondolas at floor level and then lift them through an arc by hydraulic leverage or chain drives.

Despite safety gates and interlocks, there have been lift action failures resulting in injury and death.

A few feeder builders offer rotary gon dumpers that straddle a floor bin for reduced floor space consumption. A fork lift is used to put the gondola into an overhead rotary cradle (Fig. 9). Should a dump action failure occur, the cradle will rotate and rock to a stand still; rather than drop, as a lift unit can.

Consider the rotary gon dumper for safety and floor space benefits. They are, however, about twice the price of lift units.

Stock box dumpers are used for many subtle purposes other than filling feeder bins. Long parts, shafts or tubes, with length-diameter ratios over 10:1, tend to jackstraw in bulk and resist conveyor pickup for orientation. With a stock box dumper, controlled by a level sensor, parts can be metered to the bin for better production. Jackstrawing seldom occurs where long parts are handled in small bulk quantities.

Elevator pickup action in full, large floor bins can damage parts. A metering gon dumper over a small feed reservoir can assure non-damage handling of parts in smaller quantities.

Stock box dumpers, also, provide backup parts storage.

ROD FEEDERS

Tubes, shafts and similar shaped parts can be fed with a new device, similar to a toothpick dispenser (See Fig. 10).

When diameter/length ratios are less than 50:1, a rod feeder does a good job of metering parts without damage or jackstrawing production delays.

Parts are manually loaded, diameter to diameter into a bulk reservoir. An agitator meters parts out a stub chute.

Rod feeders can be furnished with adjustable features for parts families of varying diameters and/or lengths.

Normally, parts are longer than the width of a person's fist, to permit manual loading of the reservoir. A few such feeders have had side access doors for loading short parts.

Special, remote loaded, stock boxes can be placed directly into a rod feeder for quick loading (See Fig. 11). Only a couple sources offer units to accept remote loaded, insertable stock boxes.

For example; rod feed units have been built for 1/8" diameter by 6" long fragile copper tubes and also for 2" diameter x 36" long sheared steel billets. Paper tubes, for window shade rollers, are fed at more than 100 pieces, per minute.

Two kinds of rod feeders are sold and the agitation system is the main difference between them. One style uses cam probes to isolate and roll parts. The other uses continuous motion, rotary sprockets. Continuous motion units are somewhat less expensive and can feed faster.

SUMMARY

Experience and a huge collection of trade tricks are necessary to qualify as a parts feeder builder.

Infinite variations of the described update features are being created and used, to meet the needs of competitive industry, by the parts feeding people.

Choose and get acquainted with a good parts feeding firm. You will profit, far more than they, with properly applied feeders.

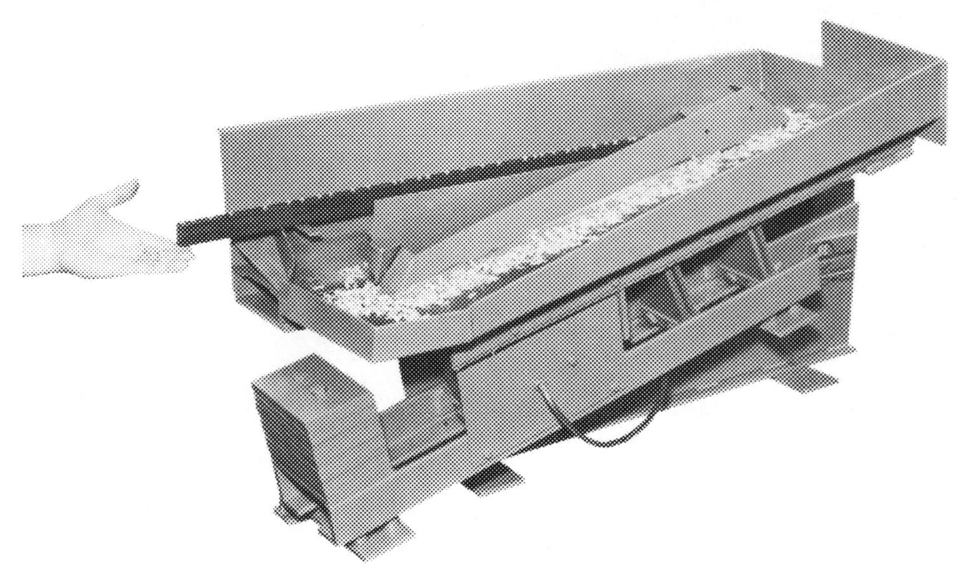

FIGURE 1

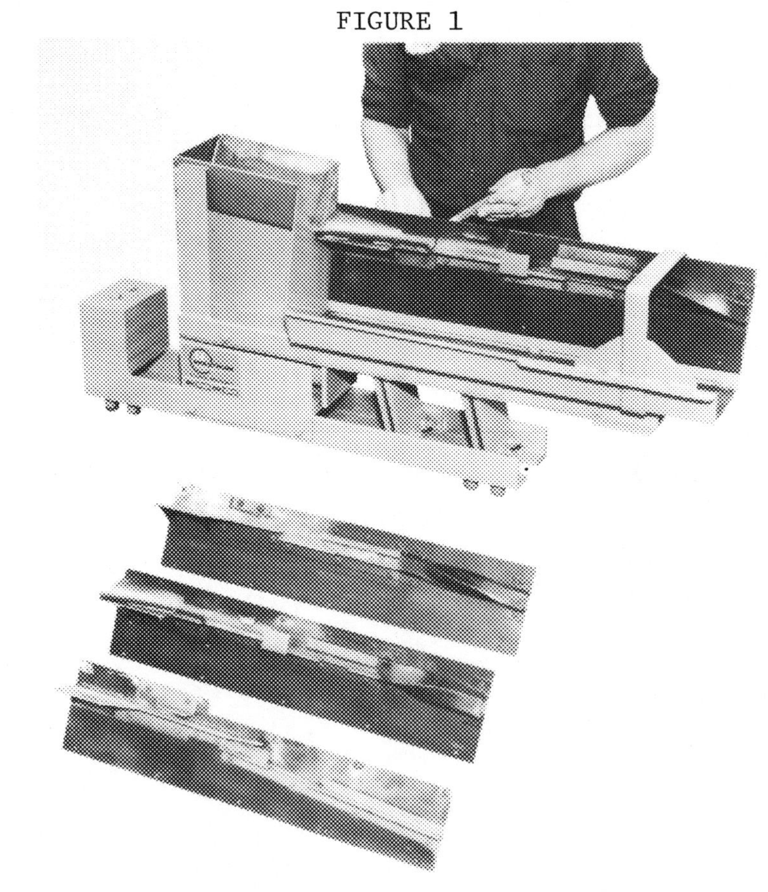

FIGURE 2

FIGURE 3

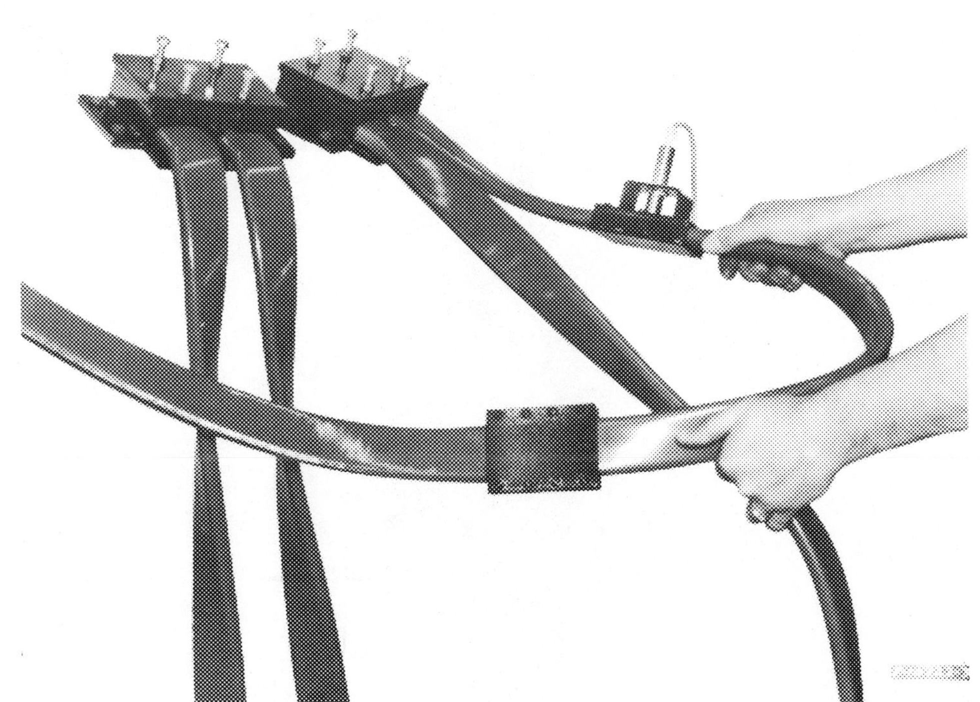

FIGURE 4

FIGURE 5

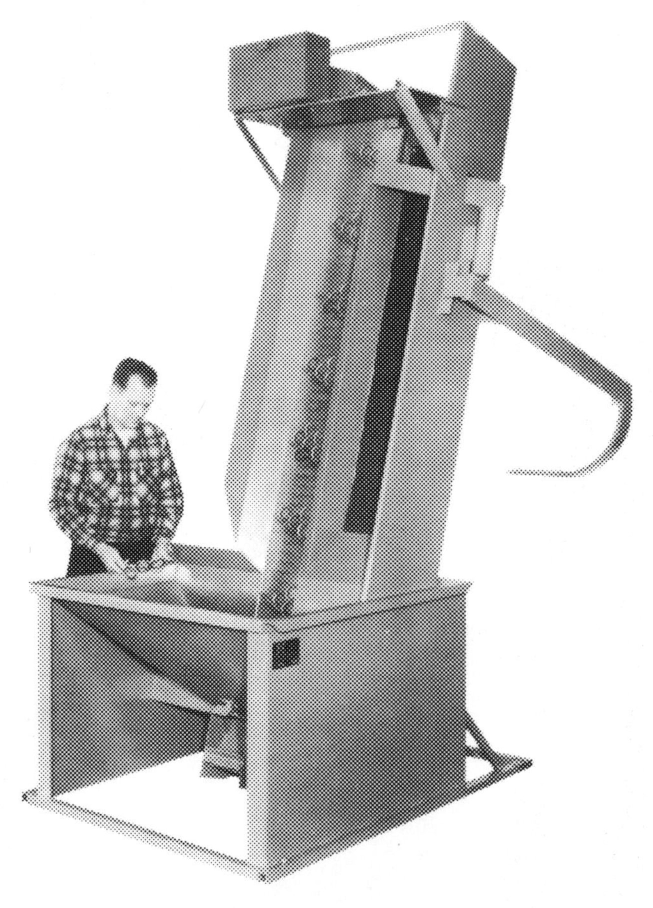

FIGURE 6

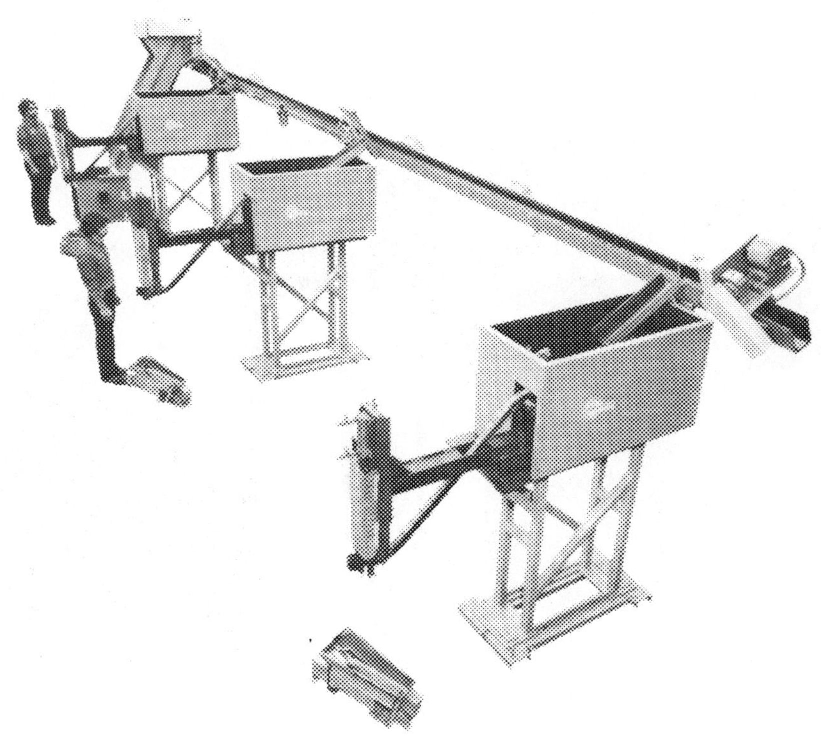

FIGURE 7

FIGURE 8

FIGURE 9

FIGURE 10

FIGURE 11

CHAPTER 4

FLEXIBLE ASSEMBLY SYSTEMS

Presented at the SME Assemblex '83 Conference, May 1983

Wire Guided Vehicles in Assembly

by Clifford T. Anglewicz
Volvo of America Corporation

INTRODUCTION

For the past 20 to 25 years, wire-guided vehicles have been used in North America primarily for material handling purposes. Companies have invested in this equipment to haul loads of material long distances in warehouses, as well as manufacturing plants. Cost justification has been based primarily on manpower reduction.

In the early 1970's, Volvo pioneered the use of wire-guided carriers for assembly purposes. Volvo's proven success with this concept has resulted in its increasingly widespread use among European manufacturers. The time has now come to look at the benefits to be derived from integrating wire-guided carriers into assembly processes. As we shall see, manpower reduction is not the sole important benefit to be derived from installation of a wire-guided vehicle system.

ASSEMBLY CARRIER SYSTEMS

Originally, wire-guided carrier assembly systems were developed as an alternative to the assembly line in an effort to improve the work environment. It was believed that a flexible assembly method would offer new opportunities for improved worker motivation and satisfaction in daily activities. Job functions could be structured in such a way to enable workers to communicate more freely, to carry out job rotation, to vary their pace of work, feel identification with the products, to be aware of quality responsibility, and to be in a position to influence their working environment. These early goals have been achieved to a large degree, with subsequent improvements in worker productivity and product quality.

Now we see carrier systems being utilized to attain a non-sequential assembly method to enhance flexibility in traditional assembly systems and to provide greater possibilities in the deployment of new manufacturing technology. If, for example, a company has a plan to introduce new robot technology into an assembly operation, most often it will be done on a limited scale, on a trial basis. Further robotization will depend on the success of the initial robot modules and the desirability of incorporating evolving robotic technology advancements. A flexible assembly line, utilizing wire-guided carriers, is a long term cost-effective means to readily modify the assembly process in steps.

Setting up assembly lines for limited production runs can be an expensive proposition with a rigid conveyor system. With a flexible, wire-guided carrier assembly sytem, changes to the line become a matter of minor guidepath variations and simple programming corrections.

In the future, manufacturers will increasingly apply CAD/CAM (Computer Aided Design/Computer Aided Manufacturing) modeling to introduce new technology to the assembly process. Wire-guided carriers are particularly well-suited to be incorporated into a graphic simulation with robotic cells and other assembly line machinery. A wide degree of variation in the movement of parts and assemblies through the assembly process is possible with wire-guided carriers. By programming in operating characteristics of carriers, robots, and other associated machinery, a working model can be developed that will achieve maximum efficiency.

MATERIAL HANDLING IN ASSEMBLY PROCESSES

Today, there is a new focus on efficient material handling methods in assembly processes. Just-in-time delivery concepts are receiving more attention. Wire-guided carriers offer a high degree of flexibility in material inventory management within the plant. Flexible carrier control systems (programmed software and operator input terminals) can be designed to insure that material is delivered to work stations in the exact amounts required and at the exact time it will be required, thereby reducing required floor space for material storage.

Or, flexible wire-guided carrier assembly lines can move through material storage areas, when desired. Such a design is particularly appropriate for use with materials with special handling requirements such as those that come in bulky, or odd-shaped packages, or those which require special care, such as acid. In addition, the carrier asembly line can move through sub-assembly build-up areas. Handling time and loss of material due to damage in transport to the main assembly line is virtually eliminated.

CARRIERS FOR CAR AND TRUCK ASSEMBLY

Since the early days of mass-produced automobiles, rigid conveyor systems have been the norm for car assembly operations. Let us look at some of the less desirable characteristics of traditional straight line conveyor systems and what the increased flexibility of carrier assembly systems can offer as an alternative.

If a throughput of 60 cars per hour is the goal of a particular line, as long as the conveyor is running and the line doesnt shut down, you will get 60 cars per hour. If, however, a problem develops along the line, with inventory, quality, manpower, etc., that causes it to be shut down, the 60 car per hour throughput cannot be maintained.

Production lines using carriers, on the other hand, can have buffer areas between individual assembly operations where one or more cars can be waiting to enter an assembly area. If a problem should arise, it can be solved in the area of the problem before the total line has to be shut down. Lost production can be made up in that area with additional manpower so that the total output is not disturbed. In addition, carriers operate at speeds of up to 200 feet per minute and can fill in voids in the production process. Theoretically, a carrier system can exceed the etablished throughput rate of a given rigid conveyor system.

Automotive manufacturers today try to provide the consumer with a complex array of model and option choices in their product line. This is a boon to consumers, but presents its share of problems on the production line. Problems with inventory control, division of labor, and tooling are inherent with such a system, not to mention the potential for confusion on the part of the assembly line workers. Space and cost limitations preclude separate lines for separate models, with rigid assembly systems.

With the right type of planning in software development and carrier guidepath design, the line can be configured in such a way, that varying models can move through their own individualied assembly patterns, thereby greatly reducing the problems mentioned above.

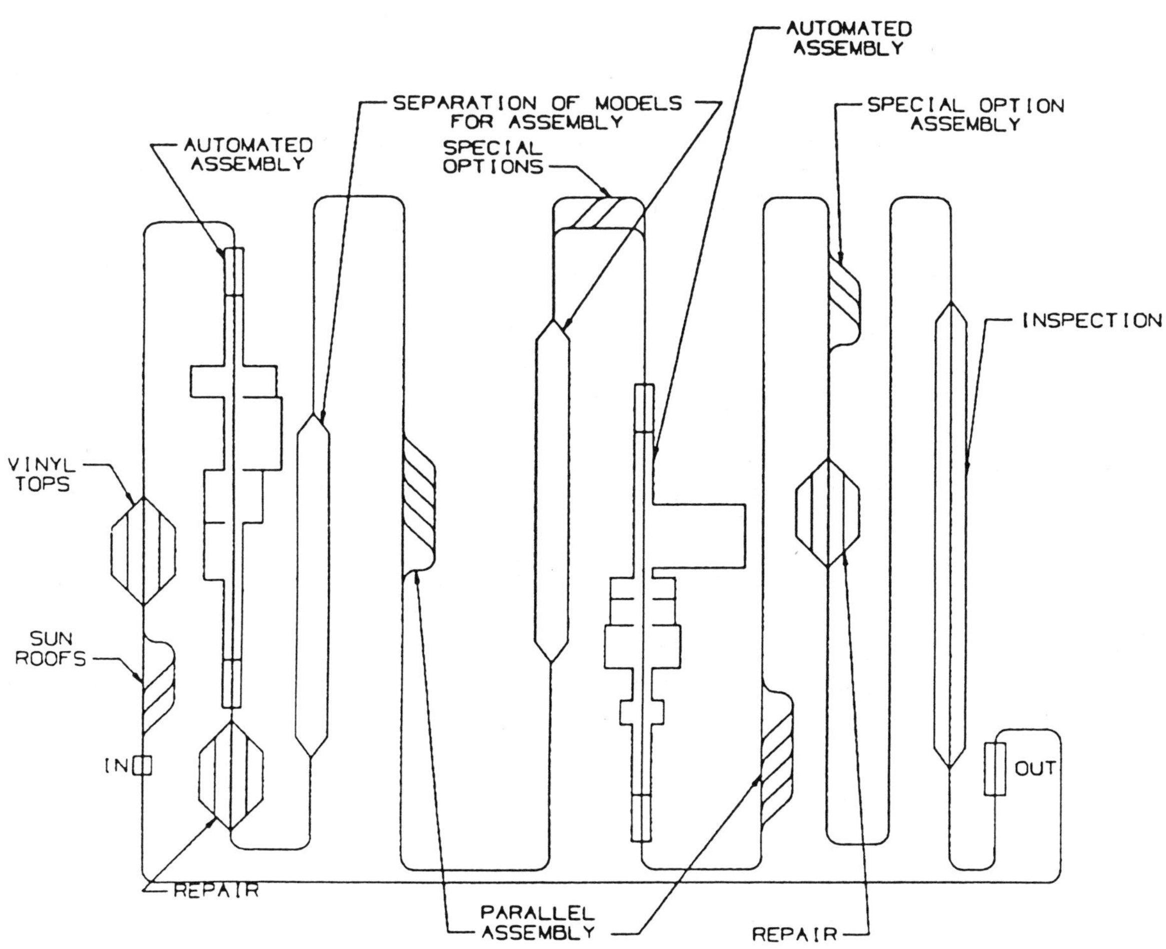

Figure 1

For example, two-door and four-door models can move along the same line, until their components begin to deviate. At that point, the two door models can be tracked along one guidepath, while the four-door models can follow another. They can then meet again when the components to be added to each are identical. This can occur any number of times down the line with models intermingling periodically in assembly areas where they are alike, and then going their separate ways. The assembly line, thereby becomes, a series of readily changeable assembly modules, with consistency built into each module.

As robots become more evident in assembly operations, another positive feature of carrier systems becomes apparent - that is, more efficient use of the robot workers. Again, it is the inexpensive design flexibility that makes this the case. Modern robots can be very versatile, capable of performing a myriad of assembly tasks. The longer a car can be left in front of a robot, without changing the final throughput time through the robot cell, the more work a robot can do during its operating sequence. One way to increase robot work time is to reduce indexing time within the cell. This can be accomplished by splitting the line through the cell. For the wire-guided carrier system, changes to increase, interweave, or reduce lines within a cell are relatively simple and inexpensive. Nearly any line scheme to reduce indexing time can be achieved easily. And as was stated above, downtime is less of a problem because the line continues to run through the cell, as the problem area is isolated for attention.

FLEXIBLE CARRIER DESIGN

Wire-guided assembly carriers have been in use for ten years. The process of designing carriers to fit car and truck assembly applications has been one of continuous development. Carrier design engineers have arrived at a point where carrier design and fixturing has been optimized for many common assembly operations, within the goal parameters of the user companies, including Volvo, Saab-Scania, and others.

In the early 70's Volvo installed a wire guided system consisting of 254 specially designed and fixtured carriers to assemble Volvo cars. Two-hundred-fourteen carriers were designed to carry the car through the complete assembly process from when it was received as a painted body to when it was shipped as a completed car.

The chassis build-up operation was done on a carrier like that shown in Figure 3.

Car body build-up took place on a carrier like that shown in Figure 2.

After the body assembly operation was completed, the two carriers met, the two assemblies were married, and the completed car was put back on the carrier (as shown in Figure 3) for completion, quality control and testing.

Figure 2

Figure 3

Later, in 1980 Volvo installed a system in the Torslanda plant in Sweden (as illustrated in Figure 4) for assembly of car bodies on carriers. This system consisted of carriers to take car bodies through welding and a material stock area where hoods, fenders and doors were installed on the car body and then transported to a metal finishing and fitting area.

When welding of assembly takes place on the car body, the carriers stop at parallel workstations instead of conventional series line type assembly. This method insures the quality of the product because the carrier is stopped so that the work can be performed properly in enough time.

Figure 4

Volvo has also utilized this assembly process in the manufacture of their 4-cylinder engines. In the Skovde facility (Figure 5) the plant is configured so that carriers can follow a predetermined course to have engine blocks put on the carriers. After they are assembled internally, they travel to the preassembly head line where heads are taken off the assembly conveyor and installed immediately on the waiting engine block. From there the carriers proceed to the external assembly area where alternators, carburetors, etc. are installed. After all assembly is completed, the carriers proceed to a test area where the engines are tested and finely tuned and inspected for emissions and quality standards. When this is completed, carriers are directed to the correct unloading area so that the completed engine can be taken off the carrier and the carrier can then proceed back to the beginning of the line. The carriers (as illustrated in Figure 5) are specially fixtured to allow the engine to be turned so that any 5 of the 6 sides can be up for ease of assembly.

Figure 5

Note how carriers have been designed to lift, tilt,and move the assembly or subassembly for maximum accessibility and ease of assembly. Ergometric and safety requirements have been carefully considered and designed into the carriers.

Future considerations in this assembly process (as illustrated in Figure 6) where a robot is preparing to do assembly work on a car body can be easily implemented into a production facility with a non-sequential carrier system.

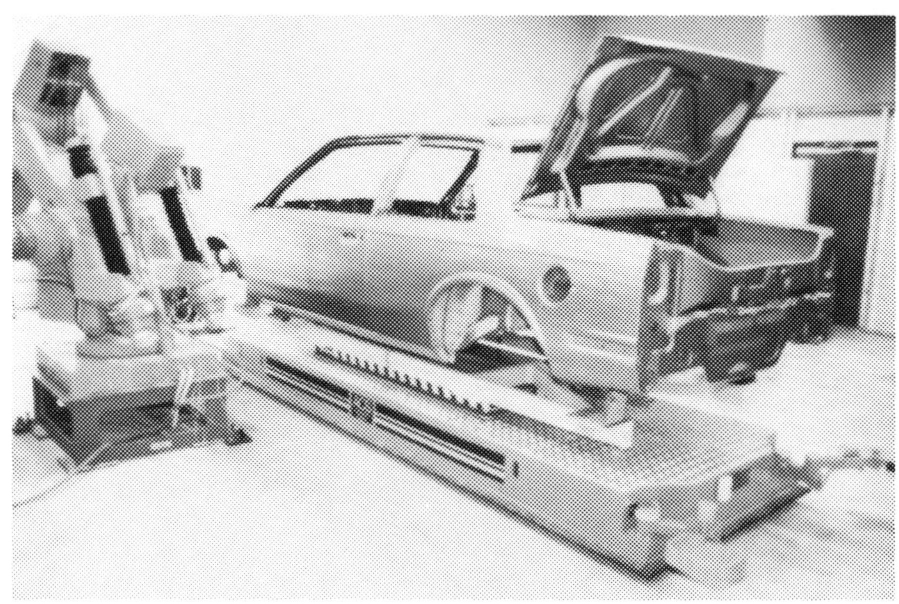

Figure 6

The control of these systems (as illustrated in Figure 7) can be done in a central location where carriers and automated processes can be tied together making full integration of the facility possible.

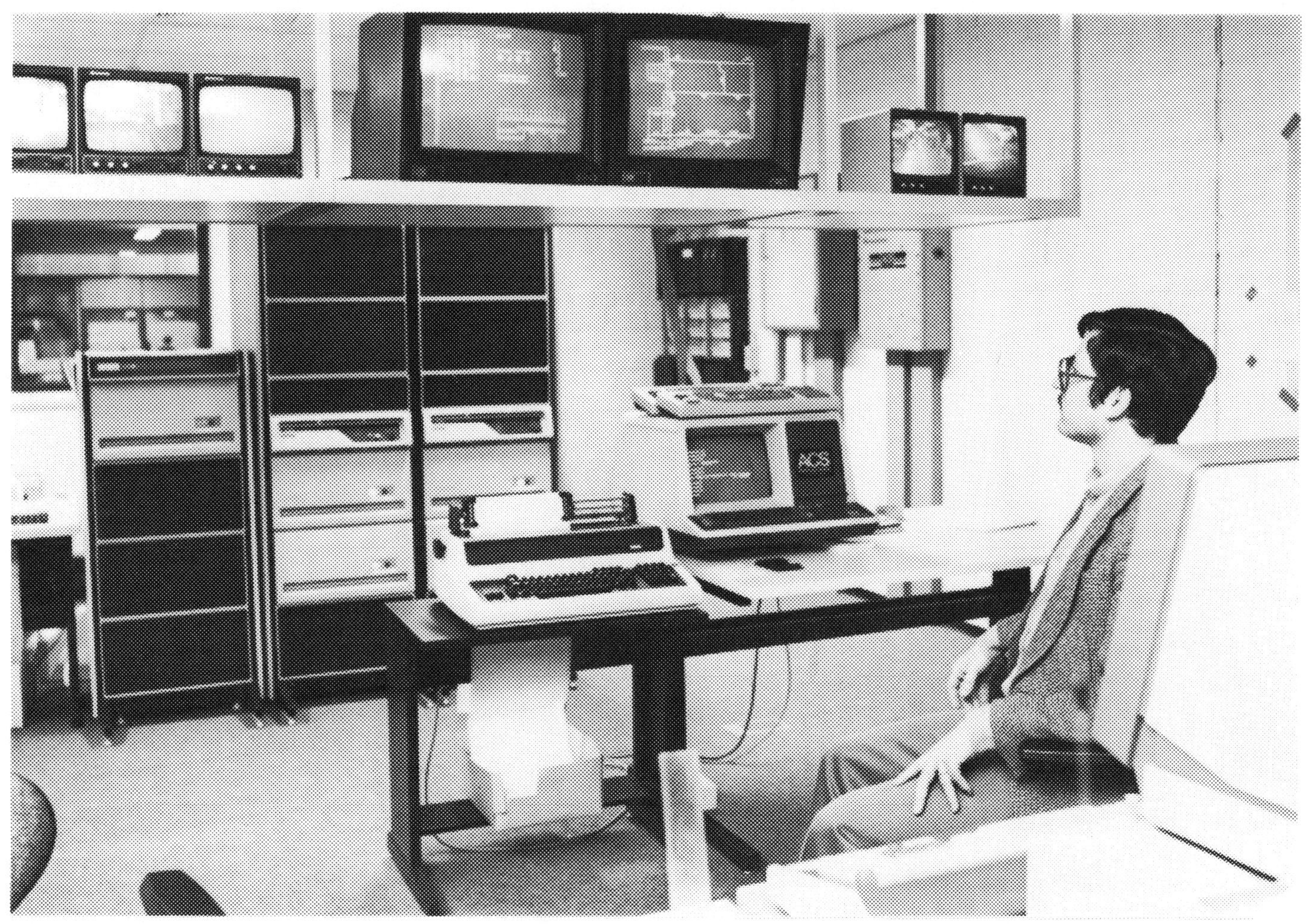

Figure 7

COST EFFECTIVENESS

Cost benefits to be derived from wire-guided carrier assembly systems are many and varied. Let us look at each of these in turn.

As with many large-scale innovations, many of the cost benefits of a wire-guided carrier assembly system will be realized over the long term. One obvious long-term benefit is the cost reduction associated with line changeovers, whether it be for the purpose of producing a new model, or or adding new technology. Not only will actual hardware and material costs be reduced, but, in addition, with the greater flexibility in test simulations, the cost of design and experimentation will be reduced, as well.

A more immediate benefit is the increased productivity to be expected with a wire-guided carrier assembly system. Down time can be expected to be less of a problem. Throughput rates can be expected to increase.

Another immediate cost-savings is inventory reduction. The just-in-time delivery capabilities of carrier assembly/material handling systems and the reduced loss of material due to transport damage can mean immediate reductions in the amount material purchased for inventory and, if floor space is money, the decreased space requirements for material storage can be looked on as a cost savings, as well.

Some other benefits to be achieved are less tangible on the balance sheet, but will have their impact also. An improved flexible working environment is likely to promote better labor relations and improve product quality, the end result being an enhancement of the corporate image and improved sales.

SUMMARY

Wire-guided assembly carrier systems have been in use outside of North America for ten years.

Wire-guided assembly carrier systems are extremely flexible at relatively low long-term cost.

Wire-guided assembly carrier systems are well-suited to blend with the new philosophy of totally flexible manufacturing.

Wire-guided assembly carrier systems are easily capable of accomodating new assembly line technology including robotics and CAD/CAM.

Wireguided assembly carriers can be combined with wire-guided material handling carriers in the same application with positive results in operating efficiency and cost savings.

Wire-guided assembly carrier systems offer a number of tangible and intangible cost benefits including:

1. Reduced line modification costs.

2. Inventory reduction and more efficient inventory control.

3. Increased productivity.

4. Better quality control.

5. Greater worker satisfaction.

Flexible assembly— a boon for short production runs

Flexible assembly systems provide the same cost-cutting benefits for short production runs that you expect from mass production. A variety of equipment and layouts can do the job; one combination may be just right for your application.

The world of "flexible" systems isn't limited to manufacturing operations that use machine tools. Assembly operations in a wide range of industries also can benefit greatly from flexible routing of work. Many such systems are operating today.

Flexible assembly systems allow work to be selectively routed to certain assembly stations, bypassing stations where the work doesn't have to go. These systems are best used in operations where more than one product or product model must be assembled.

When applied properly, flexible assembly systems can:

- Reduce work-in-process;
- Improve product quality;
- Maximize use of floor space;
- Minimize capital expenditure for equipment; and
- Provide the flexibility to change or even relocate the assembly system, at minimum cost.

In today's quickly changing marketplace, manufacturers must be able to adapt quickly and efficiently to the introduction of new products or new product models. With flexible assembly systems, changes can take minutes instead of hours or days.

Flexible assembly systems enable manufacturers to make assembly products in short production runs, with some of the same cost-cutting benefits available by using linear, mass-production assembly systems.

This changing environment is routine for many manufacturers, but perhaps even more so for electronics firms, which exist in an environment where technological breakthroughs occur often, and today's product model is outdated tomorrow. Electronic assembly operations also are characterized by mixes of many product models, and the use of automated assembly equipment, such as robots. Flexible assembly systems keep this equipment working as much as possible.

Certain kinds of flexible assembly systems have become quite popular in the last few years. For example, it's common today for electronics manufacturers to use layouts that include flexible parts routing to assembly stations set up along the length of a belt transporter. Carousels or flow-through racks often are used to store kits needed for assembly, and automatic load/

Mini-load AS/R systems can become flexible assembly systems by setting up stations at openings in the storage racks. The AS/R machines transfer containers of parts between the assembly stations and storage locations inside the mini-load AS/R system.

unload devices can interface the transporter with the carousels. But this is only one of many approaches.

Several suppliers offer mini-load AS/R systems that have openings in the walls of the storage racks. Assembly stations are set up at these openings. The AS/R machines deliver containers of parts directly to the station. This approach lets you confine several manufacturing steps —assembly, test, repair—into one assembly system.

The most recent technology to influence the world of flexible assembly systems is automatic guided vehicles, which carry automotive or other parts through assembly operations. European auto makers have used AGVS for assembly for many years. At least two U.S. manufacturers now are installing AGV systems for this purpose.

Not a new concept

Moving work through assembly operations using flexible-path routing is not a new idea. For years, automotive and appliance manufacturers have used standard power and free conveyors for assembly.

These systems are flexible, because work can be routed to specific stations. Workpieces in parts carriers are diverted to spurs off the main conveyor line, where manual or automatic assembly operations are performed.

For automatic routing, bar code scanners can be used to identify each parts carrier. Then, a computer or other controller can activate a switch in the conveyor track, and the carrier moves to the proper spur.

Car-on-track conveyors also are used in flexible assembly systems. Sections of the conveyor track can pivot automatically to selectively route the work as it moves down the conveyor.

Computer-controlled monorails, inverted power and free conveyors, and shuttle cars can do the job, too. With vertical carousels, you can install a flexible assembly system that uses more than one floor of a facility. Towline conveyors can achieve flex-

Clean room operations aren't unusual for certain kinds of flexible assembly systems. Here, a meshed belt conveyor allows air to flow through it, keeping contaminants away from sensitive electronic components as they move through the system in tote containers.

ible flow; they're used in many flexible manufacturing systems that contain machine tools.

Actually, any existing assembly line can be converted into a flexible system, but the time and cost involved may be prohibitive. If the proper equipment is specified, no converting is necessary.

What's important to remember is that the needs of each particular assembly operation will dictate what type of system is best suited for it.

Three ways to do it

Basically, there are three ways to move materials through assembly: synchronous movement, asynchronous movement and totally flexible movement.

In systems that use *synchronous movement*, work-in-process advances from one assembly station to the next, until the finished product emerges from the output end of the line. All work-in-process moves to every assembly station in the system. The work-in-process can move continuously while assembly steps are performed, or it can be indexed from one station to the next, stopping there until the work is done.

Pioneered by Henry Ford in the earliest car-manufacturing plants, synchronous assembly lines are ideal for high-volume manufacture of a single product line. Because the cycle time at each station is based on the longest time that work-in-process spends at any station in the system, the maximum production rate occurs when assembly at each station takes the same time.

With *asynchronous* systems, work-in-process moves to spurs that are set up off a central delivery path. The carousel/transporter arrangements used by electronics manufacturers, and most power and free conveyors used in assembly, are examples of asynchronous systems.

Buffers usually are required in asynchronous systems, to allow the cycle time at each station to be based on the average cycle time of all stations in the system. When new product models are added, a new cycle time must be established.

In asynchronous systems, breakdown of the central delivery path—the transporter, in the case of the carousel/transporter layouts—can shut down the entire system.

Totally flexible movement can be provided by using automatic guided vehicles or free-path deliver equipment to deliver work-in-process to assembly stations. Guided vehicles aren't restricted to moving along a central delivery path, since a guidepath can be installed in any desired geometry. Using a manual free-path delivery system trades off labor costs for totally flexible movement.

Buffers usually are used in systems that use guided vehicles, to keep assemblers and equipment working as much as possible. By using guided vehicles, stations can be bypassed, vehicles can be sent directly to repair stations, or back to stations "upstream" in the system.

Control is the key

Whatever the type of movement used in a flexible assembly system, the key to efficient operation lies in the controls used for routing and tracking work-in-process. Let's examine how a typical carousel/transporter system functions, and how the controls make the system flexible.

Usually, a minicomputer or a large microcomputer controls the overall system. Individual pieces of equipment—the load/unload device or a diverter arm, for instance—usually are controlled by smaller micros.

Upon demand from the minicomputer or a manual dispatcher, a container of parts is removed from carousel storage and placed on the belt transporter. Automatic identification equipment often is used at this stage to identify the parts container for the minicomputer. When the container arrives at the appropriate assembly station, a diverter arm or other mechanism moves the container into the station.

There, an assembler sits at a workstation and performs his or her assembly work. Workstations have many different designs; some are tailored to individual users' needs.

A typical workstation could contain a hand-held scanner or light pen

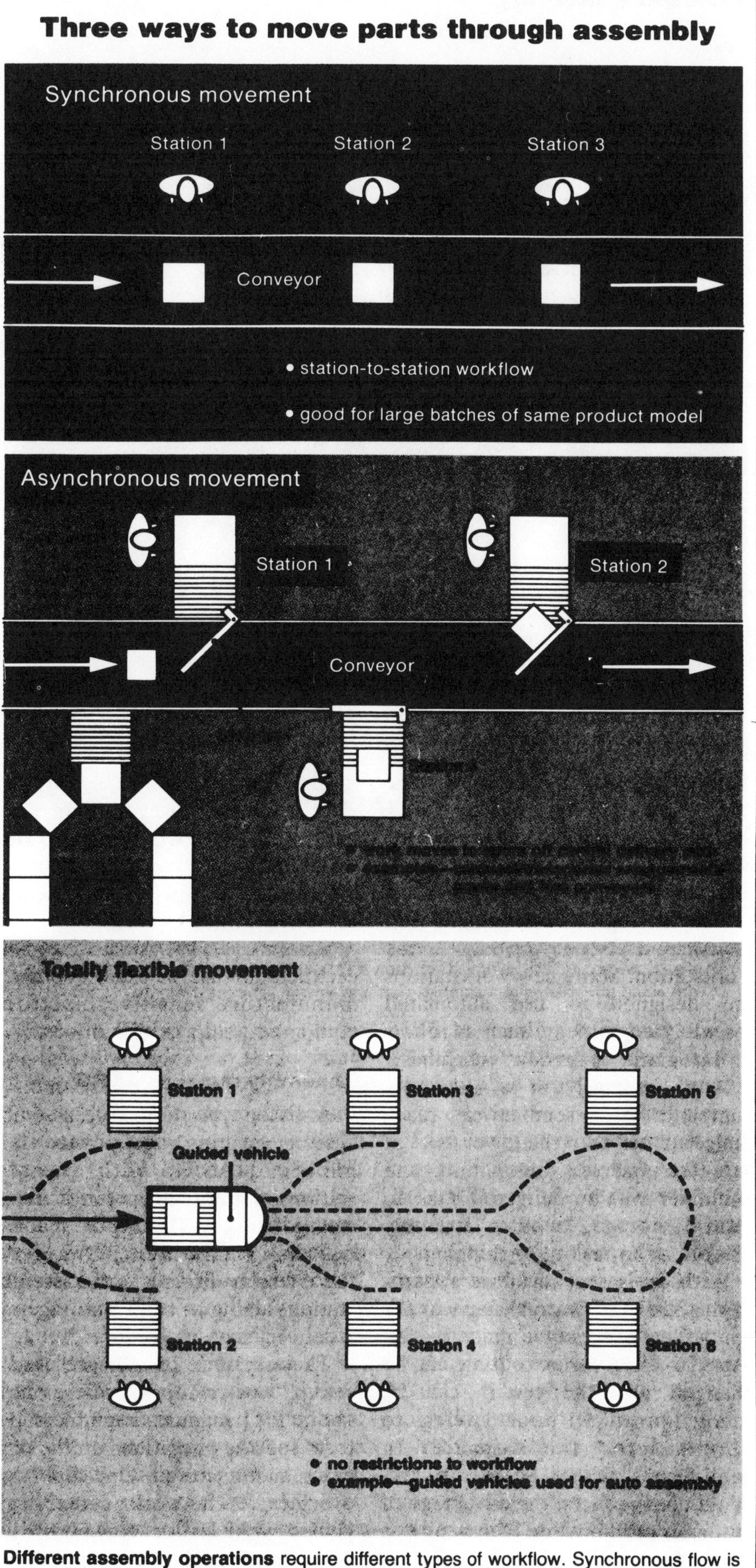

Different assembly operations require different types of workflow. Synchronous flow is best for assembling large batches of a single product model. Asynchronous or totally flexible flow is best for small batches of mixed product models.

Carousels, in background, can store work-in-process until it's needed in assembly. Then, the WIP moves down the belt transporter, and a diverter routes it into the proper station.

that the assembler uses to identify the container or parts. CRT/ keyboard terminals can be used for communication between the assembler and the control system.

Various types of assembly equipment could also be included at the workstation. Some new workstations are designed to use automated assembly equipment, such as robots or automatic "insertion" machines.

After assembly at a station is complete, the assembler can place the container onto the lower level of the transporter belt, and the container will be delivered back to central storage, another assembly station, or to testing or shipping.

With the more advanced systems of this kind, software changes to the control system can be made by the user. In other words, routings can be changed and the system can be reconfigured without having to "break into" the software. In applications where there are frequent changeovers, the advantage of "user programmable" systems is obvious.

Multiple products, moving along different delivery paths, can be built at the same time with these systems. The computer always knows what's happening, and where, in real-time. MRP and other control systems can be integrated with the flexible assembly control system.

Several of these systems have been certified for use in clean rooms, to manufacture sensitive electronic equipment, such as disc drives.

Assembly marries AS/RS

Another type of flexible assembly system gaining acceptance is a mini-load AS/RS with assembly stations set up at openings in the storage racks. The AS/R machine delivers containers or parts needed for assembly directly to the assembly station, acting as both a storage and a delivery system.

These systems usually are used to set up "work centers," where one to about fifty manual assemblers perform specific operations on the parts being manufactured. One supervisor oversees each work center, and guided vehicles or conveyors link one work center to another.

These systems can also be used for assembly of finished goods, as opposed to electronic components such as printed circuit boards or semiconductor "chips," two popular applications. When used for finished goods assembly, these systems use conveyors that run parallel to the storage racks. The workpiece moves down the conveyor, and assemblers attach components to it. The components are stored in the AS/RS, then are delivered to the assemblers by the AS/R machine or machines.

One of these systems is being installed to manufacture tape drives that are used as computer peripheral equipment. This type of arrangement may be required when the product to be manufactured is relatively large, and not suited for storage in a mini-load AS/RS.

Whether a carousel/transporter system or the mini-load system just described is best for a particular application depends on many factors, such as required throughput, available space and whether or not the system is being installed in an existing or a new facility.

Since all systems using the mini-load design operate under

Power and free conveyors have been used for years to carry car parts through assembly. The work can be moved off the "power" line to the "free" line, then be returned to the main flow when that step is complete. Inverted power and free systems can provide the same flow.

computer control, the cost of a small system can be higher than that of a carousel/transporter system where only a small level of computer control is needed.

The geometry of the two types of systems also can be a major factor. Typically, the carousel/transporter systems take up more horizontal floor space, but the transporter can be laid out in existing space in almost any desired geometry. The mini-load systems make better use of overhead space, and may be more economical when the building can be designed around the system, rather than the other way around.

AGVS as process equipment

For many years, car manufacturers in Europe have used guided vehicles to move parts through assembly operations. The benefits and operating characteristics of these AGV systems are well known.

The largest U.S. car manufacturer is installing *five* separate AGV systems to be used in assembly applications. All together, the five systems will use almost 200 guided vehicles to carry car parts through assembly operations, and to "stuff" engines into cars from below.

Another user is planning to install a small system in an existing plant, then physically move the system when the operation is relocated next year. Imagine doing this with a conventional system; the system would have to be taken apart, then re-installed at the new location. With AGVS, the vehicles are easily moved, since they aren't attached to the floor.

There are two basic ways to use AGVS for assembly. One method employs the vehicles as "mobile work platforms." The workpiece is loaded on the vehicle, which then moves from one assembly station to the next, while operators and machines perform assembly operations to the workpiece. The vehicles can deliver the work to testing or repair stations, and then finally carry it to the output end of the system. In this method, the work stays on the vehicle throughout the assembly operation.

In the second method, the vehicles are used as transport devices, moving work from one assembly station to the next, then unloading it onto fixed pickup/deposit stands that serve as assembly stations. There, assembly steps are done, another

AGVS makes its way into U.S. auto plants

Large car parts, such as this auto body, can be carried through assembly by a guided vehicle. When assembly is done, the vehicle can carry the work to testing.

Engines and transmissions often are assembled using the "team concept." Repairs can be made on-line, before the guided vehicle is released to the next assembly station.

vehicle picks up the work, carries it to the next station, and unloads it onto another P/D stand.

This method most often is used in operations that require a fairly long cycle time at each station. This prevents the vehicles from being tied up for long periods of time while assembly steps are being performed.

Both methods offer several advantages when compared with most conventional assembly systems. Because the work remains at an assembly station until it's released by the operator, the quality of work often is superior. Workers are allowed sufficient time to deal with any problems they may encounter, and usually can make repairs without sending the work to a separate station.

Also, the workpiece can be raised, lowered or rotated to facilitate manual assembly operations. Variances in worker height make no difference, and one worker can perform assembly operations to both sides of a workpiece, instead of having a worker on each side of the assembly line.

Parts can be supplied at stations set up between actual assembly operations. An operator at this station can place major parts on the workpiece or the vehicle. In some advanced systems, robots automatically place "kits" of parts on the vehicle before it moves to the next station.

Buffer queues normally are established between stations. Bumpers can stop the vehicle automatically as it touches the last vehicle in the queue, or the control system can stop the vehicle automatically before it touches the last vehicle in the queue, to allow workers to move around the vehicles.

Hand-held scanners can be used at each assembly station to allow the assembler to identify the workpiece for a computer. This allows the control system to track all work in real-time, and to be aware immediately of any problems or delays in the assembly process.

A fairly recent entry into flexible assembly is the use of light-duty guided vehicles in the electronics industry. The vehicles transport tote boxes of components or work-in-process to manual and robotic assembly and test stations. Some of these systems use optical guidepaths, which are painted or taped on the floor. No cutting of the floor is required.

The use of optical guidance, and the recent introduction of a guidance system that uses light-emitting diodes to guide the vehicles, may make AGV systems even easier to change than in the past. Rapidly evolving AGVS technology conceivably could make the use of AGVS routine in any manfacturing industry that uses assembly processes.

Presented at the SME Assemblex VIII Conference, March 1982

Flexible Assembly and Line-Feed Using Robot Vehicles

by Bruce Boldrin
Eaton-Kenway

One of today's major challenges in assembly operations is to assemble a large number of model variations with a variety of options. Many of the conventional synchronous and non-synchronous line techniques are proving cost ineffective in today's environment. This paper will summarize some of those problems, discuss assembly in the perspective of a total production system, and describe using systems of Robot Vehicles for Flexible Assembly.

There are many common problems in the effective assembly of units. Some of these problems have interesting contradictions. A sample list follows:

- High inventory levels and material stock-outs are simultaneous problems.
- Engineering changes result in substantial rework and scrapped material, when the parts are found.
- Model changeover is required without necessarily eliminating old models. It is a problem to assemble both old and new models.
- Model proliferation is becoming a common fact of life in today's market place. Meeting production goals is often difficult without resetting up the line -- a time consuming process.
- Accurate and timely information would allow management to make good production decisions, yet it is difficult to obtain.

The information system that contributes to these problems has a "dotted line" relationship between critical production components and the Host computer. The Host computer contains data bases, processes production orders, and controls inventory. The relationship between the Host computer and the major production components is not an "on-line" communication. Rather the functions of material storage, material handling, and assembly and manufacturing processes stand independently of the Host computer. Paperwork controls production. Manual data entries often result in high error rates. The basic problem with this type of system is the lack of accurate and timely information.

It has been found in batch manufacturing, and in general for job lot shops that the percentage of time that material is in a factory <u>without value being added</u> is approximately 95%. For only 5% of the materials time in the factory is the value being added, which will result in revenues to the company. The 95% is dedicated to material being either in storage or in transit. Any reduction of this queue can reduce inventory and improve overall performance of the production facility.

SYSTEM ARCHITECTURE OVERVIEW

There is an information and control system that often contributes to the solution of these problems. In this structure there is a direct line of communication between the major production components and a plant managerial computer and from the managerial computer to the Host business computer. The Host computer has major corporate data bases. It processes orders and contains the material requirements planning system. CAD/CAM, inventory planning, and group technology modules are typically on-line to the Host. The managerial computer makes best use of resources it supports, including automated storage, in-plant automated transportation, assembly, manufacturing and inspection processes. This architecture is typically used in the master planning of an overall automated factory. It is practical to implement such a system in stages.

LINE DELIVERY

Robot Vehicles are used for direct delivery to assembly lines from the work part storage systems. Figure 2 shows two types of Robot vehicles. The Robocarrier model is used for line delivery. It is battery powered, driverless computer controlled and automatically loads and unloads. It interfaces with deposit stations on the line. One such station is shown in Figure 4a. Accumulation and queueing of loads is provided by delivery to powered or gravity conveyor. This provides the continuous parts availability on the line.

KITTING

Kitting can be performed. One process integrates the Automated Storage/Retrieval Systems (AS/RS) and Robot Vehicles directly. Full pallet loads of materials are removed from the AS/RS and taken to an order picking station. The picking operator has a CRT and a hard copy terminal that allows identification of the parts to be picked and to which load they are placed. In this manner, complete kits of parts for assembly units can be picked. Partial loads are returned to the AS/RS via Robot vehicle for Automatic putaway. The kits also, can be stored in AS/RS for their appropriate manufacturing day and the Robot Vehicles can deliver the kits to an assembly station. Complete assembly and progressive assembly at a station concepts can use this technique.

AUTOMATIC TESTING OF ASSEMBLED UNITS

The guided vehicles can provide a transportation system or a testing bed for assembled units. In one application, automobile engines are transported from the end of an assembly line to a testing cell. At the cell they are deposited automatically at a dynamometer testing station. After of automatic testing is complete, the engines are transported to either finished goods if they are accepted, or to a rework area if they fail test.

LINEAR ASSEMBLY BACKGROUND

For high volume production with low model proliferation, traditional linear techniques use synchronous and non-synchronous lines. Typical linear equipment includes pallet, slat, and power and free conveyor; and tow lines. The difference between linear concepts hinges on line balancing and interstation queues (see Figure 1).

- Synchronous lines are balanced on the longest station time in the line. No interstation queues are required.

- Non-synchronous lines are balanced on the average cycle time and require interstation queues.

An analysis of linear assembly lines for 4 models with different base and options, and assembly line ranging from 30 minutes to 52 minutes is presented in Reference 3. It will not be treated in detail.

FLEXIBLE PATH ASSEMBLY SYSTEMS

Flexible Path Assembly Systems provide several distinct operational benefits compared to conventional linear systems. This section describes these benefits, after establishing a groundwork of the types of concepts, equipment and layout, and system operation.

There are two Flexible Path Assembly concepts. Both use driverless Robot Vehicles. These vehicles are battery powered, inductively guided, and automatically route and position. The typical vehicles and concepts are identified by the type of work station in the system:

1) Mobile Work Platform - Units to be assembled are transported by and remain on the Robot vehicle through all work stations. The vehicle travels in one direction, is loaded and unloaded at start and end of assembly by other equipment or manually, and is directed by on- board intelligence and local controls. See Figure 3.

2) Fixed Work Stand - Units to be assembled are picked up, transported, and deposited on fixed work stands by the Robot vehicles. Vehicles travel in two directions on the same guidepath, can automatically load and unload themselves, and

are directed by dedicated or distributed computer control. See Figure 4.

Typical parameters of the two concepts are detailed in the following table:

Parameters (Digitron Names)	Mobile Work Platform (Robomatic)	Fixed Work Stand (Robocarrier)
Vehicle (See Figure 2)		
Travel Directions	1	2
Capacity	To 700 lb.	To 4000 lb.
Load/Unload	Manual or by other equip.	Automatic
Maximum Speed	100 ft/min	200 ft/min (2000 lb.)
Positioning Accuracy	±3"	±¼"
Maximum size of Assembly Unit	36" X 48"	60" X 60"
Battery Recharge	Recharge at work station -automatically	Exchange when depleted (typ. 1 shift)
Work Station		
Type	Vehicle (Fig. 3)	One or two fixed stands, conveyor (Fig. 4)
Work Group Layout	Fig. 5	Fig. 6
Control	Microprocessor on vehicle, local microprocessor for traffic and queueing control.	Central or distributed computer control. Direct continuous communication with all vehicles.
Load Movement Command	Pushbuttons on vehicle to identify characteristics of assembly unit, pushbuttons at work stations.	Pushbuttons at work stations, CRT's at supervisor stations, and direct computer-to-computer.

Both Flexible Path concepts have a commonality of modular design:

- Work grouping. Every station in a work group is in parallel and has the same work scope. Typically, a station has 4 or more minutes of Elemental assembly time. E.g., At 4 minutes, these systems would perform the equivalent work of 4 min. ÷ 0.58 min/stations = 7 Synchronous linear work stations (see previous example). In most systems, work content per station group varies from 4 - 10 minutes. As model mix increases, the work group efficiency increases - relative to linear lines.

- Parts Provisioning. The Unit to be assembled stops at a parts provisioning station between work groups. The Parts Provisioner places major parts on the Unit or vehicle, often prepositioning parts to assist the assembly operator; this factor requires a readjustment of the assemblers work standards - to reduce materials handling functions.

- Work Station Tools and Parts. Expense stock, such as bolts, washers and fasteners are stocked and replaced at the work station. Thus, these parts are not the concern of the Parts Provisioner. Tooling is duplicated at each station in the work group, a factor that must be included in system cost.

- Bypass Capability. If assembly is not required in a work group, the group can be bypassed. This feature expedites the movement of assembly Units, expecially through the Options work stations. A second aspect of "bypass" is used for problem solution or rework. When a problem is identified that cannot, or by policy, will not be repaired at a work station, the assembly Unit can be moved by vehicle, to a grief/repair/rework station. Thus, the Unit is not passed from station to station; as on linear lines where it "steals" one assembly cycle for all downstream stations.

- Inter-Group Buffers. For throughput leveling, there is inherent intergroup buffering. Typically, several units will be queued ahead and beyond the Parts Provisioning station. Each Unit represents one cycle of line capacity. These buffers allow decoupling the work groups, so that problems or events within one work group do not immediately affect the next, downstream work group; nor do they immediately affect the upstream group. This characteristic can be used to smooth line flow - trading off with additional vehicles, accumulating conveyor, and floor space.

- Universal Vehicle. The Robot vehicle is a universal carrier for all models to be assembled. Thus, there is a standard interface to the product, such as a pallet or fixture, whose mating mechanism to the vehicle is constant. For new models, the pallet or fixture top is modified.

MOBILE WORK PLATFORM SYSTEM

The System developed for a European Auto manufacturer was designed to minimize Work-In-Process inventory and streamline material flow. Base engines are mated with transmissions and final "dress up" is accomplished. The specific application included:

- 80 Base engine types
- Over 20 transmission types
- Resulting in a total of 600 possible engine varieties.
- Assembly time range 19-35 minute/engine
- 2.1-7.0 minutes of job content per worker
- 6 work groups of 3-6 workers
- Options include air conditioner, power steering, and mounting brackets for different chasses.

This system was compared with a linear system of 2 parallel assembly loops. The system compared as followed:

Synchronous Linear Line	Flexible Path
Power and Free Conveyor: Work Platforms 180 carriers	Robomatic Mobile 100 vehicles
Conveyor length, including 2690 ft. empty carrier buffer: 2165 ft.	Guidepath length:
Required floor space 33,000 ft^2	Required Floor Space 22,000 ft^2
Hourly people 50	Hourly people 42
Delay Time (slack) 17-20%	Delay Time 4%

The installed system cost of the power and free system was 13% more than the Flexible Path System. Additional, non-quantified criteria of the Power and Free System included:

- Assembly on a hanging, moving engine
- Engines swing during assembly
- Fixed assembly height (Robomatic vehicles have 20 inch lift)
- Lack of flexibility to modify the engine model mix.

- High slack time from imbalanced work loads.

Sequence of operation (see Fig. 7):

1) Base engines are loaded onto the Robomatic vehicles by overhead hoist.

2) An induction operator clamps the engine to vehicles lift table; then programs Base model and option information into the vehicle keyboard; attaches an Engine Specifications Sheet; and releases the vehicle.

3) The engine is identified for tracking purposes with a bar code reader. This ID commands a mini-load AS/RS to retrieve the appropriate transmision (after looking up the engine record on computer disk).

4) The transmission is married to the engine with mechanical alignment assistance. An operator secures the transmission to engine with a bolt and releases the Robomatic vehicle.

5) Before entering an assembly work group, the vehicle stops at a material provisioning station. Here, the Parts Provisioner refers to the Bill of Material contained in the specification sheet and loads the appropriate major components on the engine or vehicle.

6) When loaded, a pushbutton releases the vehicle and the next vehicle automatically moves into the part loading position. The parts station has a queue capacity of 5-10 vehicles. At 0.7 min. cycle time, this results in a 3.5-7 min. buffer between Work Groups.

7) Local controls release one vehicle at a time into an open input queue position. This provides a vehicle immediately upstream of the work station.

8) When an assembler completes the Unit, the vehicle is released by pushbutton - at the work station and the vehicle moves to an output queue position on the work station spur. Simultaneously, the vehicle on input queue moves into the work station. The assembler has the required tools and fastening hardware to assemble the major components onto the engine. He has ready access to 3 sides of the engine.

9) A local controller periodically releases all vehicles on output queue positions, and the vehicles move out, simultaneously, onto the collecting guidepath.

10) Heading toward the next work group, the vehicles stop at the Inter-Group parts station. In-path accumulation

is achieved by bumper stops. When the lead vehicle moves on the trailing vehicle will start after a brief (few second) pause to provide intervehicle spacing.

11) This process is repeated through all stages of assembly, until the engine goes through a final inspection and test. Rework and adjustment stations are located on a loop that starts beyond inspection - for failed engines - and reenters the main flow upstream of inspection.

12) Bypass loops are located around the Work Groups to allow failed engines to be released directly to the rework area.

13) Engines successfully passing test proceed to the Engine off-load point. This is accomplished by the inspector reading the engine number with a hand held bar code reader - signalling completion of engine assembly. He then addresses the Robomatic vehicle to:

- An unload station for shipment to other plants, or

- The overhead conveyor linking the engine assembly area with the car assembly line.

14) After off-loading empty vehicles automatically return to the engine loading station.

In future systems of this type, there are several areas where extra capability could be installed:

1) Elimination of paperwork by carrying all engine Bills of Materials and specifications in the computer. Parts provisioners and inspectors would then be able to scan and display their information on a CRT terminal.

2) Extending the system for delivery into other inplant areas, such as the car assembly line, or a buffering In-process Automated Storage/Retrieval System. Wire guidepath is considerably less costly than overhead conveyor. The real economics of an extended system would need to be analyzed in terms of additional vehicle requirements.

3) Inclusion of hot-engine test while the engine is on the vehicle.

Fixed Work Stand System

A major European car manufacturer now has in operation a Flexible Assembly System with a planned daily output of 1500 engines, and over 100 different possible models. Other specific application parameters include:

- 4-10 minutes of job content per worker
- Individual worker freedom of ± 15%
- Group buffer capacity of 20 minutes.
- Automated handling and storage of major components and engine-in-process.

The major storage, transportation and control elements of this system include:

- Automated buffers for engine blocks, crankshafts, and cylinder heads.
- 11 work groups, with 12 work stations maximum capacity.
- 3 Automated assembly operations.
- 38 Robocarrier vehicles to service the fixed work stands.
- Bar code readers at each buffer input and output.
- 1 main and 15 microcomputers.
- 1 Backup for the main computer

The sequence of operation proceeds as follows:

1) A defined engine block is delivered from the buffer and palletized. The engine number is entered into a computer terminal and the pallet number is automatically scanned. From this induction point the engine will be tracked through assembly by the computer.

2) Pistons, rods and rings are fitted to the engine on a transfer line, followed by the crankshaft.

3) Engines enter a buffer conveyor. There are buffer conveyors between all work groups. They provide ±20 minutes between work groups, and allow the loading of heavy and larger components required in the next work station.

4) Palletized engines are picked up by the Robocarrier vehicle, and delivered when a work stand is open. In the meantime, the assembler is completing his tasks on an engine in the other work station stand. At his work station he has the necessary tools and small parts/ fasteners in a transportable module.

5) When an assemblers tasks are successfully completed, engine pickup is commanded through a pushbutton. And the engine is automatically picked up and delivered to the next buffer conveyor, or to the unload station - if complete. If the engine is defective, pushing a second button commands the engine to be automatically moved by Robot vehicle to a rework/adjustment station inside the work group. Thus, the assembler works first on one engine, then on the next without leaving the work station.

6) In this installation the Robocarriers can transport two engines at a time, thus reducing the overall travel requirements and number of vehicles.

7) When an engine is picked up from a buffer, the pallet number is automatically read by a bar code reader. Immediate updating of the engines history is performed on the computer. An assembler is notified via a signal lamp. Above their work station whether he is on or above the standard production rate.

8) Line printers at each work group can print out group and individual performance.

9) There are 25 different cylinder heads that are automatically selected and delivered to the engine pallet.

10) On the third, non-production shift, Robot vehicles pick up depleted small parts/fastener modules and replace them with a renewed kit - for the next days production.

Several important operational features have been designed into the system:

- As the Robot vehicle batteries are depleted, they automatically travel to a battery exchange area, and a standby vehicle replaces the departed vehicle. Thus, full production capability is continuously maintained for each work group.

- Software senses and diagnoses systems disturbances, breakdown of components or transmission lines. This allows rapid switch-over to backup systems components.

- Other software routines protect data in the event of system failure, and allow immediate recovery.

<u>Conclusions:</u>

1) Conventional, Linear Assembly Line concepts are best for an undifferentiated product.

2) When models and options of a product proliferate, it becomes increasingly necessary to provide a Flexible Assembly system.

3) Two field-proven Robot Vehicle systems for Flexible Assembly are available: (1) The Mobile Work Platforms, and (2) Vehicles servicing Fixed Work Stands. Case Studies are presented.

4) Bottom line justification for the Flexible systems includes balancing the factors of: Capital Expenditure, Operating Costs, Floor Space requirements, and operating Flexibility.

5) These systems apply to Assembly Units whose final configuration is within an envelope of 60" X 60" and 4000 pounds.

6) On-line tracking, minimum inventories and streamlined material flow are typical with Flexible Manufacturing Systems using Robot Vehicles.

7) Coordination of line stocking with specific model mix assembly scenarios is practical with AS/RS, AGVS and the control system architecture covered in this paper.

8) Total production systems can often be implemented in piecemeal fashion, <u>if</u> the master plan and control system structure have been properly prepared.

Capital Investment for Flexible systems can be time phased to coincide with system starting and growth. E.g., Vehicles in the system can be added to handle increases in production volume. thus deferring substantial expenditure until production level demands capacity and revenue is returned from sale of the product.

REFERENCES

1. Schneider, F., "Opel Russelsheim (FRG)." Paper presented at 1st International AGVS Conference, Stratford-on-Avon, England. June 4-6, 1981.

2. Schneider, F., "Improved Working Conditions and Productivity." Published Technical Paper.

3. Boldrin, B., "Flexible Assembly Using Robot Vehicles." Autofact III Proceedings. Detroit, November 1981.

Figure 1. Synchronous and Non-Synchronous Linear Lines

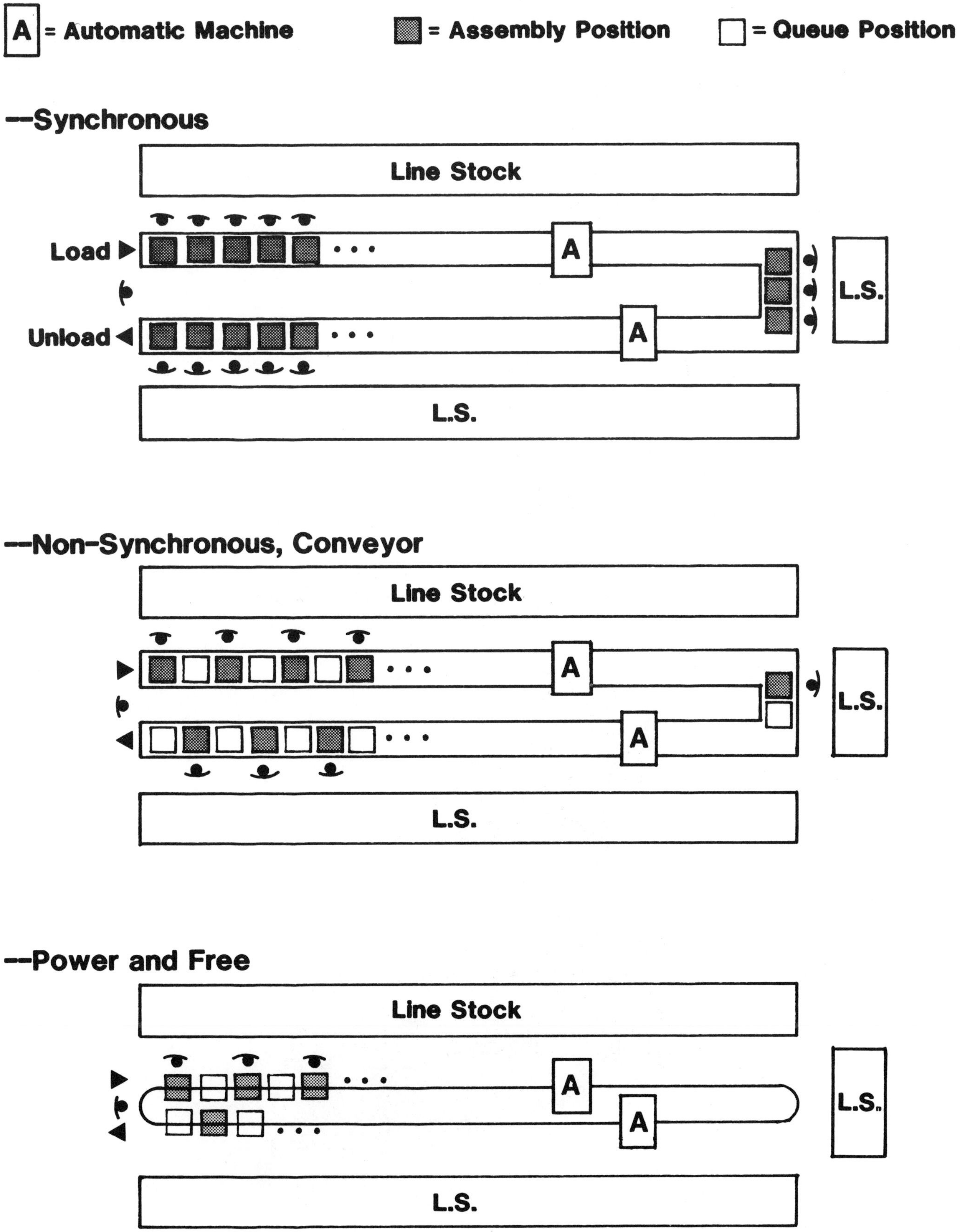

Figure 2. Flexible Path Vehicles

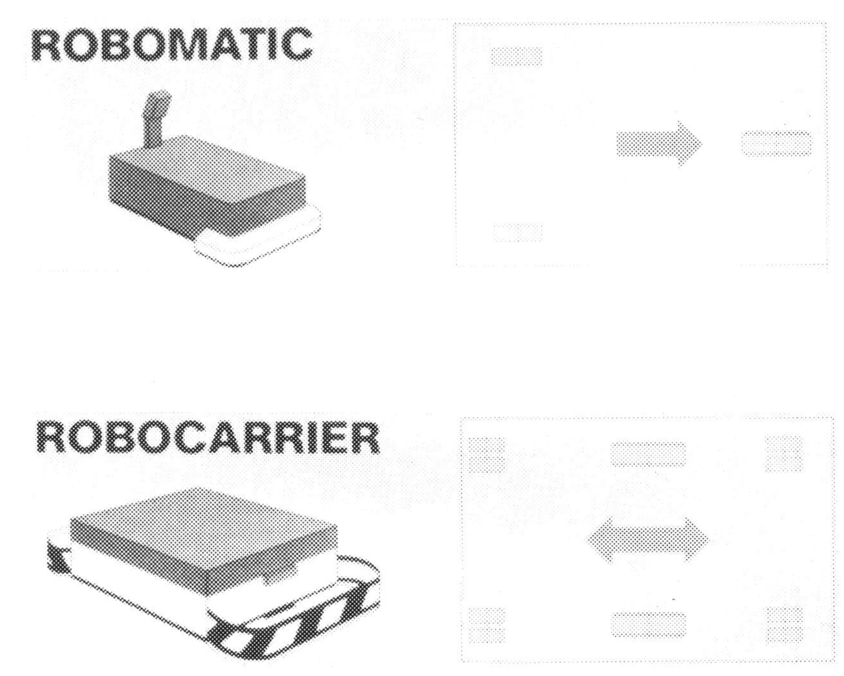

Figure 3. Mobile Work Platform with Lift

Figure 4. Work Stands—Fixed Work Stands

4a. Conveyor

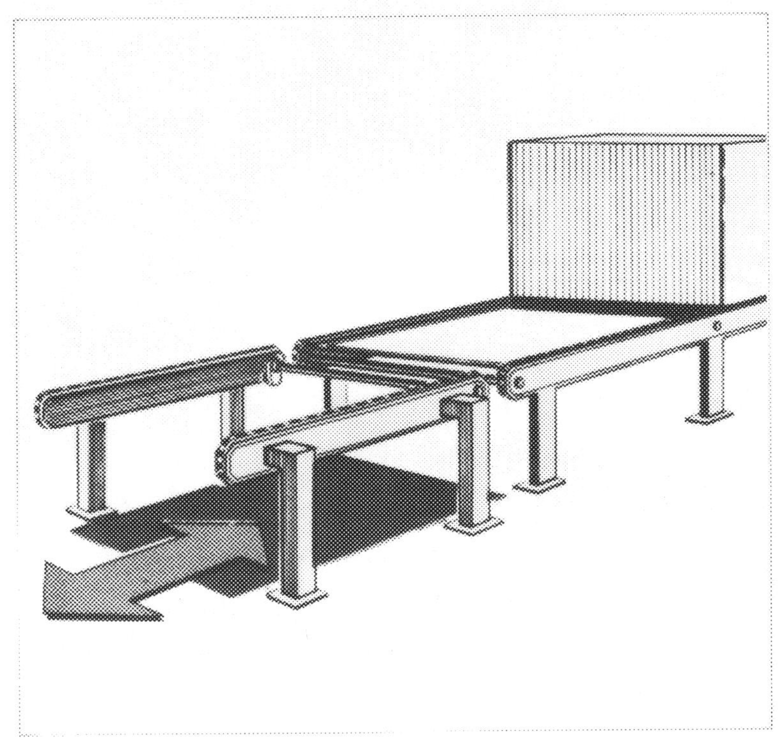

Figure 4. Work Station —Fixed Work Stands

4b. Four Post Stand

Figure 5. Work Group Layout—Mobile Work Platform

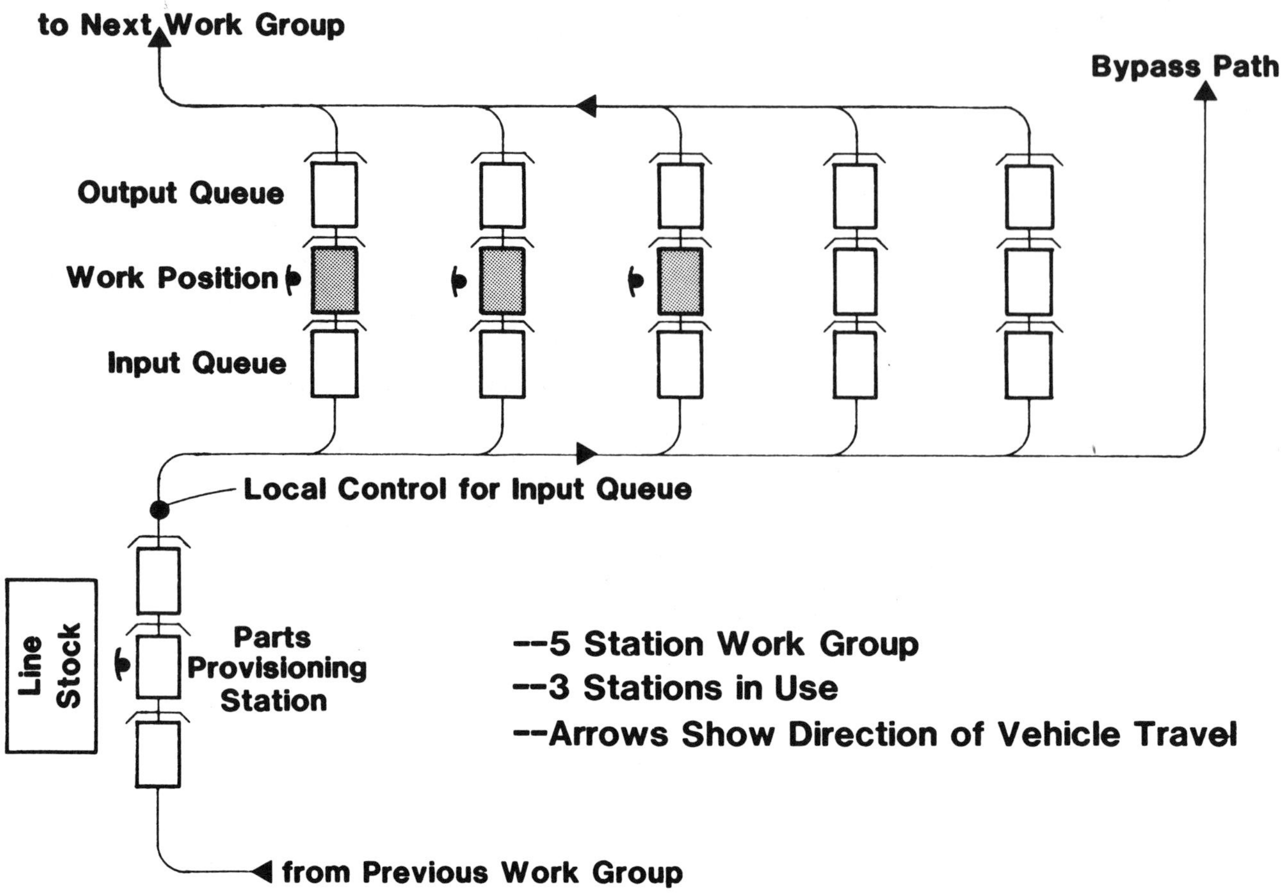

to Next Work Group
Bypass Path
Output Queue
Work Position
Input Queue
Local Control for Input Queue
Line Stock
Parts Provisioning Station
--5 Station Work Group
--3 Stations in Use
--Arrows Show Direction of Vehicle Travel
from Previous Work Group

Figure 6. Work Group Layout--Fixed Work Stand

Stock
Line Stock
Work Station
--6 Work Stations
--12 Work Stands
--5 Stations in Use
Possible Bypass Path

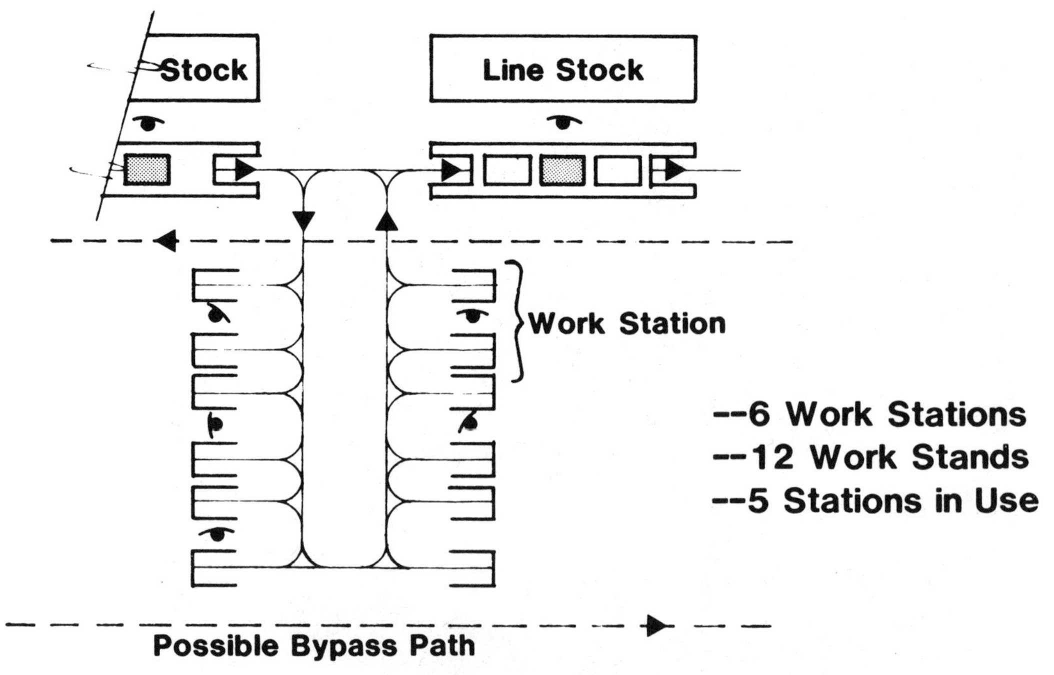

Figure 7. System Layout—Mobile Work Platform

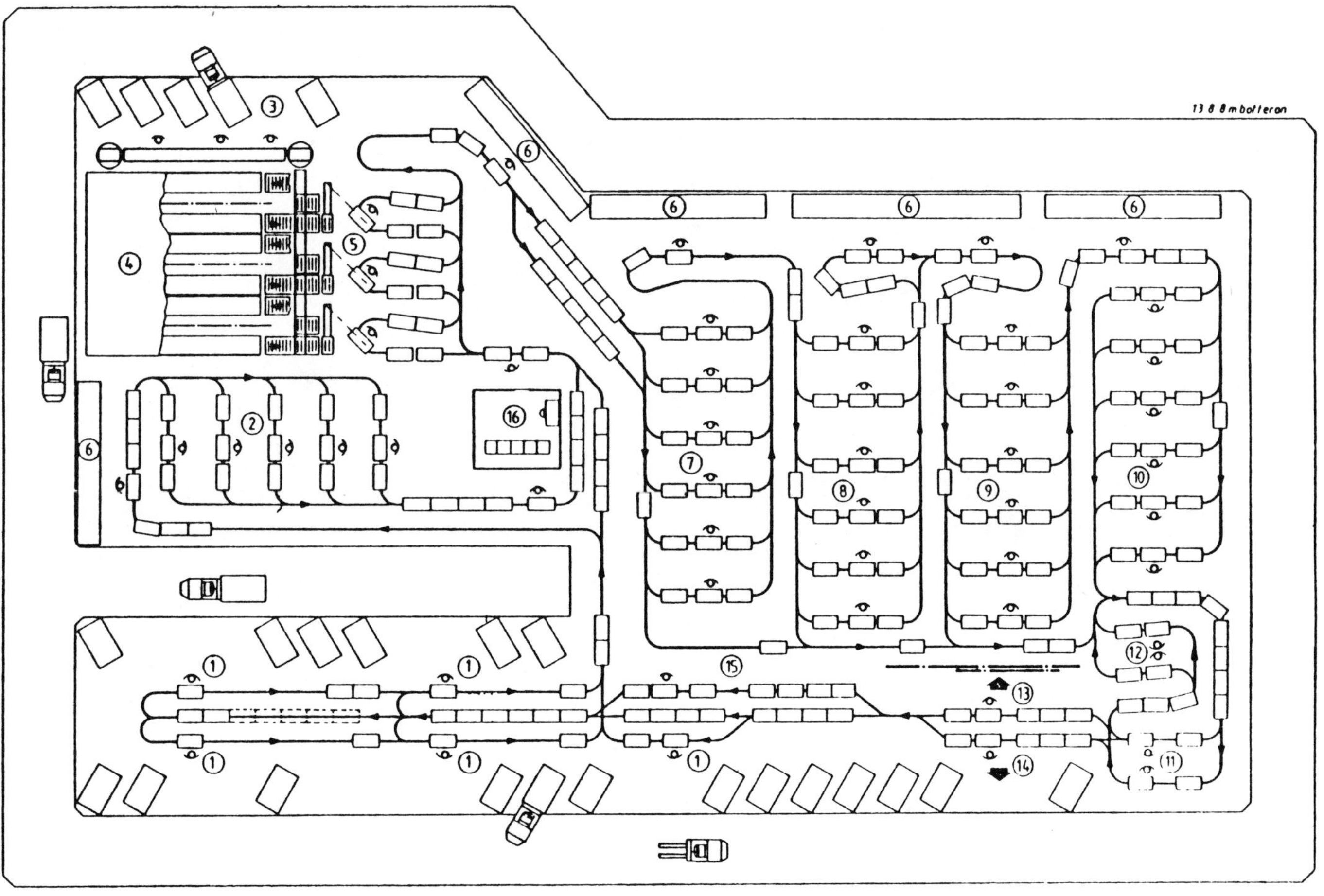

(1) Engine loading
(2) Import Engines
(3) Transmissions
(4) Mini-Load 300 Transmissions
(5) Marriage Engines and Transmissions
(6) Material preparation
(7) Transmissions
(8) Mounting brackets
(9) Generator
(10) Heater
(11) Inspection
(12) Repair
(13) P & F to car assembly
(14) To other plants
(15) ROBOMATIC Maintenance
(16) Computers

Figure 8. System Layout—Fixed Work Stand

Presented at the SME 13th ISIR/Robots 7 Conference, April 1983

A Modular Programmable Assembly Station

by Randall C. Smith
David Nitzan
SRI International

I INTRODUCTION

The automation of discrete-parts manufacturing, spurred in its practical development by both economic and social considerations, is being implemented worldwide on an ever-increasing scale. Advanced automation expands the productivity of labor, combats inflation, and makes it possible to compete effectively in world markets; it improves the everyday working conditions of the labor force and raises the standard of living of the population as a whole. Automation is especially important for batch manufacturing, which is extremely labor-intensive and accounts for the largest portion of the total cost of discrete-product manufacturing [Cook (1975)]. Unlike the application of hard automation to mass production, where the expense of acquiring special-purpose equipment can be justified by the resulting high volume of production, the automation of batch manufacturing must be programmable. Programmable industrial automation [Nitzan and Rosen (1976)] is characterized by three salient features:

* Flexibility--The capability of a machine system to perform different actions for a variety of tasks.

* Ease of Training--The facility with which a person can efficiently program a machine system to execute a desired task.

* Artificial Intelligence--The ability of a machine system to perceive new conditions (whether anticipated or not), decide what actions must be performed under those conditions, and plan these actions accordingly.

Although suitable primarily for increasing the productivity of batch manufacturing, programmable automation may also be applicable to mass production for the following purposes:

* Reduction of the setup time for manufacturing shortlived products in a competitive world market.

* Lowering the cost of production equipment by using components, such as robots, sensors, and computers, that are commercially available and recyclable.

* Producing a sufficiently large sample of new products to enable their technical and market performance to be tested before investing in a costly hard-automation system that would mass-produce them.

The factory of tomorrow will be characterized by integration of manual, hard-automation, and programmable-automation activities in proportions designed to minimize the total cost of manufacturing and servicing. The ultimate goal of programmable

----

* This work was performed as part of a program supported by the National Science Foundation, under Grant No. DAR-8023130, and by thirty U.S. industrial companies affiliated with that program.

industrial automation is an automated factory based on the advanced technology of programmable computer-aided manufacturing (CAM). Eight CAM functions are distinguished: product design, part fabrication (including metal cutting by numerically controlled machines), part storage and transportion, logistics, materials handling, assembly, inspection, and process planning. At present only the first four functions may be found in a factory employing flexible-manufacturing system (FMS). The other four are still objects of investigation in research centers worldwide.

Among these latter four research topics, programmable assembly is the most challenging for two reasons. First, programmable assembly is important; it may replace manual assembly, which constitutes the largest portion (approximately 22 percent) of the total labor cost for all durable goods [Nevins et al. (1976)]. Second, programmable assembly is complex; it includes materials handling and in-process inspection as well as programmable part presentation, trajectory planning, collision avoidance, arm control guided by multisensory feedback, part mating, and multimanipulator cooperation.

In this paper we describe recent research on programmable assembly conducted by the Robotics Department of SRI International. The paper covers two major topics:

* The characteristics of an assembly system we are developing, including hierarchical organization, modularity, distributed processing, and robustness.

* Implementation of an assembly station with the above characteristics and its use in demonstration of a programmable-assembly task.

II ASSEMBLY SYSTEM CHARACTERISTICS

The characteristics of the assembly system we are developing are as follows:

* _Hierarchical organization_--to simplify system software development by partitioning the overall control problem into manageable subsets that can be dealt with separately and simultaneously [Albus (1981)].

* _Modularity_--for rapid reconfigurability of system components, and easier modeling of those components because their interconnections and capabilities are well defined.

* _Distributed processing_--to take advantage of the opportunities for parallel computation by means of multiple processors at the points of sensing, control, and command.

* _Robustness_--to minimize the need for manual intervention in recovering from various fault conditions.

Each of these characteristics, as well as some instances of their implementation, are discussed below.

A. Hierarchical Organization

An assembly system computer will control the activities of a number of assembly stations and a transfer mechanism carrying subassemblies between them. An initial design criterion is that every station must be available for experimentation, training, calibration, setup, and debugging. For such purposes, each station will be operated independently of the rest of the system. In contrast thereto, when the entire system performs a task, some coordination of the individual stations and the transfer mechanism will be necessary. Station-level integration will be accomplished by the system computer.

We next divide each assembly station into its major functional components, called modules--e.g., manipulators, vision modules, part presenters, and support tables. A station could be composed of these modules in different ways and configurations, thus providing flexibility for batch assembly of a variety of products. Each module is encapsulated to make it self-contained, controllable by its own computer, and easily configured for coordinated operation with other modules.

A module consists of devices, each of which may be controlled by a device computer or processor. For instance, a manipulation module may consist of a robot arm with its control microcomputer,an end-effector with its microprocessor, and sensors with their microprocessors. If the control of a device is very simple, assigning a special computer to control that device may not be justified; hence, the device will be controlled by the module computer.

The outlined four-level hierarchy of system--station--module-device computers is shown in Figure 1. This computer system should be able to support a fairly large-scale assembly operation (e.g., a factory with as many as 50 modules working simultaneously).

B. Modularity

1. General

A module will

* Perform a generic operation (e.g., manipulation, image acquisition and processing, or part feeding) and control any devices used in that operation.

* Use its computer to provide an external interface to these generic operations at a high level.

* Use auxiliary sensors to verify expected local conditions; e.g., a manipulator will have a sensor to verify that it is "holding an object."

* Execute reflex actions, under predefined conditions, that can be detected by the local sensors.

* Operate independently of other modules.

Figure 2 shows the basic components of a general module with the above-listed attributes. These components are discussed in the subsequent sections.

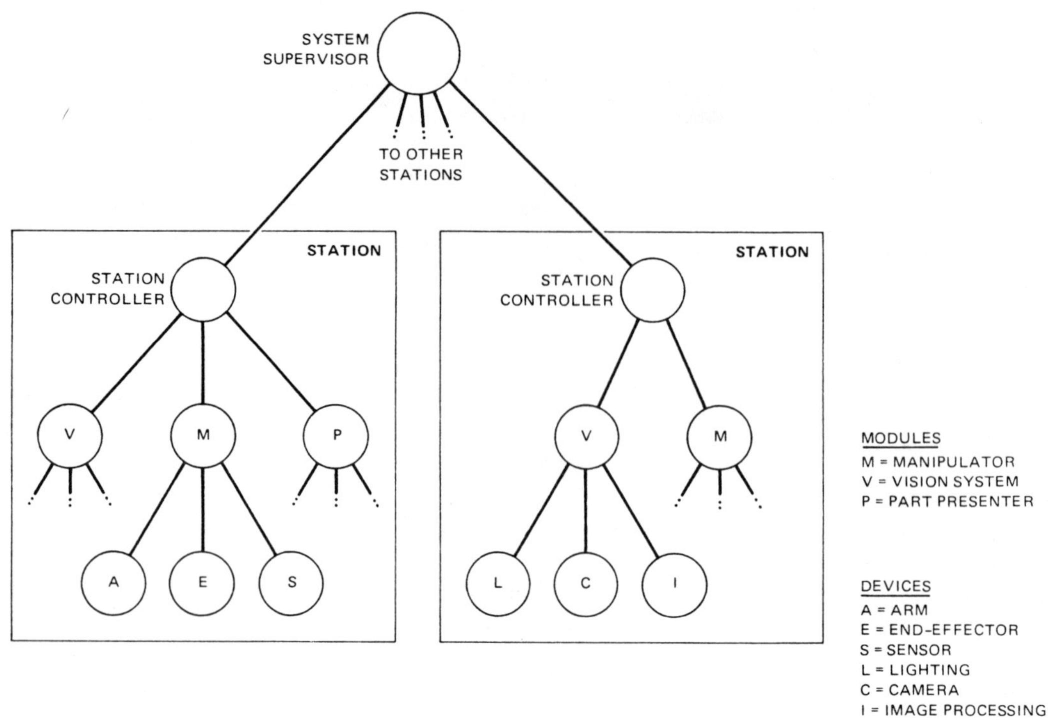

FIGURE 1 HIERARCHY OF CONTROL FOR THE ASSEMBLY SYSTEM

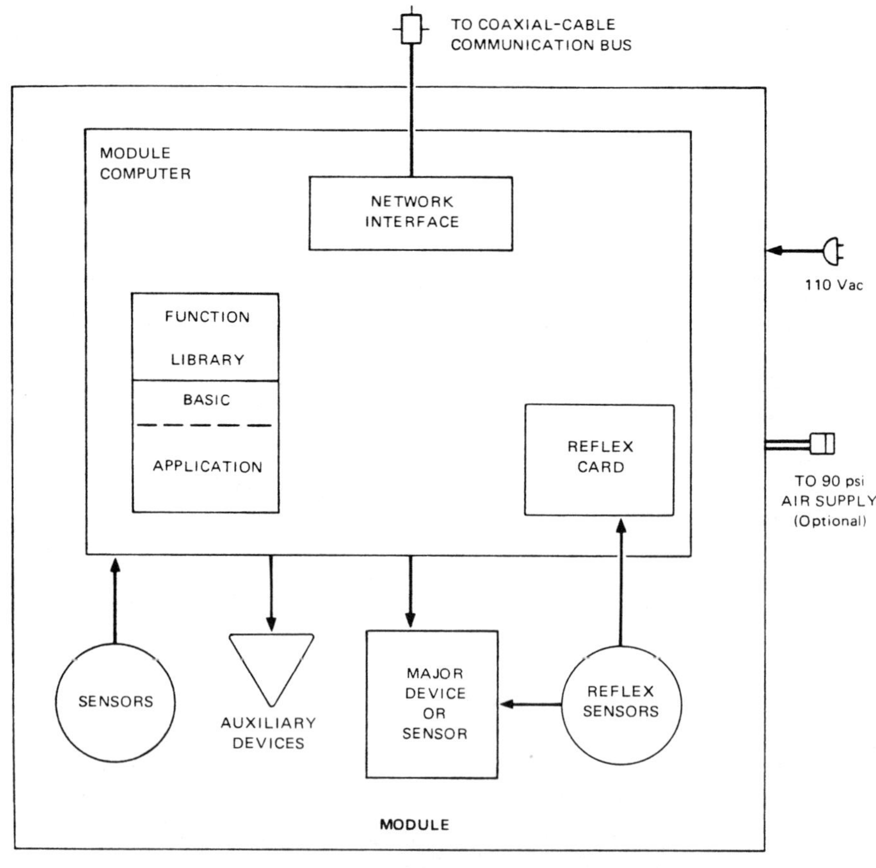

FIGURE 2 BASIC MODULE ORGANIZATION

2. Module Types

The module computer contains a processor, network interface cards, memory, and input/output interface cards for the analog and digital signals from or to the auxiliary sensors and devices of a module. Some specific module types are described below, along with examples of the basic functions they provide. In general, these functions entail intermodular communication of high-level information, rather than large amounts of raw data.

a. Manipulator Module

We have developed two manipulator modules, each consisting of a Unimation PUMA 560 robot and an end-effector. The end-effector of one arm consists of an electrically actuated two-fingered hand, a remote-center-compliance (RCC) device, and a remote-head video camera. The end-effector on our second PUMA 560 consists of a six-axis force/torque sensor mounted on the wrist, and a pneumatic two-fingered hand. The module computer and the PUMA controller for each manipulator are mounted under the arm's supporting stand. Figure 3 shows the hardware control and sensing functions associated with the manipulator module, including planned extensions for proximity and touch sensors on the end-effector. The module computer, in this case, will provide a means for controlling the hand, reading sensors in the hand and wrist, and moving the arm (indirectly, by communicating with the PUMA controller).

A few examples of the manipulator module functions are as follows:

Where(Result)
Returns the location (position and orientation) of the manipulator's end-effector.

MoveTo(Location)
Move to the specified location.

Grasp(GraspLocation)
Approach the given grasp location, open the hand, move to that location, close the fingers, and depart along the approach vector.

StopOnForceZ(Frame, StopForce, Overshoot)
Move along the z-axis of a specified coordinate frame, stopping if a contact force along the z-axis exceeds StopForce of if the manipulator travels farther than Overshoot along the z-axis.

b. Binary Vision Module

As shown in Figure 4, the binary vision module includes an SRI vision module [Gleason and Agin (1979)] with up to four 128 x 128-element solid-state cameras attached, a module computer, and illumination and camera control (to be implemented in the future).

In addition to invoking SRI vision-module functions, the module computer could be used, as needed, to

* Store the application-dependent routines developed around the basic functions of prototype training, recognition, and feature extraction.

FIGURE 3 SCHEMATIC OF THE MANIPULATOR MODULE

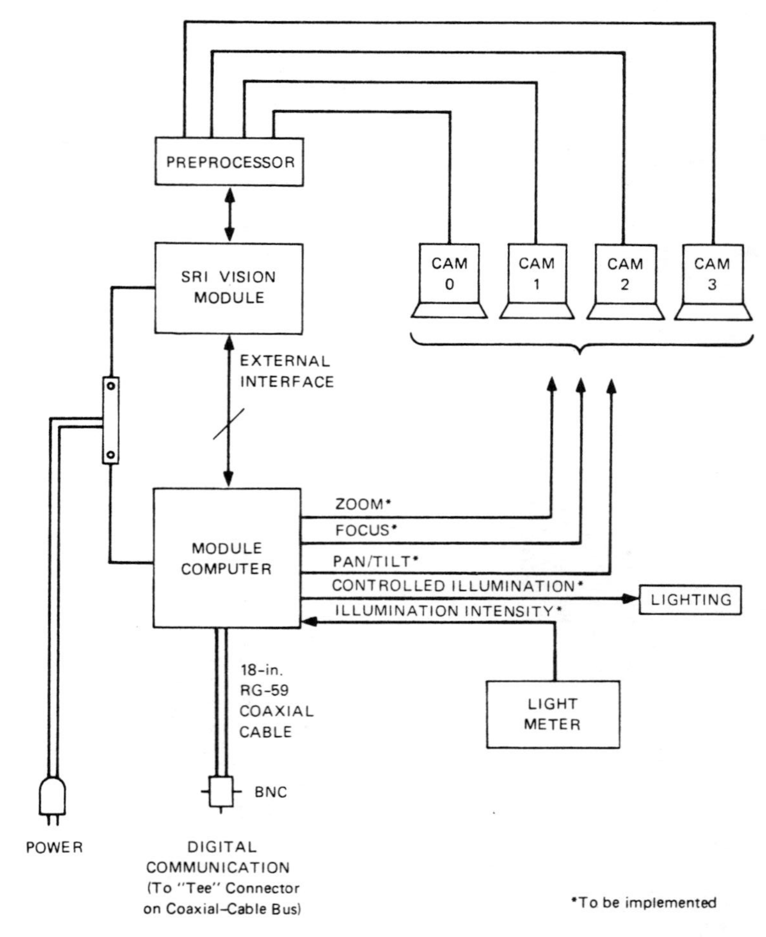

FIGURE 4 SCHEMATIC OF BINARY VISION MODULE

* Control such auxiliary devices as zoom lenses or lens turrets, and implement control of focus, directional lighting (front/back), and lighting intensity.

* Measure illumination uniformity with work surface sensors, and track intensity changes. This measurement would ensure that the lighting conditions during object recognition and object training are identical.

The electronic components of the SRI vision module and the module computer are stored inside a stand like the one on which each manipulator is mounted. The top of the stand forms a work surface and can hold the aforementioned intensity meters and/or a light table for backlighting operations. Cameras are attached to long cables and can be mounted anywhere within the assembly station. One camera is mounted on the hand of one of the manipulators.

Some examples of binary-vision-module functions are as follows:

Picture(BlobCnt)
Take a picture, perform connectivity analysis of the image, and return the number of connected regions (blobs), BlobCnt, that meet a certain blob criterion, such as minimum size and roundness.

GetFeature(BlobN,FeatN,Result)
Return the value of a blob feature (indicated by the index FeatN) of a selected blob (indexed according to decreasing size by BlobN). Features are, for example, blob area, perimeter length, and moments.

Recognize(BlobN,BestDist,NxtDist,ProtoName)
Compare a given blob, BlobN, with a set of prototypes and return the smallest distance (in feature space), BestDist, from the blob to any prototype, the second smallest distance, NxtDist, and the name of the prototype, ProtoName, which is the best match.

Find(PartType,Result)
Take a picture and look for a prototypical blob, PartType, by comparing each blob in the picture with a pretrained set of prototypes. If a good match is found, use information stored with the prototype to determine a unique set of 3-D axes in the part on the basis of its 2-D features, and return the part location.

c. Limited-Sequence Manipulator Module

We have a limited-sequence manipulator module that consists of an Auto-Place Model 50 manipulator mounted on a servo-controlled rotary table.

A limited-sequence manipulator module may be used effectively to transfer parts to and from a work surface or an inspection surface. The limited-sequence arm is less costly than servoed manipulators, is usually faster (because it operates between fixed mechanical stops), and is normally rugged enough to carry a greater payload. The limited-sequence arm, however, can operate only on parts in a fixed location (position and orientation) relative to the arm. Two examples of functions for the limited-sequence manipulator module are as follows:

AutoCmnd(RelayStates)
Set up the AutoPlace relay(s) to the requested state(s). This command actuates the pneumatics, but not the rotary table.

APTMove(Theta)
Turn the rotary table holding the Auto-Place manipulator to absolute position given by theta radians.

d. Other Modules

We have been developing or planning other modules, such as the following:

(1) X-Y-Theta Table--An x-y table whose movable surface can also be rotated about the z-axis. The table is equipped with a translucent top and a row of fluorescent lamps underneath, so we can backlight objects resting on the surface.

(2) Part Presenters--Programmable part presenters previously developed at SRI will be incorporated into the assembly system and extensions of their capabilities explored. One example is the SRI "Eye Bowl"--a standard bowl feeder that utilizes vision rather than mechanical blades for part sorting and feeding [Nitzan et al. (1982)].

(3) Programmable Jig--We wish to explore the concept of a multipurpose, computer-controlled jig that is capable of accepting and rigidly holding parts of many different shapes, which it would then present, upon request, to a manipulator or vision system in a specified orientation. This might take the form of a rugged hand and three-degree-of-freedom wrist.

C. Distributed Processing

A local-area network is a natural organization of a system in which processing is distributed among numerous computers, which are often separated from one another by, say, one meter or more. Hardware supporting various network topologies is available commercially. We use a system in which a coaxial cable forms a communication <u>bus</u> connecting all the communicating computers. The bus-network organization

* Allows direct communication between any two computers connected to the coaxial-cable bus.

* Promotes modularity of system components by requiring only a standard network interface for systemwide communication.

* Facilitates reconfigurability of components by permitting them to be connected to the network at any point on the coaxial-cable bus.

* Permits the sharing of expensive system resources, such as printers, graphics devices, and file-storage units.

Each computer connected to the communication bus contains a network interface with a unique name (a number) assigned to it. Names are used to identify both the source and the destination of a message. When a message is transmitted on the bus, every network interface compares its name with the message destination and receives the message only if there is a match. A special type of broadcast message can be addressed so that all the computers (except the sender) will receive it. This type of message is useful when a module needs help from the "system", but does not know the name of the unit that can furnish such help.

A communication software package has been written to provide flexible communication capabilities through the network interface. The selected protocol and the characteristics of the network interface are described in detail in [Smith (1982)]. Briefly, the package supports a "random-access" protocol whereby any computer may send a message to any other computer or computers if the communication bus is idle. The message is usually one of the following:

(1) A command to a module to perform one of its functions with the parameters given in the message. The command may be to supply information, request information, or to perform a specified activity.

(2) A reply to a command, containing any results or requested information. The reply also serves to confirm completion of a commanded activity, which may otherwise not have returned any results.

The communication package performs the following functions:

* Message creation, retrieval, buffering, and deletion.

* Message receipt, acknowledgment, and transmission.

* Automatic retransmission of unacknowledged messages.

* Detection of any special broadcast messages for later action.

A module will never have more than one buffered message to transmit at any time. However, it can receive and buffer numerous messages and either attend the oldest one or search in its buffer for an expected message of a certain type, from a certain source, or both.

D. Robustness

We have recently implemented three levels of processing in each module computer:

* Reflex level to detect hardware or software faults and set the module hardware devices to predetermined states.

* Bootstrap level to set the module-computer program at a predetermined state in response to reflex activation, and to notify the rest of the system about this event.

* Program level to implement the main functions of the module. A sensor-monitor routine will be implemented at this level in the future to detect conditions beyond the capabilities of the reflex level.

These processing levels are described below.

1. Reflex and Bootstrap Levels

At the lowest processing level the modules (particularly those incorporating manipulators) need self-protective mechanisms that act as reflexes--hardware responses to a set of external or internal fault conditions. Such conditions include loss of operating power, loss of program control, and human intrusion into the assembly area. If necessary, special sensors will be assigned to detect these conditions. Once enabled, a reflex device will watch for the fault condition it guards against and be triggered if that condition arises. We have designed a reflex card to implement this function and fulfill the following responsibilities:

(1) To provide a mechanism that forces the module's devices or sensors into "safe" default states when a reflex is triggered.

(2) To protect the module from power loss.

(3) To protect the module from loss of program control.

(4) To provide sensor-triggered hardware reflexes.

(5) To provide programmable-condition reflexes.

In addition, activation of a reflex may optionally force the program control in the module computer to transfer to a simple bootstrap program resident in a nonvolatile memory on the reflex card. The module subsequently executes a simple program at the bootstrap level that will

* Initialize the network interface for communication.

* Use the network to broadcast a message notifying other module computers of this module's current state.

* Provide capabilities for loading the program of this module through the network interface when commanded by another module computer.

* Supply tools for remote diagnosis of this module through the network interface.

Figure 5 depicts the relationship between the reflex and bootstrap control levels. More information about reflexes and the bootstrap functions may be found in [Nitzan et al. (1982)]. The above-listed responsibilities at the reflex level are described below.

a. Default State Conditioning

Each device or sensor connected to the module computer should be set to a safe default state whenever a reflex is triggered. It is a common practice to provide a computer bus line that carries an INITIALIZE signal to all the computer interfaces. The reflex card triggers generation of this signal when a reflex condition occurs. The INITIALIZE signal, for example, may halt any moving device if the signal is applied directly as an override or a shutdown signal to that device.

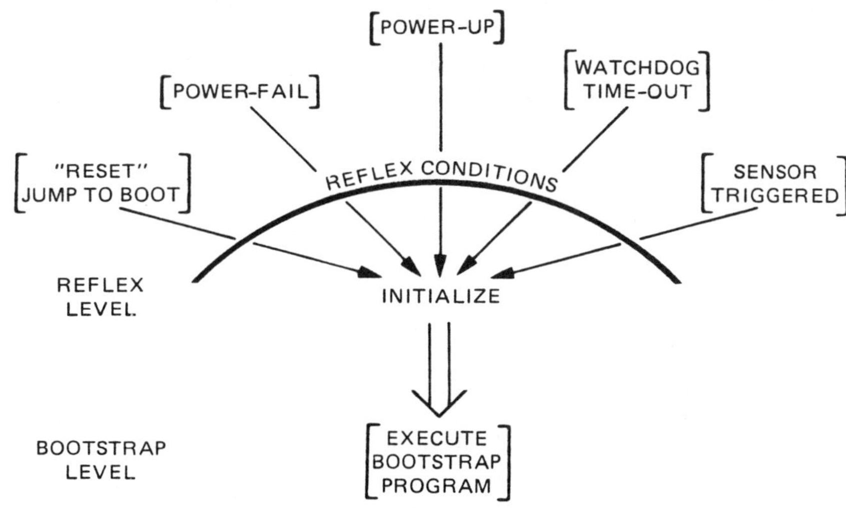

FIGURE 5 REFLEX LEVEL

b. Power Loss Protection

Like many computers, the module computer has the capability of detecting imminent power loss through a sensing circuit in its power supply. A signal indicating this event is supplied to the reflex card from the computer power supply. The reflex card triggers an INITIALIZE signal on the bus, thus initiating a command to halt moving devices, for instance, prior to the power loss in the module computer. In a "power-up" sequence, the INITIALIZE signal is generated again, and program control is transferred to the bootstrap code.

c. Protection From Loss of Program Control

The module computer's program may not always be running correctly; e.g., it may become deadlocked, halted, or contain errors. For these contingencies, an independent timer, called a "watchdog timer," is included on the reflex card. The watchdog timer must be reset periodically by a properly executing program in the module computer. If it is not reset, the watchdog timer reaches a "time-out" state. This state is a reflex condition that, like any other, will cause generation of an INITIALIZE signal via the reflex card and transfer the module computer's program control to the bootstrap code.

d. Provision for Sensor-Triggered Reflexes

Certain events detected by sensors connected to the module computer may require an emergency response from the module. The reflex card provides such a response capability by accepting binary signals from these sensors. The binary signals could, for example, indicate the state of contact/noncontact or proximity/nonproximity of objects with respect to a manipulator's end-effector. One reflexive response might be to halt a moving manipulator anytime an intruder is detected in the work space. As another example, a proximity sensor on a manipulator's hand may be used to trigger a "stop-arm" reflex to prevent collision with unexpected obstacles; however, sometimes this reflex must be disabled to permit the hand to reach a target object. The computer program may disable any reflex circuit on the reflex card by transmitting a special code word. This encryption reduces the possibility that the reflexes may be disabled accidentally. Reflex devices will always be placed in the "reflex disabled" state following the INITIALIZE signal, so that the module will react to the fault condition once rather than repeatedly. The triggering condition should be determined by reading the reflex status and then be removed before the reflex is enabled again. When the module program begins, it enables those reflex sensors that should be active at that time.

e. Provision for Programmable-Condition Reflexes

The reflexes just described are directly triggered by simple binary sensor signals. Certain more complex conditions detected by the module-computer program may also warrant initiation of an orderly shutdown. Loss of communication with other devices is one instance of a potential shutdown condition; another example is detection of anomalous conditions computed from local sensor values and internal program states. A RESET instruction should be supplied by the module computer to activate the INITIALIZE bus signal from software. Utilization of this provision will be based on the estimated urgency of the situation. After it carries out the RESET command, the program should transfer execution to the bootstrap program.

2. Sensor State Monitoring at Program level

a. Program level

The reflex and bootstrap levels are concerned with initializing the module, getting it loaded and running, and providing a uniform method for detecting and reacting to local anomalous conditions. The main functions of the module are performed at the program level, i.e., controlling the module's main device or sensor. The program level has the following responsibilities:

* Reset the watchdog timer periodically.

* Provide full, flexible intermodule communication via the network interface.

* Implement the generic functions defined for the module type to control the module's devices or sensors.

* Provide periodic sensor-state monitoring.

Examples of module functions and a description of module inter-communication capabilities have been given in Sections II-B-2 and II-C, respectively.

b. Sensor State Monitoring

A background process for monitoring the numerous sensors associated with a module is under development; it is presented as a principal component of the program level of each module. The monitoring process is intended to read the values of local sensors periodically and to compare these values with their expected range, which is given in a table. When an actual sensor value is found outside the expected range, a programmable-condition reflex may be executed or, less drastically, a message reporting the anomaly may be broadcast. Range entries in the sensor table will be made in one of two ways:

* Explicitly, through a MONITOR command given to the module, indicating which sensor to use and the expected range values of that sensor in the next interval.

* Implicitly, by execution of a generic function that imposes known constraints upon a sensor's values.

A MONITOR command can be used, for example, to instruct a manipulator module to monitor the forces and torques acting upon its end-effector and to assure that they remain within a specified range for a given application. Monitoring will then proceed independently until the MONITOR command is withdrawn.

In some cases the effect of a generic function on a set of sensors is known a priori and sensor monitoring can be initiated implicitly. For example, let us consider two functions for a manipulator hand: GRASP and RELEASE. The value of a binary touch sensor on the hand's fingers after execution of the GRASP command is expected to be ON. The GRASP routine itself will verify this condition and enter the tolerance range for the sensor value (unnecessary in this instance) in the sensor range table. Execution of the RELEASE routine will generate an entry in the table corresponding to the OFF value for the contact sensors (if, indeed, they were off). The sensor value corresponding to the ON or OFF state can be checked repeatedly by a sensor monitor routine. Thus, an external application program directing the manipulator module need not verify continuously that an object in the hand is still there, but will instead be notified immediately if the object is dropped.

Other operations similarly impose anticipated constraints on associated sensors. Should a discrepancy occur between the sensor table range and the actual value of a sensor, the safest approach will be to execute a "programmable-condition" shutdown reflex and enter the bootstrap level. Since the reflex does not destroy the resident program, state information can be retrieved from the module by means of the diagnostic routines available in the module's bootstrap program. Such information may be used in future work to determine why a module failed and to direct the recovery of that module or the entire station accordingly.

III ASSEMBLY STATION

A. Station Configuration

Figure 6 shows the current configuration of our assembly station. It consists of a station controller, a binary vision module with three video cameras, and two manipulator modules whose black and white arms are called Arm 1 and Arm 2, respectively. The end-effector of Arm 1 consists of a plastic remote-center-compliance device, the front end of a video camera (the back end is mounted on the manipulator arm), and a two-fingered hand. The end-effector of Arm 2 consists of a six-axis wrist force/torque sensor and a two-fingered hand. Not included in this configuration are the limited-sequence-manipulator module and the other modules described in Section II-B-2; these modules may be incorporated into the assembly station as needed during the performance of other assembly tasks.

This station configuration includes two support surfaces. The first, located between the manipulators, is the assembly area (the binary vision module, excluding its cameras, is mounted beneath it). The second surface, located near Arm 1, supports

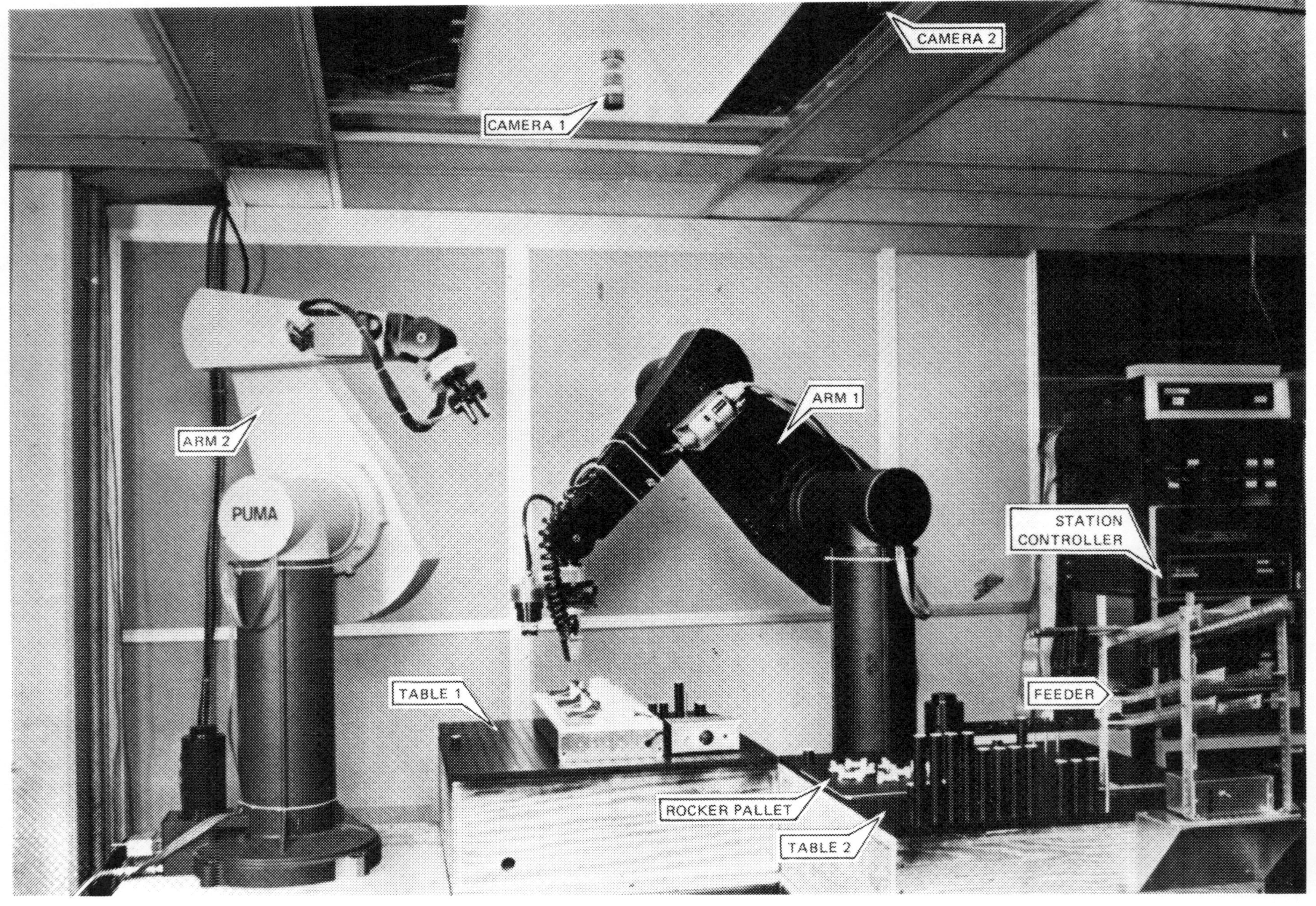

FIGURE 6 ASSEMBLY STATION

general-purpose part feeders and pallets within reach of that arm. Both surfaces have fixed cameras mounted above them, and both have a grid of holes, which are used to provide mechanical support components and to aid in calibrating the coordinate frames of the manipulators, the support surfaces, and the cameras relative to one another.

The station controller, a small computer, is used to control the sequential and parallel operations of the modules by means of commands on the communication network. The station controller also controls a printer, a speech output device, and two disks. Messages from the modules to the operator are printed or spoken; files stored on the disks are sent to the module computers upon request. In addition, the station controller has menus of module commands that can be executed interactively by the user, as well as a cross-network debugger that allows him to examine, alter, and set breakpoints in the programs of the module computers.

B. Calibration of Coordinate Frames

Communication of object locations among modules is facilitated if there is a common coordinate system, or reference frame, in which to define these locations. For instance, a camera may be used to determine the location of a part; that information, expressed in terms of the coordinates of the reference frame, will enable any manipulator within reach to access the part. Each manipulator module will need to know only its relationship with the reference frame rather than all its pairwise relationships with other coordinate frames in the assembly station. We, therefore, calibrate all the station modules to a station reference frame.

In Figure 7 we define the base coordinate frames of two manipulators, R1 for Arm 1 and R2 for Arm 2, the coordinate frames of their respective end-effectors, E1 and E2, and two coordinate frames, T1 and T2, on Table 1 and Table 2, respectively. Camera 1 overlooks Table 1 and Camera 2 overlooks Table 2. Frame T1 is chosen to be the reference frame and the other frames must be related to it, as indicated by the dashed arrows, by means of transforms--4 x 4 homogeneous coordinate-transformation matrices [Paul (1981)].

We use the following general notation. The position vector (x,y,z,1) of a point P in an arbitrary homogeneous coordinate frame F is denoted by P(F). The same point may be described by P(F1) or P(F2), where F1 and F2 are two different frames. We denote the transform from Frame F1 to Frame F2 by [F1/F2], where P(F1) = [F1/F2] * P(F2). Using this notation, note that P(F2) = [F2/F1] * P(F1), where [F2/F1] is the inverse of [F1/F2]. We depict [F1/F2] by an arrow pointing from the origin of Frame F1 to that of Frame F2 (see the examples in Figure 7); note that the inverse of [F1/F2] would be depicted by an arrow in the opposite direction.

Using the above notation, we obtain P(R1) = [R1/T1] * P(T1). The elements of Transform [R1/T1] can be determined by mounting a pencil-like tool on the wrist of Robot 1 and leading that robot so that its tool tip touches three points in Frame T1--its origin, a point on its +x axis, and a point on its +y axis. Reading the robot positions in Frame R1, the coordinates corresponding to the origin yield the translation vector of Transform [R1/T1], while those of the other two points yield its rotation matrix. A similar sequence using Arm 2 and the same calibration points in T1 may be performed to derive [R2/T1]. At this stage, the position of a part given in Frame T1 can be transformed into the frames of both arms.

Next, points in the 2-D image plane of each camera are related to a 3-D space. Consider first Camera 1, which is calibrated directly to the Table 1 grid it views. Pegs of varying lengths are inserted into the grid holes, which are easily identifiable integer coordinates in the grid framework. The white tops of the pegs can be seen as bright spots by the camera. A set of known peg-top (x,y,z) positions in Frame T1 and the corresponding set of (u,v) image coordinates in a 2-D Frame C1 are utilized, using a least-mean-squares fitting method, to produce the camera calibration matrix [Bolles et al. (1981)]. The camera calibration matrix can be used to compute the image coordinates (u,v) of a given point (x,y,z) in the calibration frame. Generally, however, we do the reverse--the x and y coordinates of a point in the calibration frame are obtained as a function of the camera calibration matrix, the image point (u,v), and the known z-coordinate of that point. For Camera 1, the calibration frame is equivalent to the reference frame (Frame T1).

Calibration of Camera 2 is similar to that of Camera 1, except that the relation between the calibration frame (Frame T2) and the reference frame (Frame T1) must be derived. The relation [T2/T1] is determined by making the tool tip of Arm 1 touch points on the origin, the +x axis, and the +y axis of Frame T2 to derive [R1/T2], and computing [T2/T1] = [T2/R1] * [R1/T1].

The hand-held camera is calibrated similarly to the other cameras, except that only one peg is used and the camera is moved to view it from different locations. The camera is rigidly attached to the tool-mounting flange on the wrist of Arm 1; hence, points seen by the camera will be initially referenced to a frame, FL, fixed in that flange. Assume that the peg is placed in a grid hole on Table 1 and that its position in Frame T1 is known. After each time the arm is moved, we compute the peg-top position in Frame FL, using the relation Peg(FL) = [FL/R1] * [R1/T1] * Peg(T1). From the resulting list of (x,y,z) positions in Frame FL and the corresponding (u,v) image coordinates, the camera calibration matrix can be computed as for Camera 1 or Camera 2. Relations between the hand-held camera and other frames are illustrated in the next section.

C. Example of Intermodular Communication

Most of the communication that occurs in the context of our assembly station is between the station controller and the modules it commands or queries. However, one important example of direct communication between the modules themselves involves the use of a camera mounted on a manipulator's hand.

A stationary camera can supply information about the location of an object it views in a fixed coordinate frame, such as Camera 1 over Table 1 in Figure 7, because the relationship between the camera and the table is constant. A mobile camera, on the other hand, can supply information about the position of a part relative only to the camera's viewing location. If the part location is desired with respect to a fixed frame, such as T1, then the location of the viewing camera must be known. Figure 8 shows schematically the transforms between frames associated with finding a part, PARTX1, in a pallet on Table 1 by means of a camera attached to a flange, FL, that holds the end-effector of Arm 1. The following sequence of commands illustrates a direct communication between a binary vision module using a hand-held camera and the manipulator module.

MoveTo (ARM1, AbovePallet)--The station controller commands Arm 1 to move to Location AbovePallet above a pallet with a desired part. Since AbovePallet is described with respect to T1, Arm 1 converts AbovePallet to a location in its own frame, using the relation [R1/AbovePallet] = [R1/T1] * [T1/AbovePallet], and moves to that location. After the move, Frames E1 and AbovePallet will coincide.

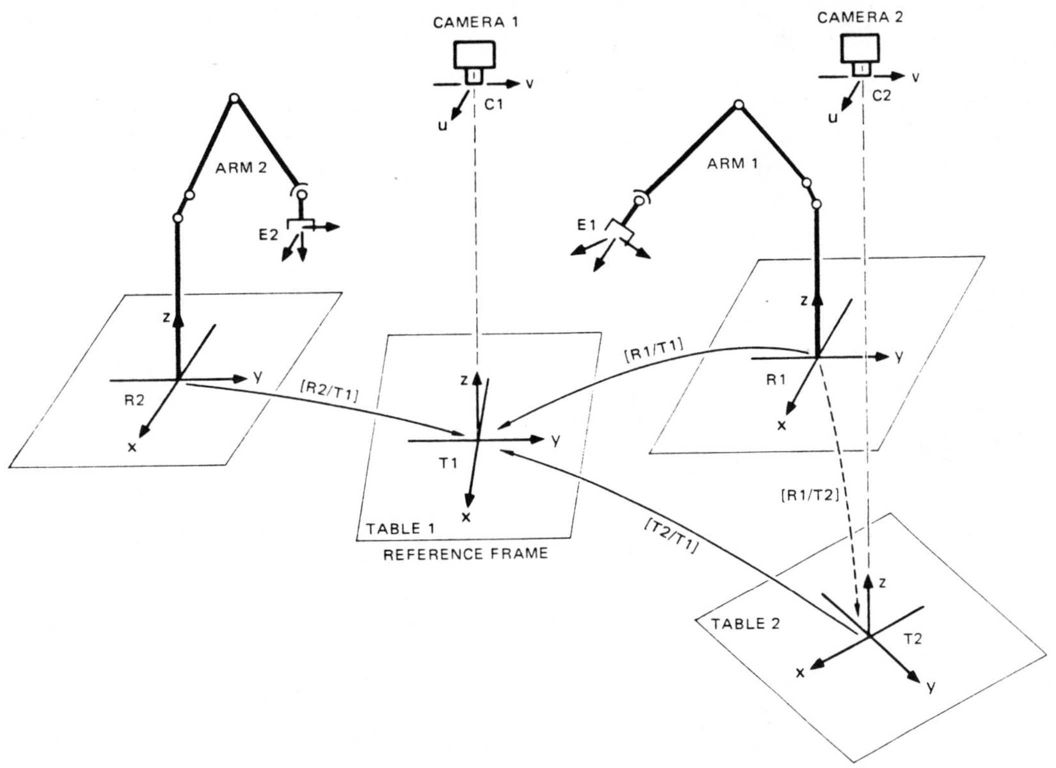

FIGURE 7 STATION COORDINATE FRAMES AND TRANSFORMS

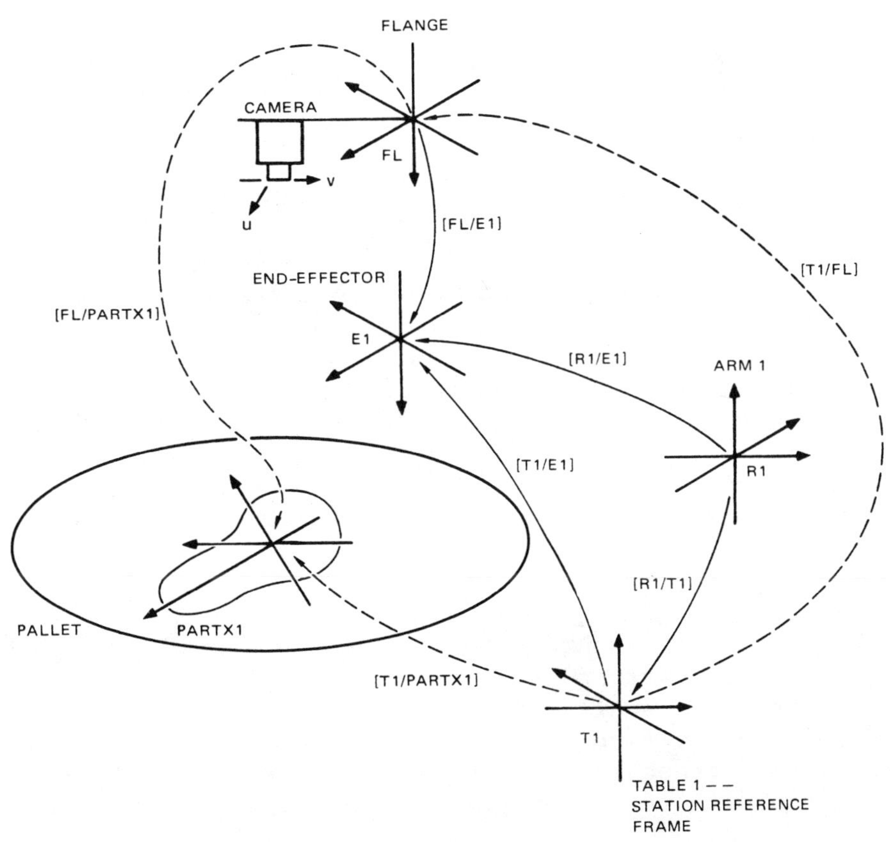

FIGURE 8 USING THE HAND-HELD CAMERA

SetCamera(VM1, HandCamera)--The station controller commands Vision Module VM1 to select the hand-held camera for input.

Find(VM1,PARTX, Result)--The station controller commands VM1 to take a picture with the selected camera, compute the location of a part of Type PARTX in reference coordinates, and return this result.

The task of VM1 is thus to compute Transform [T1/PARTX1]. Since [T1/PARTX1] = [T1/FL] * [FL/PARTX1], VM1 will first ask and obtain from Arm 1 the value of [T1/FL]; it will then find PARTX1 and compute its transform, [T1/PARTX1], as follows:

WhereFL(ARM1, Result)--The vision module asks Arm 1 for the location in T1 of its tool-mounting flange, FL, to which the camera is attached. The camera has been previously calibrated so that points it views will be referenced to a coordinate system fixed to the flange.

Reply(VM1, Result)--Reading its current value of [R1/E1], Arm 1 computes [T1/FL], using the expression [T1/FL] = [T1/R1] * [R1/E1] * [E1/FL], and gives its value to the vision module. The latter then takes a picture with the hand-held camera, recognizes an instance (PARTX1) of PARTX, if present, and determines [FL/PARTX1] on the basis of 2-D image features and additional prototype information.

Reply(StationController, Result)--The vision module tells the station controller where PARTX1 is in T1, using the relation [T1/PARTX1] = [T1/FL] * [FL/PARTX1].

D. Assembly Demonstration

A few demonstrations of the assembly station have been performed, the most recent one involving the assembly of part of a DEC LA-34 printer carriage [Nitzan et al. (1983)]. That assembly contains four part types (see Figure 9):

(1) A friction shaft with four plastic rocker arms, each containing a hole.

(2) Four plastic rockers, each of which snaps into the hole of one of the rocker arms.

(3) Two small rollers, each of which snaps into the front end of two adjacent rockers.

(4) Two large rollers, each of which snaps into the back end of two adjacent rockers.

All the parts are acquired by Arm 1 from their locations on Table 2 (see Figure 6). The shaft and the two rollers slide into fixed pickup locations on three feeders, each consisting of two inclined rails. The rockers are presented on a sticky pallet under Camera 2 in one stable state (upright), but at arbitrary locations.

Table 1 supports a simple fixture with "V" notches for aligning cylindrical parts, such as the friction shaft and the rollers. The fixture is attached to a light table that backlights it and a small surrounding area. Camera 1 is used to locate the fixture visually from above by recognizing and locating a reference target on it; the locations of a few fixture components, such as the holding support for the friction shaft, are computed according to their relative locations with respect to the target.

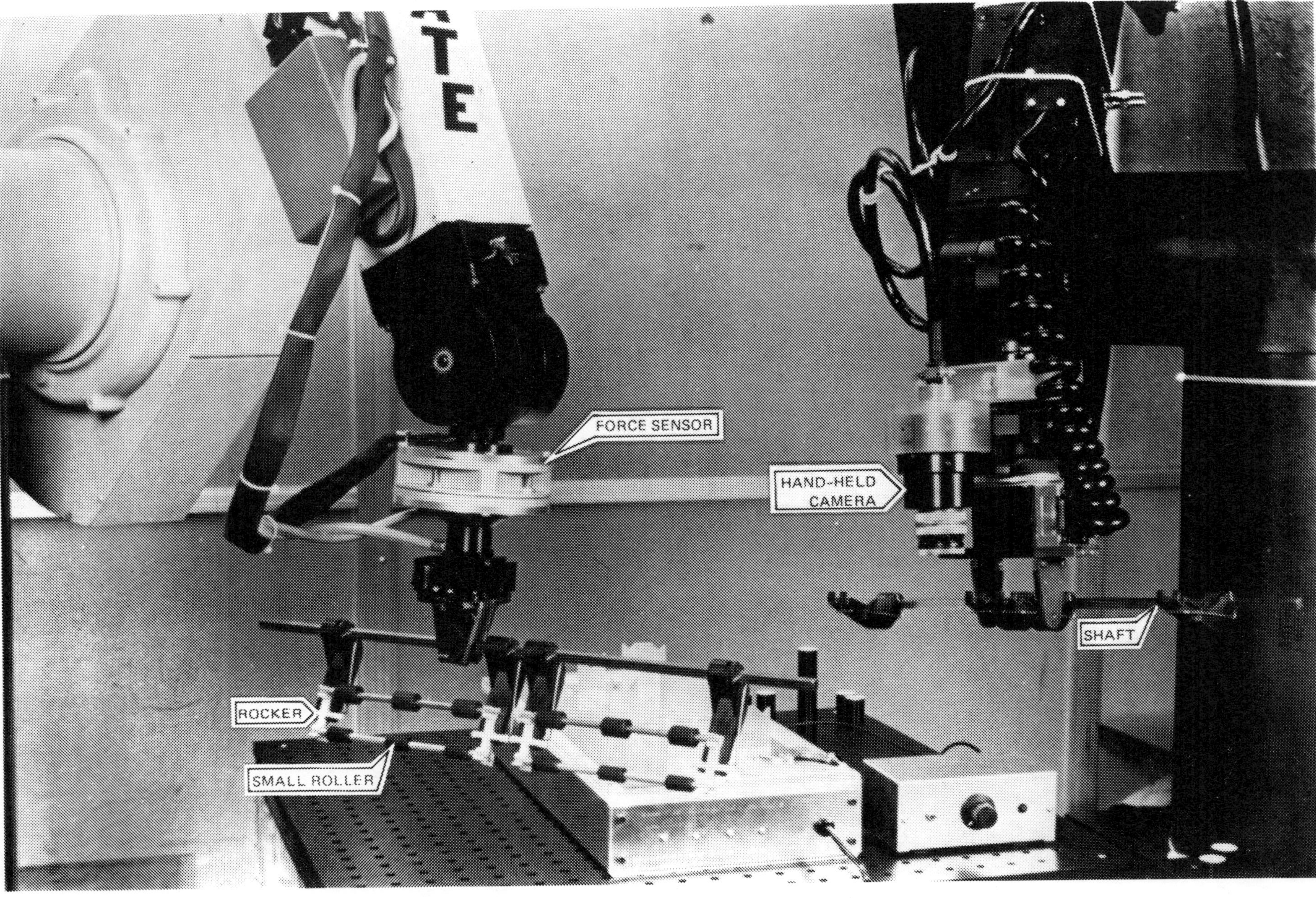

FIGURE 9 SIMULTANEOUS PERFORMANCE OF INITIAL AND FINAL ASSEMBLY STEPS

Meanwhile, Arm 1 acquires a friction shaft from its feeder and places the shaft on its support (after the latter has been located). After Camera 1 locates the fixture, Camera 2 is used to locate a rocker on the pallet. Subsequently Arm 1 acquires the rocker and places it beside an empty rocker arm whose hole has been located by the hand-held camera on the arm. That camera then determines the precise location of the rocker.

Part mating with force feedback is performed next by Arm 2 because Arm 1 has no wrist force sensor. Given the locations of each rocker and its destination hole, Arm 2 grasps the rocker and places it in the hole while making small corrective movements based on information from its force sensor. That information is used to determine when the rocker should be set into place, as well as to protect the assembly from excessive forces that Arm 2 might accidentally exert. An increase in the applied force followed by a sudden drop in that force provides confirmation that the two parts have been snapped together successfully. While the rocker is being inserted by Arm 2, Arm 1 acquires one of the four rollers needed and places it behind the shaft support on the fixture.

The above cycle is repeated until all four rockers have been inserted in their respective rocker arms and all four rollers have been placed in their single fixture. Subsequently, Arm 1 turns the friction shaft subassembly over onto the fixtured rollers and holds the shaft in place, as shown in Figure 10. Arm 2 then presses every rocker down until it is snapped onto the corresponding front and back rollers; force sensing is used again to verify this operation. Finally, Arm 1 releases the friction shaft and Arm 2 removes the completed assembly from the assembly area.

E. Conclusions

The assembly station and assembly demonstration described in this paper have exhibited the following capabilities:

* No special-purpose fixtures--some parts are presented to the manipulator on a pallet, others are on a slide. The assembly is assisted by a simple fixture with "V" notches for centering cylindrical workpieces.

* Utilization of medium-level module commands, such as MOVETO, FIND, and GRASP.

* Binary vision to locate and identify parts on a pallet for acquisition. Three cameras are used during the assembly--two fixed in the ceiling and one attached to the end-effector of one of the manipulators.

* Force feedback to actively control compliance and insertion forces in snap-together, part-mating operations.

* A stationwide calibration scheme for two manipulators and multiple cameras, enabling part locations to be given with respect to a common reference frame.

* Concurrent operation of two manipulators in a central assembly area, while the binary-vision system recognizes and locates parts elsewhere.

* Synchronization of manipulator motions to avoid collisions in the central assembly area.

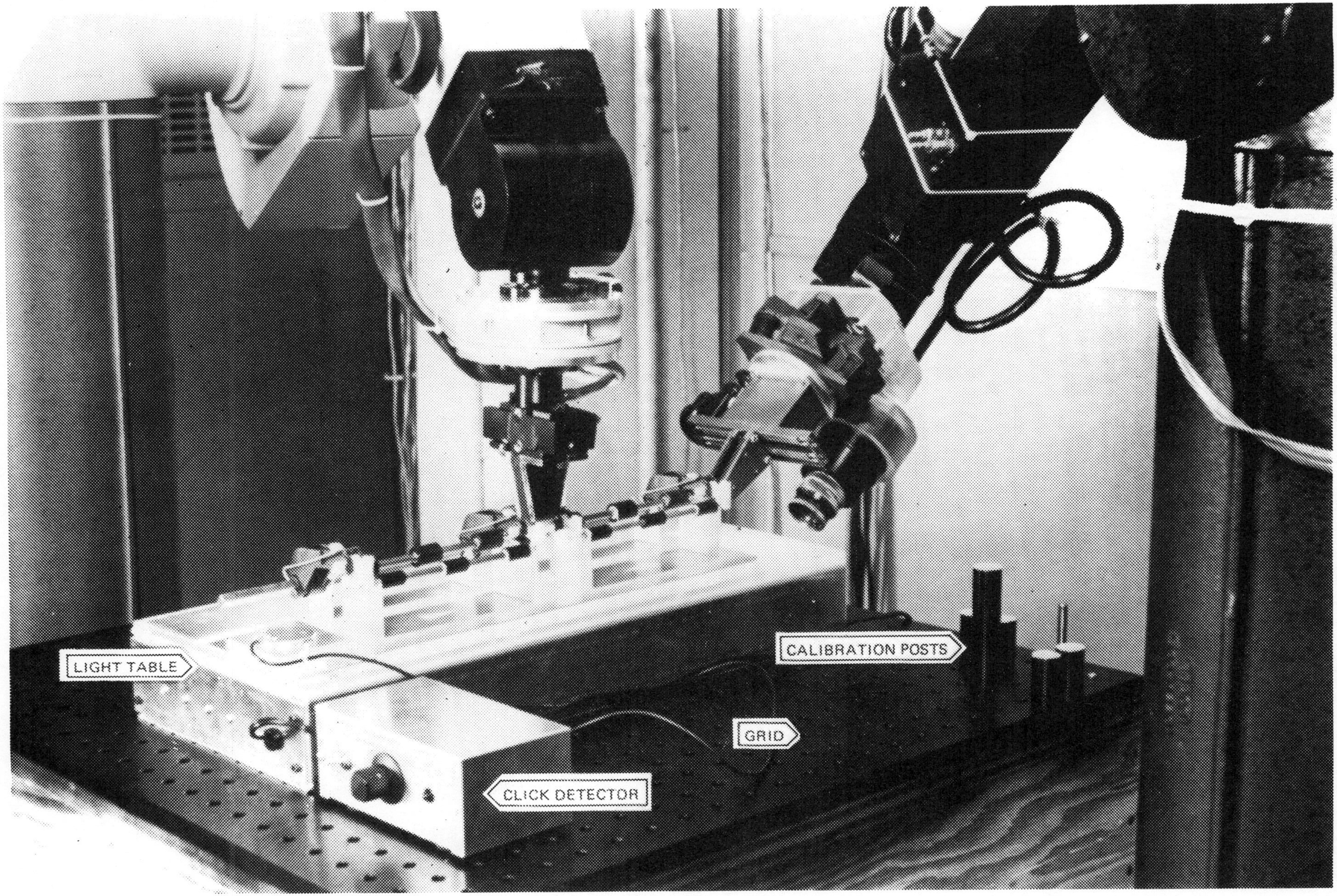

FIGURE 10 PRESSING ROCKERS ONTO ROLLERS

* Simple sensors to verify operation, including the use of finger separation to ascertain the presence of a part in the hand, and a click detector (employing a microphone mechanically coupled to the assembly area) to verify that semirigid parts have snapped together successfully, or to detect a drop of a part.

These capabilities are important for the development of not only programmable assembly but also of programmable automation in general.

REFERENCES

1. J. J. S. Albus, Brains, Behavior, and Robotics (BYTE Books, Subsidiary of McGraw Hill, Peterborough, New Hampshire, 1981).

2. R. C. Bolles, J. H. Kremers, and R. A. Cain, "A Simple Sensor to Gather Three-Dimensional Data," A.I. Center Technical Note 249, SRI International, Menlo Park, California (July 1981).

3. N. H. Cook, "Computer-Managed Part Manufacture," Scientific American, Vol. 232, pp. 86-93 (February 1975).

4. G. J. Gleason and G. J. Agin, "A Modular Vision System for Sensor-Controlled Manipulation and Inspection," Proc. of 9th Int. Symposium and Exposition on Industrial Robots, Washington, D.C. (March 1979).

5. J. Nevins et al., "Exploratory Research in Industrial Modular Assembly," NSF Grants GI-39432X and ATA74-18173A01, Reports 1-3, Draper Laboratories, Cambridge, Massachusetts (June 1, 1973 to August 31, 1975).

6. D. Nitzan and C. A. Rosen, "Programmable Industrial Automation," IEEE Trans. Computers, Vol. C-25, pp. 1259-1270 (December 1976).

7. D. Nitzan et al., "Machine Intelligence Applied to Industrial Automation," Eleventh Report, NSF Grant DAR80-23130, SRI International, Menlo Park, California (January 1982).

8. D. Nitzan et al., "Machine Intelligence Applied to Industrial Automation," Twelfth Report, NSF Grant DAR80-23130, SRI International, Menlo Park, California (January 1983).

9. R. P. Paul, Robot Manipulators (MIT Press, Cambridge, Massachusetts, 1981).

10. R. C. Smith, "Design of a Modular Programmable Assembly System," Technical Note, Robotics Department, SRI International, Menlo Park, California (December 1982).

CHAPTER 5

COMPLIANCE DEVICES IN ASSEMBLY

Reprinted from *Autofact West Conference Proceedings*, Volume 2, 1980, Copyright SME

Precision Assembly Using Compliance Devices

by Jack D. Lane
Robotic Integrated Systems Engineering, Inc.

Abstract

The Remote Center Compliance (RCC) device is a mechanical device which acts as a multi-axis float to correct for misalignment errors encountered when mating parts during precision assembly operations. The devices utilize reaction forces which are encountered during assembly to cause elastic deformation of its structural members. This deformation results in rotational or translational movement of the device and the part being assembled so as to result in alignment with minimal force contact. Numerous potential applications have been evaluated to determine the capabilities of the device to achieve alignment. Several of these applications will be discussed with regard to the results which were obtained under laboratory conditions. Potential production applications will be reviewed.

Introduction

Compliance is a new concept and term which, until just recently, has not been used or applied for performing assembly operations. In the past, when parts were assembled and difficulty was encountered in attaining alignment, the solution which was provided was to "float" or allow free movement of one or both of the parts. Although this approach does reduce the problems inherent from misalignment, it can be extremely expensive and does not completely reduce the tendency toward part jamming, wedging, and plastic deformation to the components. The use of passive compliance devices to assist parts mating during assembly greatly reduces the problems associated with misalignment. So long as chamfers are provided on one or both of the parts being assembled, engagement or insertion of the parts within these chamfers can be achieved without physical damage to the components and without jamming and wedging. The compliance, which can result between parts, can be achieved through the use of passive compliance devices. These devices, due to the designed principles through which they operate, are called Remote Center Compliance (RCC) devices and are the subject to be discussed and evaluated in this paper.

Assembly Problems

At the present time, assembly operations can be performed either manually, automatically, or integrated in an optimum manner. If manual assembly is employed, an operator can adapt to changing conditions such as those brought about by part variation, mislocation, and product model mix. An operator can compensate for these changing conditions and, as a result, does not require elaborate tools and fixtures to perform the assembly tasks. However, operator error and fatigue can result in quality control problems.

When production volumes are high enough, assembly operations can be performed automatically on special purpose machines. These automatic assembly machines consist of work stations grouped along some type of transfer system for part conveyance. Each station performs one task through the aid of dedicated station equipment, jigs, and fixtures. Part variation, misalignment, and product model mix are not readily adapted to since sensors cannot always be employed efficiently or economically to guide or monitor the assembly process. Therefore, part variations and slight misalignments can result in jamming, incomplete operations, and excessive machine downtime. However, automation can still be justified when the production volumes are high, product life is long, and assembly tasks are simple. For an assembly operation to be performed successfully on a repetitive basis, it is absolutely essential that part variation and location be minimized and consistency in dimensions and location be maximized. To achieve this in a mass production environment requires elaborate and costly tooling, fixtures, and the employment of expensive production controls. Therefore, many assembly operations are performed manually to resolve some of the problems in mating parts with variations or mislocations, but this results in increased assembly costs and lower productivity. The potential application of a mechanical device such as the RCC device would greatly facilitate automatic assembly and could be an aid to an operator performing manual assembly operations.

Description of the Remote Center Compliance Device

The Remote Center Compliance device is an entirely mechanical device which has the designed capabilities to act as a multi-axis float to provide lateral and angular alignment of components being assembled. Due to the inherent designed elasticity of its structural supporting members, these members deform elastically when acted upon by forces encountered during assembly. The response of these structural members to these reaction forces results in a unique compensation to variations in angular or translational misalignment. The device can translate movements to parts to which it is attached to correct for lateral misalignment without generating moments or causing rotation of the parts being assembled. It can also correct for angular misalignment to bring about parallelism between the center line of a part to be assembled and the center line of the hole in which the part is being inserted. The action is accomplished without lateral movement which, if it occurred concurrently, would tend to increase insertion forces.

Design Variations of the Commercially Available Compliance Devices

At the present time, three different companies or organizations are producing compliance devices. All of these operate in principle as described above. However, each is configured differently to respond to the reaction forces which are generated when parts make contact.

Draper Laboratory Device

The Charles Stark Draper Laboratory, Inc., in Cambridge, Massachusetts, were the originators of Remote Center Compliance devices. Through a research program entitled "Exploratory Research in Industrial Modular Assembly" [1, 2, 3, 4], they evolved the concept of compliance in assembly as a result of studying the phenomena of parts and how they are mated in assembly. The design of an RCC device was evolutionary in that several different design configurations were developed. General Motors purchased five RCC devices which were designed to be lightweight and easily disassembled. Figure 1 and Figure 2 show this device which achieves compliance through the three high-strength 17-4 PH stainless steel wires and the side and top aluminum structural members. This device was designated as Model 4-B and is the Draper unit that was used for evaluation purposes.

Astek Engineering, Inc.

The Astek Engineering Company in Watertown, Massachusetts, has developed a compliance device which utilizes elastomeric shear pad elements

(Figure 3). These elements are constructed by bonding alternating layers of metal shims with an elastomeric material. In this manner, tensile, compressive, and shear properties can be accurately controlled by varying shim spacing and elastomeric materials. As a result, specific properties can be supplied to accommodate interference or clearance fits for intended applications. Maximum force requirements need to be known for a given application. If the design force is exceeded during operation, the elements will collapse and compliance capabilities would be lost.

Lord Kinematics

The Lord Kinematics Corporation of Erie, Pennsylvania, also produces a compliance device which utilizes elastomeric mountings within an aluminum or steel structure (Figure 4). The Lord Corporation has had extensive experience with shock mounting and dampening equipment and has applied their knowledge of this technology to the design of a compliance device. Their devices operate in principle in a manner similar to that previously discussed with the Astek device. Basic designs have been developed which would be suitable for most assembly operations. However, specific applications can also be accommodated by altering the properties of the mountings to achieve desired characteristics.

Research Activites

In addition to the organizations discussed, which are producing compliance devices, research activities are being pursued in other assembly areas using compliance techniques. The Draper Laboratory is developing an active type of compliance mechanism which will act as a feedback device to provide information to robots or other positioning equipment to achieve alignment between mating components [5]. The Japanese are working on the development of robots which can achieve their own compliance during assembly [6]. These robots would be capable of much higher insertion forces to accommodate the assembly of heavy components. Further research work is being performed in England and New Zealand to develop compliance devices to assist robot assembly operations [7]. Devices using cylinders, springs, and rigid links have been developed to be interfaced to an assembly robot.

Advantages of Compliance During Assembly

The purpose of the Remote Center Compliance device is to automatically provide compliance between two parts being assembled to accommodate for any angular or translational misalignment which might exist. The misalignment might exist between two parts, a tool and a part, a tool and a fixture, or any other components being assembled with close clearance or interference fits. As a result of its capability to provide alignment, it offers the following potential advantages:

1. Interference fits and close clearance fits can be achieved with minimal force during the assembly of two components.
2. Part damage as a result of jamming and wedging can be minimized or eliminated during assembly operations.

3. Expensive floating workheads and floating pallets can be eliminated on manual assembly stations or automatic assembly machines.
4. Tooling, fixtures, feedtracks, escapements, and other component alignment mechanisms can be designed and built with wider tolerances.
5. Assembly tolerances can be decreased to reduce tolerance stackup problems and to improve assembly component functioning.
6. Machine downtime resulting from component jamming and wedging can be reduced.
7. Rework operations resulting from jammed, wedged, or galled components can be reduced.
8. Manual assembly operations that previously required operator force feedback and wiggling to perform successfully can now be done automatically.
9. Manufacturing costs of tools, fixtures, pallets, indexing mechanisms, workhead mechanisms, and other associated assembly components can be reduced.
10. Assembly operations can be performed with less costly and less accurate industrial robots and pick-and-place mechanisms.

In general, increased productivity can be achieved with the implementation of a rather low cost device requiring low maintenance and upkeep and increased flexibility in alignment in performing precision assembly operations.

Reprinted from *Autofact West Conference Proceedings*, Volume 2, 1980, Copyright SME

Some Insertion Compliance Options

by Jack Rebman
Lane R. Miller
Lord Corporation

ABSTRACT

Applications of insertion compliance devices have many variations. Users should have choices, through design flexibility and device adjustability. The remote center compliance (RCC) is an optimal solution to the insertion problem. The design challenge is to project the center of compliance to a suitable remote location, provide sufficient compliance in translation and rotation, and encompass these in a compact, serviceable package. Design alternatives available include RCC style, resilient materials, device geometry and dimensions, and detail construction. Also, a means for adjusting the remote center position, after the insertion compliance device is in service, is another alternative.

INTRODUCTION

The remote center compliance (RCC) is rapidly gaining acceptance as a solution to the classical pin insertion problem. The mechanics of pin insertion have been well described by the work at the Draper Laboratories.[1,2] Out of that work came the recognition of the RCC as a passive means for automated pin insertion. There are several ways to achieve a remote center of compliance. Inherent in those several ways are the options available to the designer for meeting different job demands.

A variety of application requirements can dictate a variety of performance capabilities. Let's focus on two areas of capability. One is design flexibility, and the other is device adjustability. In this paper, we will describe several approaches for obtaining a remote center of compliance, significant parameters and their effect on performance, and what they can mean for design flexibility and device adjustability. The user can have some choices.

It should be noted that the utility of a remote center compliance is based on two sets of parameters. The first governs the location of the center of compliance. An optimal design would place that location at or very near the lead end of the pin being inserted. This first set also governs the magnitude of compliance in its several modes. The design challenge for this set is achieving sufficient projection of the compliance center along with appropriate compliance in each of the modes. The objective is to achieve insertion without wedging or jamming and with loads and moments within acceptable limits.

The second set of the important parameters governs the many aspects of practical use. These include such things as size, weight, envelope, and capacity. The second set also governs the stability, repeatability, dependability, and versatility of a particular remote center compliance in use. The design challenge for this set is achieving a compact design with sufficient capacity and, in addition, stability, repeatability, dependability, versatility, etc.

In this paper, we will be concerned primarily with the first set of parameters. At this point, it should be helpful to review some characteristics of the system center of compliance.

CENTER OF COMPLIANCE

The center of compliance has these properties:

1. It is the point in space at which a system of unit compliances appears to be concentrated.
2. When a force is applied to a body through the position of the center of compliance, the body supported by the system of compliances will move in pure translation.
3. When a moment is applied to a body supported by a system of compliances, the body will move in rotation about the center of compliance.

Fig. 1 shows a model of a focalized system of unit compliances. There are four units in the system, equally spaced around the mounting circle. Each unit is a strut with a rigid center section and two elastomeric springs at each end. The system is focalized; that is, each strut is pointed toward a common point along the axial center line of the overall system. This one set of four unit struts provides both the rotational and translational compliance of the system. The supported cylindrical body is shown free of any external load or moment. Please note the reference line aligned with the center line of the supported body. The position of the center of compliance for the system is clearly marked on the body.

With the application of a force through the position of the center of compliance, as in Fig. 2, a pure translation of the body will result. The parallel displacement of the body center line from the reference line so indicates. With the application of a moment, as in Fig. 3, by means of a couple, a pure rotation of the body results about the center of compliance. The displacement now shows an angular relationship of the body center line and the reference line, with the reference line intersecting the center line at the center of compliance.

A combination of the two effects results from the application of a force at some position separated from the center of compliance, as in Fig. 4. At the center of compliance this appears as both a force and a moment. The center line of the supported body indicates a translational displacement relative to the reference line, as well as a rotation. The reference line and the center line of the body now intersect at a point away from the position of the center of compliance. If the vertical load in Fig. 4 were applied at a point on the opposite side of the center of compliance, there would also be a translational displacement of the body and a rotation such that the center line and the reference line would intersect at a point between the center of compliance and the support ring.

These characteristic deflections under a force and/or a moment are essential to the operation of a remote center pin insertion device.

IMPORTANT SYSTEM PARAMETERS

In Fig. 5, important dimensions are defined for the system shown in Fig. 1. The formulae[3] which express the center of compliance projection and the compliance values in the several modes, rotation and translation, are listed in Table 1.

Special note should be taken of the important parameter L, which relates the unit compliance magnitude in the several principle directions. The L values and the angle at which the unit compliances are focalized are primary determinants of the distance which the center of compliance is projected to its remote location. The third major parameter is m, the radius from the center line of the system to the individual compliance units. A plot of the relationship of the projection distance, α, the radius to the unit compliances, m, and the focal angle, Φ, are shown in Fig. 6.

As already described, the compliance system in Fig. 1 includes four unit compliances with each unit a strut having a rigid center section and two elastomeric springs at each end. With this arrangement, the individual unit exhibits relatively high compliance in directions perpendicular to the strut and relatively low compliance in a direction along the length of the strut. The ratio of the two spring rates, along the length of the strut and normal to the length, establish the value of the parameter L. Please note that the spring rate is the inverse of the compliance. In Fig. 5, parameters L' and L" are shown. Consider that the reference direction for these L values is tangent to the circle in the plane of physical support which passes through all of the unit compliances. We can then identify three primary spring rates for the unit compliance:

- k, in the direction of the tangent
- L'k, in a direction normal to k and intersecting the system centerline
- L"k, in a direction normal to both k and L'k

DESIGN FLEXIBILITY

Now we can begin our discussion of design flexibility. Each of the major parameters, m, Φ, L, and k, is an avenue for providing options to the designer. Each of these is an independent variable. There are, however, important tradeoffs among them when considering overall system behavior. For example, reference to the curves in Fig. 6 would reveal that a larger m and/or a larger L would mean greater projection of the center of compliance. The projection distance is linear with m and non-linear with L. In a practical design, it is simpler, for example, to double the projection for lower values of L than it is for higher values of L. The curves of Fig. 6 also indicate that there is an optimum focal angle, Φ, for maximum projection. With a high value of L, the projection, α, is quite sensitive to changes in the focal angle, Φ.

The parameter L, as a function of L' and L", characterizes the style of the compliance unit. These styles or configuration types are important alternatives which are available in achieving desired system performance. Over the years, a variety of configurations has been developed which provides choices of both L' and L". Fig. 7 shows a plot of L versus L' for different values of L". The relative effectiveness of different styles in their ability to project the center of compliance can be readily seen. For example, given an L' of 100, the value of L would be approximately doubled in changing L" from a value of 2 to a value of 0.5.

Another aspect of configuration style provides further design choices. The system in Fig. 1 has L' equal to 50 and L" equal to 1. Remember that the unit compliance for the system in Fig. 1 is a rigid strut with an elastomeric spring at each end of the strut. This configuration is similar to many that have been used widely as aircraft engine suspensions. In many situations, this particular style can serve very well as an insertion compliance device. The unit compliance in this system would behave much the same as the wire flexure used by the Draper Laboratories[1] in their insertion compliance devices. In this system, however, a single set serves both the translational and rotational modes.

In some instances, the system design with a single set of unit compliances for both translation and rotation does not offer sufficiently high compliance along with a large projection of the center of compliance. This difficulty can be overcome by adding another set of unit compliances to soften the system in the particular mode which might be deficient. This leads to another style currently popular for insertion compliance devices.

Please refer to Fig. 8. The system shown includes two sets, three each, of unit compliances. Each unit is a high capacity laminate construction of elastomeric springs[4] which, over recent years, has been developed as a rugged elastomeric bearing for oscillating motion. For these units, the L' equals 230, and the L" equals 1. The effective L has a value of 114.5.

One of the two sets of unit compliances shown in Fig. 8 is focalized similar to the system in Fig. 1. This focalized set is termed the "rotational sub-system." It is soft in the rotational mode and stiff in the translational mode. The other set is not focalized (angle $\Phi = 0$), and this set is soft in translation in a plane perpendicular to the center line of the system and quite stiff in translation in the axial direction along the center line of the system. This second set is termed the "translational sub-system."

The two sets in series are soft in lateral translation and rotation and are stiff in the direction of the system axial translation. This is the desired combination of system compliances for an insertion compliance device. Therefore, this is another design option which will meet the first design challenge for an insertion compliance device, to reach well out with the center of compliance and have the required softness in the several system modes.

Suppose we use the design in Fig. 8 to examine further the question of design flexibility. Within a given general envelope and given a particular configuration style, what range of system performance can be available? Without changing envelope at all, the particular elastomeric compound can be modified and the number of shims in the high capacity laminate can be changed. If slight modifications in the detail of the envelope configuration can be incorporated, then variations in the diameter of the unit compliance can be sized up or down. Further, little or no change in the gross envelope would be made with changes in the focal angle, Φ.

For an insertion compliance device such as that shown in Fig. 8, Table 2 lists changes in system characteristics as a result of changes for the several major system parameters. Six cases are shown in the table. The first serves as a reference. In the second, the shear modulus of elasticity of the unit compliance is doubled. The third case provides for an increase in the diameter of the unit compliance. The load area is doubled. The fourth case shows the effect of removing approximately half of the shims in the construction of the high capacity laminate unit compliance. In the remaining two cases, the focal angle for the system is doubled and then halved.

It should be clear from these results that an appreciable latitude of design choice is available, even with a relatively fixed envelope constraint.

DEVICE ADJUSTABILITY

Insertion compliance designs already described are fixed in performance once they are fabricated. Properly engineered, they should perform well. Many times, however, system parameters are not fixed. Not all objects to be inserted have the same length. Therefore, optimal insertion would not be achieved for pins having significantly different length from that length for which the RCC was designed. A compliance device with a continuously variable position of center of compliance should be useful in such cases.

An insertion compliance that is conveniently adjustable can be available in the near future. The manner by which such a device projects the remote center is different from those already described in Figs. 1 and 8. Even so, similar projection of the compliance center and desired system compliance patterns can be achieved. Fig. 9 shows a schematic drawing of the concept. It is comprised of several links and a concentrated compliance. A patent application has been filed.

Link (A) is a member which represents the pin to be inserted, the gripper which would hold the pin, and the insertion device member to which the gripper is attached. Member (A) in turn attaches to both links (B) and (C), each of which in turn attaches to the insertion machine. Pivot attachments of links (B) and (C) to the machine are shown respectively at points a and e. Link (B) can rotate freely about pivot a, as can link (C) about pivot e. Link (B) can also slide vertically on pivot a. Link (A) is attached to link (C) through the pivot, d. Likewise, link (A) is attached to link (B) at the pivot, b. Each pivot, b and d, permits free rotation of (A) about the pivot point. A slot in link (A) at pivot b permits sliding motion of the pivot b within link (A) in the vertical direction. This is not true for pivot d. At pivot d, rotation only is permitted. The slot in (A) is elongated to provide for one of the possible ways for adjusting the location of the remote center.

The two links, (B) and (C), are tied one to the other with the concentrated unit compliance at point c. Of the three components of compliance for the unit (D) in its three principle directions, only the lateral component need be considered. This is the case because link (B) can slide vertically on pivot a. Many forms of spring element can be used for compliance (D). For our purposes, it is assumed to be a bonded elastomeric sandwich. A concentrated compliance (D) clarifies description of the device. The compliance could be distributed in various places throughout the linkage.

For this design, Fig. 9, the dimension which best describes the location of a remote center of compliance is df. So long as dimension df is greater than dimension de, the center of compliance can be considered to be in a remote location. Given the definition of a center of compliance, stated earlier, the dimension of df can be established as a function of the known geometry of the rest of the system and the concentrated unit compliance, (D).

If we assume that the center of compliance is located at point f, a lateral load F applied to the supported member at that point should produce pure translation laterally as the movement of link (A) under the influence of load F. This would mean that pivot point b would have a lateral component of movement equal in direction and in magnitude to that for pivot point d. This condition would, in turn, prescribe the rotations for links (B) and (C) and the deflections of unit compliance (D). The result is an expression defining dimension df.

$$df = \frac{bd}{[Q-1]} \tag{1}$$

where: Q is a function of compliance values of the unit compliance (D) and also the geometry of the linkage.

The center of compliance will be projected for Q>1, that is for positive values of df. By examining more closely the motions of pivot points b and d, the necessary conditions for projection can be identified.

Projection occurs if the lateral deflections of b and d, in Fig. 9, meet the following conditions:

-Both are in the same direction.
-Deflection at b is greater than deflection at d.

These lateral deflections are the summation of deflections due to the linkage alone and the deflections due to straining of the compliance (D). A simplifying assumption will help explain the action. Let the compliance (D) behave like a pivot connecting links (B) and (C). Assume that the pivot is fixed relative to (C). Also relative to (B), assume it can rotate freely and slide freely in the vertical direction. The expression for df then becomes:

$$df = \frac{bd}{\frac{\left(\frac{cd + de}{de}\right)}{\left(\frac{ab + bc}{ab}\right)} - 1} \tag{2}$$

Further, if links (B) and (C) are equal in length, ac = ce, the expression reduces to:

$$df = \frac{bd}{\left(\frac{ab}{de}\right) - 1} \tag{3}$$

By pinning link (B) to link (C), as described, the deflections of points b and d are forced to be in the same direction. Also, if ab is greater than de, df is a positive number. Therefore, projection is achieved.

This projection is fine, but such a linkage with the pin joint replacing the compliance, (D), would not serve well as an insertion compliance. A lateral load applied at the center, thus established, would meet very stiff resistance. If the links were rigid, the compliance would be zero, at that center.

Now let's provide some "give" in the system in (D). Lateral loads across (D) would then produce lateral deflections at points b and d which are opposite in direction. They would combine with the deflections of b and d due to the linkage alone. By introducing a proper amount of compliance for (D), the system becomes a practical means for achieving not only compliance center projection but also a necessary combination of system compliance values.

The potential for adjustability in this design becomes apparent. To change the dimension df, a simple change in one or another of the pivot locations on links (B) and (C) would be needed. For example, suppose we change the position of pivot b. The results are shown as Case I of Table 3. Another possibility is to move only pivot a. Results for this condition are shown as Case II of Table 3. Finally, another option would be to move both pivots b and d while maintaining the distance bd constant. Results for this situation are shown as Case III of Table 3.

The values of Table 3 were computed using equation (3). The same equation applies for the use of a resilient component for compliance (D). This holds true so long as the only load developed across (D) is in the lateral direction. One way to accomplish this is to permit sliding in the vertical direction for link (B) relative to pivot a, as well as rotation, as indicated in Fig. 9. If vertical components of force develop across compliance (D), non-linearities become involved, which can seriously degrade performance.

CONCLUSION

The user of remote center compliance devices for insertion operations has a wide range of options for best meeting application requirements. Design flexibility is available with alternatives among compliance system configuration styles, resilient materials, system dimensions and geometry, and detail construction. Further, designs are available which can be adjusted to meet changing requirements while in service.

REFERENCES

1. "Exploratory Research in Industrial Modular Assembly," by Nevins, Whitney, Dunlavey, Drake, Killoran, Kondoleon, Mogged, Seltzer, Simunovic, Wang, and Watson, Fifth Report Covering 1 September 76 to 31 August 77, C. S. Draper Laboratories Report R-1111.

2. "Using Compliance in Lieu of Sensory Feedback for Automatic Assembly," by Samuel H. Drake, M.I.T. Doctor of Science Thesis, September 1977.

3. "Report on Predetermination and Control of Vibration in Aircraft Originating from the Engine," K. A. Browne Wright Aeronautical Report 353, dated June 10, 1937.

4. "Compliance: The Forgiving Factor," by Jack Rebman, Robotics Today, Fall 1979.

TABLE 1

Focalized System Performance Formulae
(See Figure 5 for definitions and dimensions.)

- Projection of Center of Compliance = α

  where: $$\frac{\alpha}{m} = \frac{1}{2} \cdot \frac{\sin 2\Phi}{\frac{L''+1}{L'-L''} + \sin^2\Phi}$$

  and for use with Fig. 6, $L = \frac{L'-L''}{L''+1}$

- System Compliance Values

  - Compliance for rotation about X axis $= \frac{1}{R_x}$

    where: $R_x = nkm^2$ in-lb/rad.

  - Compliance for rotation about Y or Z axis $= \frac{1}{R_z}$

    where $$R_z = \frac{n}{2}km[L'\cos\Phi(m\cos\Phi - \alpha\sin\Phi) + L''\sin\Phi(m\sin\Phi + \alpha\cos\Phi)] \quad \text{in-lb/rad.}$$

  - Compliance for translation along X axis $= \frac{1}{K_x}$

    where $K_x = nK[L'\cos^2\Phi + L''\sin^2\Phi]$ lb/in.

  - Compliance for translation parallel to Y or Z axis $= \frac{1}{K_z}$

    where: $K_z = \frac{n}{2}K[L'\sin\Phi + L''\cos\Phi + 1]$ lb/in.

Table 2

Focalized Compliance System Characteristics With System Parameter Changes

	Cases					
	1	2	3	4	5	6
Shear Modulus	45	90	45	45	45	45
Number of Shims	17	17	17	8	17	17
Unit Compliance Diameter (in.)	0.47	0.47	0.66	0.47	0.47	0.47
Focal Angle Φ (deg.)	13	13	13	13	26	6.5
L'	230	190	460	35	230	230
Projection, α (in.)	5.3	5.0	6.2	2.2	3.2	6.2
Translational Spring Rate, $K_{y\&z}$, (lb/in.)	70	140	150	30	80	60
Rotational Spring Rate, $R_{y\&z}$, (in-lb./rad.)	3500	6600	8000	900	1050	8000
Longitudinal Spring Rate, K_x, (lb/in.)	9500	16000	38000	1000	9000	10000
Torsional Spring Rate, R_x, (in-lb./rad.)	120	240	240	75	120	120
Load Capacity (lb.)	350	700	700	150	350	350

Remarks: Case 1 - Reference values
Case 2 - Elastomer shear modulus doubled
Case 3 - Unit compliance diameter increased
Case 4 - Number of shims reduced
Case 5 - Focal angle doubled
Case 6 - Focal angle halved

TABLE 3

Center of Compliance Projection Changes
With Changes in Device Geometry

		Dimensions					
		ab	ac	de	ce	bd	df
Case I:	a	0.50	1	0.50	1	1	∞
Move pivot b	b	0.53	1	0.50	1	0.97	16.7
only	c	0.55	1	0.50	1	0.95	9.5
	d	0.60	1	0.50	1	0.90	4.5
	e	0.65	1	0.50	1	0.85	2.8
Case II:	a	0.50	1	0.50	1	1	∞
Move pivot a	b	0.53	1.03	0.50	1	1	34.3
only	c	0.55	1.05	0.50	1	1	20.8
	d	0.60	1.10	0.50	1	1	11.0
	e	0.65	1.15	0.50	1	1	7.7
Case III:	a	0.50	1	0.50	1	1	∞
Move pivots	b	0.53	1	0.47	1	1	7.8
b & d only,	c	0.55	1	0.45	1	1	4.5
with bd	d	0.60	1	0.40	1	1	2.0
constant	e	0.65	1	0.35	1	1	1.2

FIGURE 1:

FOCALIZED COMPLIANCE SYSTEM

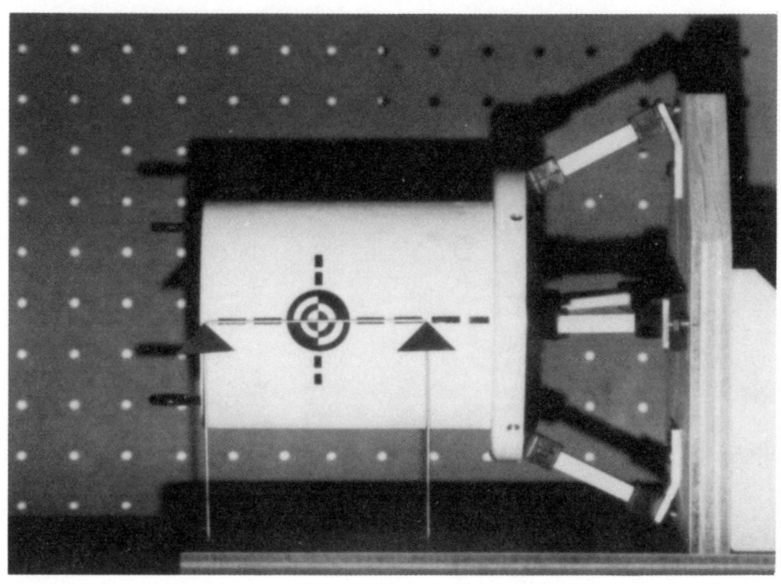

FIGURE 2:

DEFLECTION WITH LOAD AT CENTER OF COMPLIANCE

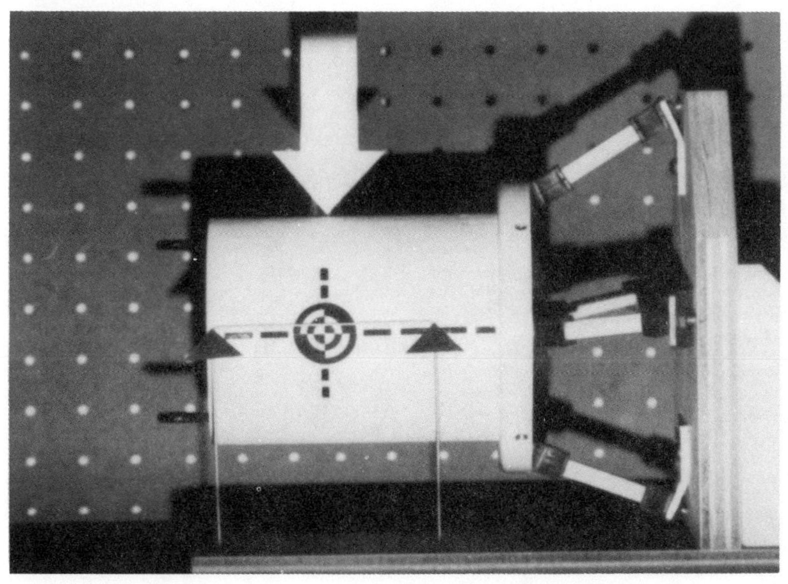

FIGURE 3:

DEFLECTION WITH MOMENT APPLIED

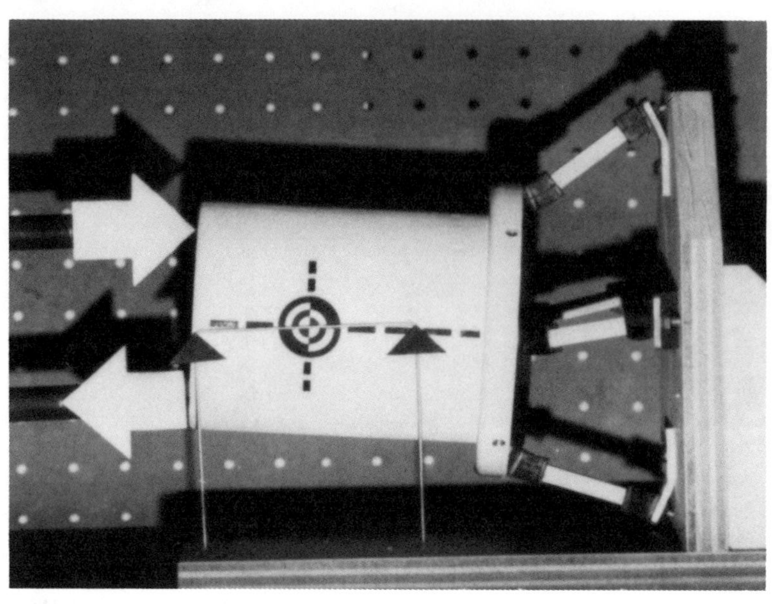

FIGURE 4:

DEFLECTION WITH LOAD AT POINT SEPARATED FROM CENTER OF COMPLIANCE

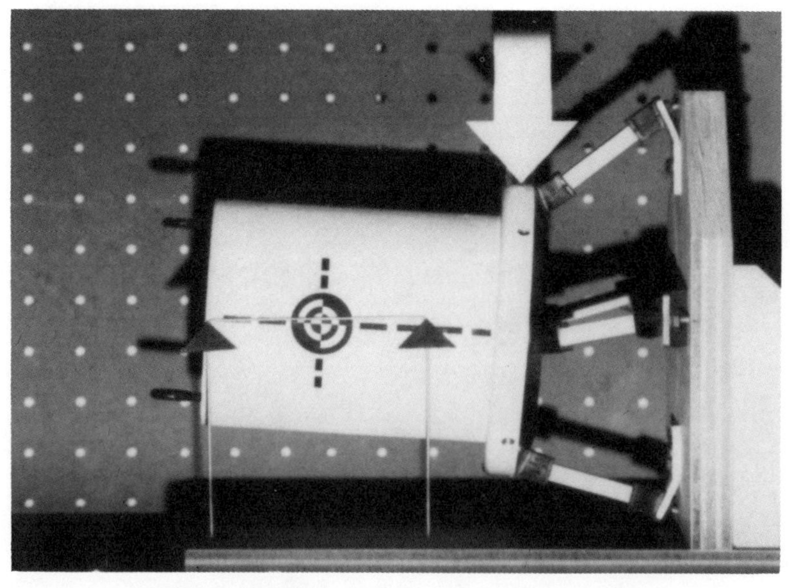

FIGURE 5:

FOCALIZED COMPLIANCE SYSTEM DIMENSIONS

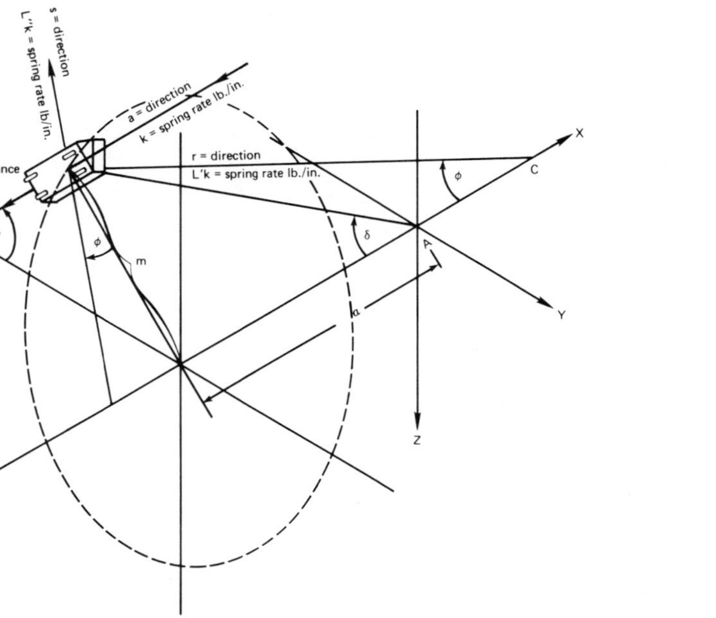

FIGURE 6:

REMOTE CENTER OF COMPLIANCE PROJECTION CURVES FOR FOCALIZED SYSTEM

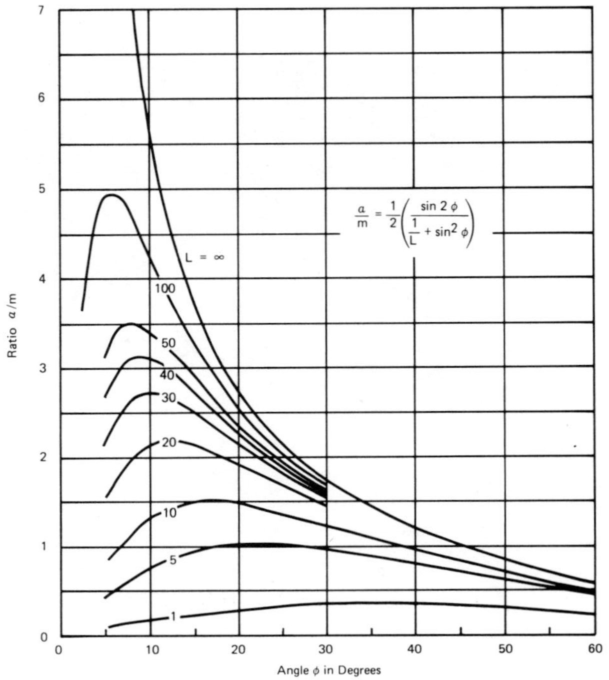

$$\frac{\alpha}{m} = \frac{1}{2}\left(\frac{\sin 2\phi}{\frac{1}{L} + \sin^2\phi}\right)$$

FIGURE 7:
VALUES FOR VARIOUS L′ AND L″

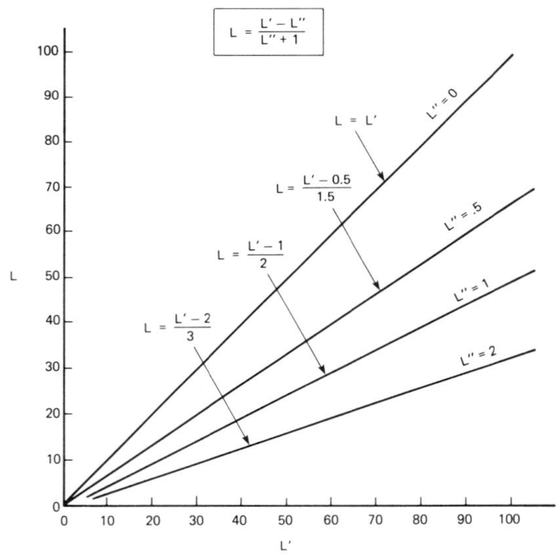

FIGURE 8:
LORD ELASTOMERIC RCC

FIGURE 9:
LINK-TYPE ADJUSTABLE RCC

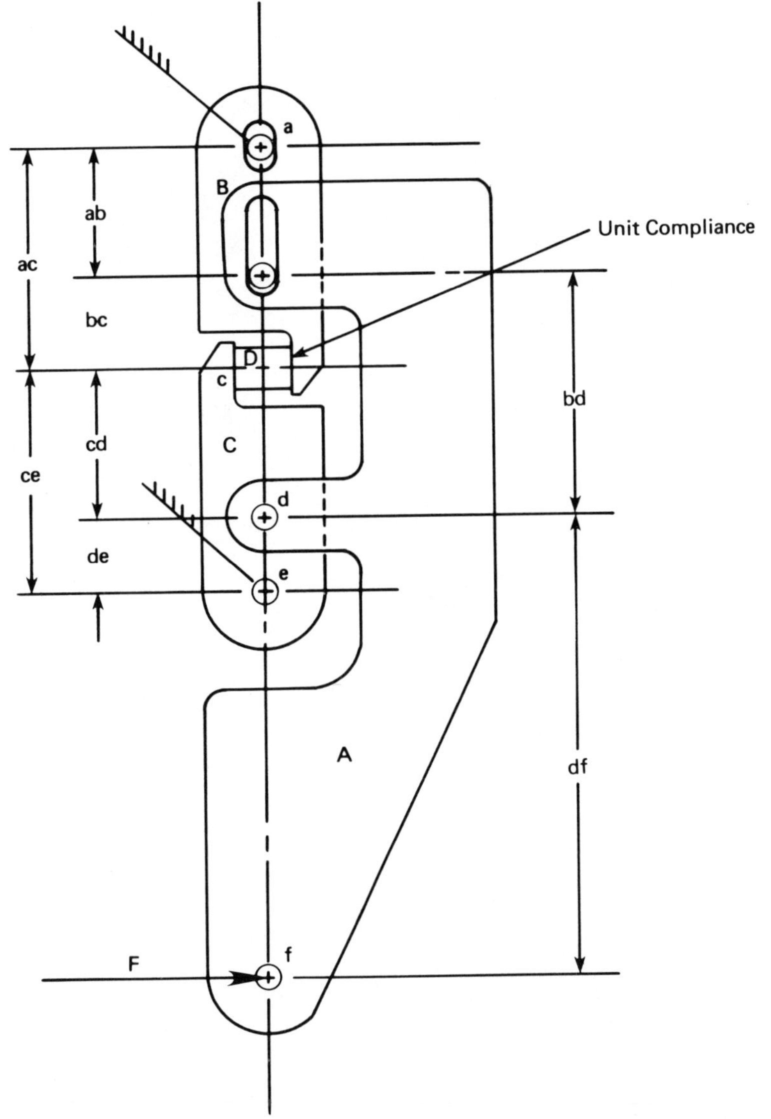

CHAPTER 6

ROBOTS IN ASSEMBLY

Reprinted from *Modern Machine Shop*, December 1982

Chrysler's Productive Assembly Weapons

Robots, programmable controllers, and new computer-aided quality concepts are some of the new weapons in Chrysler's fight to regain economic health.

By John Martin

One of the linchpins in the Chrysler strategy for a return to economic health was the introduction of the Chrysler LeBaron and Dodge 400 passenger cars. The first of these front-wheel-drive, mid-sized luxury sedans rolled off the St. Louis assembly line, with Lee Iacocca at the wheel, on September 9, 1981.

The 75-million-dollar conversion of the plant involved a major renovation of the body shop, replacement of assembly tooling, rearrangement of conveyor systems to accommodate the new assembly process, extension of the paint spray line, and a 44,500-square-foot expansion of manufacturing floor space.

The showcase of the plant is its body assembly operation—95 percent of all the 2500 spot welds the car gets are made by welding machines and robots. The robot population (all Unimate 4000 series) was increased from 14 to 64.

Control for this operation is supplied by two of the Allen-Bradley programmable logic controllers

Thirty-two Unimate 4000s weld the Chrysler body on the St. Louis automatic respot line.

(PLCs). All told, there are 26 programmable controllers in the plant, tied into a data highway. This data link is used for monitoring and information gathering. The link compiles downtime, generates information forms, finds out how many parts are being built, how many of which type car, how many with or without air conditioning, and so on. In the future, additional software will be written to extend control of the process and to introduce new ideas into the assembly operations.

One of the PC's sole function is to monitor the 25 others—it prints out problem messages on Teletype. A future goal is to put even more specific diagnostics into the software.

For example, if a limit switch goes bad somewhere, it would be nice to have a graphic display of the machine with the faulty part come up on a monitoring screen. Then the particular switches could be indicated by a number, and the location of the faulty switch could also be pointed out.

Welding

The first welding operation involves joining the underbody/floor pan assembly. This is done with automated fixturing supplied by Progressive Tool and Die, out of Detroit. One of the lessons learned from Chrysler's Belvidere, Illinois installation and put to good use here was to put all the fixturing controls (valves, solenoids, and so on) on a level above the actual welding to facilitate repair and maintenance.

After welding, the underbodies go to a buffer. Each one is optically read to see if it has air conditioning, is a component for a two-door or four-door model, and so on. Then they are stored, accordingly, in the buffer area. An automatic crane-and-hook system, also under computer control, picks up the respective underbodies as they are needed to fill the day's orders.

Next step is the so-called "toy-tab" operation (basically folding over some sheet metal protrusions into slots to temporarily hold the assembly together). Here, the underbody and side apertures are tabbed together on a slat conveyor and transferred to the centerpiece of the automated welding—the robogate.

Basically, the robogate is an automatic framing fixture where the assembly is locked in position for the 22-station automatic welding line. The gates close around the car body and clamps secure it in place. At this first station, six robot welders give the body 26 stabilizing welds (critical welds to such areas as the side rail, wheel house, headers, roof and driprail) to assure consistant car width and dimensioning.

The robots get the nod for their particular welding program from photoelectric switches, which read the car model type, put the information into the computer's shift registers, and send it through the system to the weld guns.

The automatic single-gate framing provides a constant dimensional control. Flexibility is also important, in that Chrysler only has to build a new gate for model changes. The robots can then be reprogrammed for new positioning. Chrysler builds two sets of gates for every car model. The second one is for use as a simulator or to replace the first if repairs are necessary.

Next stage of the welding is the automatic respot line, where 32 robots do all the outside sheet metal welding. Two inspectors at the end of this line check welds for visual qualities. They have the prerogative to shut down the entire line if the welds look faulty.

A body buffer follows (the automatic welding has a capacity of 104 bodies per hour—plant output of finished cars is 60 per hour). Then there are several-hundred-feet of workers with manual weld guns to get the difficult and inaccessible spots the robot couldn't hit. This line also permits taking a robot off the automatic line to do some work on it and still make sure the car is properly welded.

Assembly

After welding is completed comes the main assembly line—paint, trim, testing, and so on. In the paint department of the St. Louis assembly plant, a major expansion of facilities involved lengthening of the former color and prime spray booths by 70 feet to provide greater capacity. New automatic paint spray equipment also was installed, reducing manual painting to only isolated areas of the car body.

In the trim, chassis, and final assembly departments, facilities were revised to accommodate a new engine dress-up area, new roll test machinery for checking transmission gears and driveability, and five new machines to adjust wheel alignment. A new technique for installing the engine/transaxle assembly replaces the former body drop with a pedestal loop. The

Car body is secured in robogate fixture and welders move in for stabilizing welds.

tradition of lowering the car body over the engine and drive train is reversed, and the chassis, supported by a pedestal, is lifted into the body.

On the south side of the plant a 44,450-square-foot building was constructed for tire and wheel assembly, as well as storage for incoming shipments of tires and engines.

The electrical system on all Chrysler LeBarons and Dodge 400s is checked out with equipment developed by Chrysler's Huntsville, Alabama, Electronics Division. The computer checkout stations inspect the instrument panel and steering column circuits before these major subassemblies reach the final assembly line. Instrument panels are checked out in stations that inspect up to 100 circuits each in a 90-second test cycle. At the end of the steering column subassembly operation, on-line computerized stations check up to 45 steering-column circuits, depending on the number of options.

For additional quality assurance, electronic test systems in the final assembly area inspect all circuits in each car, including the previously tested instrument panel and steering column circuitry.

Quality Control

If an automobile isn't welded right, it will fall apart when the safety division crashes it or it will have a rattle that can't be fixed or a squeak that can't be taken out.

Every day, three cars are torn apart after welding. Workers take a chisel to the weld, wedge it in, and start hammering. If the weld is good, it will hold. Twice a day these inspection routines are checked by plant supervisors, who examine documentation sheets which indicate any welds that were not up to par. The same is done to the car underbody. It is put on a fixture, turned over, and beat upon to search for any defective welds.

The Quality Assurance Department also "plates" one car daily. They put the automobile on a large flat plate, and use QA inspection equipment to read and locate all the points on the car to make sure they conform to specifications.

The sheet metal is constantly checked to make certain everything matches up. There will be occasional variations in stamping lots and, when that occurs, adjustments must be made to the weld tooling account for the variation.

Chrysler Corporation has its own weld auditor, who shows up once a month from Detroit on a surprise visit. He spends a week at the plant beating apart some 20 cars.

The welding equipment itself has some adaptive features. For example, current is monitored through silicon controlled rectifiers that fire the weld. If the weld is pulling less than 11,000 ampers, a low-current alarm will be sounded.

Additionally, the weld timer increments become hot as the weld tip wears. It will also stop the welding after a certain number of welds if the tip hasn't been changed.

Plant Startup

The welding, and the plant in general, had a fairly smooth startup, which is particularly impressive given the fact that before the retooling most of the welding was done by hand guns. Chrysler, of course, was also able to draw from their experience at their three other plants that have robogate setups.

One thing they did was to assemble and train a multi-skilled welding maintenance crew, rather than just rely on plant specialists in different areas. Welding maintenance encompasses many disciplines: fluidics (water and oil), pneumatics (clamps), electrics and electronics (the controls), and tooling (locators, blocks, and so on). This requires the skill of electricians, electronic technicians, toolmakers, pipefitters, and millwrights.

What Chrysler elected to do was assemble a team and cross-train within the group. This included an extensive stint at the Chrysler Institute in Detroit. Unimation and Parker-Hannifin (two major suppliers) also provided extensive training. The team of workers were also sent to the Newark, Delaware plant for on-the-job training. Over 300,000 man-hours of classroom, on-the-job training, and specialized training were logged for the entire St. Louis workforce.

There were problems, of course, much of them having to do with the accelerated retooling schedule. But remember that an entire plant was stripped and rebuilt in three months. There were some tempers and words behind those closed doors.

Unbroken chain of precision-welded bodies dramatize the new computerized system of robot welders. The computer-programmed welding machines assure precise fits and improved alignment of trim features and accessories, as well as stronger body components and greater structural integrity.

Jim Foreman, Plant Superintendent, explains the defect/demerit system Chrysler uses to rate its car assembly process.

Sometimes outside contractors, who came in to put a lot of the equipment together, didn't meet their deadlines or Chrysler's specifications. Some of the equipment that Chrysler purchased was brand new (one piece was even a prototype) and some massaging and improvising were necessary.

Programming the PCs also took some time. The PCs use everyday ladder-diagram logic on their CRT screens—five rungs of a wiring diagram. Some of the workers, although they were good electricians, took a while to get used to the more formal conceptual requirements that are involved in programming.

A number of the workers had done programming before, but principally with the Unimate, which combines an on-board computer with a hand-pendant teach mechanism. But PCs, especially in a network system, require some experience and expertise to massage and manipulate the program just so, especially when a complete change from one type of tooling to another is involved.

The experience Chrysler gained at its other plants paid off, though. For example, the St. Louis plant knew they would be welding double-galvanized sheet steel, and selected adequate transformers to generate the temperatures needed at the weld points. At Belvidere, however, the plant didn't start out with that material; they changed over to it. They then had to retrofit their power sources to generate enough heat to make the weld hold properly.

The St. Louis plant also developed a floor-pan multi-welder, as opposed to the seven-stage machine used in Belvidere. At St. Louis, they subassemble prior to welding, whereas the Belvidere plant used the seven different machines sitting one after another on a shuttle bar to complete the operation.

Eliminating unnecessary transfers was a big goal of the new setup. To this end, the St. Louis facility direct-linked the toy-tab/robogate/automatic-respot sequencing; the other Chrysler plants require transfers between the different steps.

Lots of little improvements were made, too. These include, for example, increasing the amount of light, air supplies and the size of oil reservoirs in certain areas.

The banking system (workpiece buffer storage area) is an excellent means to handle an assembly line when there is one model and one type firewall.

But when there are three different body styles, two different floor pans, two different tops, and a special firewall for the vehicles that have air conditioning, the banking system does become complex. Especially since Chrysler has less than a two-hour bank for the two major systems (side aperture and automatic respot welding).

Scheduling problems can be terrific, and they sometimes outrun the software. It's not a question of hardware memory capability, but rather the algorithms for feeding the data through the system and arriving at decisions.

A worker on the test line for electrical components takes time out to explain his test procedures.

There are two possible approaches for this predicament. One, of course, is further refining and rewriting the supervisory software. The other involves product design. The AC air distributor box includes a heater core and an AC evaporator (essentially just a radiator) while the heater box air distributor (without AC) is just half

Tom McTernan, General Foreman for Maintenance and Welding, at the PC control station for the robotized welding.

the size. If AC air distributor boxes were used in all the cars (whether they had air conditioning or not) it would standardize the box size and cut scheduling costs by one-half (by eliminating buffering for that component).

The Human Side

There was a significant management change three years ago at Chrysler's St. Louis plant. Management became much more people-oriented—the new managers felt a job could be done with a minimum of aggravation. The greatest strength of the organization is its people.

In the early days of the auto industry, brute force was often used to "persuade" the workers to do their jobs. Chrysler is in the vangard of the new management, which realizes that you have to give people a good place to work. People today know there is a better and an easier way.

Reprinted by permission of IFS (Publications) Ltd. from *Assembly Automation*, May 1982

Car-body construction and assembly

S. Muller, KUKA Schweißanlagen + Roboter GmbH, Augsburg, W. Germany.

More than half of the industrial robots throughout the world are used in the automobile industry. The following deals with the most important application so far, spot welding in car-body construction, and one with a promising future, namely, assembly.

THE AUTOMOBILE industry has had a profound effect on the development and operation of industrial robots. Whereas the development of this modern manufacturing technique is now proceeding at a measured pace, its utilisation is increasing rapidly. The automobile industry with its high output, high investment levels and also in its worldwide competitive situation provided the necessary prerequisites for today's extensive use of industrial robots.

Industrial robots are used in the automobile industry in a number of different manufacturing sectors – often as flexible replacement for rigid, fully automatic transfer systems, or in order to supplement such systems. The main fields of application are in car-body construction and painting. So far their use in the manufacture of gear systems and axles and in stamping plants has been less significant and there has been practically no use of robots for assembly to date.

Spot welding in car-body construction is used for assembling individual sheet metal parts into a car body. Depending on the car model, 500 to 1,000 stampings are joined into subassemblies – e.g. front ends, sides, floors – and are welded to the car body in the next job step, 4,000-6,000 spot welds are required for each car body.

Up to 1960, car bodies were primarily assembled on carousel type assembly lines. The individual parts were brought to the main production line via conveyor systems. This was the first mechanisation of car-body construction. The output per shift was about 500 car bodies.

Since the sixties custom-made transfer lines have been used employing presses with shuttle transfer. In these transfer lines, the loading operations are triggered automatically and the spot welds are placed automatically and precisely.

All the motions and welding operations carried out on the line, as well as the call-up and loading of the individual parts are monitored by a sequence control. These systems have a production rate of 800-1,000 parts per shift. Cycle times of 15s are possible.

Since 1970, there has been increased usage of flexible manufacturing systems with industrial robots. At first this was out of interest for a new means of production, and then later because the application of flexible manufacturing techniques proved to be economically viable.

Cycle times

So far, these new manufacturing systems have been primarily suitable for medium range production rates – corresponding to 600-800 units per shift and thus slightly longer cycle times.

The difference in cycle time between transfer lines and flexible manufacturing systems with industrial robots, which have cycle times below 30s, is that the amount of time required for positioning the parts is relatively long. As a result, there is so little time left for spot welding that the robot is not used to its optimum capacity.

However the automobile industry also wishes to benefit from the obvious advantages of flexible systems for higher production rates. Consequently, the welding system manufacturers, in close co-operation with the automobile industry, continue to develop and improve overall system designs within which the spot welding rates are optimised and the robots operate at full capacity, and thus economically.

The transition of the manufacturing method from the rigid transfer line to the flexible system, has no effect on the labour number basically required for operation and maintenance of the production line. Of course, the qualitative requirements demanded from these men are higher.

In flexible automatic manufacturing, the welding equipment, such as welding guns, are moved by industrial robots. These flexible systems fulfill essential objectives set by the automobile manufacturers[1]:

- Retooling production lines for different models or component alterations.
- Better adaptation to capacity changes, i.e. to increase or reduce production rates.
- Mixed production of different models on the same production line.
- Greater manufacturing efficiency, e.g. by means of stand-by equipment.

The economic efficiency of an automobile manufacturer and thus the manufacturer's competitiveness increases with the degree to which these objectives are fulfilled. This is the reason for the automobile industry's demand for flexible automatic

Footnote: This article is based on a paper from the International Conference 'Robots in the automotive industry', held in Birmingham, UK, 20-22 April, 1982.

manufacturing systems. It is easy to understand why this demand is gaining more and more importance, bearing in mind the increasing shortages of energy and raw material supplies, the rapid development of new techniques, the growing demand for large ranges of model variations and the increasing need for quick reactions to the demands of the market.

KUKA

A few application examples for KUKA industrial robots are as follows:

- *Driver's seat support* – an IR 601/60 with a transformer gun welds the driver's seat supports for a small truck. The components to be welded are loaded by hand into a dual-station turntable which then positions them in front of the robot for welding. The turntable is equipped with a turn-over device. When the robot has welded all the spots on one side, the component is turned over 180° and is then finish welded.
- *Side members of pickup trucks* – two IR 601/60 welding the righthand and lefthand side members of a pickup truck in a four-station turntable. The various side members are loaded into two separate loading stations. The robot works at two adjacent welding stations. The appropriate program is called up automatically (Fig. 1).
- *Floor and front end* – robot-equipped and conventional welding systems are used alternately in the transfer line for welding the floor and front end assembly. First of all three IR 601/60 track-weld the vertical plates to the side and cross members in order to produce a transportable assembly. At the next station, the front panel and the radiator partition are connected to the basic unit by means of a multispot welding press. In the third station, a further five IR 601/60 robots – three of them on the left and two of them on the right – weld the engine compartment. At the same time two IR 601/60 weld the vertically clamped rear floor in a rotating welding fixture in a welding line of right angles. The front end, the mid-floor and the rear floor are then welded in a conventional welding press. The production line opertes at a cycle time of 36s; the robots weld 6-16 spots during this time. The robots are equipped with welding guns with integrated 25 kVA welding transformers[2] (Fig. 2).
- *Spot welding car sides* – four trunnion mounted fixtures with three IR 601/60 spot welding robots each are used for welding car-body sides. Two weld on each side. Each trunnion mounted fixture has two welding stations and an automatic loading and unloading station. Accordingly, each trunnion is equipped with three fixture mounts; in each of these mounts, six clamps position one component. An indexing system turns the trunnion by 120° at each step. Indexing is sufficiently accurate to obviate the use of indexing or locking pins. The complete part is automatically unloaded and a new one is automatically loaded, the robots weld a total of 194 spots using transformer guns. The cycle time is approx. 150s. The flexible design of the system makes the use of a fourth robot possible with each trunnion and enables the installation of interchangeable fixtures for different components on the mounting faces of the trunnions.
- *Complete car body* – IR 601/60 are also used for spot welding the complete car body.
  Example 1: In this case one IR 601/60 each on the righthand and lefthand of a turning device are welding approx. 80 spots which could not be welded during the previous operations due to clamping devices preventing access. The car body transfer is by overhead conveyor. The cycle time is approx. 170s. Three of these systems operate in parallel.
  The shape and attitude of the parts and access conditions required that small welding guns were specified for the robots. Moreover, all the welding operations had to be made by means of the same welding gun type.
  Example 2: In this case, the car body is being welded with the aid of a carrier system based on a so-called respotting line. The carriers convey the car bodies to the IR welding stations. The individual car parts are first tack-welded in a conventional welding station.
  Different vehicle models, e.g. with extended wheel base – are detected by means of mechanical sensing of the carriers. The signal automatically triggers the appropriate robot program[3].

Industrial robots equipped with welding guns cannot be used for welding large-surface or deeply curved stampings or at any rate only to a limited degree. This is why the KUKA industrial robot IR 200 was developed as a 'flexible overhead tool'. This robot welds against backing elecrodes with a single pivoting thruster, an

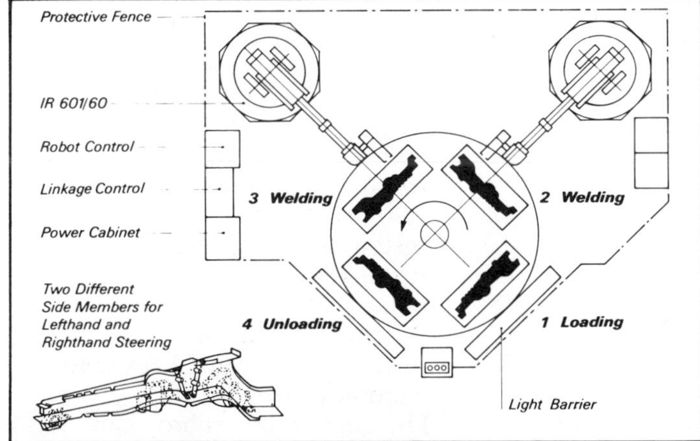

Fig. 1. Spot welding station with IR 601/60 welding side members for pick-up trucks.

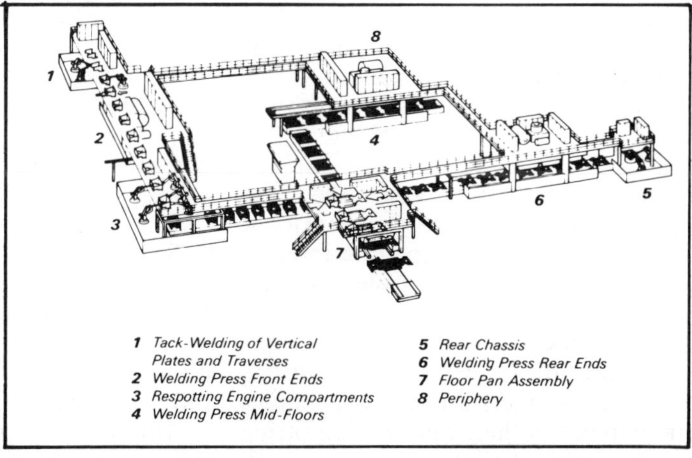

Fig. 2. KUKA industrial robots in the transfer line for floor assemblies.

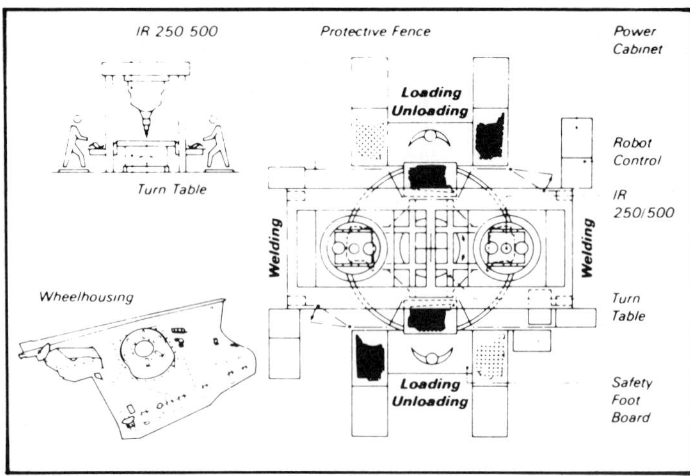

Fig. 3. Spot welding station with IR 200 welding wheelhousings.

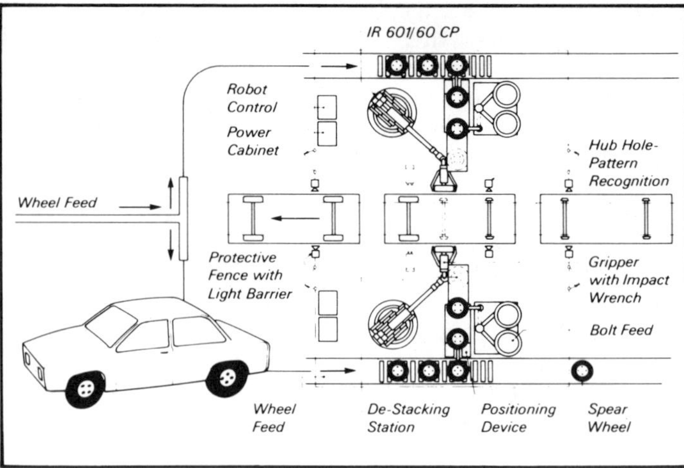

Fig. 4. Assembly station with IR 601/60 CP for car wheels.

electrode gun or a spring-backed electrode.

The IR 200 is a portal-mounted robot which requires no floor space apart from that taken up by the support columns. Its maximum welding force if 500daN. The robot is designed to provide the necessary rigidity required for this.

An application where two IR 200 robots could be used, is during direct welding of lefthand and righthand car wheelhousings from both the outside and inside. This requires a turntable with four stations which are equipped with different backing electrodes. The wheelhousings clamped onto the backing electrodes are swivelled into the working zone of each of the two IR 200 by alternative clockwise and anti-clockwise rotation of the turntable by 90°.

Direct welding is more suitable for safety critical applications, improves clamping ability and helps to save space. The wheelhousings are welded in a small area. Various small parts, such as weld nuts and reinforcement panels are welded to both sides of the main component. The 56 spots of the wheelhousing are welded in 140s; some welds are made with two thicknesses and others through three (Fig. 3).

For welding floor assemblies, the system consists of a conventional loading and tack-welding line and two respotting lines with flexible assembly systems.

In the tack-welding line, the front ends, the mid-floors, the rear floors, the side members and the rear panels are joined together. The two respotting lines are parallel and are both equipped with three robot stations. This enables each robot to weld more spots per cycle. Another advantage is that the manufacturer still achieves 50% of output if one of the respotting lines stops.

A total of 20 industrial robots are used in this production line – eight of them floor mounted models and 12 portal-mounted robots[4].

Production

As demonstrated by these examples, KUKA has considerable experience with regard to the development and application of flexible manufacturing systems with industrial robots. The prerequisites for their use in car-body manufacture are the following:

- Creation of reference points for all models.
- Robot-adapted car-body design.
- Assembly-adapted car parts.
- Accurately made stampings.
- Centralised production control.
- Qualified maintenance personnel.

The most important elements for flexible automatic car-body manufacture are the following:

- Positioning devices for the welding tools, i.e. industrial robots.
- Welding tools, i.e. transformer gun, push guns, etc.
- Carrying and conveying fixtures, i.e. lifting shuttles, turntables.
- Data storage systems.
- Sequence controls.

All of these elements form part of KUKA's delivery program. They are the result of long-standing close co-operation between the automobile industry and KUKA.

Automobile manufacture is characterised by the following apparently contradictory concepts: on the one hand highly mechanised manufacturing processes scarcely allow further rationalisation, and on the other, labour intensive manufacturing sectors, such as car-body and final assembly.

Approximately 40% of the manufacturing time of an automobile is taken up by assembly[5]. This is why this sector is regarded as a major focal point for rationalisation. The reasons for the low level of mechanisation in automobile assembly are easy to understand:

- Assembly work is complicated and requires more manual skill than sheer physical force.
- The more car models there are, the greater the variety of necessary assembly operations.
- Positioning the individual parts is often difficult and requires an accurate eye.

In order to promote mechanisation in assembly effectively and cost-efficiently, two conditions should be fulfilled in advance:

- Assembly-adapted automobile design
- Pre-assembly of as many sub-units as possible or automatically.

Consequently, the following sectors are suitable for the use of flexible assembly lines:

- Installation of engines and gear systems.
- Installation of rear axles.
- Installation of seats, doors, fuel tanks, batteries.
- Assembly of wheels.
- Assembly of front, back and side window glasses.

The industrial robot can fulfill essential complementary functions – de-stacking and separating parts.

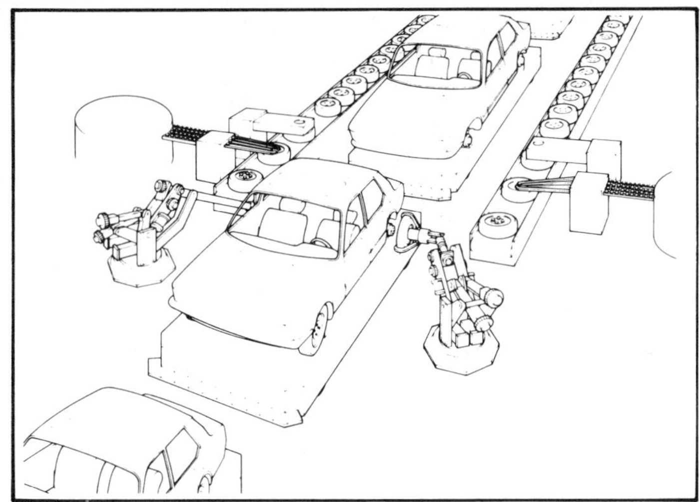

Fig. 5. IR 601/60 CP mounting car wheels.

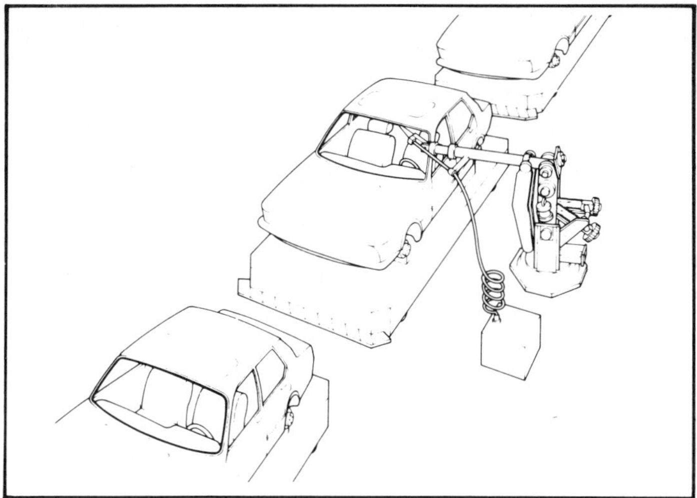

Fig. 6. IR 601/60 CP applying adhesive for car windshields.

Complex

As a rule, assembly operations are highly complex procedures. For this reason pilot studies are carried out on selected assembly operations prior to the installation of industrial robots for assembly on site of the user. Test facilities are required for this.

Three such test facilities, equipped with KUKA industrial robots IR 601/60 CP and IR 250/500, will be described.

Two IR 601/60 CP are used to fit front and rear wheels in pairs on suspended or floor-conveyed cars. The essential functional units of this assembly system include the following:

- ○ Two high-capacity industrial robots with continuous path control and conveyor synchronisation.
- ○ Television sensor system for synchronising car and robot motions and for hole pattern detection.
- ○ Robot wrist with integrated wheel-gripping device and multiple impact wrench.
- ○ Position control system for determining the car conveyor speed.
- ○ Reloading station with wheel-aligning device and feeding device for wheel bolts.
- ○ Centralised control system (Fig. 4 and 5).

Windshields are fitted in two stages in the test facility. In the first stage an IR 601/60 CP with conveyor synchronisation applies the adhesive. In the second stage an IR 250/500 picks up the windshields and presses it into the windshield frame. The essential functional elements of this application are the following:

- ● Long-reach, floor mounted industrial robot with continuous path control and conveyor synchronisation.
- ● Television sensor system to synchronise the car and robot motions and for detecting the position of the windshield opening.
- ● Portal-mounted industrial robot with vacuum pick-up device (Fig. 6 and 7).

Such a pilot facility is used to lift a fully assembled rear axle into the car body from below. The fastening operations required for this are carried out by two additional devices. The positioned rear axle is bolted into place from above by an IR 601/60 CP. The essential functional elements of this pilot application are the following:

- ○ One or two industrial robots, depending on the cycle time, with multiple impact wrench for bolting back the shock absorber legs.
- ○ Conveying facility for the rear axles.
- ○ Lifting and positioning device with multiple impact wrenches and bold feeding devices. (Fig. 8 and 9).

The main application sectors for industrial robots is in the automobile industry. In future the assembly of vehicles will be one of the main application fields for robots. □

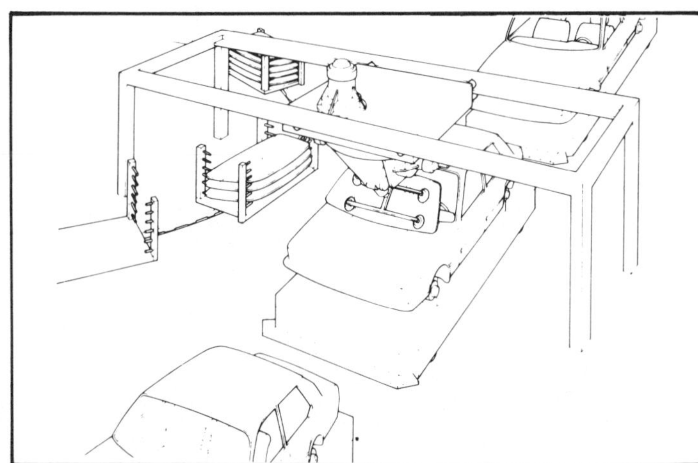

Fig. 7. IR 250/500 mounting car windshields.

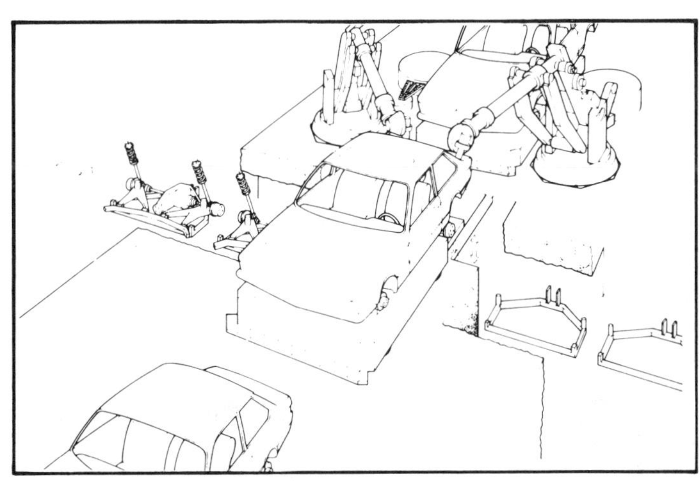

Fig. 8. IR 601/60 CP mounting car rear axles (general view).

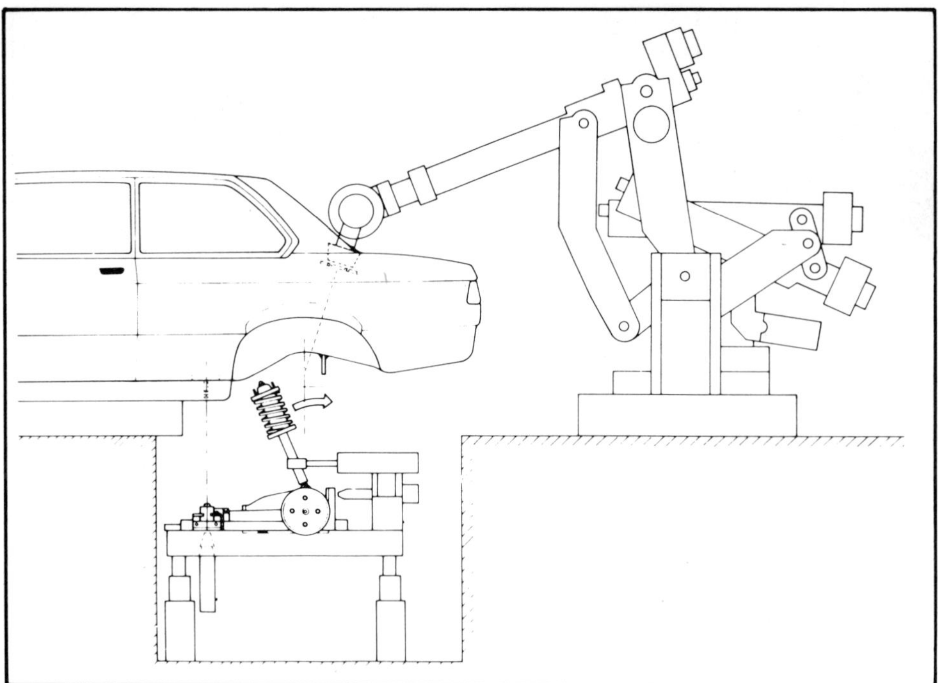

Fig. 9. IR 601/60 CP mounting car rear axles (side view).

References

[1] J. Kraus: Autoflex-System (The Autoflex System). Brochure published by KUKA Schweißanlagen + Roboter GmbH, Augsburg, 1980.

[2] H. C. Minhöfer: Roboter in der Schweißstraße – zentrale Kontrolle erhöht die Verfügbarkeit (Robot in welding lines – centralised command facility increases control). Moderne Fertigung 5/1981, page 42-54.

[3] H. Weule, W. Pollmann: Industrieroboter in Karosserieschweißanlagen (Industrial robots in car body assembly plants). Automobilindustrie 4/1980, page 55-61.

[4] D. Udelhofen: Roboterapplikationen in der Schweißtechnik – Darstellung einer neuen Anwendung (Robot applications in welding – portrayal of a new application). Conference on 'industrial robots – the strongest link in the manufacturing chain', 29-30.10.81, Böblingen, Verlag Moderne Industrie page 1-28.

[5] F. Köhne: Montageautomation als Rationalisierungspotential der Zukunft (assembly automation as a rationalisation potential of the future). Special conference at the 4th German Assembly Congress, 1981. Verlag Moderne Industrie.

Reprinted with permission from *Electronic Design*, Vol. 30, No. 9, copyright Hayden Publishing Co., Inc., 1982

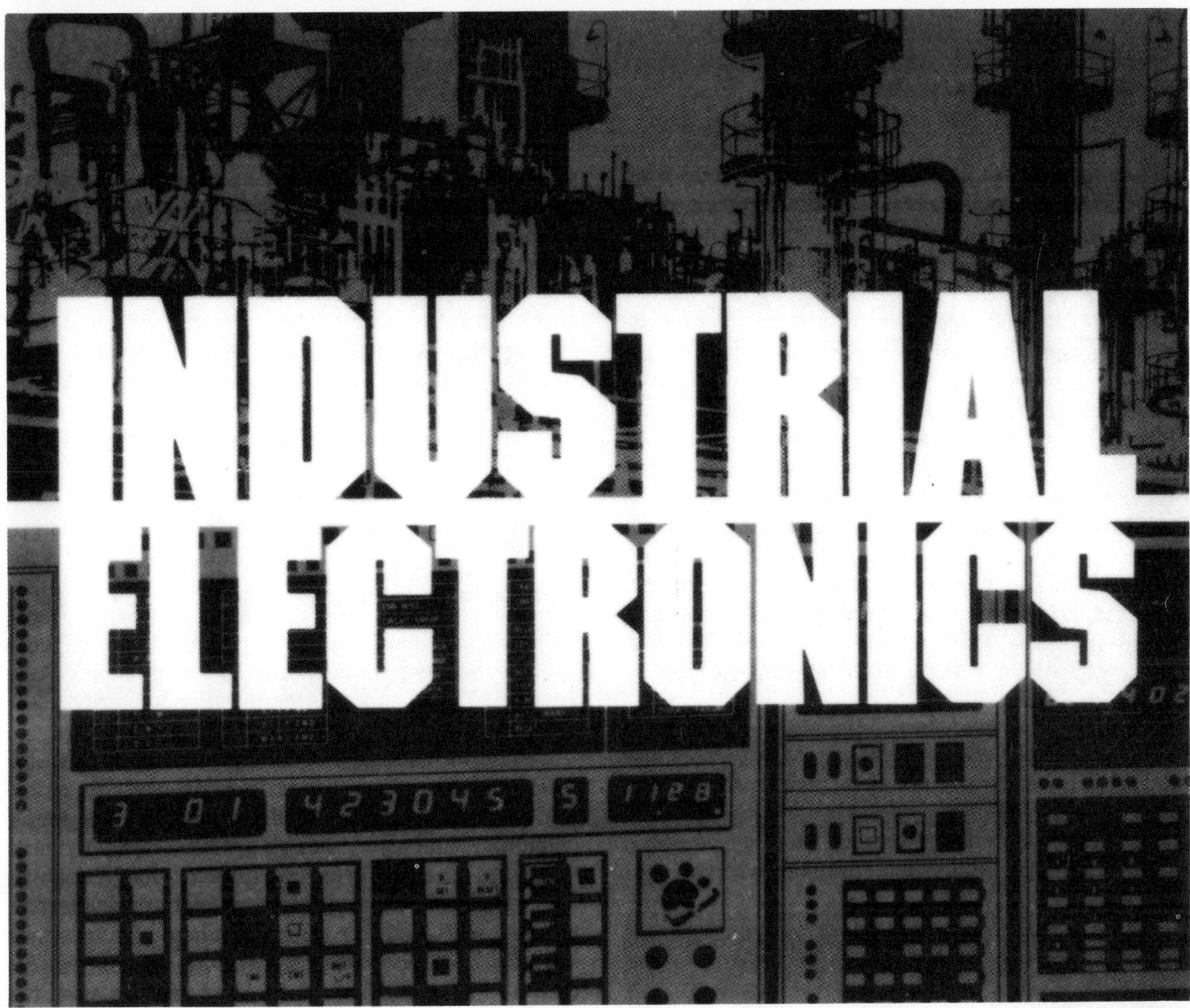

Robotics comes into its own with improved vision, sense perception, teaching languages

Roger Allan

Robotics is no longer merely a technological field of the future. Today it is not uncommon to see industrial robots loading and unloading machinery, handling materials, spray-painting, die-casting, or performing a multitude of assembly tasks. Improved sensory perception, movement, speed, and programming languages have yielded robots that are capable of performing many jobs now done by human beings. In fact, the prospect of using robots to perform mundane, repetitive assembly tasks has prompted many industries to move strongly into the field of robotics.

Future robots will gain in sophistication, intelligence, and dexterity. They will have three-dimensional vision systems that operate at split-second speeds. Furthermore, tactile and force array sensors will be attached to their arms and wrists (with the latter becoming increasingly flexible). Finally, off-line software-programming languages will supplement on-line methods to increase the robots' productivity in assembly applications.

Today's vision-sensing systems generally perform better in robotic applications than their predecessors did. Many have greater gray-scale capabilities (now at about 16 levels) and higher resolutions. Still, even better gray-scale capabilities are needed, as robots are increasingly being called on to see all images regardless of background lighting or overlapping, something most vision systems cannot do.

Researchers also are attempting to come up with more effective and powerful vision-sensing algorithms to tap the resources of present computer hardware. They note that the current 256-by-256-element array cameras have insufficient resolution and also require tremendous amounts of computer power, making their use in continuously operating vision systems both uneconomical and impractical. In addition, researchers are looking at higher resolutions—1000 by 1000 elements in an array—and at processors that can analyze complex scenes in about 100 ms (compared with 2 s currently). One solution is the integration of low-level image-processing operations like edge detection on the chip.

At Stanford University (Palo Alto, Calif.), researchers have designed a general-purpose VLSI NMOS chip suitable for parallel implementation of computer vision algorithms. It has a two-dimensional array of processors, with each processor connected to four neighbors, handling 32-bit internal storage in three shift registers, and performing arbitrary Boolean and serial-bit arithmetic operations (Fig. 1).

Ultimately, the researchers are hoping to develop one processor per vision-system pixel, which would require a very dense VLSI chip. That chip would likely take up an entire wafer and need fault-tolerant hardware to deal with the fabrication errors that are inevitable in such large circuits. (Redundant links provided between processors would be routed around faulty units.)

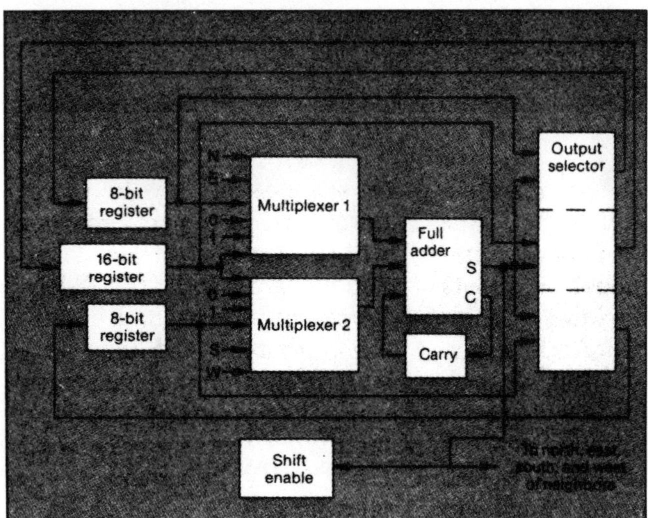

1. Researchers at Stanford University have designed a computer vision system that uses a two-dimensional array of serial processors on a single VLSI NMOS chip. Each processor is connected to its four neighbors and has a data path like the one shown.

As a benchmark, the researchers used a Digital Equipment Corp. KL-10 computer, which contains 6000 ECL chips and performs 8.7 million 8-bit additions/s. To match that performance, they say, a wafer with 16,384 processors would be needed, each processor doing 530 such operations/s at a clock rate of 4.2 kHz. They are currently testing a prototype chip with a single processor and redesigning it for more compact, regular layouts. They estimate that their chip could be placed on a wafer that would have a 10-MHz clock rate, yielding a 2500 times increase in speed.

From global to local

The first robotic vision systems detected differences between objects only by recognizing global features. Thanks to greater intelligence, gray-scale capability, and color, today's systems can detect and recognize fine features two dimensionally, making them well suited for quality-control and inspection applications—with or without robots. These newer systems perceive depth through light beams reflected off objects that contrast sharply with the background and that do not touch or overlap.

Like radar, some systems time the reflection of a laser or ultrasound beam off an object to measure its distance and discern its shape. Other systems project controlled light onto the object and determine the distance by triangulation, deducing the shape from the pattern formed at the intersection of the light beam and the object's surface.

At Carnegie-Mellon University's Robotics Institute (Pittsburgh, Pa.), researchers are working on a different method of depth perception for vision systems. They use a circle of fixed and focused LEDs that flash in sequence. When an object appears in the LED focal plane, an optical sensor detects a point of light. For objects further or nearer than the focal point, the sensor sees a circle or an ellipse of determinable size that rotates either clockwise or counterclockwise, depending on whether the image is nearer or further than the focal point. So long as an object is within 5 cm of the sensor, the sensor can compute its distance and orientation and determine its shape and dimensions.

Greater vision-system speed

One trend in vision systems is toward faster systems that will meet the demands of the high-speed conveyor belts and transfer lines they must

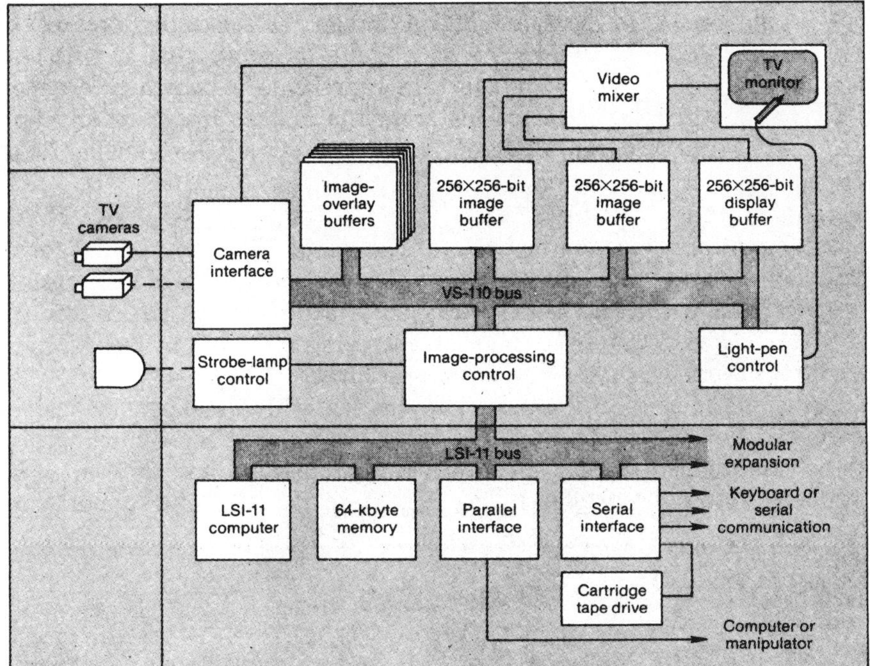

2. **The VS-110 vision system (a) from Machine Intelligence can inspect up to 900 precisely indexed parts/min for small dimensional defects. Its LSI-11–based circuit (b) views images and transforms them through masking, adding, or differencing operations.**

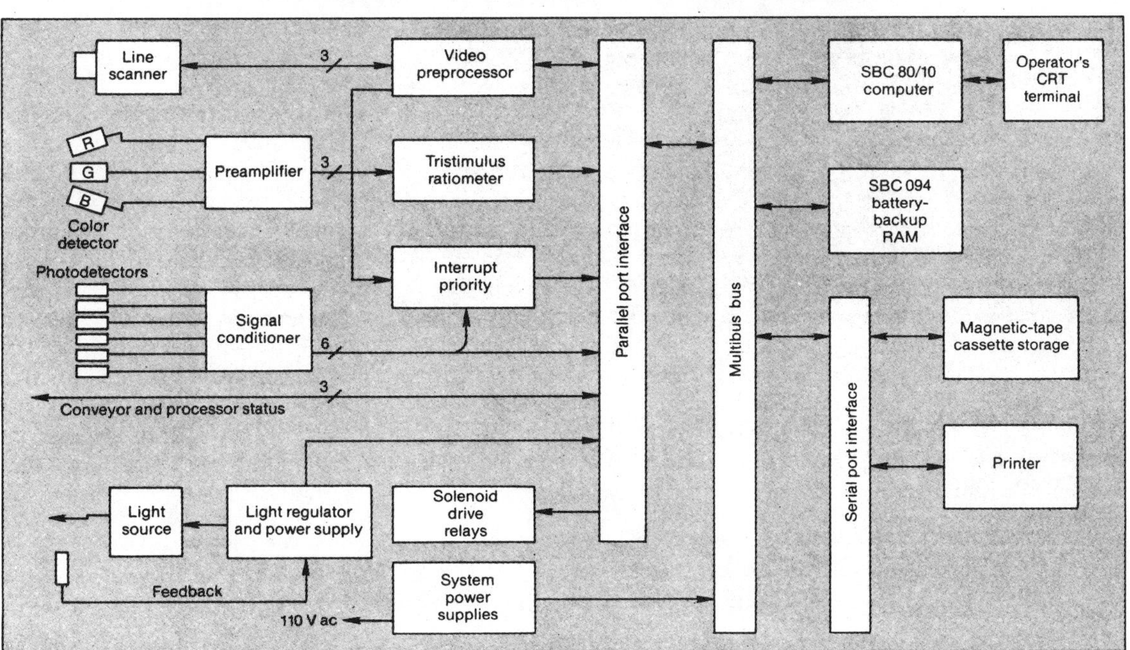

3. **Using the circuitry shown, Optical Recognition Systems' high-speed vision system can sort out 330 cigarette packs/min merely by recognizing their graphics content.**

often work on. One such system is the VS-110 from Machine Intelligence Corp. (Menlo Park, Calif.). The device (Fig. 2a) recognizes and inspects images of complex objects in real time against a contrasting background.

Designed for applications in which an object's position and orientation remain constant, it can visually inspect 900 precisely indexed parts/min for small dimensional defects. It does that by cross-checking an object's image with a master image stored in the system memory. The LSI-11–based system (Fig. 2b) views images and transforms them by masking, adding, or differencing operations.

Masking blocks out portions of the viewed field, allowing any unwanted background to be removed. Adding eliminates unnecessary fine details by filling in portions of an image. Differencing creates a pseudo-image of a scene by comparing it with an overlay image that represents a known reference. Any deviations from the master image or any apparent differences due to motion in the viewing field can thus be readily detected.

Optical Recognition Systems Inc. (New York, N.Y.) is another manufacturer of relatively inexpensive high-speed robotic vision systems. One of its systems (Fig. 3) currently sorts 330 cigarette packs/min in a tobacco plant, recognizing each pack by its graphics content with 98% accuracy.

Still another sophisticated and inexpensive vision system comes from Automatix Inc. (Burlington, Mass.). Autovision II operates on 360 moving parts/min, has 16 gray-scale levels, and supports up

Classifying industrial robots

The Robot Institute of America defines a robot as a "reprogrammable multifunctional manipulator designed to move material, parts, tools, or specialized devices through variable programmed motions for the performance of a variety of tasks."

The key words that distinguish industrial robots from robotlike machine tools are "reprogrammable multifunctional manipulator." Despite that distinction, typical industrial robots are little more than advanced, highly flexible, manipulative machine tools that generally accomplish their work through one or more precisely controlled articulating appendages.

Robot manufacturers classify their products in three ways—the type of motor-driving power, the degree of robot-arm freedom, and the method of control—all of which affect a robot's performance level and intended application.

The most common robot classification is by the type of motor-driving power, specifically pneumatic, hydraulic, and electric in ascending order of average cost and weight-lifting capability. (Hydraulic robots operated in a closed-loop control system, however, can lift the heaviest weights.) Pneumatically driven robots are generally poorly suited for applications that require precise manipulation of parts and materials, as air is difficult to control for speed and position.

Most industrial robots use hydraulic drives, which afford a lot of force and power in relatively compact sizes and within relatively tight spaces. These drives give a good balance of high accuracy and weight-lifting capabilities but require periodic maintenance, usually done by pipefitters. For hazardous and potentially explosive environments where electrical contacts may be a problem, hydraulic robots may be the answer.

Electrically driven robots, though comparatively more expensive than pneumatic or hydraulic ones, are beginning to appear for more assembly applications, particularly those that involve parts weighing a few pounds or less. With all three drives, the robot's positioning accuracy generally decreases as its weight-lifting capability increases.

Regardless of drive power, robots are also classified by the number of axes through which their arms or manipulators work, usually expressed in degrees of freedom (or axes of articulation). As shown, robot-arm or manipulator motions are

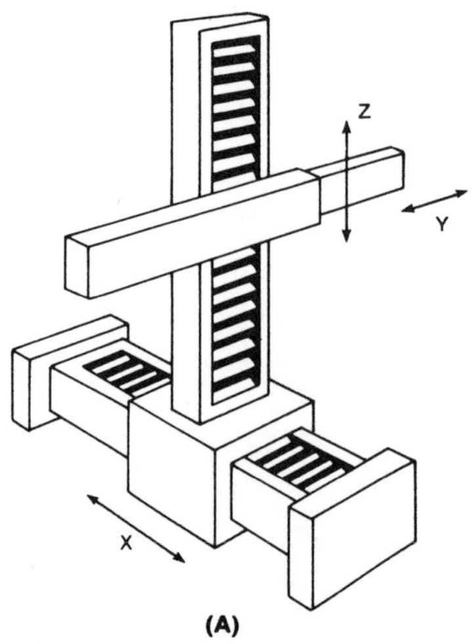

(A)

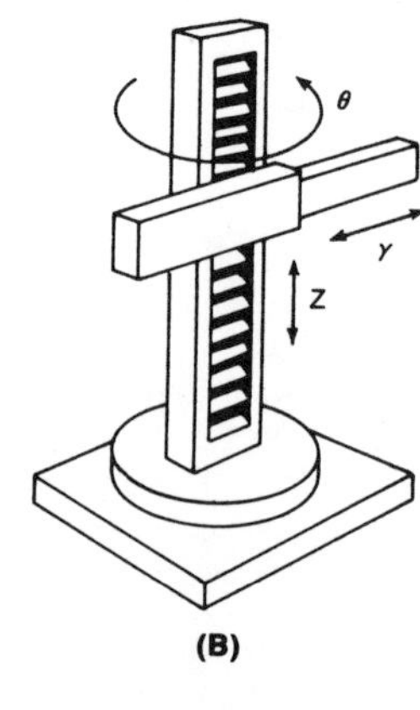

(B)

to four cameras (Fig. 4). It visually inspects, identifies, orients, and locates parts. The system is useful as a stand-alone or robotic system and is taught by the firm's conversational language, called Rail.

High-speed versatility

One of the most advanced vision systems is the versatile Optomation II from General Electric Co.'s Optoelectronic Systems operation (Syracuse, N.Y.). The microcomputer-based product performs 100% noncontact on-line inspection of randomly oriented parts at 900 parts/min. Its main components are the PN230Y decision processor and the TN2500 charge-injection device (CID) camera (Fig. 5), which is used in many other commercial robotic vision systems. The system employs three microprocessors, one for image analysis and two for system and display control. It is taught using VPL, a high-level language similar to Basic, and can handle up to four cameras.

Other real-time noncontact vision systems, like the Optocator from Selectronic Electronic Inc. (Valdese, N.C.), are pushing robot vision to new heights. The Optocator is teamed up with a robot from Asea (Troy, Mich.) to gauge, profile, and inspect components with a resolution of 0.00004 in. and a maximum error and nonlinearity as low as 0.00008 in.

The system comprises a modulated laser or infrared light source, a photodetector, and a processor. At a rate of 16,000 times/s, the light source projects a spot of light having a diameter of 0.04 to 0.25 in. onto any object with measuring ranges of 0.3 to 10 in. The photodetector reads the light spot in one

classified in four ways: rectilinear, cylindrical, polar, and anthropomorphic.

In rectilinear, or Cartesian, motion, a robot's arm moves along three Cartesian axes—X, Y, and Z (A). This is the simplest kind of robot-arm motion. A cylindrical robot arm is a point-to-point device mounted on a rotation axis that creates a cylindrical work envelope (B). Polar motion permits a robot arm to reach all points within a spherical envelope (C). In polar-coordinate robots, the arm can be programmed for both continuous-path and point-to-point motion. Anthropomorphic motion, also known as revolute motion, is the most sophisticated. It gives a robot's arm a total of six degrees of motion freedom in the shoulder, elbow and wrist (D). However, an anthropomorphic arm generally cannot lift as much weight as one involving fewer axes of articulations.

Robots are probably most widely classified according to whether they are controlled by nonservo or servo drives. That classification is the simplest way to separate less sophisticated robots from more intelligent ones.

With nonservo drives, the robot's tool-center point stops only at the fixed-end points of each axis of motion. Many sequential motions are possible along each axis, but they can be accomplished only between two specific end points; thus there is no motion control between points. Furthermore, nonservo-controlled robots cannot accelerate or decelerate their arms. Simple pick-and-place robots are in this class, as they are controlled by stops, peg drums, or plug boards that define the end points of motion. These point-to-point robots, however, are consistently more accurate in their tool-center-point positioning than servo-controlled ones. Also, they are usually smaller in size and less expensive, though they cannot lift as much weight as most servo-driven robots.

Servo-controlled robots offer continuous-path arm motions and can generally be programmed to stop at any point within the arm's range of motion. Control of acceleration and deceleration is also possible. These robots typically use internal sensors to gather data on the velocity, position, acceleration, force, and torque status of the manipulator; they then compare that data with predetermined operational parameters in the control program. When deviations from those parameters are detected, the robot's servos start corrective action.

The most sophisticated servo-controlled robots can be instructed not only to operate in a continuous path but also to branch off at a specific point in that path and go into a subroutine of motions. They then return automatically to the point where they branched off once they complete the subroutine.

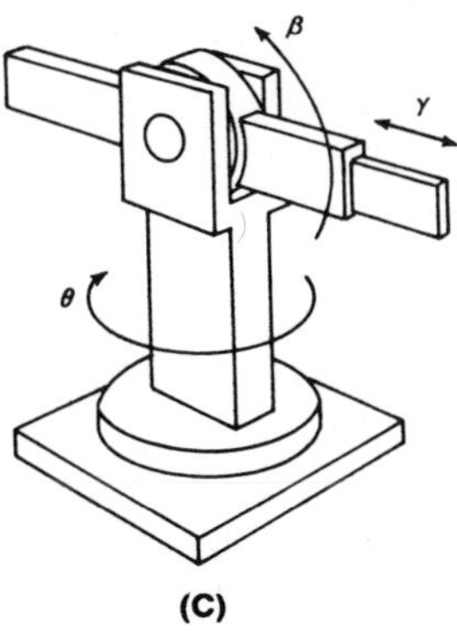

(C)

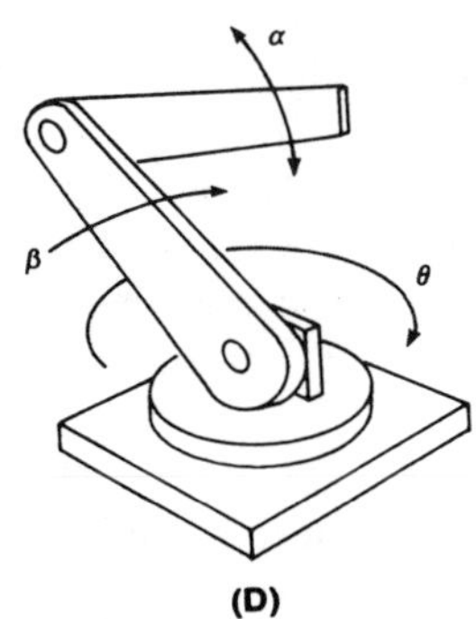

(D)

4. Enclosed in a rugged case, the Autovision II system from Automatix can inspect up to 360 moving parts/min. The system has 16 gray-scale levels, supports up to four cameras, and is programmed by the company's conversational language, called Rail.

dimension and delivers a signal to the processor, which then produces an output signal, displayed as numerals, that accurately indicates the light spot's position on the object under examination.

Plug-in board versatility

Vision systems like the Ermac (electronically reprogrammable modular automation control) from Everett/Charles Automation Modules Inc. (Rancho Cucamonga, Calif.) can be easily upgraded using plug-in printed-circuit boards.

Geared for real-time visual inspection and control of parts, the basic system (Fig. 6) uses the Ermac 1100 controller and accepts several single-board plug-in modules. For example, an expansion chassis with up to 15 slots for options can be added to make the Ermac 1200 system, which acts as a stand-alone package. Connecting a CRT display and keyboard assembly results in the Ermac 1300, and incorporating a closed-circuit TV camera and proprietary video-processing hardware produces the Ermac 1500 system.

Taught by the Direct Logic language, which is an English-like high-level language, the Ermac systems also provide a variety of I/O features such as slots for STD, IEEE-488, or serial interface buses.

Though high reliability in a vision system is a must, many of today's vision systems do not necessarily provide that sense of security. Manufacturers often leave the control of such variables as background lighting, moving and randomly oriented parts, external noise—as well as the subsequent complex software programming—to the user. What is needed, some believe, is a vision system that is simply plugged in and ready to go.

One system that may solve these problems is the MV100 from Perceptron Inc. (Farmington Hills, Mich.). Designed to work on a variety of conveyor-belt configurations, it permits a robot to identify, count, and apply finishing paint to different parts merely by viewing their silhouettes.

Functionally, the system comprises a proprietary high-speed image preprocessor that interacts with an image buffer (both are controlled by an LSI-11 16-bit microcomputer), a solid-state photosensitive camera (128 by 128 elements), and a 4-by-4-ft lighting unit, which creates the silhouettes. The microcomputer also monitors the console that displays the part images.

Information on the parts to be identified is programmed into an internal cassette-tape recorder and then fed to both the robot and a programmable controller through an I/O port. With the system, a robot can work on up to 60 different parts/min, storing the count in a non-resettable internal counter. Up to 32 color flags can issue appropriate commands through the system's serial or parallel ports in case the parts colors change.

Fiber optics—a low-cost solution

Dolan-Jenner Industries Inc. (Woburn, Mass.), a manufacturer of robot joint drives and fiber-optic components, proposes a scanning fiber-optic vision system for robots as a low-cost solution to vision sensing. The firm is designing a programmable multipoint robotic vision-sensing system that uses coherent fiber optics as well as noncoherent light guides and photoelectric controls. The system, which the company says should be relatively inexpensive, will perform a number of functions similar to those of programmable controllers.

Optical fibers are used to transmit signals from a robot's moving hand to its stationary arm (Fig. 7). When mounted at every axis of the hand, such fibers could provide complete information on the hand's location with respect to its environment. Furthermore, the fibers can be molded to the outline of the object being viewed, so that strategically located fibers can determine the object's shape by picking up its discrete characteristics.

Stereo vision coming

In order to perform such tasks as picking specific parts out of a bin full of different parts, robots will require three-dimensional vision systems. Fortunately, various universities—the Massachusetts Institute of Technology (Cambridge, Mass.), the University of Rhode Island (Kingston, R.I.), Stanford University, the Jet Propulsion Laboratory at the California Institute of Technology (Pasadena, Calif.), and Carnegie-Mellon University—are conducting research on stereo vision that should give robots the ability to locate and grasp randomly oriented objects, regardless of background lighting, and to negotiate objects in their path.

At MIT, researchers are working on practical hardware that will implement a stereo-matching circuit for real-time three-dimensional vision. Two 1024-pixel linear-array scanning cameras developed by MIT are used for vision-system input, and a pair of serpentine memories buffer the video outputs from the cameras (Fig. 8). Signals are fed first to matching digital convolvers, which compute the video signals at a pixel rate of 1 MHz, and then to matching zero-crossing encoders whose outputs are fed to a stereo-matching circuit. The encoders detect zero crossings in the convolution raster signal and code the orientation and gradient of the image contour at that location. The matching circuit then computes both disparities of unique-image matches as they occur. It also computes information on the absence of any matches at locations where they are expected.

The MIT researchers soon hope to have a prototype three-dimensional image system that can produce a depth map of 1000-by-1000-pixel images within 1 to 4 s.

Geometric shapes as cues

At Stanford University a research team has developed the Acronym three-dimensional vision system for inspecting and picking parts out of a bin. Simple geometric shapes are stored in the system's memory and act as cues for a robot to represent objects in three dimensions. Acronym represents objects as partial or whole graphs composed from generalized cylinders like cones or polyhedra.

Cal Tech's Jet Propulsion Laboratory (JPL) is also developing a stereo vision system with 188-by-244-

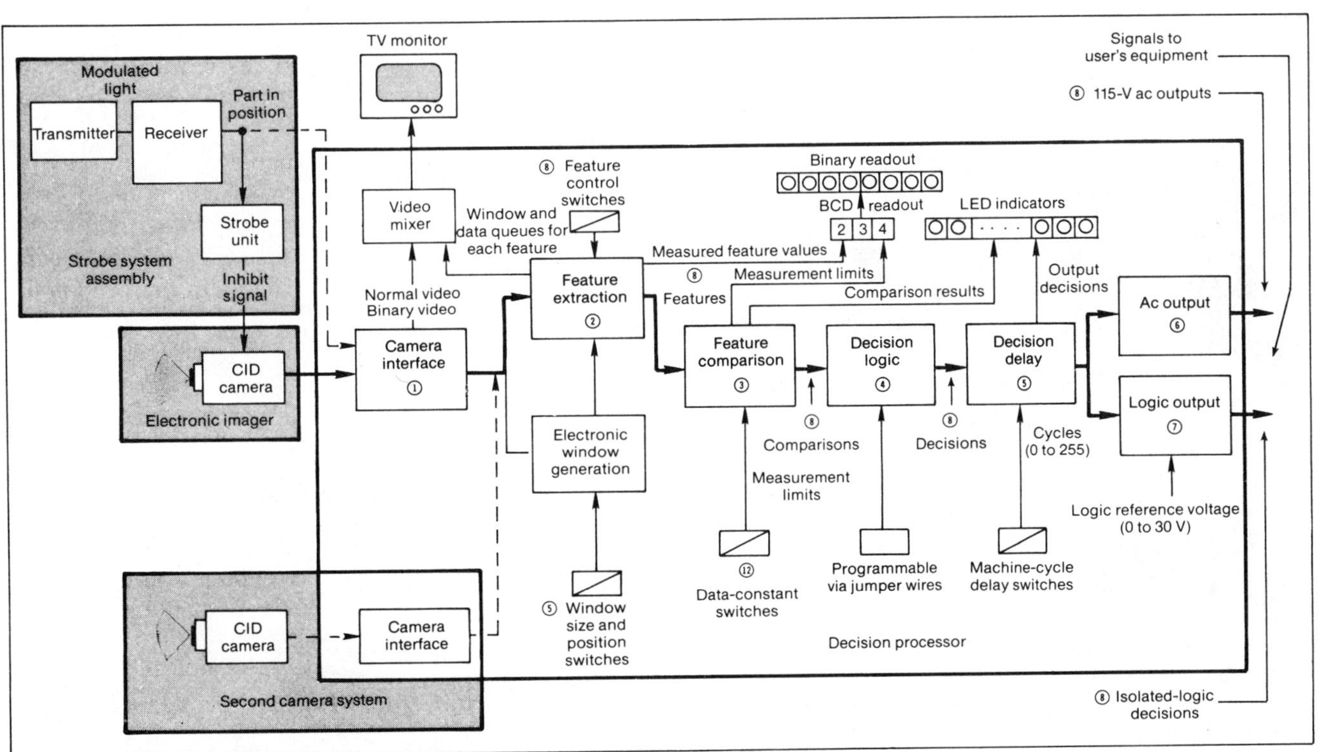

5. General Electric's Optomation II vision system can inspect up to 900 randomly oriented parts/min and support four charge-injection device (CID) cameras. The circled numbers indicate respective processing steps for video data.

pixel solid-state cameras. It will generate a gradient image to discern edges and thresholds, thin the edges to obtain a binary image, and chain-code the information to reduce the data set that needs processing. Real-time vision is achieved through pipeline processing (i.e., the hardware is segmented so that gradient images, thresholds, and thinning are processed concurrently). With that method, the lower-level features of a complete image can be processed 30 times/s.

JPL researchers built a custom hardware preprocessor that processes television images in real time. Imfex (image-feature extractor) uses a pipeline architecture to process video input signals by fitting straight lines to edges (Fig. 9). The JPL researchers have also demonstrated an autonomous robot manipulation system that employs vision and force-torque feedback while the robot places a pin in a hole.

SRI International (Menlo Park, Calif.) is working on a program sponsored by the National Science Foundation that would enable robots to recognize printed-circuit boards and their contents. Researchers there are considering using all types of sensors, as well as three-dimensional vision.

6. **Automatic 100% inspection of small parts is possible with the Ermac modular vision systems from Everett/Charles Automation Modules. The basic Ermac 1100 controller can be upgraded with such equipment as an expansion chassis or a closed-circuit television camera. Users can easily program the system with the firm's English-like high-level language, Direct Logic.**

Many industrial-robot manufacturers are realizing that certain tasks, like assembly, require more of the robot than just vision sensing. To be effective machine tools, industrial robots must also have tactile and force- and proximity-sensing capabilities and more dextrous appendages.

Tactile sensing gains ground

In addition to algorithms and processing power that operate on sensor information, inexpensive tactile and wrist sensors must be developed. Most sophisticated wrist sensors currently cost $6000 to $7000, but the user must usually develop the sensor.

Nevertheless, inexpensive wrists and tactile sensors are not far in the future. Cincinatti-Milacron (South Lebanon, Ohio) has already developed its three-roll wrist to simplify the design of wrist-drive mechanisms. The device uses three roll axes to get pitch, yaw, and roll motions that coincide at a single point within the wrist.

Modular and flexible robot wrists like the Hydrowrist from Bird-Johnson Co. (Walpole, Mass.) have servo valves and feedback transducers for each of those three motion axes. Thus each axis may be regulated individually or with other axes for maximum flexibility.

At MIT, a 16-by-16 touch-sensitive array of pressure-sensitive switches was constructed from pantyhose material sandwiched between a flexible PC board and graphite-impregnated conductive silicone rubber. About the size of a human fingertip, the 256-point array fits on a robot's finger, enabling it to discriminate between six different objects (screws, washers, cotter pins, and so forth).

When a robot using the fingertip sensor touches an object, the array switches make contact in a pattern that is detected and analyzed by a minicomputer, thus providing the robot with the ability to recognize objects as well.

The U.S. Naval Surface Weapons Center (Silver Springs, Md.) is currently developing high-performance, rugged tactile and force feedback sensors that are cost-effective and relatively simple. In the program, magnetoelastic/magnetostrictive sensors are retrofitted to a typical robot or machine-tool gripper (Fig. 10), giving it the sensing feedback to detect grip, slip, and torque conditions. (A material becomes magnetoelastic when changes in its physical length, or in the physical forces applied to it or by it, influence its magnetic field. The magnetostrictive effect works in the reverse: changes in a material's magnetic field cause changes in the material's physical length.)

For such simple assembly tasks as inserting one part into another, the Remote Center Compliance

Robots as assemblers

Despite the fact that only a small fraction of today's industrial robots are used to perform assembly tasks, robotic designers and researchers are leaning toward assembly applications as the next big area in which these programmable manipulators can be put to work.

Many assembly operations are performed on either a vertical or horizontal axis and can best be served by a Cartesian robot moving along X, Y, and Z coordinates. (Cartesian robots are easier to control and program than anthropomorphic ones.) One example, the Cybervision assembly robot, comes from Automatix Inc. (Burlington, Mass.). The turnkey system comprises an AID600 Cartesian robot, which lifts up to 17 lb; an AI32 controller; the firm's conversational teaching language, called Rail; its Autovision inspection system; and equipment that presents the parts and transports the material.

The AI32 controller has a 32-bit processor that can monitor multiple robots, sensors, and processors simultaneously. Its expandable memory (up to 2 Mbytes) has enough storage capacity to handle many software programs as well as user-defined functions.

Bendix Corp.'s Robotics Division (Southfield, Mich.) has developed its own compact assembly robot. The AA-160 computerized numerical-control (CNC) industrial robot has six axes and an all-electric design that permit it to move in polar coordinates and carry up to 45 lb. The robot's telescoping arm—reaching to 102 in. vertically and 60 in. horizontally—can literally bend over backwards within a compact spherical work envelope, thus preventing interference with other closely spaced machines in a tight area. The robot can position a part to within ±0.004 in. and is consistently accurate to within ±0.002 in. All six articulation axes can be used at once.

Bendix' Dynapath system 5AR, a computerized numerical controller, automatically allows the robot to traverse the shortest path between any two programmed points at a record speed of 170°/s for robot-base rotation.

The trend toward using robots in assembly applications was well demonstrated at last month's Robots VI Convention in Detroit, where many small and nimble electrically driven assembly robots went about their paces. These included the six-axis T3-726 from Cincinnati-Milacron (South Lebanon, Ohio), the six-axis Puma 260 from Unimation Inc. (Danbury, Conn.), the six-axis Series 1000 and 2000 from Westinghouse Electric Co. (Pittsburgh, Pa.), and a four-axis robot that is part of the 7535 manufacturing system from IBM Corp. (Boca Raton, Fla.).

Also at the convention were multiple-arm assembly robots like the Allegro A12 from General Electric Co. (Bridgeport, Conn.). Two years ago the firm obtained licensing rights from Digital Automation SpA (Turin, Italy) to manufacture the latter's Pragma A3000 robot, upon which the A12 is based.

Driven by dc servomotors, the A12 has up to four arms, which operate independently or in unison, and employs tactile sensors that can be installed on any or all of the arms' three sliding axes. The tactile sensors attached to the grippers allow the robot to react to missing or substandard parts, with an accuracy of within ±0.001 in. and a repeatability to within ±0.0001 in. The robot has three standard Cartesian axes movements, as well as two or three optional rotations for wrist roll, pitch, or yaw.

The robot can assemble a variety of electromechanical, electronic, or automotive components (weighing up to 14 lb) at 225 ft/min at the arm tip. It also can accelerate at 20 ft/s² in its Y axis and 13 ft/s² in its X and Z axes.

To get a better idea of the robot's versatility, speed, and capabilities, General Electric offers the following examples: A three-arm A12 can assemble as many as 450 20-part odometers with five-digit drums in one hour; a two-arm A12 can assemble 320 12-part compressor valves in one hour; and even a single-arm robot can insert 1000 integrated circuits onto a printed-circuit board in one hour, choosing from among 100 different chips.

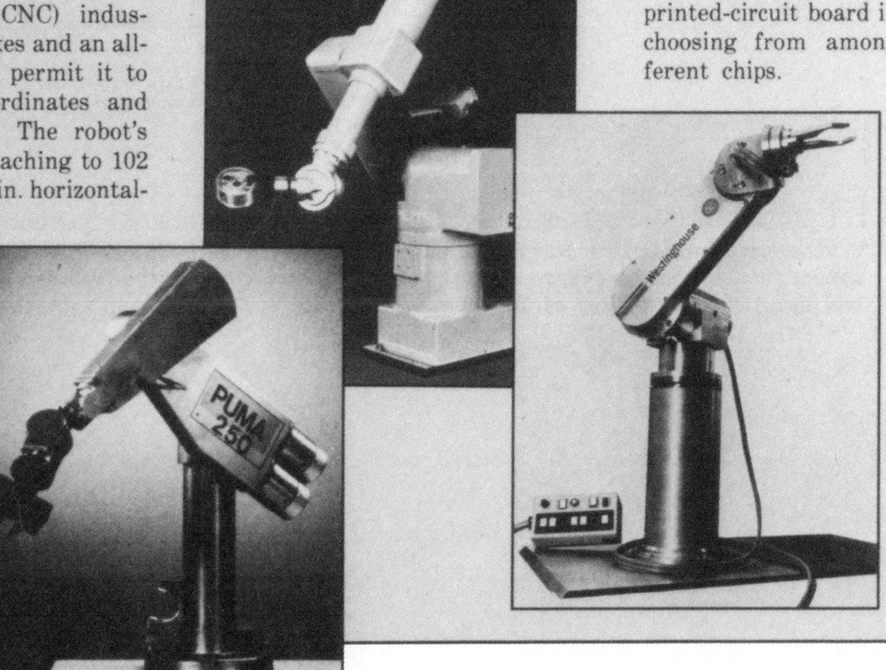

(RCC) device developed at MIT's Charles Stark Draper Laboratories has proved to be an inexpensive way to make a robot's arms more flexible. Mounted on a robot's wrist, the mechanical device lets the machine mate contacting parts together (like a peg in a chamfered hole), even if the parts are misaligned by 1 or 2 mm. The operation is based on the principle that the center of compliance for the mating parts is absorbed and shifted by mechanical springs on the RCC device, thus giving more flexibility to a robot's arm to correct mismated parts. Recently, MIT used the RCC concept to make the insertion of chamferless parts possible.

Lord Corp. (Erie, Pa.) has taken a different approach with the RCC concept, one that is said to be used in many industrial robots today. Instead of springs, it uses elastomeric materials to give the robot's arm flexibility.

The company has also just unveiled a tactile array sensor for robots that uses elastomeric materials. The 8-by-8 array has elastomeric compliant tips spaced 0.3 in. apart. When a robot fitted with such an array touches an object with a 1-lb force, the appropriate compliant tips are depressed about 80 mils; an optical, magnetic, or other sensor fitted behind each of the tips detects the degree of depression. With appropriate computer circuitry, the robot can determine a part's contour, outline, orientation, and even any force and slip sensing.

Ac servomotors are lighter

General Electric Co., Siemens AG, Toshiba Ltd., and several other Japanese companies are investigating using ac servomotors to control robot joints instead of using the conventional dc servomotors. An ac servomotor is less expensive and lighter than a dc motor, so that lighter-weight arm materials can be used.

For the same horsepower rating, an ac servomotor weighs 30% to 40% less than a dc servomotor; when used to drive a lighter robot arm, the cumulative effect is about 60% less weight. Also, an ac servomotor does not create sparks, so that it is safer to use in hazardous environments. Unfortunately, today's ac servomotors are difficult to control and cannot be stopped accurately, though those problems may be corrected soon.

General Electric's Industrial Electronics Develop-

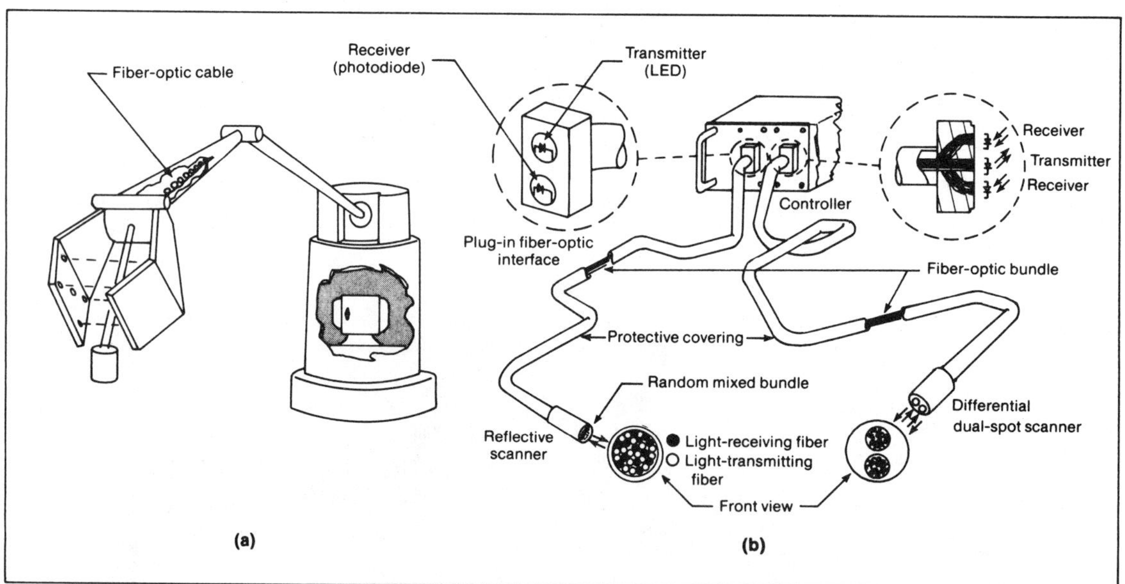

7. Fiber-optic cables can form an inexpensive vision-sensing system in industrial robots. Coiled through a robot's arm (a), the fibers can transmit signals from a rotating hand to the stationary arm as well as extend and retract when the arm moves. The receiver-transmitter system uses dual-spot scanners (b).

ment Laboratory (Charlottesville, Va.) is developing control algorithms in hardware that will control ac servomotors more accurately.

Most industrial robots use either optical encoders or resolvers and resolver-to-digital converters in their joints for determining position. A resolver placed in a robot joint, for example, produces a cosine signal indicative of the angle subtended by that joint. Twelve-bit resolvers with r-d converters and 11-bit optical encoders are used most commonly.

Sophisticated assembly robots, like the Puma of Unimation Inc. (Danbury, Conn.), use absolute optical encoders, whereas less sophisticated robots employ incremental optical encoders. The latter are generally half the price of absolute encoders for a given resolution level but are more susceptible to external electromagnetic interference. They may lose bit counts as a result of such interference, in which case a robot would not know where its arm is at a specific time.

Resolvers and r-d converters—manufactured by such firms as Analog Devices Inc. (Norwood, Mass.), Natel Engineering Co., Inc. (Canoga Park, Calif.), and ILC Data Device Corp. (Bohemia, N.Y.)—are now mounting an offensive against optical encoders. More rugged and easier to use, they can be wired into a robot system. In addition, their signal levels are not susceptible to electromagnetic interference. Optical encoders, on the other hand, are generally fragile devices and therefore susceptible to inaccuracies caused by vibrations.

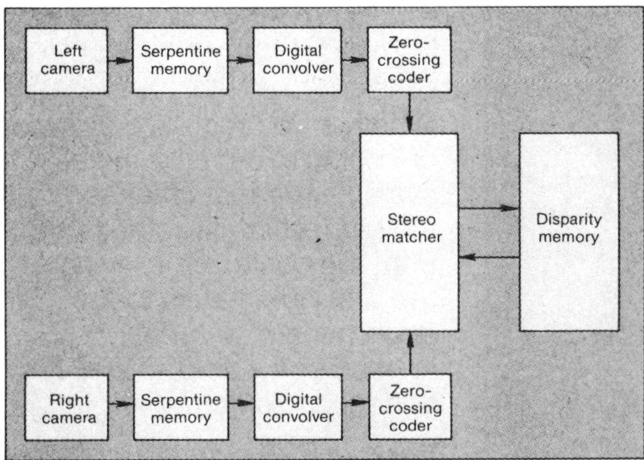

8. Stereo matching of images from two cameras is a problem for most commercially available vision systems. But researchers at MIT have developed a stereo-matching circuit that provides robots with real-time three-dimensional vision.

Teaching the robot

Most industrial robots today are programmed on line, meaning they are either led through the desired motions via a remote box or literally walked through those motions. Since playback speeds are independent of teaching speeds, teaching on line can be done slowly at first for more accurate positioning, with the robot operating at faster speeds later. On-line programming, however, has proved to be too time

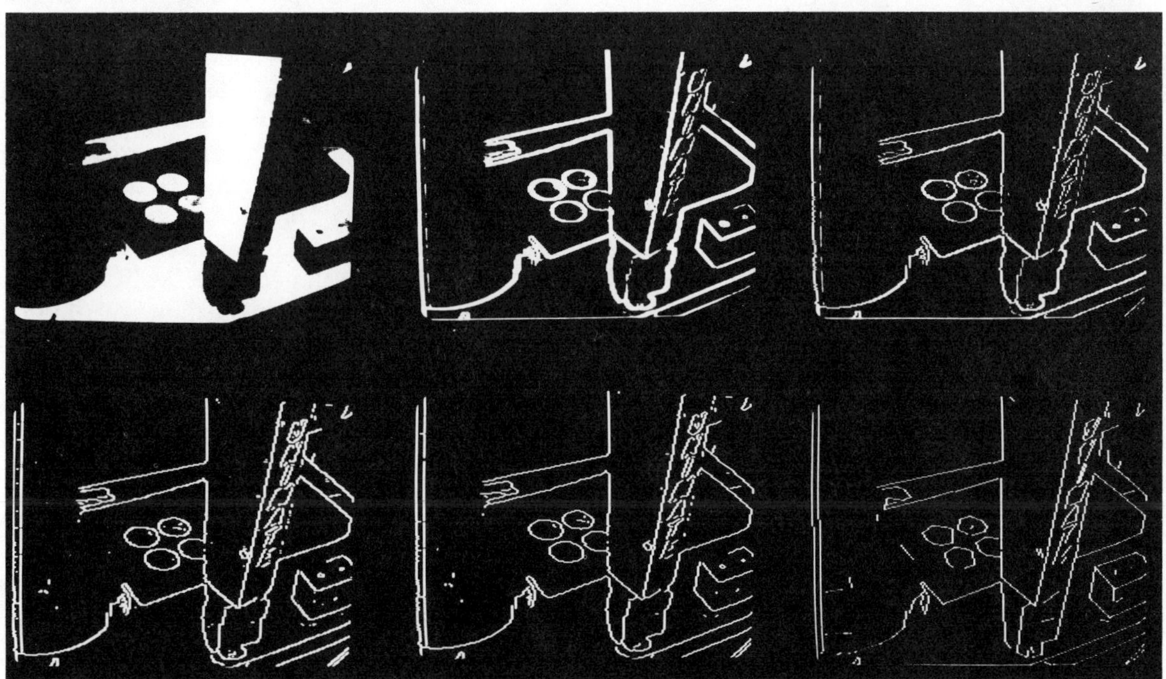

9. A preprocessor for three-dimensional vision, designed at Cal Tech's Jet Propulsion Laboratory, processes video input in real time. A pipeline processor, called Imfex (image-feature extractor), fits straight lines to edges in a series of steps. From top left to bottom right, the stages are raw video, gradient image (spatial differentiation), primary-edge thinning, edge thresholding, secondary-edge thinning, and line-segment fitting.

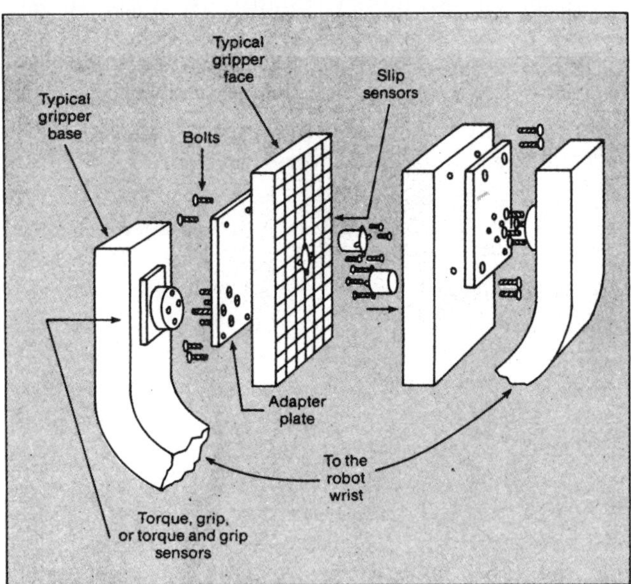

10. Magnetoelastic sensors developed at the U.S. Naval Surface Weapons Center have been retrofitted to robot wrist grippers. The high-performance sensors provide torque, grip, and slip information.

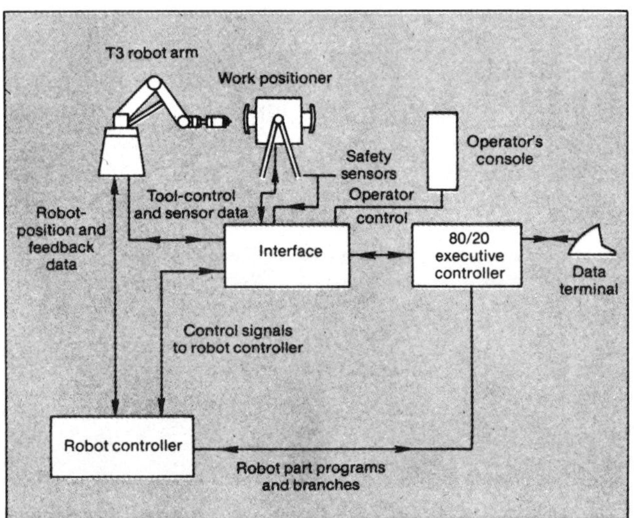

11. A robot work station at General Dynamics Corp. integrates a Cincinnati-Milacron T3 robot into a system controlled by a CNC and an Intel 80/20 microcomputer.

consuming and unproductive for small- and medium-batch assembly jobs.

The greater use of multiaxis robots capable of complex manipulations has brought with it a need for more sophisticated programming. Newer off-line teaching methods using explicit or world-model high-level languages are beginning to surface.

With off-line teaching, a robot is taught either explicitly (a programmer specifies every individual motion with a statement) or implicitly (a programmer specifies only the general task to be performed). Implicit programming assumes that the robot has a large data base in which such things as workpiece and work-tool characteristics and their spatial relationships are described. Implicit languages thus require more complex software compilers and more sophisticated data-base management systems than explicit ones.

Though off-line languages generally reduce machine tool setup times and permit better software documentation, they also require greater computer power than on-line teaching languages.

Integrating the robot with a system

Robot interfacing is an important issue that has yet to be tackled by robot manufacturers to any large degree. As yet, there are insufficient standards, given the many robot interfaces that control-system manufacturers provide; many of these interfaces are not readily compatible with one another.

General Electric is currently working on a program to interface robots with other automation tools, robot-teaching languages, and robot positioning and control. The firm is also working on the direct downloading of software programs from one of its own Calma CAD terminals to a robot controller.

Thermwood Corp. (Dale, Ind.), a major manufacturer of industrial robots, would also like to see controllers designed specifically for robots. It cites one particular problem, however: many numerical control systems cannot handle five or six axes of robot articulation, and when systems are modified to handle that, the cost becomes prohibitive.

Control Automation (Princeton, N.J.) has taken an innovative approach to robot networking (see the news story in this issue) with an assembly robot that is controlled by any computer. Instructed in English-language commands, the robot interfaces with other automation elements via an RS-232 communications line.

For robot applications to take off, they must be totally integrated with computers, machine tools, sensors, and inventory-control systems. At General Dynamics Corp. (Fort Worth, Texas), for example, the industrial robot's role in a large closed-loop control system has been well defined (Fig. 11).□

J.G. Aceti | C.B. Carroll

Robotic assembly of the VideoDisc player's arm drive

Electromechanical assembly—an elusive task for robots, and a lucrative area for manufacturing—has begun at RCA.

In April 1982, the VideoDisc Player Manufacturing Technology group at RCA Consumer Electronics and the Electromechanical Systems Group at RCA Laboratories proposed a prototype subassembly as a candidate for robotic assembly. The subassembly was a VideoDisc arm drive to be used in RCA's prototyped J-line player, scheduled to be assembled in the Bloomington plant. This project would become the first large-volume robotic assembly system within RCA. The obstacles to implementation present a double-edged sword, both technological and economic. But, correspondingly, the rewards for meeting the challenge will be a two-pronged solution to lagging U.S. productivity.

Robotic assembly today

Electromechanical assembly presents a formidable task to robotic system designers.

Abstract: *Competitiveness in manufacturing allows only the progressive and flexible companies to survive. Robots will play a role in the factory of the future as the flexible muscle driving the manufacturing process. This is the account of RCA's first introduction of a robot into the production environment of electromechanical assembly. It involves interesting economic and technological issues, and shows cooperation among groups at RCA Consumer Electronics and RCA Laboratories.*

Final manuscript received February 6, 1984.
Reprint RE-29-2-2

Programmable mechanical arms have accomplished many industrial tasks—such as spray painting, welding, and part handling—where the task does not require the exacting accuracy of assembly. Assembly requires more of the system—an additional degree of flexibility, adaptability, precision, and in general, sensory feedback mechanisms similar to a human assembler on a production line.

Assembly and inspection of product account for 40 percent of direct labor in the radio and television industry.[1] Many of the "simple" tasks performed by these workers actually require a complex interaction among sight, hearing, and touch that is difficult if not impossible for an electromechanical system to integrate and mimic. Moreover, robotic system designers must pay particular attention to the part-feeding methods that provide the robot with oriented and accurately positioned parts (see box, page 12). And even before that, careful consideration should be given to the design of parts for automatic handling. A more extensive review of this subject is covered in an article by Malcolm Cobb, elsewhere in this issue. Understandably, application of robots for assembly-line work has been minimal—amounting to a little over 5 percent of the robotic systems operating nationwide. A recent survey conducted by *Assembly Engineering* magazine showed that 13.3 percent of the 1,000 assembly plants participating in the study used robots and that 28.3 percent of those robots were part of the assembly line. The other 71.1 percent were used for part handling, electronic component assembly, arc welding, stand-alone operations, inspection, spot welding, fastening, wire harness assembly, and calibration.[2]

Economic and political issues

With the impressive array of tasks being performed and the potential economic advantages, why are only 13.3 percent involved in robotics? There are several reasons:

- Risk associated with new machinery
- Labor conflicts
- Inadequate project proposals

Understanding the risk

Those involved with providing, justifying, or using assembly automation equipment have associated the risk of using robotics with that of using uncompromising hard automation—machinery uniquely designed for a specific product, with little possibility for future modification. To a great extent, the difficulties associated with hard automation are appropriate, but the flexibility, reprogrammability, and reduced cost of modern robots have diminished the uncertainty of bringing automation into the factory. Robots can reduce both economic and technological risk—the economic risk of having automated assembly costing more than manual assembly and the technological risk of being inflexible to product improvements. An understanding of the economics when using truly reusable "off the shelf" equipment proves profitable by shortening the payback for each application. When all avenues of savings are consid-

Orienting and feeding parts

Part feeding is a technology that has lagged behind the advanced automation systems it supports. Neither flexible nor sophisticated, part-feeding equipment is usually constructed by artisans working in small specialized shops with welding torch and hammer.

The most common feeding method, bowl feeding, provides the builder with versatility to handle many different parts, and provides the user with many years of reliable service. Being the lowest-cost feeding method, designers should prepare for automating the assembly with bowl feeders by following some guidelines for part design.

Not all parts can be bowl fed. Delicate, sticky, light, tangled, or abrasive parts are generally not bowl fed. For most parts the overriding concern is geometry, and in particular, symmetry. If a part is either symmetric or grossly asymmetric, then feeding will be easier and more efficient.

Dr. Geoffrey Boothroyd, of the University of Massachusetts in Amherst, has researched this subject quite thoroughly and has produced a handbook called "Feeding and Orienting Techniques for Small Parts." Although the text is all-encompassing, three rules can be stated to cover most design problems:

1. Avoid projections, holes, or slots that will cause tangling with identical parts when placed in bulk in the feeder. This may be achieved by arranging that the holes or slots are smaller than the projections.
2. Attempt to make the parts symmetrical to avoid the need for extra orienting devices and the corresponding loss in feeder efficiency.
3. If symmetry cannot be achieved, exaggerate asymmetrical features to facilitate orienting or, alternatively, provide corresponding asymmetrical features that can be used to orient the parts.[6]

Providing the bowl builder with well-designed parts simplifies the bowl design and supplies the factory with a more reliable and efficient system.

ered, a window appears on the production volume scale (typically between 0.25 million and 2 million parts per annum) where robots compete favorably with manual labor and hard automation.

In addition to the typical costs associated with getting a robot up and running, an economic analysis includes the varying costs associated with auxiliary equipment (that is, fixtures, part-feeding mechanisms, vision systems, automatic tools) essential to support the assembly robot. For assembly applications, these costs can amount to three times that of the robot itself. With knowledgeable foresight, robotic assembly can still be had with minimal risk. Considering the robot by itself:

> "Robot manufacturers estimate that it will cost (average) about $6 per hour to operate a robot, based on a two shift operation and a useful equipment life of 8 years. If direct labor costs are around $16 per hour in a five-day, two shift operation, then justification is no problem . . . . This will give a payback period of less than two years for most robots . . . ."[3]

Labor conflicts

Due to insignificant unemployment relating directly to robots, few real conflicts have erupted between labor and management. Many robotic applications, especially in the metal-working industry, have replaced hazardous or tedious jobs without much complaint. But eventually, as large numbers of robots begin working on the plant floors, serious problems could erupt. Proper training programs can benefit the workers and the factory by channeling personnel into new and more appropriate jobs as system operators, programmers, and maintenance experts.

Inadequate proposals

When proposing the use of robots, it is essential that manufacturing engineers outline the potential risks and savings for management. Increased use of robots will occur not by the justification methods of the past but by outlining the additional savings beyond reduction of labor and increased quality. The value of robots as reusable tools must be considered, as must the value of increased flexibility gained by manufacturers who must constantly deal with consumer demand, part quality, and seasonal production variations.

In addition, the proposal should adequately define the need for maintenance personnel capable of supporting robotic systems, and show an increased awareness of the safety issues involved.

Robotic assembly within RCA

To dispel some of the reservations and to consider the possibilities, groups at RCA Laboratories are investigating various phases of flexible manufacturing. Flexible manufacturing, in a practical sense, is the means to vary the manufacturing process to accommodate changing market demands, product changes (within a family of products), without additional capital expenditure or labor. In particular, the area of robotic assembly is being studied for application in RCA's manufacturing facilities. What better way is there to learn and apply new-found knowledge than by doing a "real" job?

Designing the product for automation assembly

The initial task was to consider the feasibility of robotically assembling a 24-piece subassembly, the VideoDisc arm drive. After consideration, our first response was that the product could indeed be robotically assembled, but not economically. Several small external retaining rings, set screws, and a 3/8″ long extension spring presented formidable assembly tasks for the mechanical arm. Thus, we attempted to modify the design to one that is easier to assemble.

Applying a disciplined set of guidelines (Fig. 1), we designed a simplified product. The principal improvements were:

- Fifty percent reduction in part count
- Layered assembly
- Elimination of set screws and retaining rings
- Elimination of extension spring

Of course, there was no reduction in the capability or quality of the product (Fig. 2).

Today we would have taken the new design and quantitatively analyzed the improvements. Using a software package called *Design for Automatic and Manual Assembly,*[4] an assembly efficiency would be derived for comparison among any suggested improvements. Design-for-assembly cannot be overemphasized as a critical point in the design-for-manufacturing process. Not only is there a savings in part cost but also a drastic reduction in assembly time.

> "Experience shows that it is difficult to make large savings in cost by the introduction of automatic assembly in the manufacture of an existing product. In those cases where large savings are claimed, examination will show that often the savings are really due to changes in the design of the product necessitated by the introduction of the new process. It can probably be stated that, in most of these instances, even greater savings would be made if the new product were to be assembled manually. Undoubtedly, the greatest cost savings are to be made by careful consideration of the design of the product and its individual component parts."[5]

Less inventory, increased reliability, and higher quality are also realized without any additional capital investment if assembly is considered during the design phase. In addition, it is appropriate to design for automatic assembly by specifically allowing for automatic feeding/handling of components. Simple modifications may be made at this point for the purpose of robotically or automatically handling the part.

Modeling the assembly

Having a workable model, a strategy for assembly must be considered within the limits imposed by manufacturing. For the gearbox, the production rate was naturally a critical issue. One system working one shift had to produce 1,400 assemblies—one about every 20 seconds.

Knowing approximate execution times for various robot maneuvers, a machine-time schedule can be readily constructed. The user must creatively determine the sequence of events and must discover if any auxiliary equipment (that is, fixtures, vision, tactile feedback, etc.) will be employed. For maximum flexibility, a minimum amount of tooling should be considered. On the other hand, additional tooling can be used effectively to "buy time" by assisting the robot. Typically, dedicated hardware is required to feed parts to the robot (that is, bowl feeders, magazines, pallets). Unlike the robot, dedicated hardware is not easily reusable, and therefore, is less economical for medium-volume applications.

After many attempts, a scenario, or sequence of events, for assembling the gearbox was produced that called for an assembly time of under 20 seconds, and required a minimum amount of dedicated hardware (Fig. 3). In this case, hardware was used only to apply grease, accurately position parts, pre-mesh the three gears, and automatically apply screws. The scenario also required that:

- The potentiometer, lever, and gear be manually assembled—the design did not allow efficient use of the robot for this assembly.
- The three reduction gears be premeshed and a palnut accurately located below the cluster (a palnut is a formed sheet metal nut that is pushed rather than turned on a shaft).

1. Strive to minimize number of parts.
2. Select a base to which other parts will mate, that is, design the product in layers to allow each part to be assembled from above.
3. Ensure that the base part has features that facilitate locating and securing of mating parts.
4. Use liberal chamfers, tolerances, tapers, and rounded edges where possible.
5. Avoid expensive and time-consuming fastening operations, such as screwing, soldering, and so on.
6. Assure that parts remain in an oriented form from vendor to factory floor, or be certain that parts can be easily fed by conventional means.

Fig. 1. *General rules for designing a mechanical product for manual and automated assembly.*

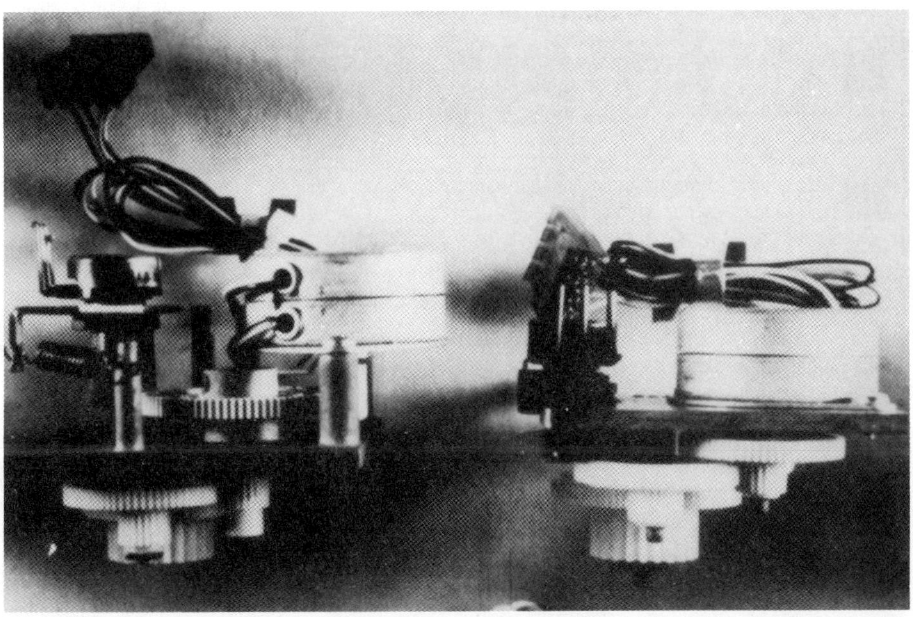

Fig. 2. *Following the design guidelines for assembly, the prototype design (left) was simplified, thereby making automated assembly feasible and manual assembly much faster.*

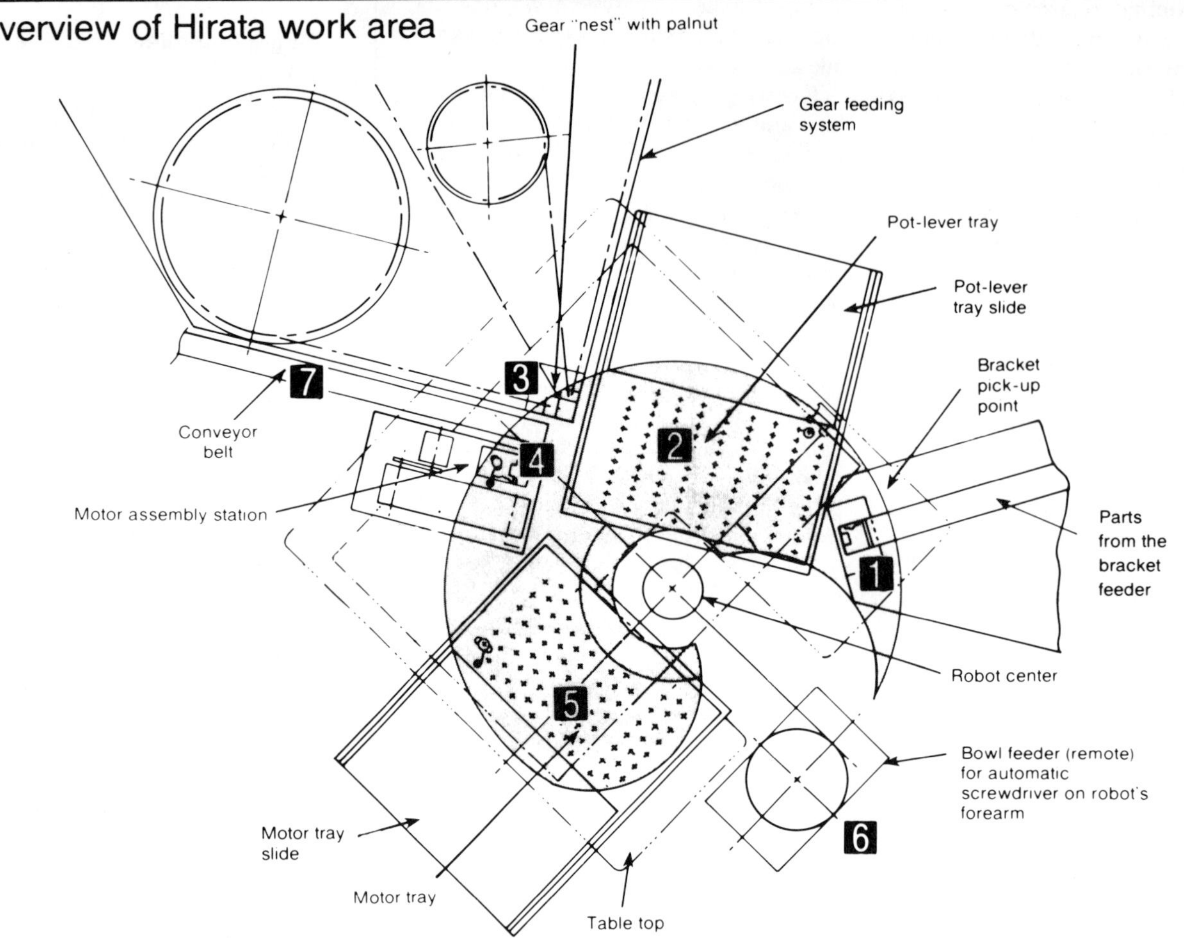

1 The robot picks up a bracket from the bowl feeder's nest. The nest is instrumented to detect the part and test the bracket's three gear shafts for straightness. Failed parts are removed by the robot and rejected. A small amount of grease is also applied to each shaft for lubrication between the rotating gear and metal shaft. The gripper raises the part from the nest and a sensor on the gripper checks that the part is properly held.

2 The arm moves to remove one of the 108 pot-lever assemblies. Each pot-lever is accurately located and retained by a plastic clip in the tray. The arm is lowered until the bracket is snapped onto a pot-lever.

3 This station contains three premeshed gears and a palnut (a type of fastener which is pushed on). The system interrogates four proximity sensors for positive location of each part before the arm lowers the bracket into the assembly. Each shaft on the bracket moves into its mating gear until the longest shaft pierces the palnut.

4 Having captured the gears and the palnut, the robot positions the partial assembly in the motor assembly station. This station is equipped to sense that the pot-lever was properly installed and that one of the two screws is applied in the next sequence.

5 The gripper is moved over and selects one of 108 motors from a tray similar to the pot-lever tray and moves back to the motor assembly station. The motor is moved in a way that meshes the pinion smoothly. While placing the part on the bracket, the system actuates a clamp that will hold the motor in place while the robot positions the screwdriver and applies two screws.

6 A Weber automatic screwdriver mounted on the forearm of the robot is fed screws from a remote bowl feeder via a pneumatic feed tube.The system controller positions the arm and actuates the driver. After fastening a motor to the bracket, the robot swings around to the bracket station to begin the sequence again.

7 Simultaneously, the motor assembly station flips the completed assembly onto a belt conveyor, which moves the part away from the robot. If the robot has successfully completed the assembly, it will travel down the conveyor to an operator for visual and electromechanical inspection. Incomplete assemblies are directed to a rework bin. Incomplete assemblies are caused by parts missing in trays or nests or the robot failing to pick up a part. If a missing part causes a jam, the system will stop and notify the operator via a red rotating beacon. Inconsequential misses are ignored by the robot and sorted on the conveyor, thus minimizing downtime. At the completion of 108 assemblies, the system automatically stops and turns on the rotating beacon. The operator simply replaces the empty trays with filled ones and presses a resume button.

Fig. 3. *Assembly sequence and overview of Hirata work area and ancillary part-handling equipment. The nautilus-shaped area encloses the work area of the robot. The actual robot, the robot controller, and the Weber screwdriver are not shown. Several bowls, for example, the 40" diameter bowl for bracket feeding, are not shown.*

- The motor and its pinion be pre-assembled—the motor manufacturer provided this service.

Prototype testing

Prototype fixtures and software were then developed for the IBM RS-1 robot (Fig. 4), which is an advanced assembly robot. Concepts, cycle times, reliability, and robot performance were all evaluated for the selected scenario. The sophistication of the RS-1 provides the flexibility to modify programs quickly and to experiment with different techniques (See also, J. Baldo's description of Astro-Electronics' IBM robot application elsewhere in this issue). At this point we tried, as an exercise in "devil's advocacy," to identify and work with imperfect parts that statistically are hard to find but, when present, cause an outstanding percentage of problems. Also we looked for handling problems. In this case, we discovered a particular way of wrapping the motor's leads in small bundles to prevent the possibility of their getting caught on a fixture during assembly.

From this stage, with the initial boundary conditions satisfied, a proposal was developed for management that outlined:

- Which robot was required?
- What hard tooling was required?
- How much money was required?

For this case, the IBM RS-1 would have been excessively sophisticated in terms of both the robotic skills required and the cost.

A Hirata AR-300 robot was proposed due to its repeatability and low cost (Fig. 5). Also, the Hirata robot is found in several RCA facilities including Lancaster, Indianapolis (Sherman Drive), Princeton, and now Bloomington. This type of robot is called 'SCARA' (selective compliance assembly robotic arm) and is less costly and more suited for certain assembly tasks than the rectilinear-type RS-1. While spherical or cylindrical coordinate robots such as the Puma 600 have roughly isotropic rigidity, the SCARA-type robot is designed to move anisotropically. The SCARA-type robot, therefore, is designed to respond selectively to loads applied at the gripper. The kinematics of the Hirata, which moves in a nautilus-shaped work area (Fig. 3), allows compliance in the X-Y (horizontal) plane, which permits compensation that is helpful when the robot is locating a part in a fixture or hole having a tolerance smaller than the repeatability of the robot, as when pressing a peg into a chamfered hole. But the robot is rigid in the Z (vertical) axis, for pressing, as required for screw driving or press fitting.

IBM 7565 (RS-1 shown)
Coordinate system: Cartesian
Repeatability: ±0.005
Actuators: All hydraulic
Maximum velocity: 40 in/sec
Payload: 5 lbs
Controller: IBM Series 1 computer
Cost: $120,000 typical

Fig. 4. *The IBM 7565/RS-1 robot (Series 1 computer and operator terminal not shown) is one of the most advanced systems available today. Using the system's advanced optical/tactile sensors, the robot is capable of executing complex tasks that require iterative part positioning for precision assembly.*

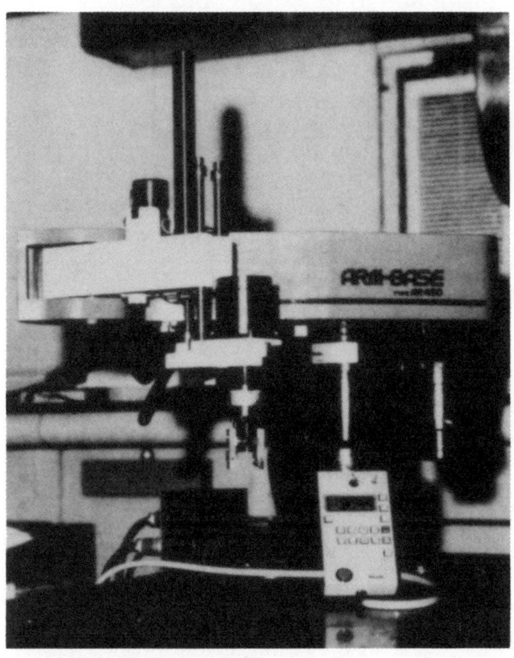

Hirata AR-300 (AR-450 shown)
Coordinate system: SCARA type
Repeatability: ±0.002
Actuators: dc servo motors
Maximum velocity: 55 in/sec
Payload: 4.5 lbs
Controller: Hirata program controller
Cost: $18,000

Fig. 5. *Hirata AR300/AR450 robot was first manufactured for Hirata's own customized automation equipment. Noted for its low cost and repeatability, it has become favored for small electromechanical assembly tasks.*

Programming the Hirata

The Hirata AR-300 system consists of the robot, controller, and a pendant. The pendant is the user's interface to the system providing teach, programming, and two run modes (step and continuous). The user basically inputs all position points (up to 1,000) and then writes his program in sequential steps (up to 256). For this system a control panel with four control pushbuttons, a counter, and eight status lights was added to simplify the operator-robot interface. The panel also eliminates the ability and need for the factory operator to use the pendant, thus preventing accidental changes in the program.

Position points can either be taught or programmed in, but our experience was that taught points were more accurate and quicker to input than those programmed. Most robots have this peculiarity in that their accuracy is dependent on a buildup of mechanical tolerances, but that once a point is taught the robot can repeatedly move to that same position.

Feeding parts

Hard tooling, as identified in the prototype stage, became the greatest expense in developing this system. The handling, orienting, and feeding of parts as they come from the vendor is a formidable job. The design

The robot-CAD/CAM connection

Feeding an assembly robot accurately positioned parts is a major concern for robotic system designers. Parts sometimes cannot be fed/oriented automatically due to some delicacy of the part. You must then rely on tray feeding, which provides a queue of manually placed and accurately secured parts accessible by the robot. Two such trays were required for the arm-drive assembly system.

Both the motor, with its 16-inch-long wires, and the potentiometer-lever assembly were considered too delicate and too difficult to feed automatically. With the assistance of CAD/CAM facilities available at the RCA Laboratories, trays were developed for the components mentioned. The designer had to both determine and avoid interference problems and "dense-pack" as many parts onto a 16- by 24-inch tray as possible. On the Computervision system, a design was produced, verified, and then downloaded to a Spindle Wizard (computerized numerically controlled milling machine) where two stacks of six trays were machined. A pot-lever tray consists of 108 positions that secure the part and allow clearance for the extended shafts on the bracket (not visible).

The same number of plastic spring clips, which retain the pot-lever, were designed on the same system and downloaded to a Graziano NC Lathe for production.

Software is now available for some CAD systems that not only allows modeling of parts and fixtures but also shows the robot's kinematics. An entire workcell can be animated for analysis before any hardware is purchased. In the future, robot software will be developed in this manner and fed directly to the workcell, eliminating robot downtime for programming. The CAD/CAM-robot system will be a valuable tool in the factory of the future.

of a reliable feeding mechanism is usually best left to those companies who are skilled in the art. If the parts are relatively small and are not delicate, a vibratory bowl feeder is commonly used. This type of feeder has a helical track that passes around the wall of a bowl. When a driving vibration is applied to leaf springs mounted under the bowl, the effect is to cause parts in the bottom of the bowl to climb up the track to an outlet at the top of the bowl. Various devices are welded into the bowl that allow only properly oriented parts to reach the outlet. Delicate parts or parts that tangle, such as motors, are better fed either by magazine or tray.

In summary, the proposal presented for this system included a Hirata AR-300 robot and the following feeding systems:

- Vibratory bowl for feeding brackets
- Vibratory bowl for feeding 3rd-reduction gears
- Vibratory bowl for feeding 2nd-reduction gears
- Vibratory bowl for feeding lst-reduction gears
- Vibratory bowl for feeding palnuts
- Tray for holding 108 motors (with spares for reloading)
- Tray for holding 108 pot-lever assemblies (with spares for reloading)

In addition, a Weber automatic screwdriver, a nest for meshing the three gears, and a motor assembly fixture had to be built/purchased.

An overview in Fig. 3 shows the placement of feeding mechanisms into the robot's work area. Starting at the bracket feeder and then working counterclockwise the robot follows the scenario presented in the figure.

Safety

Safety of the operator is paramount in the system's design. For this application, the robot's work area was totally enclosed in removable plexiglass sheets. These lightweight shields are easily removable but, if removed, cause all power to be dropped from the arm. Feeding equipment is outside the enclosure and can be replenished without interrupting the system. Additionally, the motor and pot-lever trays automatically move out from the work area on a slide for replacement.

ADAM

To ensure a safe, efficient transfer of the system from Princeton to Bloomington, the engineers and the technicians from the plant visited the Laboratories before shipment. Seeing and working with the equipment in the lab allowed the present users to become familiar with the equipment as well as with programming the robot. They nicknamed their first robot ADAM (arm-drive assembly machine).

After almost two years of development and six months of actual construction, the Hirata system was dismantled and shipped to Bloomington in early November to begin production on the factory floor (Fig. 6).

What we have learned

As of this writing, the arm-drive assembly system has only been operating for a short time. We intend to keep track of the system's performance, and in particular, the reliability of the Hirata robot. Our future work in robotics and design for robotic assembly will be based on our present experiences with this system. In a nutshell, we found our greatest effort and cost was in part feeding, not in programming or using the robot itself.

Part feeding curtails flexibility, increases costs, greatly increases floor space required (as compared with the robot itself), and lengthens the concept to delivery time. Bowl feeding requires that parts remain exactly

John Aceti graduated from Stevens Institute of Technology in 1977. He received a BE degree in Mechanical Engineering and, in 1982, was awarded a Professional Engineering License by the State of New Jersey.

Since 1977, Mr. Aceti has worked for RCA Laboratories, Princeton, N.J., as a Member of the Technical Staff in the Electromechanical Systems group. His work as a Project Leader includes a solar-energy concentrator, computer-based automatic parts-inspection devices for VideoDisc caddies, and flexible manufacturing systems. In 1983, he received an RCA Laboratories Outstanding Achievement Award for his designs of automatic inspection systems for delicate plastic parts. He also holds a U.S. patent for this work.

Contact him at:
RCA Laboratories
Princeton, N.J.
TACNET: 226-2800

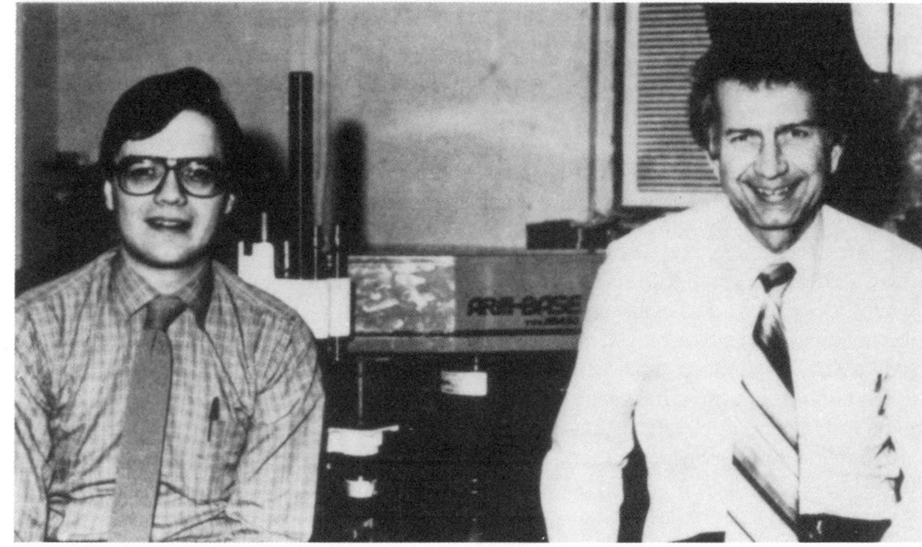

Authors Aceti (left) and Carroll.

Charles Carroll is Head of the Electromechanical Systems Research group in the Manufacturing Systems Research Laboratory. He graduated *cum laude* from the University of Florida in 1960, with a BS degree in Mechanical Engineering; in 1966, he received the MSME degree from Drexel University. In 1960, he had joined RCA Astro-Electronics, Princeton, N.J., as a design engineer developing sensor and camera components. After serving as a Lieutenant in the U.S. Army Artillery from 1962 to 1964, Mr. Carroll returned to Astro, and then transferred to RCA Laboratories, Princeton, in 1966.

In 1973, Mr. Carroll transferred to the RCA Palm Beach Division, Palm Beach Gardens, Florida, and worked on products using mini- and microcomputers for hotel/motel management. He rejoined RCA Labs in 1975, and he received an RCA Laboratories Outstanding Achievement Award for work on equipment for optical scanning and readout of the RCA VideoDisc. He then led the mechanical design effort that succeeded in producing early prototypes of a flat-panel television system. Mr. Carroll holds seven U.S. patents.

Contact him at:
RCA Laboratories
Princeton, N.J.
TACNET: 226-3030

the same, thus making modifications to the product impossible—modifications for which the robot could easily be reprogrammed. Part feeding accounted for almost 70 percent of the purchased part costs for the system, leaving only about 20 percent (the robot) reusable, thus decreasing the savings. Bowl feeding requires two to four months for construction while a robot can be programmed in a day or two. However, with all the drawbacks, bowl feeding remains the most cost-effective method for orienting and feeding bulk parts.

In the future, the use of robots for assembly will depend on an improvement in the methods to deliver oriented parts to the robot workcell. Vision and tactile feedback systems are presently too expensive and slow for most applications. In the near future, these systems should be available at reasonable costs and will be fast enough to assist the robot in handling unoriented parts. Still, most applications will require ingenuity on the part of designers, manufacturing engineers, and purchasing department personnel. Parts should be delivered already oriented, such as in egg crates or pallets. Disciplined methods such as the *Design for Assembly* software, previously mentioned, will assist engineers in making parts

Fig. 6. *The Hirata gearbox assembly system, without its protective Plexiglas envelope, is shown in the Laboratories environment.*

that can be efficiently fed by ordinary means.

Conclusion

From television receivers to VideoDisc players, electromechanical assembly is a vital factor in RCA's future. Competition in this arena is fierce and the winner will provide the consumer with what he wants when he wants it. As flexible manufacturing becomes more of a reality, and robots play a vital part, the competition will become only more aggressive. We must continue to plan and implement effective modern techniques in our engineering and manufacturing.

Acknowledgments

The authors would like to acknowledge Jack Drake, Manager of Player Manufacturing Technology, and Del Baysinger at RCA Consumer Electronics for their continued contributions from inception to completion of this project. Also, Scott Feeser, Associate Member of Technical Staff, Bob Schneller, Senior Technical Associate, and Paul Smalser, Associate Member of Technical Staff, all from the RCA Laboratories, were the driving force behind the systems programming and construction. And finally, Bernie Parambo's suggestions and aid greatly eased the entry into production; he is in the Manufacturing Methods Engineering department at Consumer Electronics.

References

1. J.L. Nevins and D.E. Whitney,"Assembly Research" *The Industrial Robot*, p. 29 (March 1980).
2. "Assembly Robots—Better Than Ever," *Assembly Engineering*, p. 57 (May 1983).
3. "A Systems Approach to Staying Competitive," *Assembly Engineering*, p. 48 (May 1983).
4. G. Boothroyd, "Design for Assembly," University of Massachusetts, Amherst, Massachusetts.
5. G. Boothroyd, C. Poli, and L. Murch, *Automatic Assembly*, Marcel Dekker, Inc., p. 255 (1982).

Robotic PC Board Assembly

Frank C. Romeo
FARED Robot Systems, Inc.

Printed circuit board (pc board) assembly and testing must be approached as any other small parts assembly task. Simply stated, it represents a significant instance of Boothroyd peg-in-hole and multi peg-in-hole problems. Viewed as such, the task should not be relegated to a special assembly category. The problem can be solved using traditional automation techniques and robotics.

Robotics has brought us new programmable machinery that is especially suitable to pc board production. The advantage of robotics as applied to pc board manufacturing lies in the intelligence inherent in the robot. The equipment can be directed to make decisions concerning size, weight, lead placement, process and function. Most importantly, it can decide electronically that a solution is not possible for a particular attempt at an assembly task, and thus take remedial action, such as passing that component or pc board to a human operator. These intelligent decisions can help conquer traditional automation's two greatest limitations, assembly quality and run time efficiencies. High assembly quality is a goal in all assembly approaches. Robotics offers many unique advantages over former alternatives. One can develop an assembly pedigree for low volume, high part value assemblies. Such a tracking device or pedigree, permits a complete record of individual parts in the assembly, and also any anomalies that occurred during its assembly or processing. Robotics offers unerring record keeping no human can match. And through the nature of its microprocessor controls is exercised individual action and retention control not available in dedicated automation.

This gripper incorporates sensors to differentiate between good and bad insertions.

Total run time efficiencies are again important because of the need to maximize yield. Humans have the mechanical ability to provide unparalleled yield, but sociological problems often result in poor performance. On the other hand, dedicated automation runs at the highest assembly speeds but has limited or no error recovery capability, therein requiring a high degree of human intervention.

Flexibility is the hallmark of robotics. We feel that with a simple tool changer design or with easy-to-replace tooling, any robot on a line can install all but the most elaborately-configured components.

Robots also offer ease of programming. IBM offers a high level AML language with fewer than 40 commands, Adept offers VAL II, and Seiko offers DARL, programming languages that are Basic-like with robotic additions. All are extremely easy to use and do not require previous programming experience. Both are tailored for factory floor implementation and modification. Software driven robots allow decision-making throughout the assembly process. And more importantly, well-written software allows the robot to solve rudimentary problems.

Error recovery may be the single most important reason for utilizing robots in pc board assembly and testing. While robots are slower, we have typically seen fewer errors with robot error correction and process correction. Thus, error recovery alone may justify the machinery purchase. It is important, however, to note that robotic installations are not yet wide spread and actual long-term experience may alter this expected efficiency. Further, we have found our results are more dependent on component specifications than robotic machinery.

With the above as background, an explanation of a design philosophy in developing an actual pc board assembly line follows. This is an insertion system consisting of seven robotic work-stations which install 23 components, including DIP's, radials, axials, and odd-lead components. Line operation is non-synchronous with a pallet transporter. The system interfaces with tote stackers on each end and is controlled by a personal computer (PC) and a programmable logic controller (PLC). The line functions with an average cycle time of eight seconds per component insertion and allows the passing of only 100 percent quality boards. It is also capable of running a batch size of one on an exclusion basis.

Board handling starts with a tote stacker system. The totes are inserted manually. The boards are indexed automatically into a board removal area. At that location a robot reaches forward via a rod extension on the end effector and probes for a board. If a board is not present, the tote stacker is commanded to the next position. Upon finding an available board, the robot loads it into a custom gripper and holds it flat by gripping all four sides. The board is then loaded on a pallet. The pallet consists of an aluminum frame with steel locating pins and hold-down devices to eliminate board warp (a common cause for misinsertion). A shot pin receptacle is machined into each pallet for precision pallet location at each station. A pallet carrier was chosen over bare board transportation because on bare board transport systems, the misregistration potential occurs at each stop station, but a board, once located on the pallet, is secure from any further misregistration. Such a method also permits utilization of a large board area for support as opposed to the thin edges on the side of a board available for the transport belts. We have also shyed away from light duty adjustable conveyors now sold as pc board transports. Rather, we use a dual belt system with custom modifications for board handling. This system is heavy duty, and provides stop-station accuracy of ± 0.002 in. At the end of the system, the completed boards are loaded on a short belt conveyor for transportation to the wave solder unit or loaded into storage totes if wave solder is unavailable.

A line for inserting odd form electronic components. Several robots are in view.

Component feeding is determined on an individual basis. We use tubes, vibratory bowls, linear vibrators, magazine feeders and tape transporters, as needed, based on price and availability of component packaging. Regardless of the feeding method used, each is fully instrumented to minimize machine stoppages. No sequenced tapes are used because of cost and because their use eliminates practical error recovery if a component will not insert.

Component lead preparation is critical to the success of robotic component insertion. At each component feed location the leads are mechanically straightened and a special preform is utilized to eliminate clinching. We do not clinch components on the board saving equipment expense, programming and maintenance. The preforming of component leads effects acceptable part retention.

All components are tested in the lead alignment station for polarity and in some cases for function. These tests are simplistic but cost effective and can eliminate many faulty components from the insertion cycle.

All our gripper designs incorporate two important design features. The first is force-controlled insertion. Most dedicated automation systems insert with cam-controlled mechanisms to achieve high speeds. This eliminates insertion verification and increases scrap and rework costs. By using force controlled insertion and position detection, we can insure proper placement. If insertion is not achieved we can remove the part and place it in a bin for later analysis. The robot then attempts to insert a new part. If a second insertion attempt fails, it searches the immediate area in a pattern to detect an offset drill pattern or to accommodate partially blocked holes. If neither recovery technique is successful, the board is electronically flagged and delivered to a manual repair station.

The second gripper design features lead control. All of our grippers grasp the leads and not the component body. The mechanical accuracy of component bodies is notoriously poor and therefore unreliable as a reference for orientation and position. This limits component spacing in some cases, but we have designed grippers for the most common components.

Our system control is unique, and involves a three-level hierarchy. In our approach, the robot controls everything inside its work area. If any control functions are required that the robot controller cannot supply, the cell's needs are supplied by the PLC or by the host PC. Communications to these devices are handled serially via RS232 to the PC, and digitally to the PLC. If service is needed from the PLC or the PC, such is coordinated by the robot controller. The robot stations are designed to be as autonomous as is practical. If a PC fails or is taken off-line, the system continues to operate. All serial communications are handled by a multiplexer. Each line is buffered and the multiplexer functions as a cross compiler. This allows all serial communicating devices to be networked with error messages and diagnostic information. The host also records operating statistics such as system run time, number of parts produced, the performance of each cell, the number of rejects cell-by-cell, the average cycle time of the board repair operations, etc. The multiplexer enhances the system by allowing orderly start-up and shutdown. The user can address each robot or ancillary device individually, or simultaneously, providing the console operator complete control. The system is designed for a batch of one to insure maximum flexibility, but is generally not operated in that mode. The primary goal is 100 percent quality output at any quantity.

The PC board testing line presents new problems for robotics. Because of the large size of the components to be interfaced; i.e., functional testers, shorts testers, and large transport totes, the robots must cover their full working envelope and still maintain repeatability and accuracy. With larger boards, we must also contend with a part that does not easily provide a flat plane. Board warp and planar distortions are often the most difficult problems to solve.

All boards are loaded into a custom tote after wave solder and inspection. All totes from the solder station are directed to the storage rack. The storage rack can function in a first-in, first-out (FIFO) or last-in, first-out (LIFO) method. We are using FIFO in this instance. From the storage rack which can hold 300 totes with 14 to 30 boards each, a PLC directs totes to work stations on demand. Tote identification is not necessary because of the zone control feature used by the supplier of the conveyor equipment. It allows the user to define work zones and allows single totes to pass through a zone at a time. This permitted considerable savings over a tote ID system. Each work cell on the distribution system side is identical regardless of function. Each contains a filled tote with boards that are to be tested and an empty tote for rejects. The available slots in a reject tote are electronically stored and it is released when full or released if an operator at a repair station requests more work. The totes supplying boards to be tested are assumed full, but only a small cycle time penalty is incurred if a slot is empty.

Bulk feed, lead form and insertion of capacitors.

One of the key features of this system is the tote used. The tote consists of two distinct elements. An outer plastic box open on the top allows transportation without any board contact. The inner frame is completely independent from the outer shell. This frame also contains four shot pin receptacles. These seats allow each tote to mate with corresponding shot pins driven by the conveyor stop stations. Shot pinning the frame is the only sure way to guarantee repeatable tote location. The milled slots also insure that all boards are spaced on precision centers. Our design goal was to locate boards within ± 0.015 in. We found using such precision less costly than a vision system interface for slot location. Without this precision control, an unacceptable number of misses will occur.

The robot support unit is a precision casting. The transport, the testers, and other peripheral equipment are also attached to this stand. The use of a precision casting permits clear definition of the robot's coordinate system. It allows maintenance of dimensional accuracy and the exchange of components without reteaching the entire robot program. The robots used in this system offer the accuracy and repeatability

Power resistors are bulk fed to lead prep equipment. This sort of handling is standard for axial components supplied in bulk.

necessary for the task. We modified the robot's "Z" stroke to allow for the 15 in. of movement required for board removal from the tote.

The gripper used is a propriety design. It consists of two moveable side guides. These guides are driven by our precision parallel gripper to insure proper centering of the board. Another stationary guide is placed at the top of the board. An independently driven loading tang pulls the board from the tote and into the side and top guides. When pulled into the guides, end sensors indicate proper board placement. The side guides then close, centering the board. To facilitate the unloading of the board into the tester, we made two modifications to the bed of nails test head. The guide pins were extended one inch to allow the robot to load the board directly onto the pins and the bed of nails was fitted with a small elevator platform. This provides a consistent placement and pickup location for the robot. To unload the board into a tote, side guides are released, but maintain their guiding position and push the board into the tote slot until fully seated. The gripper uses many sensors to confirm each action and the presence of the board. It is mounted on a tool changer that also interconnects all air and electrical connectors. In this test system two boards are run concurrently making the tool changer mandatory.

A typical cycle for testing is as follows: the conveyor system indicates a tote is shot pinned in the working/good location and a tote is shot pinned in the reject location; the robot searches the supply tote for a board and upon locating one, removes it; a wrist action changes the board from vertical to horizontal and locates it on the tester's extended pins; the robot then signals the tester to lower the platform, actuate the vacuum and initiate the test; if the board tests acceptable, it is reloaded into its original slot and the cycle repeats; if the board is unacceptable, it is removed and the robot presents its Code 39 label to a stationary bar code reader before placing it in the reject tote. The serial number is transmitted to the SCI-1033 and the error code is requested from the tester. The SCI-1033 merges the two pieces of information and both are sent to the PC host for retrieval during repair. The load cycle is 12 seconds; the unload cycle is also 12 seconds; the scanning cycle is 19 seconds. If a board is misplaced in the grippers or will not load into the tester or a tote, it is placed in a holding area. All opportunities for machine stoppages are designed out if practical.

Acceptable boards are stored until needed. Rejected boards are also stored during first shift lunch and breaks and any second-shift or third-shift operation. This allows three shift testing and single shift debugging and repair. When shipping personnel are available, the system sends qualified boards from the storage area to a bracket installation station where the product is removed from the tote and a mounting bracket is installed. After the bracket is installed, a processing tab is sheared from the face of the board assembly and the unit is then sent to a packaging station.

The control system architecture is the same as that used in the pc board assembly system. Every precaution has been taken to keep the system running at all times.

This system and others like it represent the vanguard of new options available to electronics manufacturers. As the electronics and communications industries continue to explode, fierce competition is forcing all profit margins down, therefore you must be an effective manufacturer to succeed. Robotics offers the cost-effective edge necessary to make the high quality/low cost products the marketplace demands.

Presented at Tower Conference Management Company's Second Annual International Robot Conference, 1984

Criteria for Selecting an Assembly Robot

by Patrick Nicholson
Aerojet Electrosystems Company

The purpose of this talk is not to provide all the answers--but rather to prompt each of you to ask yourselves all the right questions. There is no one best robot for all situations--but there may be one best robot for a single, given situation. By thinking about all the contributing factors, and all the criteria that must be met, it is possible to narrow the choice down enough so that some sort of rational decision can be made. Keep in mind that this presentation addresses the choice of assembly robots--not welding robots, or painting robots, or machine load/unload robots.

1. First, how do you expect to use your robot? Before you answer that you want to eliminate the people on your widget assembly line, the first consideration is: do you want to create a robot assembly cell, with one robot doing a series of different tasks, or do you want to place robots along a conveyor or indexing line, with each robot doing one specific task? This questions is fundamental, and must be answered before you can get much further.

The Robot Assembly Cell

The Robot Assembly Cell is a miniature assembly line where one robot is surrounded by feeders, process machines, and fixtures. The one robot does a series of tasks in turn, taking parts from several feeders, perhaps applying adhesive, placing parts, installing screws, then placing the completed assembly on the outgoing transport mechanism. Several characteristics are required of a robot working in a cell:

(1) The robot will probably require a large, powerful controller, in order to store all the necessary tasks, calling up each program as needed.

(2) The robot will probably require more than simple X-Y-Z motion in order to accomplish a variety of tasks.

(3) The robot will probably require tool changing capability, so it can change from a handling gripper to an adhesive dispenser, to another handling gripper, to a power screwdriver.

(4) The robot will probably require motion-control programming, so that it can avoid obstacles as the assembly is built up, and so that the hand can trace a complex contour, in case it is required to apply adhesive to a complex shape, for example.

(5) The robot will probably require some type of sensor, so that missing parts, assembly errors, cross-threads and the like will be detected, to avoid the robot mindlessly continuing an assembly operation that was botched in the first step.

(6) The presence of a sensor and an intelligent controller permits the programming of interactive error correction, such as correction for mislocated holes, rejection of defective screws, etc., thus reducing the need for human intervention.

What kind of robot do you see emerging from this series of considerations? Well, it appears that a robot destined for an assembly cell has to be a fairly complex robot. It will probably be a 5 or 6 axis, jointed arm type, servo-controlled, with a "smart" controller capable of interfacing with tactile or vision sensors. Even though we haven't selected a robot yet, notice that the simple series of questions and answers has revealed the following: either we will have to simplify the task as described, or we are forced to use a robot in the upper tier of complexity--probably in the price range of $45,000 to $150,000.

The Robotized Indexing Line

The robotized indexing line, with simple, 1-function robots placed along an indexing conveyor or rotary table, attacks the same problem in a totally different way: the robots can be simple pick and place robots--but there will be several of them. Let's examine some of the criteria:

(1) Each of the robots can probably be simple, with 3 or 4 degrees of freedom, a single, fixed end-effector, and a simple controller that only has to store a single program. Each will perform a single function when the conveyor or rotary table indexes, bringing work into its position.

(2) The range of motions required of each robot is very restricted. Obstacle avoidance, for example, may be accomplished simply by programming in a couple of intermediate points between the "pick" position and the "place" position in order to go up and over, or around, an obstacle.

(3) Coordination of the indexing conveyor and several "pick and place" robots will probably require a separate system control computer.

(4) Interactive control between the robot and the assembly--such as, correcting for a mislocated hole--is incompatible with the single-function, indexing line concept, because of the "smart" controller requirement. Presence-sensing and error detection will probably be under the control of the system-control computer, with errors resulting in defective assemblies being placed in a "defective assembly" chute, to be handled off-line by humans.

(5) This type of assembly requires that cycle time be roughly equal in each of the various stations, with the indexing rate of the conveyor (or rotary indexing table) determined by the slowest cycle time in the series.

What kind of robot emerges from the robotized indexing conveyor concept? Essentially, a "pick and place" robot, with a single function, a simple controller, and only 3 to 4 axes. Such a robot will cost somewhere in the $5,000 to 13,000 price range. However, several robots are required. In case 10 robots are required, the cost of the 10 "simple-minded" robots is roughly the same as the cost of the 1 "smart" robot used in the robot cell arrangement.

How to Choose

How to choose between the robot cell and the robotized indexing line? Each case will have to be considered on its merits, but the following items should be considered:

(1) In general, the robotized indexing line will have a faster cycle time. Essentially, it is faster to index the line one step than for the "smart" robot to change tools or end-effectors to do the next task. Furthermore, the addition of sensors and interactive programming inevitably takes more time.

(2) Programming the robotized indexing line will be simpler and faster. As a rule of thumb, the smarter the robot, and the more interactive decisions it has to make, the longer it takes to program.

(3) End-effectors and tool holders can be much simpler (and much cheaper) for the "pick and place" robots on the robotized indexing line--after all, each end-effector is fixed, and has a single function. On the other hand, several end-effectors are required. The robot assembly cell requires either a "quick-change" concept, whereby the robot can change end effectors quickly, or it requires a very complex, multi-function end effector to perform all the functions needed in the cell.

(4) On the other hand, the robot cell is much more flexible, and much closer to being a direct replacement for a human operator. The sequence of events can be changed by a simple programming change, for example, where it would require physically relocating stations on the robotized indexing line. All robotic applications include a considerable amount of "hard" automation: feeders, escapement mechanisms, fixtures, special tools. However, the robotized indexing line is much closer to a true "hard" automated assembly line, and as such, has only limited flexibility.

In summary, the first decision to be made--robot work cell vs. robotized indexing line--will establish several of the most important parameters to be considered in choosing a robot--level of intelligence, number of axes, compatibility with external sensors.

2. Second, having decided how you want to use your robot, the next step is to define--precisely--what you expect your robot to do. Let's assume you are going to use your robot to install the bottom plate on a small transmission.

That's simple, you're probably thinking--I'll just pull out my shop order and list the operations. If that is what you're thinking--think again! It is <u>not</u> sufficient to list steps as follows:

A. Apply self-forming gasket material to mating flange.

B. Place bottom cover in position.

C. Install nine socket-head 6-32 machine screws.

D. Torque 9 socket-head screws to 20-24 ft-lbs.

Although the steps listed above are more detailed than the typical shop order, they are woefully inadequate for analyzing a robot application. The most important step--turn the assembly upside down--was not even included, because it was obvious to a human operator. To analyze a robot application, you will need the type of micro-motion analysis used by industrial engineers to assign standard hours to operations. The steps above might start out as follows:

1. Receive partial assembly on locating pallet on indexing table.

2. Grasp partial assembly, lift straight up to clear locating pins. Continue lifting until the housing clears the locating pallet by a minimum of 6 inches. Then rotate the assembly 180 degrees in roll, lower onto the pallet and position housing over location pins. Release assembly.

3. Reach to right edge of work area. Grasp self-forming gasket dispenser head. Lift head free of support bracket, move to a position above the mating flange of main housing, lower to a position 1/2" to 3/4" from flange, and depress foot pedal to dispense gasket material.

4. Move gasket dispenser at a rate that leaves a continuous bead 3/16" to 1/4" in diameter on the mating surface, making a continuous circle around mounting holes and locating pins.

5. Release foot pedal. Move dispenser to right edge of work area and replace in support bracket.

You get the idea. Even though it appears simple-minded to make that kind of analysis for human assemblers, you cannot define your robot without it. For example:

(1) How much weight does the robot have to lift? Let's assume that the partial assembly, without cover, weights 4 pounds. Does that mean that a robot with a 5-pound payload can do the job? That depends. How much does the end-effector weigh? Most robot parameters apply to the mounting flange where the end effector will be installed. If your end effector is a complex, multi-purpose device, its weight, added to the assembly to be moved, could easily exceed the rated payload of the robot.

(2) Consider the motion requirements. Let's assume our bottom cover is bullet-shaped, with three straight sides and one curved side. Can the robot you are considering execute a smooth parabolic curve? Can that curve be defined by the formula for a parabola with the variables defined? If not, you may find yourself programming that curve with an inordinate number of point-to-point moves.

Robot Selection Criteria

Let us examine the criteria for selection of a work-cell robot to perform the assembly function described above (applying form-in-place gasket material and installing the cover plate on a small transmission). The following steps describe one way to develop enough characteristics to permit an informed selection of the assembly robot:

1. Lay out the work area and have a human operator actually perform the task required. This is a very large step, involving the following:

    a. Select the source of motion to bring parts into the station, and take them away. We have assumed an indexing conveyor.

    b. Determine the means of providing repeatable location and orientation of the parts to be assembled. We have assumed that parts will arrive on a pallet, which provides both location and orientation.

    c. Determine how parts to be added to the assembly will be presented. We have assumed that the bottom cover will be presented by a gravity feeder, with a slot escapement. Screws will be bowl-fed directly to the automatic screwdriver.

    d. Lay out the assembly area, providing sufficient room for all operations. It is important to reduce space to that actually required, which will usually be less than the space required for human operators.

    e. Determine end-effector requirements and come up with at least a "such-as" design, in order to determine weight and size. We have assumed a multi-purpose gripper, capable of picking up three items: the assembly, the dispenser head, and the power screwdriver.

Selection Criteria

(1) Work Envelope: The first consideration is: how far must your robot reach? Typically, a robot intended for use on an indexing line will require a smaller envelope than one that is intended for a robot assembly cell. But, in either case, if you simply take the dimensions of a human work station, you will probably be overstating the envelope requirements. Robots require less work space than humans. There are some precautions to be taken when you figure out the work envelope, however:

1. Provide some over-reach. Accuracy, resolution and maneuverability all drop dramatically in a jointed-arm robot, when the arm is fully extended.

2. Consider the expected dimensions of the end-effector and its probable position--it may not be favorable to the envelope.

3. Take into account the angle of entry and any special movements required to acquire any part or tool. For example, in the simple case we have been talking about, the plastic case of the gasket dispenser tends

to cling to the bracket. The robot has to rock the hand slightly to free it, and it must then move straight up for some distance. Note: A fully-extended jointed arm robot cannot move its hand straight up.

(2) Payload Requirements: This is the one everybody talks about--but it's still easy to get tripped up. Let's assume the partially-assembled transmission mentioned above weighs 4 pounds. Is it safe to assume that it can be handled by a robot with a 5 pound load capacity? Maybe not--let's consider:

1. The end-effector--If the end-effector is a complex, multi-function device, as it is in most robot work cells, it may very well weigh over a pound. Right away, you have exceeded the rated capacity of your robot. The payload rating normally applies to the mounting flange where the end-effector mounts; therefore, the end-effector is considered to be part of the payload.

2. Method of measurement--Most robot manufacturers rate payload conservatively, with payload calculated with the arm at full extension, and moving at maximum rate. Other manufacturers calculate payload under more average conditions, and "de-rate" the payload, based on % of maximum extension, speed and direction of movement. Read the fine print. Better still, get a demonstration, at a trade show, or at the supplier's headquarters. I would be reluctant to deal with a supplier who is unable or unwilling to arrange for a demonstration.

(3) Number of Axes: As mentioned earlier, the simpler robots used on the robotized indexing line need fewer axes--possibly only three. For example, to install screws through the cover of our simple transmission--assuming that it is located horizontally--would require just three axes. Don't be too conservative, however. Close observation often reveals subtle but complex motions. In applying sealant, for example, it may be necessary to go through a lift-twist-rotate motion to avoid making strings. the problem is similar to pouring wine without letting it drip--watch the motion used by an experienced waiter to get a clean cut-off. In short, analyze your requirements carefully before reducing the number of axes. Specifying too few can be an expensive mistake.

(4) Power Source: The choice is hydraulic, pneumatic or electric. Hydraulic power offers high payload and high speed, at the expense of possible contamination with oil, noise, and high energy requirements. Pneumatic power offers low power and speed, and generally lower accuracy--but an attractive low price. Electric power offers moderate to high payload, good speed and excellent control. Electric power is the cleanest of the three, and is therefore the source of choice for electronic assembly, foods, hospital goods, or other applications where contamination is a concern. In general, electric power is steadily gaining ground as motors become smaller and more powerful. If an electric-powered robot is available that meets the speed and payload requirements, it will usually turn out to be the best choice.

(5) Speed: Speed is usually not a prime consideration. Most industrial robots can, in fact, move the part or tool from point "A" to point "B" in a fraction of a second--but that blinding speed is seldom used. Let's consider what's involved:

a. What are the requirements? Moving a power screwdriver in position to put a series of screws through a cover to hold it in place appears to require high speed. After all, the screwdriving operation goes fast, so a slow move has a direct effect on cycle time. However, the distances are often short, so that the robot is always accelerating and decelerating--it may never reach its top speed. The difference in cycle time between a high speed and moderate speed robot may not be significant.

b. Speed may be process-dictated rather than robot-dictated. Dispensing gasket material, for example, is process-related. The speed is one of the variables that determines the size of the bead of sealer applied. In this case, ultimate speed is less important than good control. Once again, careful analysis of the specific case is mandatory.

(6) Controller Capability: The controller may be a stand-alone computer, capable of controlling all aspects of the robot work cell--or it may be a relatively-simple microprocessor system at the low end of a hierarchy of computers. This is an extremely complex subject, so it will be considered only in a broad-brush fashion. What are the considerations?

a. Do you require continuous path generation, or is point-to-point programming sufficient? In the case described, dispensing the sealant requires continuous path generation, since steady, programmed motion is required, following a curved path at a continuous rate. Do not assume a given robot can generate a continuous path--many, perhaps most, cannot.

b. Do you require coordinated movement of all axes? Coordinated motion is graceful and elegant--but if you don't need it, it's expensive. To drive screws straight down into an assembly, for example, independent motion of the axes is adequate. This is the job for a scara or cartesian robot with a simple controller.

c. Will the robot be required to control other parts of the automation system? Controlling the conveyor, the screwdrivers, the dispenser and interfacing with preceding and following automation elements--all this requires increasing complexity; complexity that you pay for:

d. How about more exotic capabilities--program down-load from a CAD/CAM master computer; off-line programming; error-detection and reporting; error-detection with decision-making capability...Here, as elsewhere, it is mandatory to do a thorough analysis of the requirements, so that the controller will be neither too complex or too simple.

(7) Programming Language: Most analyses of robot installation costs say that the robot itself represents about 30% of the total costs, with tooling, fixturing and other mechanical interface requirements accounting for about 35% and programming costs, 35%. These figures are starting to change, with programming costs taking a larger and larger share--up to 50% and 60% in some cases. It's a straightforward situation: the smarter the robots get, the most difficult it is to program them. The programming language can have a powerful effect on those costs. Once again, a complete discussion is beyond the scope of this paper. Some points to be considered are:

a. Does it use English-language statements? Most languages do--but some are more user-friendly than others. The simplest robots tend to be programmed in basic, which is a widely-known language, but not ideal for describing robotic functions.

b. Does it have the capability of describing a trajectory in mathematical terms? That capability is mandatory for direct down-load of CAD/CAM inputs. It is a virtual necessity for programming a robot to follow a complex contour, such as in adhesive application, seam welding or de-burring.

c. Does it provide means of specifying all the necessary parameters in simple terms--location at start of move, location at end of move, acceleration, deceleration, maximum speed, obstacle location?

d. Can programming be done in modular form, with provisions for calling up modules (subroutines) as they are required?

e. Can sensors such as vision or tactile sensors be integrated into the program by simple means?

Again, the list is not comprehensive, but the object is to stimulate the thought processes. The ideal would be to learn to program several robot languages. Barring that, the only feasible approach is to formulate a list of characteristics, and then ask the various robot suppliers if they robot language has this or that characteristic. Demonstrations are also extremely valuable, if the opportunity is provided to ask meaningful questions.

Conclusion

This paper is not comprehensive. A number of significant considerations have been omitted: number of robots of the model considered already in use; what type robots your company may already own; service and support provided; spare parts availability; training availability; the strength and stability of the manufacturer. It is hoped that the criteria here presented will--at the least--stimulate the thinking of those responsible for robot selection.

CHAPTER 7

LASERS IN ASSEMBLY

Optical I/O Hybrid Microelectronics Packaging

by Gary J. Peterson
Hewlett-Packard Company

ABSTRACT

This paper gives an insight into new concepts used in optical I/O hybrid microelectronics design and packaging. The scope of the paper is limited to a solid state laser hybrid package. Package parameter requirements are discussed in depth, especially the optical and thermal parameters. The package design was influenced by the dual heterostructure $Ga_{1-x}Al_xAs$ laser diode requirements: hermeticity in the 10^{-8} STD cc He/sec leak range, optical flat lens, and inert encapsulated atmosphere. Manufacturing methods presented include assembly, substrate fabrications, reflow solder processes, precision component alignments, interconnection wire bonding, and encapsulation.

INTRODUCTION

Applications of hybrid microelectronics in today's industries are virtually innumerable. Two primary reasons why hybrid microelectronics are used so predominately are circuit performance applications and package density. Optical I/O hybrid microelectronic packages are devices that either transmit or receive optical signals or do both as a part of the circuit's primary function.

An electronic-total-station surveying instrument, recently developed to measure distance, vertical and horizontal angles, as well as combinations thereof, united state-of-the-art design in three areas; electronics, optics, and mechanics. This instrument is illustrated in Figure 1. Design requirements for a light weight small sized instrument suggested the need for high density electronics packaging. These requirements led to the development of unique hybrid circuits that combined solid state optical elements with the electronics; consequently, precision tolerances were imposed upon both the optical components and the package structures.

Optical I/O hybrids are used in the instrument's distance and angle measurement systems. These hybrids will be briefly introduced. Functional applications of the hybrids are explained in Reference No. 2. The laser design is introduced separately, followed by discussions on the package design and manufacturing methods. The scope of this paper is limited to the laser hybrid package.

HP3820A ELECTRONIC TOTAL STATION

Figure 1.

OPTICAL I/O HYBRID CONCEPTS

Visible light emitting diode (LED) sources (700 nm wavelength) are used in seven areas of the angle measurement system. These illuminators are made by mounting a GaAs LED die on a specially designed black ceramic (94% Al_2O_3) header. The header assembly is optically aligned to a lens cell inside a steel cylindrical assembly and then epoxied into place. Figure 2 illustrates the application of the illuminator in the angle measurement system. Light from these illuminators is projected through encoder disc elements and then detected by discrete digital sensors and angle sensing hybrids in order to measure angle.

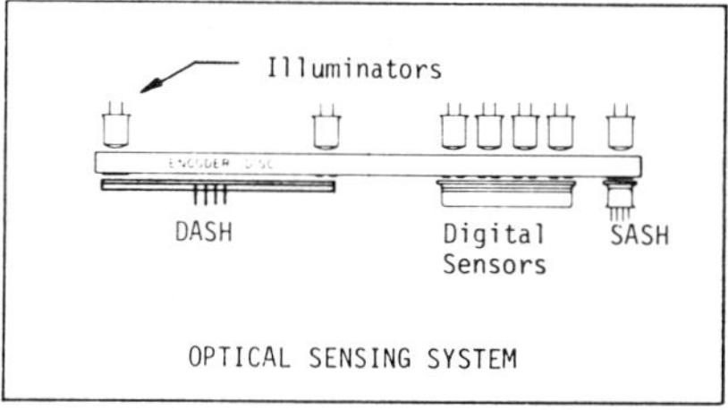

Figure 2.

The single angle sense hybrid (SASH) uses an array of four closely matched silicon photodiodes with very precise geometry to interpolate the optically-encoded angle information to better than one part per million (less than one arc second). The photodiode array (PDA) is fabricated by integrated circuit bipolar processes.

The hybrid lid for SASH must have a transparent opening over the PDA that is free from dust, scratches, inclusions, etc. Other areas of the lid surface must be opaque to prevent errors caused by the photosensitivity of the differential operational amplifier chips. In addition, the lid must be very thin and mounted closely to the PDA so that the distance between the circular encoder disc and the PDA is minimized. These lid requirements are met by selectively screening an opaque epoxy "ink" on a .007 inch thick transparent soda-lime glass lid. An opening in the screened lid provides the transparent window over the PDA chip. A .020 inch thick ceramic spacer separates the lid from the thin film metallized ceramic substrate. The structure is bonded together with .004 inches thick epoxy preforms (Ablefilm 550-1-004 with silane).

The dual angle sense hybrid (DASH) combines the circuit of SASH together with another different type PDA with similar circuitry to detect coarse angular position. This hybrid also incorporates all of the package features and constraints of SASH. The two PDAs are positioned on the substrate at locations equal to the separation distance between the fine and coarse encoder track. These photodiode arrays are aligned with respect to each other within ± .002 inches and one degree of relative rotation. This insures that when one of the photodiode arrays is aligned to the encoder track the other array will also be within positional range to the other track. Alignment keys incorporated into the photodiode array metal mask and into the thin film Mo-Au pattern on the Al_2O_3 substrate aid the precision alignment operation used in chip mounting.

Another hybrid used in the angle measurement system is the level sense hybrid (LSH). This hybrid uses a transparent soda-lime glass substrate which has a metallized thin film Mo-Au pattern on one side. The thin film pattern is used for both the component chip mounting and circuit interconnections, as well as for the optical mask pattern geometries. Two photodiode array chips are mounted on the substrate with epoxy preforms (Ablefilm 517-1); the phototransistor is mounted with conductive silver filled epoxy as illustrated in Figure 3. Level sense hybrid component

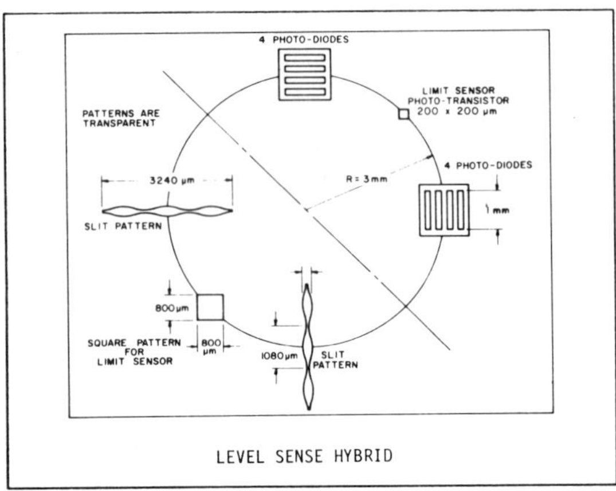

LEVEL SENSE HYBRID

Figure 3.

alignments are more critical than those for DASH; however, the most demanding areas of fabrication are keeping the glass substrate and glass lid free from scratches during processing and also the removal of particulates larger than 10 microns from the photodiodes and glass surfaces. The hybrid is encapsulated with the same type of methods as those described for SASH and DASH; ceramic spacer, epoxy preform, and glass lid.

The optical layout[2] of the level sensor system is shown in Figure 4. Visible light (700nm wavelength) from an illuminator, accurately positioned behind the hybrid substrate, is defined by the metallized pattern openings on the substrate. Light patterns are focused by a series of collimator lenses, reflected off of a gravity sensing mercury pool and is then imaged on the two photodiode arrays and phototransistor. If the instrument is leveled within the range of the level-compensation circuitry, the square shaped light pattern reflected from the mercury pool strikes the phototransistor to provide an "in range" signal that allows the level sensing circuit to be activated. The sinusoidal shaped light patterns detected by the photodiode arrays are used in a level sensing circuit to determine the instrument's position relative to the earth's gravitational field, which then compensates all distance and angle measurements for an instrument mis-level condition.

The optical I/O hybrid concepts discussed have been briefly introduced. In the manufacture of these devices, non traditional materials were used, new fabrication processes were developed, special handling and environmental control methods were established as well as precision assembly techniques.

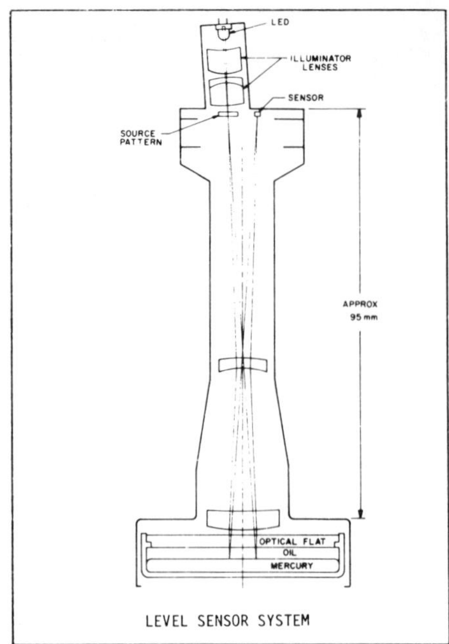

LEVEL SENSOR SYSTEM

Figure 4.

LASER HYBRID PACKAGE

Another optical I/O hybrid concept is used for a laser hybrid package. This hybrid emits infrared lasing light which is used in the instrument's distance measurement system. A solid state laser diode is the key component in this hybrid package. The diode is a double heterostructure continuous wave $Ga_{1-x}Al_xAs$ laser with an output wavelength of ≃840nm at 25°C. The output power is controlled at 4mW. Light is emitted from both ends of the cleaved mirror faces of the diode. The light flux from the front face is emitted out from the hybrid package for the transmitted light source used to measure distance. Light flux from the back face illuminates a silicon photodiode in the feedback control circuit. The hybrid circuit contains a pass transistor which controls the laser current obtained from a 3.6 volt source. This transistor is, in turn, controlled by a feedback network that incorporates the silicon photodiode which sinks the hybrid input current. Thus if the laser flux increases, the photodiode sinks more current allowing less input to the amplifier. The pass transistor bias is then decreased and the laser current falls. The laser circuit exhibits a linear power - current characteristic and is stable with a fixed output power over a wide temperature range. The circuit block diagram is outlined in Figure 5.

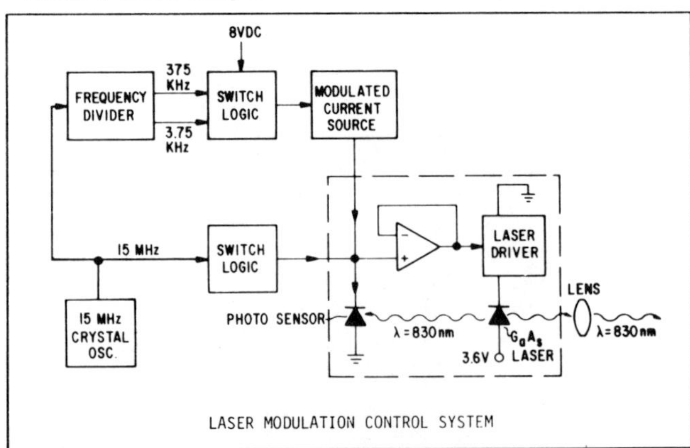

LASER MODULATION CONTROL SYSTEM

Figure 5.

The laser diode and hybrid circuit are packaged in a special TO-8 type package with a soda-lime glass window. The package is hermetically sealed with inert atmosphere inside. The header base provides a mechanical reference surface for aligning the emitted laser beam to the instrument's optical system as well as providing a heatsink for the laser diode. A photograph of the hybrid is shown in Figure 6.

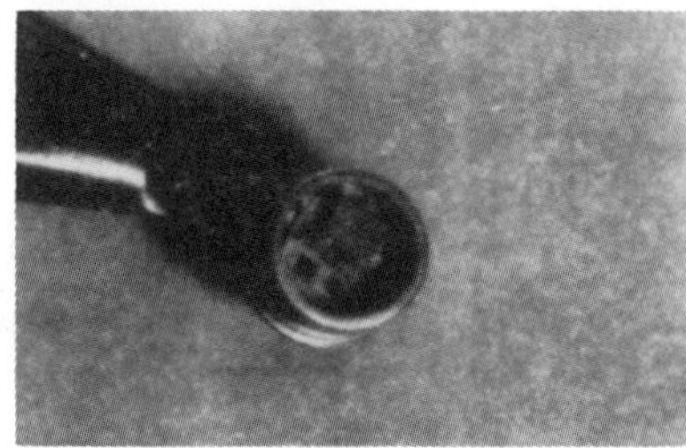

Figure 6.

PACKAGE REQUIREMENTS

The hybrid package requirements are listed in Table 1. In addition to these requirements, the hybrid had some special optical and electrical parameters that influenced both the package design and manufacturing processes. Window properties such as birefringent effects, surface defects, et al, directly affected the quality of the laser beam and its far field pattern. Also, any film deposited on the light emitting mirror surfaces of the laser die will attenuate the laser beam and may also cause unwanted reflections, depending upon the film thickness and type. The thermal efficiency of the metallic bond between the laser die and its heatsink affected the performance of the laser at elevated temperatures. Excessive heating caused the lasing efficiency to decrease and the wavelength to exceed its 840 ±15 nm limits.

Table 1.

PARAMETER	REQUIREMENTS
Hermeticity	Leak rate ≤ 9.9×10^{-8} STD. cc Helium/sec.
Encapsulated inert environment	Nitrogen or Argon (no organic epoxies, etc.)
Environmental	-40°C to +75°C storage temperature. +40°C at 95% relative humidity for five days
Thermal	Heatsink dissipation ≈200mW, TA + 25°C Package dissipation ≈400mW, TA + 25°C
Light beam orientation	Center of beam ≤ ±2° with respect to the mounting surface of the package
Light beam location	Center of laser beam concentric to center of package within ±.010 inches
Package geometry	TO-8 style header compatible with bracket mounting or with cylindrical mounting.
Optical lens (light transmission area)	1. Flat within 5 X 10^{-6} inches
	2. Irregularity within .5 X 10^{-6} inches
	3. Maximum defect size ≤ .0005 inches not to exceed 2 defects
	4. Non polarized transparent material
	5. Thickness = .040±.001 inches

PACKAGE DESIGN

In the early design stages, the laser hybrid was an integral part of a much larger hybrid which used epoxy methods for assembly and encapsulation. The laser diode and its associated circuitry was separated from this hybrid so that it could be used as a device and because of the required inert atmosphere and efficient thermal heatsinking. Hybrid circuit performance was dominate over cost effectiveness. The laser hybrid package is illustrated in Figure 7.

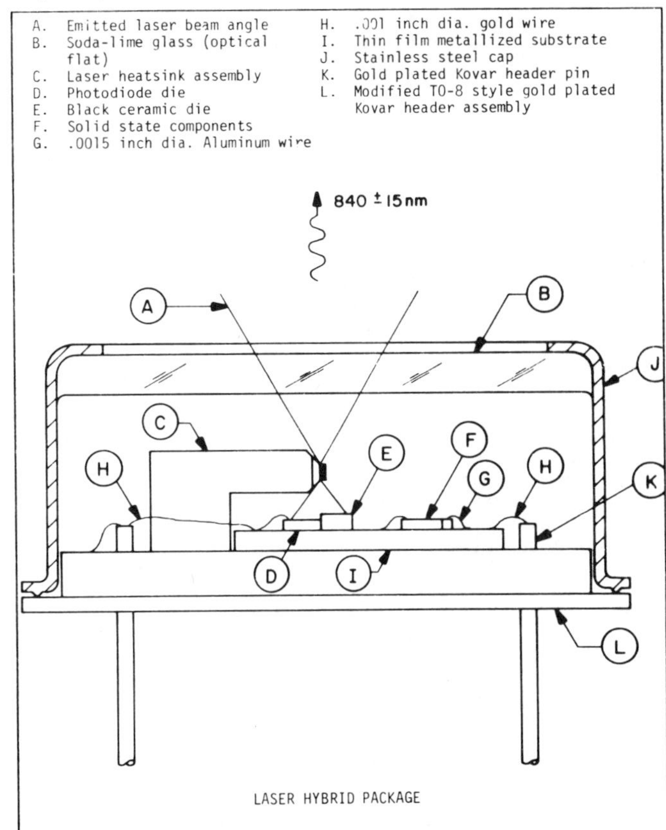

Figure 7.

A modified TO-8 type header was used as the building block for the hybrid; six leads instead of the standard 12, and the gold plating thickness was increased to 80 microns. These modifications made it possible to use reflow solder techniques for component assembly. Encapsulation was made by welding a 300 series stainless steel cap with a fused soda-lime glass window.

The hybrid circuit was built on an alumina substrate with Ta_2N-Mo-Au thin film circuit pattern with add on die components. Five thin film Ta_2N resistors are incorporated into the circuit pattern. All of the die components were Si-Au eutectically bonded to the substrate. Component interconnections were made with .0015 inch dia. aluminum wire. The back side of the substrate had thin film metallization which was soldered to the header.

The laser die was attached to a copper heatsink by an indium-gold bond. Gold wire bonds (.0007 inches dia.) connected the laser die to a metallized alumina standoff on the heatsink. Silicon dioxide was evaporated onto the mirror surfaces of the laser die to protect the active light emitting area. Another set of gold wires (.001 inches dia.) connected the standoff to the hybrid substrate. The heatsink provided a structure to position the laser die above the photodiode and non-reflective black ceramic chip.

It also provided a conductor for the other connection of the laser die. Of course, the primary purpose of the heatsink was to conduct the heat from the laser die to the base of the header. The base of the heatsink was plated with gold which was soldered to the header.

Consideration was given to the development of a new LSI device that would integrate all of the hybrid circuitry including the photodiode. This remains a viable consideration for future laser hybrid packaging because the package would then become smaller as well as less expensive. The concept was abandoned because of the project schedule.

Interconnections were made between the header pins and the hybrid substrate by .001 inch dia. gold wires. Gold wire was also used to connect one of the pins to the base of the header; however, changes have been made to directly connect the pin to the header by a copper braze.

MANUFACTURING METHODS

Established microelectronics manufacturing processes were used as much as possible in the fabrication of the laser hybrid. The following sections provide an insight into some of the manufacturing methods that were developed for the hybrid.

Assembly

The hybrid assembly sequence had two production stages as illustrated in Table 2. These stages were separated by a hybrid circuit test operation. The laser heatsink assembly was assembled and tested separately because of a required "burn-in" test which eliminated early life failure.

An exploded view of the laser hybrid assembly is shown in Figure 8. Some of the assembly methods are discussed in subsequent sections. Figure 9 depicts the method used for the laser heatsink assembly.

Table 2.

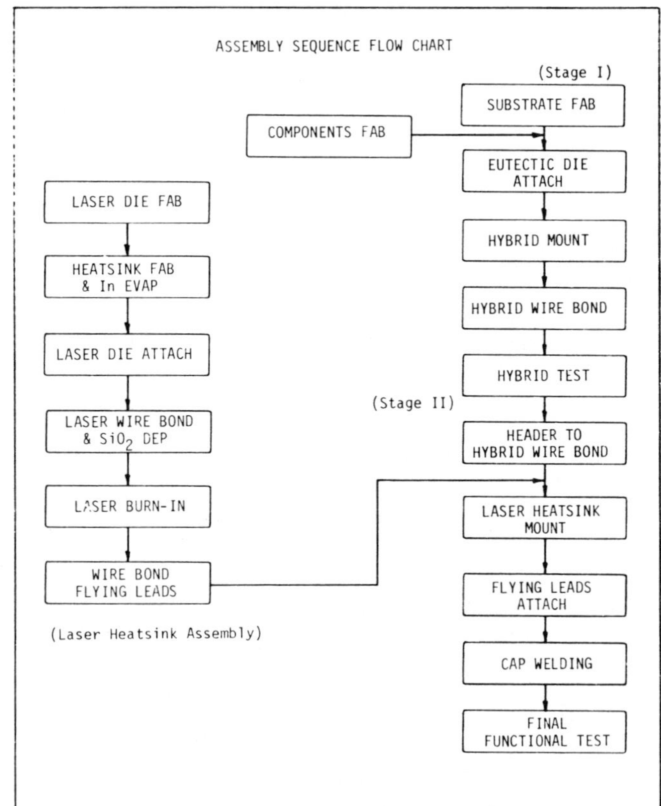

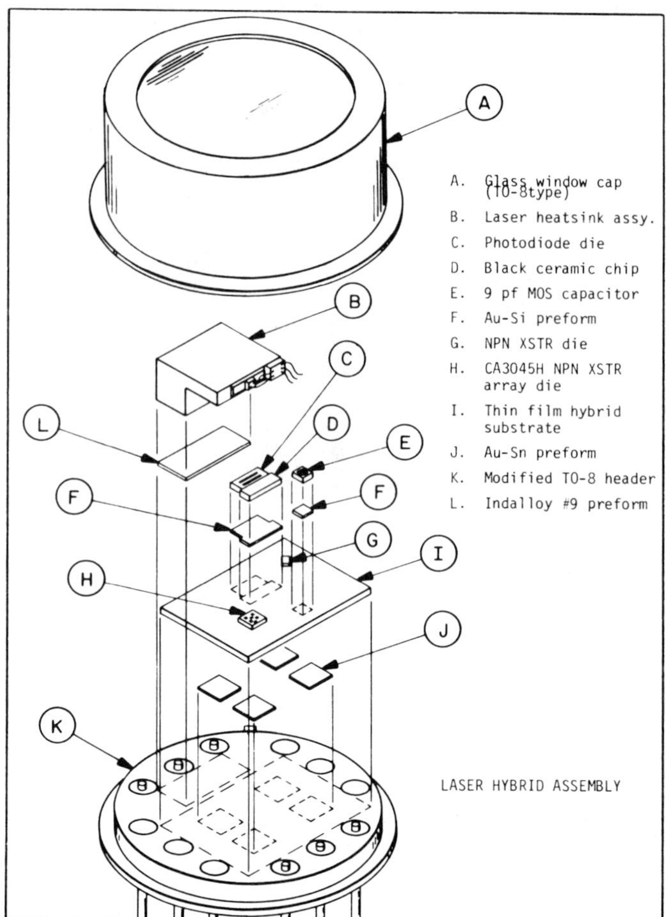

Figure 8.

Figure 9.

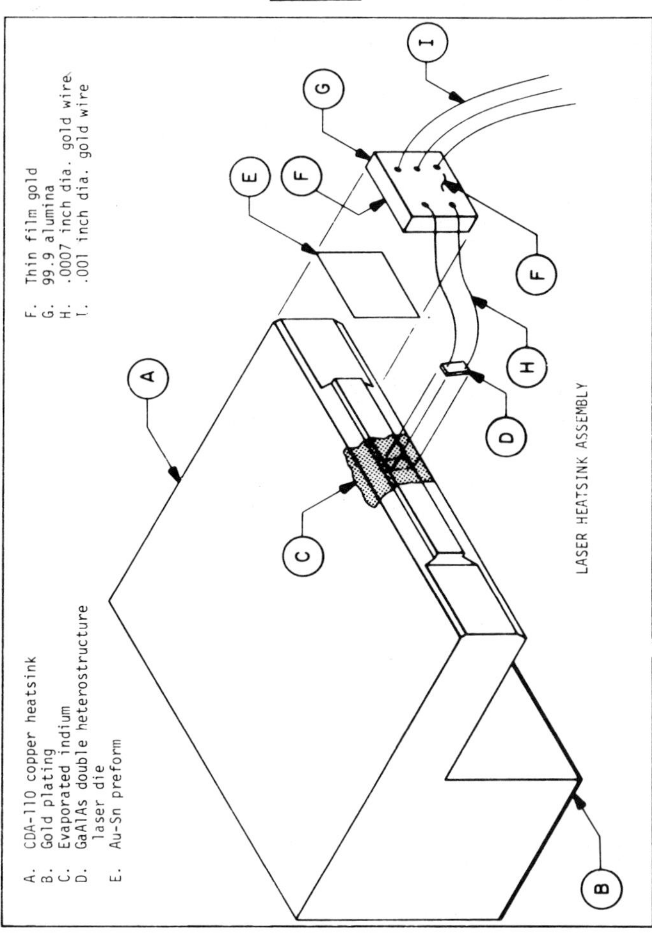

Substrate Fabrications

Three different types of substrates were fabricated for the hybrid package. Standard established thin film metallization processes were used in the fabrication of the substrates, except for some of the unique requirements imposed upon them.

The hybrid substrate was fabricated on a 99.9% alumina substrate with dimensions of 1.00 x 1.00 x .025 inches. Both sides of the substrate were metallized. Also, the "circuit side" had twice the "standard 2 microns" gold thickness which aided the eutectic die bonding processes. Both the top and bottom surfaces were sputtered with tantalum nitride (standard 25 ohms per square). This was followed by a molybdenum-gold sputtering over the tantalum nitride.

During the circuit definition processes, the "backside" metallization was protected with a continuous coating of photoresist. The standard gold etch time was increased because of the 4 micron gold thickness. The masks had an array of 9 circuit patterns. Each circuit pattern had 5 tantalum nitride resistors. All of the resistors in the array were aged together on the substrate. During the aging process the resistors were selectively probed and measured. After the aging process, two resistors on each individual circuit pattern were measured for acceptance ($\pm$5% of design value). Those that are not acceptable were identified. A wafer sawing process was used to saw the substrate into individual hybrid substrates. A 90 micron thick diamond composite blade was used in the sawing process. The final hybrid substrate size was .250 x .375 x .025 inches.

In the fabrication of the black ceramic chip, a 94% alumina substrate was used, with dimensions 1.00 x 1.00 x 0.15 inches. The substrates were sputtered with a metallization of tantalum nitride, molybdenum, and gold on one side only. Tantalum nitride and molybdenum were used for adhesion purposes. The substrate was subsequently waxed onto a "reject" silicon wafer and then sawed into individual chips. This waxing allowed the 90 micron wide wafer saw blade to saw all the way through the substrate, leaving a required burr free surface. The chip size was .038 x .085 x .015 inches.

Another type of substrate was used to fabricate the ceramic standoff for the laser heatsink assembly. This substrate was only .010 inches thick. The material was 99.9% alumina with dimension 1.50 x 1.50 x .010 inches. This substrate was metallized on both sides using the previously mentioned sputtering processes of tantalum nitride, molybdenum, and gold. The substrate was then sawed into individual chips (.030 x .035 inches) using the same method stated in the fabrication of the black ceramic chip.

Reflow Solder Processes

Since the laser hybrid package requirements prevented the use of epoxies for assembly, reflow solder methods were developed for the package. Solders of different melting points were selected to be compatible with the sequence of assembly. The solders are depicted in Figures 8 and 9. They are listed in the table below.

Table 3.

ASSEMBLY AREA	SOLDER TYPE	M.P.	COMPOSITION (% by weight)
Component die to substrate	Au-Si Eutectic	370°C	97Au, 3 Si
Substrate to header	Au-Sn Eutectic	280°C	80Au, 20 Sn
Ceramic stand-off to heatsink	Au-Sn Eutectic	280°C	80Au, 20 Sn
Laser die to heatsink	Indium	157°C	99.9 In
Heatsink to header	Indalloy #9	162°C	70 Sn, 18 Pb, 12 In

Components were die attached to the hybrid substrate in two separate stages. The two XSTR die and MOS capacitor were Au-Si eutectically die bonded to the substrate on a semi automatic Hughes model HDB 980 Die Bonder. A unique feature on this bonder was the dead weight quick-change die collets and the folded optics for viewing the bond and die pickup areas. Each collet was designed for a particular size die. A Au-Si preform was used for the gold backed capacitor die, the other silicon backed die quickly formed a Au-Si solder during the bonding process. The bonding parameters were optimized for the three sizes of die; however, the largest die size dictated the scrub cycle time. Heated nitrogen provided a quasi inert atmosphere during die bonding. A critical parameter was the total time that the thin film substrate was at the 385°C bonding temperature. At this temperature the tantalum nitride thin film resistors continued to oxidize (standard aging temp. for Ta_2N resistors is 400°C). Also, since the gold substrate pads were only 4 microns thick, all of the gold diffused to form the Au-Si solder in less than 2 minutes. After 14 minutes at 385°C the metallic bond between the molybdenum and the Au-Si was lost. An average of 3 minutes of die bonding time has been experienced for this first stage of eutectic die bonding.

During the second stage of die bonding, the photodiode chip and black ceramic chip were Au-Si eutectically die bonded on a specially tooled die bonder with a water cooled baffle to protect the operator and heated nitrogen to retard oxidation. The photodiode was manually scrubbed into the substrate pad area with tweezers to form the Au-Si eutectic solder. The gold backed black ceramic die was then set into solder and positioned to the pad. Following this, the photodiode was moved into position against the black ceramic chip. The physical contact between these two components required the separate manual die bonding operation. Average time for this operation was 4 minutes; therefore, the total time that the substrate was at the 385°C bonding temperature was $\simeq$ 7 minutes. At this average total time, strong metallic bonds were realized.

Gold-tin preforms were used to bond the hybrid substrate to the header and the standoff chip to the copper heatsink. These reflow solder processes were manual operations that required a strip heater with a gas (90% N_2, 10% H_2) purged bell jar for assembly. Special fixturing held the assemblies during the reflow. Strong metallic bonds were also achieved.

The laser die was bonded to the evaporated indium surface on the copper heatsink after the die had been aligned. A Hughes HVT-100 strip heater was used for this 160°C reflow process. During the reflow, a diffused Indium-Gold intermetallic phase was formed between the gold plated bonding surface on the laser die and the indium which provided the metallic bond. A critical parameter was the slopes of the temperature vs. time curves. These parameters were controlled by the cooling rate of the strip heater electrodes, flow rate of the (90%N_2, 10% H_2) forming gas and the electrode current rate. Thermal stresses introduced to the laser die directly affected the life of the die.

The heatsink assembly was also bonded to the header with a reflow solder process. An Indalloy #9 preform was used to bond the gold plated heatsink base to the header. An alignment holding fixture was used during reflow. The preform melts at 160°C; however, the indium on the leading edge of the heatsink where the laser die was previously attached melts at 157°C. It was important to quickly cool the solder bond at the heatsink base, otherwise the heat would transfer out to the leading edge and reflow the indium. The aforementioned strip heater was also used for this process. The Indalloy #9 provided a strong structural bond as well as a thermally conductive interconnection with low circuit impedance.

Assembly Alignments

One of the more challenging areas of the manufacturing processes were the assembly alignments. Typically, component placement positional accuracies of less than .010 inches are very uncommon for most hybrid assemblies. Four of the alignments that were performed at various stages are explained.

The photodiode die together with the black ceramic chip were aligned to a substrate pad within .002 inches. Tantalum nitride alignment keys, incorporated into the thin film metallization substrate circuit pattern, provided a visual reference surface. Alignment was performed manually during the eutectic die attachment of the components with the use of tweezers, stereozoom microscope and a special tooled eutectic die bonding machine.

When bonding the substrate to the header, the substrate alignment keys were positioned to the center of the header within .001 inches to one axis and within .004 inches to the other axis. A 10X microscope was used for the alignment. A special microscope reticle was designed for the alignment. The reticle pattern had a cross hair centered on three concentric circles of 11, 12, and 13mm diameters. During assembly, the reticle was superimposed on both the circumference and the terminals of the header. The substrate was then manually positioned with tweezers so that its alignment keys were also superimposed on the cross hair.

Another alignment occurred during the laser die attachment. The laser die was rotated 6° $\pm$ 1/2° with respect to a reference surface on the heatsink. This alignment was done just prior to reflowing the indium for die bonding. Alignment tooling provided a reference surface for positioning the angle of rotation. The alignment process was viewed with a 30X monocular zoom microscope with a reference reticle. A vacuum pickup tool and a hand probe was used to manually position the laser die.

The laser heatsink assembly was aligned to the header such that the center of the laser die was located at the package center within .004 inches. Special fixtures were used to hold the assembly in position during the reflow solder bonding process. The same type of special reticle aforementioned was used in the alignment microscope. The reticle pattern was centered to the header within .001 inches. The center of the laser die was then superimposed at the center of the reticle cross hair within .002 inches by manually positioning the laser heatsink assembly.

Heatsink Fabrication

The heatsink was manufactured from CDA #110 copper material. As an intermediate step during the machining operations, it was plated with a gold strike followed by a .005 inch thick gold plating. This provided a gold surface for the mounting base. The critical machined surface was on the lapped area where the laser die was attached. A surface finish of 4 microinches AA was required.

After the heatsink had been fabricated, the ceramic standoff chip was attached. This attachment was reviewed in a preceding section on reflow soldering. Special preparation of the lapped surface was required prior to indium evaporation. The heatsink was cleaned in a sequence of hot organic solvents and then heated in a reducing atmosphere at 600°C to remove any cuprous oxides from the lapped copper surface. Immediately following, a thickness of 4 microns of indium was selectively evaporated on the lapped surface area for laser die attachment. Of particular interest was that laser die attachment had to occur within 24 hours from the indium evaporation process; otherwise, the indium oxides which were formed impeded the die attachment process.

Wire Bonding

Many different types of wire bonds were made for the assembly of the laser hybrid package. Standard .0015 inch dia. aluminum wire was ultrasonically bonded to interconnect the hybrid components to the thin film pads on the substrate with a Tempress ultrasonic wire bonder. This method was used because it was faster than the subsequent gold ball wire bonds used for the other interconnections.

Gold ball wire bonds (.001 inch dia. wire) were used to connect the substrate to the header base and pins. The longest bond was .140 inches. A Hughes HPB 360 pulsed thermocompression gold ball wire bonder was used for all of the gold wire bonding operations for the laser package.

The laser die was very sensitive to externally applied strains; therefore, ultrasonic methods of wire bonding on the laser die was prohibitive. To minimize the applied bonding forces transmitted to the laser die, .0007 dia. gold wire bonds were used to interconnect the die to the ceramic standoff on the heatsink.

After the laser heatsink assembly (laser die) had survived a 96 hour burn-in test, three .001 inch dia. gold wires were attached to its ceramic standoff. These "flying leads" were subsequently used to continue the connection between the laser die and the substrate circuit. The length of these leads were approximately .125 inches. The pulsed thermocompression gold ball bonder was very suited for this type of bonding because of its manual operation mode and unusually long search height range. The attachment of these "flying leads": to the substrate after the laser heatsink assembly had been bonded to the header required some operator skill. Two of the three flying leads were laced down from the ceramic standoff and onto the hybrid substrate bonding pad.

The third lead was used for insurance only and was removed if it was not needed. The laced leads were wedge bonded to the substrate pad with two wedge bonds made by the bonding capillary, after first removing the wire from it.

Encapsulation

Encapsulation was accomplished by welding the stainless steel cap rim to the base of the gold plated Kovar header by the resistance welding method, most commonly referred to as "spot welding". The cap welder was enclosed by a nitrogen filled dry box. During the welding operation, nitrogen was entrapped inside the hybrid which provided a quasi inert atmosphere for the laser die. The nitrogen also enhanced the weldability.

There were several parameters that affected the weld: notably, assembly parts material and geometry; electrode material and geometry; applied electrode force; amount and type of current pulse; cycle time before, during and after the current pulse; electrode cooling rate; contact resistance and surface area between the two mating parts. After final characterization, successful welds were made using tungsten-copper electrodes, 500 pounds applied force, a high current pulse (AC) for 30 milliseconds and a quench time of 1.40 seconds. It was important that the header base temperature remained lower than the reflow solder melting points during the welding progess. Leak rates after encapsulation were in the 10^{-8} STD. cc He/sec range.

SUMMARY

The packaging concepts presented were primarily developed from established hybrid microelectronics principles and manufacturing methods used in today's electronics manufacturing industries. These concepts were integrated into the design of hybrid packages that had unique optical I/O requirements and new applications.

Optical I/O hybrid microelectronic packages developed for the HP 3820A Electronic Total Station have contributed to the overall success of the instrument's design and accuracy performance. The laser hybrid package has proven to be a viable device as a continuous wave lasing light source for the measurement of distance using light.

ACKNOWLEDGEMENTS

Acknowledgement is given to those who substantially contributed to the development of this paper: Don Bradbury of HP Labs; HP 3820A Project Manager Alfred Gort; Device Technology Dept. Manager Tom Christen and development team members Hal Chase, Ken Gilpin and Tom McDonald. Appreciation is also expressed to Diana Roberts for the manuscript typing and to Glenda Dixon for the graphic illustrations.

REFERENCES

1. Alan Thompson, John Williamson, Kent Garliepp and Charles Stolte. HPL/SSL TECHNICAL MEMO No. 77-11 "Life Testing GaAlAs Double Heterostructure Laser Hybrids". Hewlett-Packard Company. 1977.

2. Alfred F. Gort. "The Hewlett-Packard 3820A Electronic Total Station". Civil Engineering Division, Hewlett-Packard Company. 1977.

Reprinted from *Electronic Packaging & Production*, June 1984

Lasers Tackle Tough Soldering Problems

Surface-mount component assemblies sometimes pose unique soldering problems to which the laser currently offers the best solution.

By Earl Lish, Martin Marietta Aerospace, Orlando, FL

Numerous technical problems must be addressed when attaching surface mounted components to a substrate. Frequently, components are difficult if not impossible to solder to a PWB with conventional mass soldering methods. This is because heat must be applied to the component side of the PWB. Solder paste normally used for attachment can commonly form solder balls when melted. Often solder mask must be used to prevent formation of solder bridges. Sometimes IR, hot belt, and vapor phase soldering subjects the entire assembly to soldering temperatures for long periods of time, causing components or PWBs to be damaged or destroyed. Furthermore, mass soldering techniques may not work well on assemblies containing heat pipes or heat sinks because of the heat absorbing characteristics of such devices. Differences in thermal expansion properties between ceramic devices and PWBs during thermal cycling can result in cracking of solder connections.

The laser offers a means of potentially overcoming many of these microsoldering problems. Laser microsoldering eliminates solder ball and solder bridge formation, and thus the need for solder mask. Thanks to the laser beam's unique and controllable nature, heat that is highly localized for an extremely short time duration can be applied, thus minimizing or eliminating component or PWB damage (Reference 1). Furthermore, laser soldered PWB connections are more ductile than conventionally formed connections since the short duration laser pulse does not form as much brittle intermetallic compound. When soldering SMDs to PWB assemblies containing heat pipes or heat sinks, the laser is an ideal tool since the short duration pulse forms the solder connection on the PWB surface before the heat can be conducted through the PWB into the heat pipe or heat sink.

Special considerations

Applying a laser beam to solder connections presents a number of special geometric and thermal problems. Infrared radiation from a YAG laser at a 1.06 μm wavelength is almost equally absorbed by metals and PWB materials, whereas the 10.6 μm wavelength radiation from a carbon dioxide (CO_2) laser is mostly reflected by metals but completely absorbed by PWB laminate materials. However, both the YAG and CO_2 lasers are quite capable of forming acceptable microsoldered connections. Primarily because of availability, a 20 W continuous wave (CW) flowing CO_2 gas laser was selected for use in studying the attachment of SMDs and connectors to polyimide PWBs.

Beam power levels and durations used to solder SMDs ranged from 5 to 10 W focused into a 0.010 in. diameter spot for 0.2 to 0.5 sec. This is equivalent to 300 KW to 600 KW per square inch. Since energy equals power times duration, CO_2 laser soldering was accomplished with 1 to 5 W-sec (Joules) of applied energy per soldered connection.

Heat transfer

Since the laser beam is applied only once to a given area, and then for only 0.2 to 0.5 sec, an important factor to consider when laser soldering is the heat transfer between the device and PWB. Because of the consistent formation of solder balls that can be neither readily nor reliably removed, solder paste is unsuitable for laser soldering. This is unfortunate because solder paste acts as both a heat transfer media and a final connection material.

Solder balls created during laser soldering are due to the partial expulsion of solder paste beyond the laser beam heat zone by the instant vaporization of flux before all solder particles can fuse on the connection. Further work with solder paste was discontinued in favor of reflow soldering of solder coated devices to thick, unfused tin-lead plating on PWB solder pads.

The decision not to use solder paste not only eliminated formation of solder balls but also formation of solder bridges. Since laser soldering can be accomplished without formation of solder balls or solder bridges, the need for solder mask is thereby eliminated.

Solder dipped or thick tin-lead plated surfaces with a film of rosin-base soldering flux on top of and between each solderable surface were found to be easily soldered without the addition of more solder.

Burns of Apollo Laser (Ref. 2), reported that rosin base flux readily absorbs CO_2 laser radiation and transfers this energy into the solder coated device-to-PWB interfaces to form reliable solder connections. In fact, without rosin base flux present on the surface exposed to laser radiation, solder melt does not occur, indicating that most of the laser energy is reflected.

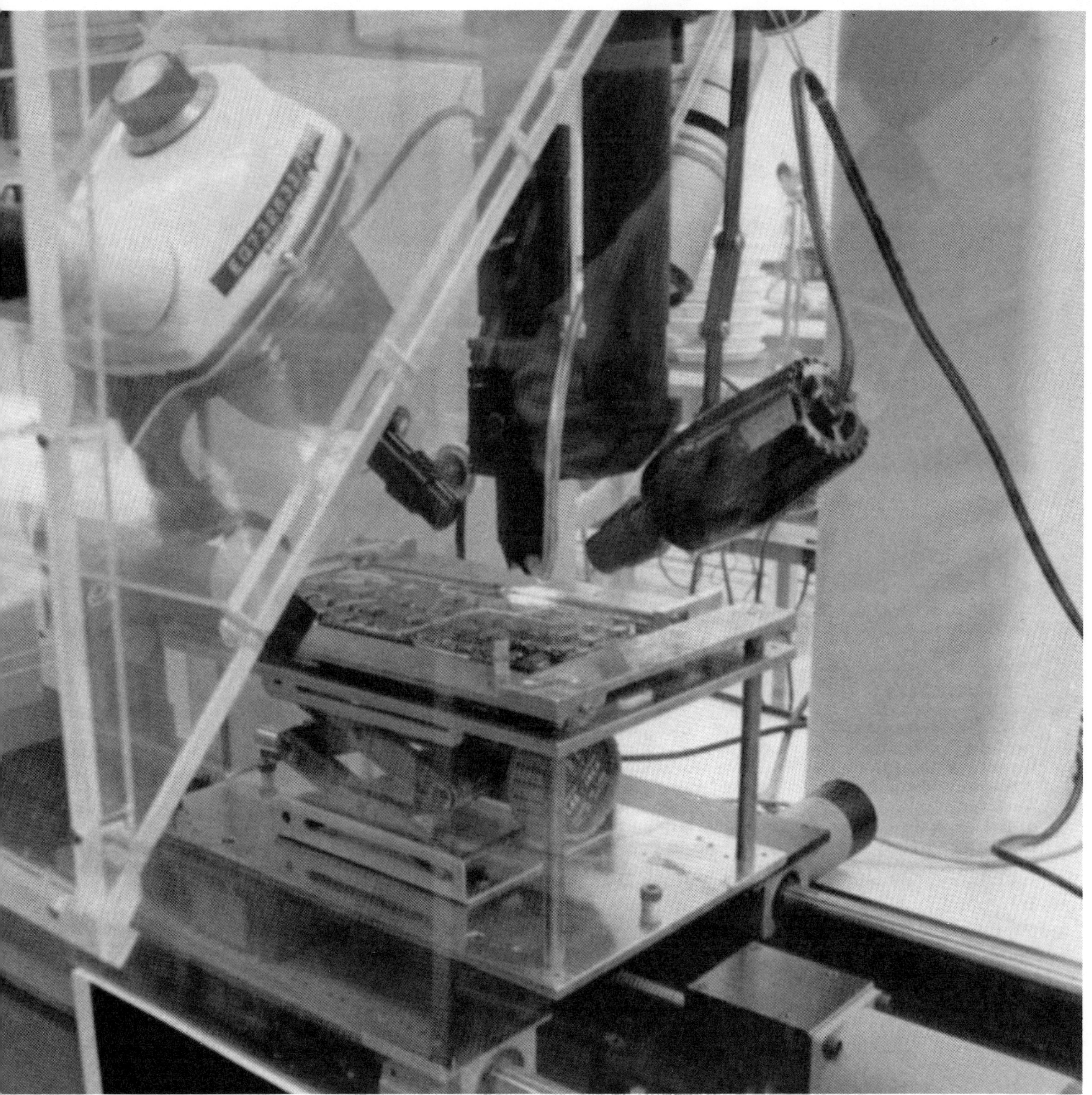

*1. **Lasers sometimes offer** the best way to tackle tough soldering problems. Pictured is a CO_2 laser with a computer numerically controlled X-Y table soldering an edge connector to a PWB and heat sink assembly.*

Angle critical

The angle of the laser beam relative to the device is critical to successful laser soldering. A vertical beam (Fig. 2A) will melt but not flow the small quantity of solder contained in a solder dipped HCC crenelation because of surface tension of the fluxed, molten solder. Even when a beam is applied at a 45 degree angle to the HCC (Fig. 2B), the aspherical tinned pads do not flow, thus a solder connection is not formed unless the parts are in intimate contact. However, a solder connection is formed by the same surface tension that causes flat areas of solid solder to assume raised, aspherical configurations when flattened solder tinned pads on HCCs are melted. These raised areas touch each other, then join and form the connection.

Flat solder coating is attained on HCC pads by means of a hot, moving belt that flattens the aspherical solder-dipped coatings into the same plane under the influence of heat and HCC weight. Figure 3 illustrates the aspherical configuration normally obtained on solder dipped HCC pads, while Fig. 4 illustrates the desired

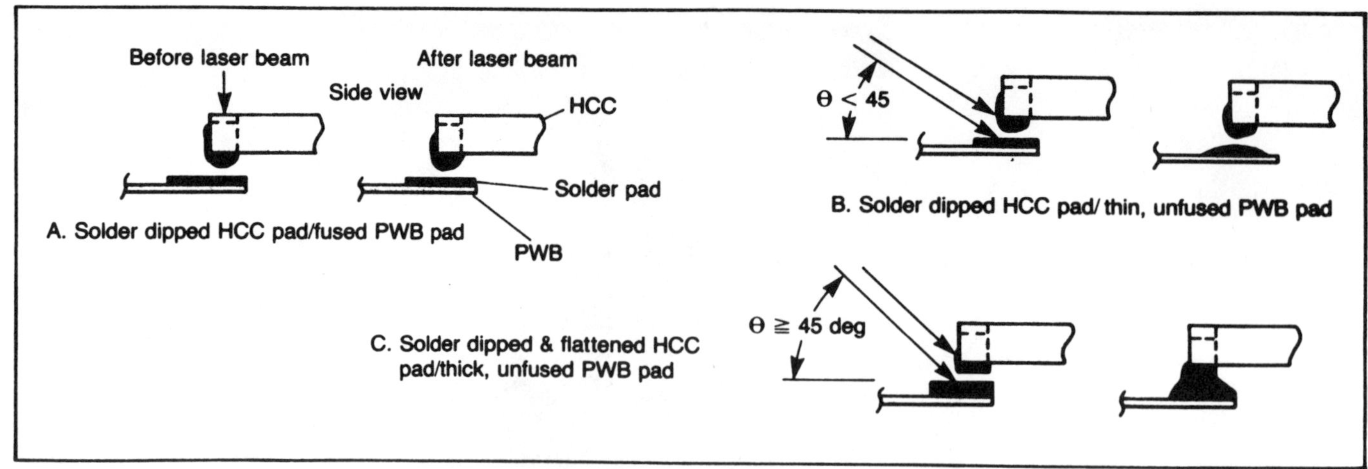

*2. **Effects of the laser beam** angle and solder coating configuration on the solder joint formation.*

planar configuration attained by heat flattening.

Joint cracking

Much concern has been expressed about the potential for cracked solder connections between relatively low expansion ceramic HCC devices and higher expansion polyimide-fiberglass substrate material after thermal cycling (Ref. 3, 4, 5, 6). Because most commercial processes heat the entire PWB and devices to at least 419 degrees F (215 degrees C) for 10 to 30 sec, the solder connections are increasingly stressed beginning at the moment of solidification and continuing as the assembly cools.

Laser soldering, however, does not subject PWBs or devices to elevated temperatures except in the immediate solder connection area, and then for only 0.2 to 0.5 sec. For this reason, laser-soldered connections have less initial stress and therefore withstand more thermal cycles before cracking than do connections formed by more conventional processes. Furthermore, laser-soldered connections are more ductile because there is less formation of brittle copper-tin intermetallic compound as a result of faster solder fusion and cooling.

Figure 5 shows the difference in the cross section of a laser soldered connection as compared with a wave-soldered connection. The most striking difference is the almost complete absence of copper-tin intermetallic compound in the laser soldered section that is always quite prominent in conventional wave soldered sections. This difference is attributed to the period of time in which molten solder is present on the copper PWB pad, i.e., the shorter the time (0.25 sec by laser) the less formation of intermetallic compound.

It should be pointed out that extreme care must be observed during section preparation to avoid heating the sample during sawing, mounting, grinding and polishing because such heating has been demonstrably shown to cause formation of additional intermetallic compound that was not present after initial cooling of the original solder connection. Use of low exotherm casting resins, and low grinding and polishing pressures in conjunction with copious liquid cooling, are therefore a must in order to achieve a truer measure of the copper-tin intermetallic compound thickness.

Insulation burning

Inexplicable burning of the polyimide PWB insulation was observed through a microscope at the moment of firing the vertical CO_2 laser beam onto a flat PWB solder pad. Investigation revealed that this burning was not due to reflection as had been suggested, but to the fact that the focused laser beam was larger in diameter than originally believed.

The beam energy distribution was stated to be Gaussian, and the spot diameter was stated as 0.010 in. This spot diameter was defined as that

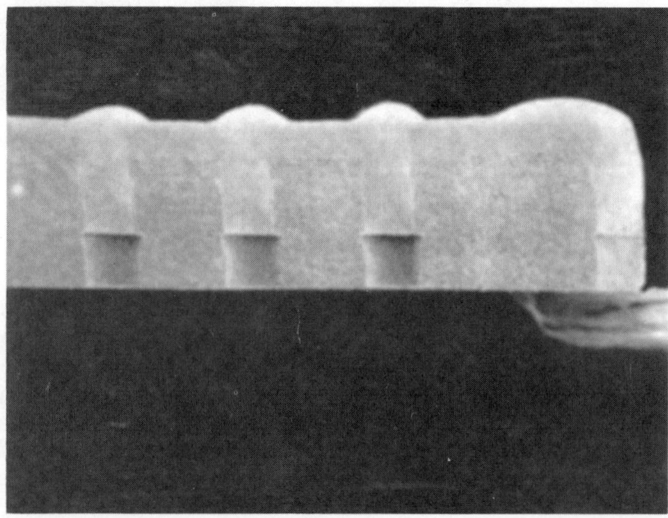

*3. **SEM photo shows** how the solder-dipped pad of a hermetic chip carrier appears at 100X magnification.*

*4. **SEM photo** shows how the solder-dipped pads of an HCC appear after heat flattening at 100X magnification.*

Table 1. Solder inspection - connector terminations per MIL-STD-454, requirement 5, soldering*			
Defect	Group		
	1 Hand tinned, Laser soldered	2 As-received Laser soldered	3 Hand tinned, Hand soldered
Insufficient solder	2/23 = 9%	18/18 = 100%	1/20 = 5%
*Solder dipped PWB No additional solder RMA flux			

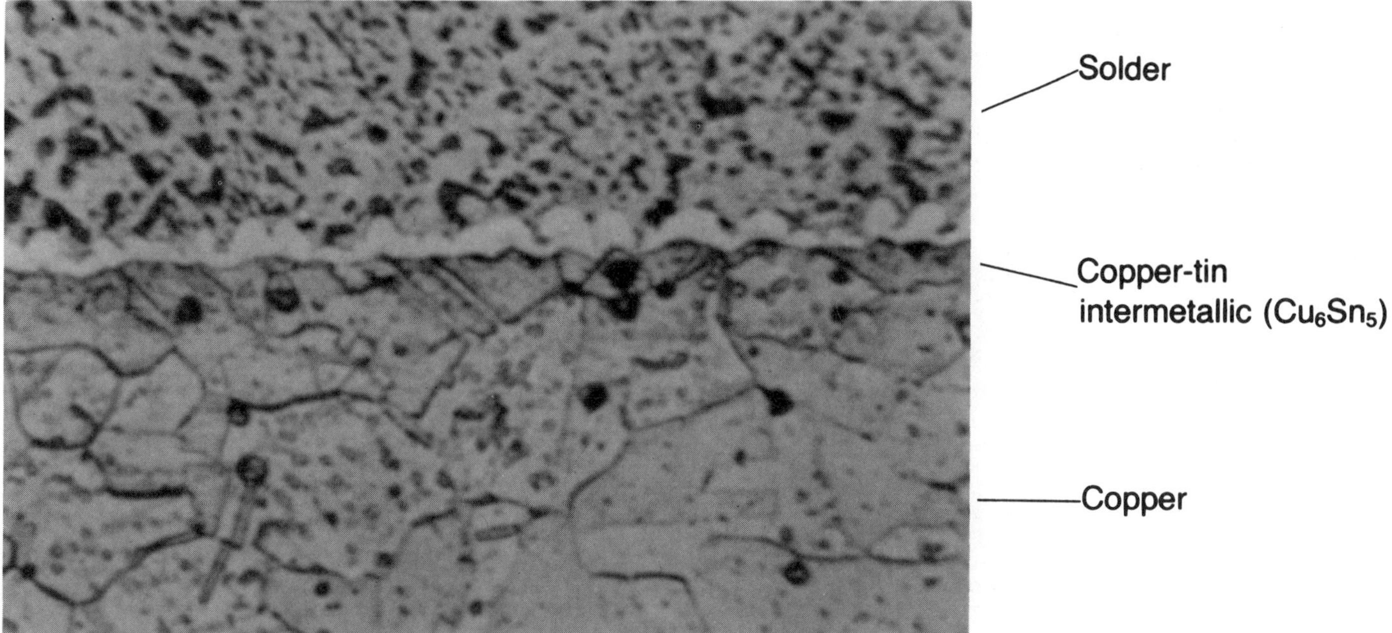

Wave soldered for 3 sec at 500 F

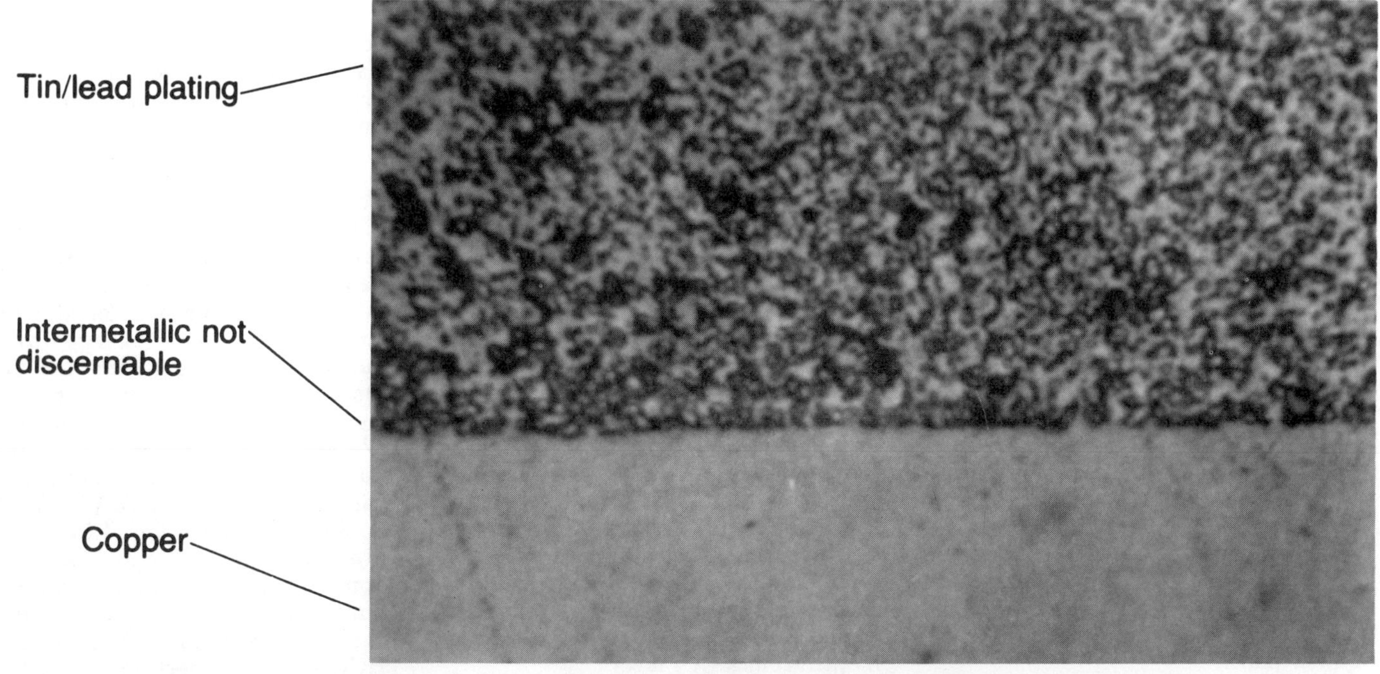

Laser soldered for 0.25 sec at 8 W

5. ***Cross sections*** *of a wave soldered solder-copper interface (a) revealed much more copper-tin intermatallic formation than found at the laser soldered solder-copper interface (b).*

*6. **Hand tinned terminations** that have been laser soldered are shown in this 100X SEM photograph.*

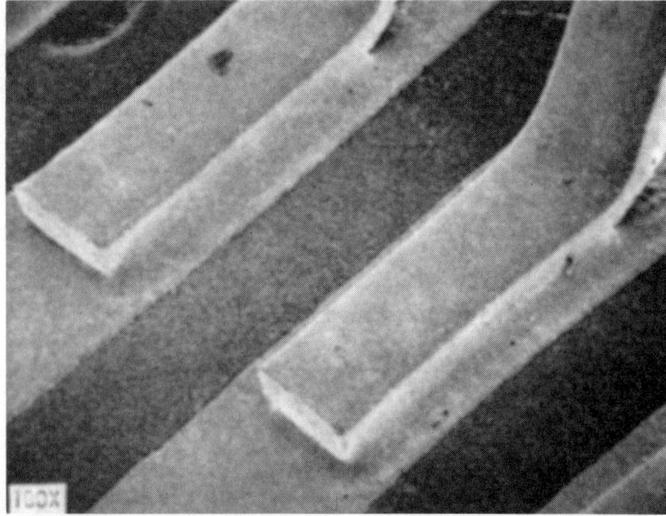

*7. **Terminations that were laser soldered** as received without any special preparation appear like this when viewed by an SEM at 100X magnification.*

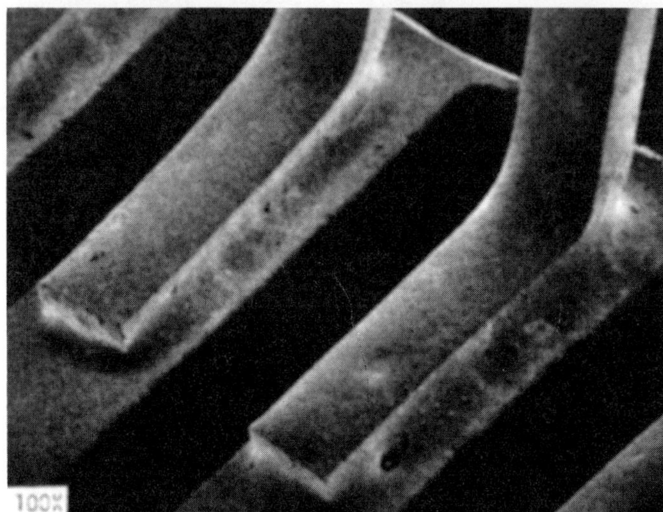

*8. **When viewed by an SEM** at 100X magnification, hand tinned terminations that have been hand soldered looked like this.*

central circular area of the laser beam which contained 86 percent of the total beam energy. Since a Gaussian distribution forms a straight line when plotted on probability paper (in contrast to the more familiar bell-shaped normal distribution curve when plotted linearly), this line was used to graphically determine the total beam diameter of the area enclosing the remaining 14 percent of the laser energy, known in laser circles as "fringe energy," that extended beyond the central spot area. The total beam diameter, consisting of 100 percent of the energy, was found to be 0.025 in. From this data, it was determined that the relatively low power, large diameter fringe energy of the CO_2 laser beam could burn insulation on either side of a PWB pad less than 0.025 in. wide even if the central 0.010-in. diameter spot were centered on the pad.

However, it was determined that an optical method of beam expansion (up-collimation) could reduce the focused spot and beam diameters. It was found that a 2.5X beam expansion in conjunction with the existing 2.5 in. focal-length lens would produce a central spot diameter (86 percent of the energy) of 0.004 in. and a total beam diameter (100 percent of the total energy) of 0.010 in., which would be safely usable on PWB pads down to 0.010 in. wide.

Reflected energy damage

Examination of CO_2 laser-soldered HCCs and connector terminations showed that solder-coated surfaces could reflect some of the laser energy and damage nearby surfaces such as PWB insulation and vertical ceramic sides of HCCs. As will be shown later, the 1.06 μm wavelength energy from a YAG laser is better absorbed by solder and therefore does not present as severe a reflection problem when compared with the 10.6 μm wavelength from a CO_2 laser.

Tin-lead coated circuit conductors that were 0.010 in. wide by 0.0028 in. thick were found to fuse only about 0.005 in. beyond the PWB pads after being laser soldered to a HCC; as will be illustrated later, this supports the claim that laser soldering does not appreciably heat surrounding PWB areas.

Power levels and pulse duration

Laser soldering, unlike cutting or welding operations, is more time dependent than total-energy dependent. Time is needed for the highly concentrated heat energy to flow outward, melt the solder, and for the molten solder to form the solder connection. For example, a single 100 W continuous wave pulse (CW) applied for a duration of 0.024 sec contains the same energy as an 8 W CW pulse applied for a duration of 0.3 sec (both are 2.4 Joules). The 100 W pulse, if applied to a PWB pad, would probably burn a hole through the pad, whereas the 8W pulse will result in an excellent HCC solder connection. For soldering connector terminations, a power level of 10 W applied for 0.5 sec (equal to 5 Joules) was found more effective than higher power having shorter pulse duration.

Connector terminations

Three sets of surface mount connector terminations were soldered on the same polyimide PWB and then submitted to Quality Assurance for determination of compliance to MIL-STD-454, Requirement 5, Soldering. The first set consisted of 23 hand-tinned terminations that had been laser soldered to the PWB. The second set consisted of 18 as-received, unfused, bright acid tin-plated terminations that had also been laser soldered to the PWB. The third set consisted of 20 hand-tinned terminations that had been hand soldered to the PWB. An RMA flux from the same bottle was used throughout all soldering operations. Results of the soldering inspection, summarized in Table 1, indicated that acceptable laser-soldered connections were achieved on hand-tinned terminations, but not on as-received, unfused, bright acid tin-plated terminations.

After visual inspection of these three sets of soldered terminations, SEM photographs were made of terminations from each group (Fig. 6, 7, 8) to illustrate the solder fillets formed. Some of the terminations selected also have minor cosmetic blemishes, such as pin holes and voids, that were acceptable per MIL-STD-454, Requirement 5, since the bottom of these areas could be seen as being coated with solder.

Connector peel strengths

After SEM photographs were made of these three sets of soldered terminations, five terminations from each set were peel tested to destruction by means of a Unitek Model 6-092-03 pull tester operating at a peel speed of 4 in./min. The results of this test, listed in Table 2, indicated that laser soldering of connector terminations did not degrade the adhesive bond strength of the copper circuit pad to the polyimide PWB when compared with the peel strengths obtained with hand soldered terminations.

The peel strength test also revealed the weakness of soldered connections having insufficient solder as listed in Table 1 for Group 2 terminations (as-received). These terminations had an unfused, bright acid tin-plate finish per MIL-T-10727 not exceeding 0.00025-in. thick, a finish that DeVore of General Electric has shown will readily de-wet due to the outgassing of co-deposited organic brightener materials in the tin plate (Ref. 7). It is therefore possible that the low peel strength of Group II terminations was due more to outgassing effects of co-deposited organic material than to insufficient solder.

After performing the peel strength tests, an additional five terminations were pull tested to destruction at a pull speed of 4 in./min. Results of this test, shown following the peel test data in Table 2, indicated that a higher range pull test gauge should have been used since all pull tests except one (Term. No. 2-3) pinned the gauge needle beyond the gauge's full scale range before termination failure occurred.

The effect of CO_2 and YAG lasers on materials

Samples of polyimide PWB insulation materials and unfused tin-lead plating were taken to the Air Force Wright Aeronautical Laboratory (AFWAL) for determination of their absorbance, reflectance and transmittance at 10.6 and 1.06 μm wavelengths of CO_2 and YAG lasers respectively in an attempt to explain: (1) why polyimide PWB insulation was burned by fringe energy from a CO_2 laser, and (2) why tin-lead plate appeared to reflect CO_2 laser energy to a greater degree than that from a YAG laser.

Fig. 1 and 2 are composites of the reflectance data obtained. According to Fig. 1, the reflectance of polyimide PWB insulation at 10.6 μm wavelength is 2 percent (equivalent to 98 percent absorbance and/or transmittance), whereas Fig. 2 shows the relfectance at 1.06 μm to be 27 percent (equivalent to 73 percent absorbance and/or transmittance). The practical result of this data is to show that the polyimide PWB material is one-fourth less likely to be burned by YAG laser energy than by the same energy from a CO_2 laser.

In similar manner, Fig. 1 shows that the reflectance of unfused tin-lead plate at 10.6 μm wavelength is 74 percent (equivalent to 26 percent absorbance), whereas Fig. 2 shows the reflectance at 1.06 μm to be 21 percent (equivalent to 79 percent absorbance). The practical result of this data is to show that only one-third as much energy is required from a YAG laser to melt solder as that required by a CO_2 laser.

One conclusion that can be made from this data is that the YAG laser is potentially less damaging to devices and PWB insulation by direct or reflected laser energy than is the CO_2 laser. Another conclusion is that the YAG laser is more efficient for soldering. However, these advantages of the YAG laser tend to be offset by factors such as the much higher initial cost (twice as much), and lower efficiency (1 percent versus 10 to 15 percent), leading to higher operating and maintenance costs than for the same power CO_2 laser. With proper attention to system design, either type laser can effectively perform microsoldering operations, and a cost trade study is recommended before selecting one type in preference to the other.

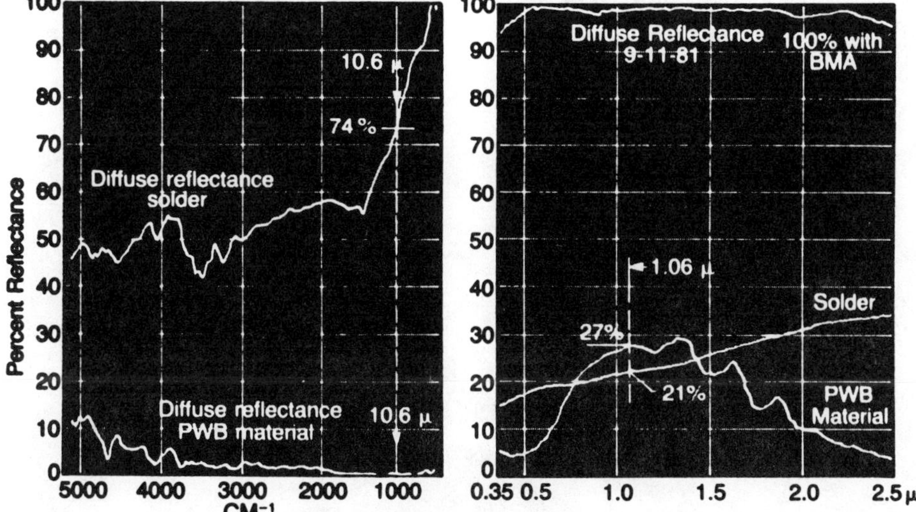

1. The reflectance of 10.6 μm radiation from solder and polyimide PWB material.

2. The reflectance of 1.06 μm radiation from solder and polyimide.

Table 2. Peel and pull test data - soldered terminations

Group								
1. Hand tinned, laser soldered			2. As-received, laser soldered			3. Hand tinned, hand soldered		
Term no.	Lb pull	Fail site*	Term no.	Lb pull	Fail site*	Term no.	Lb pull	Fail site*
Peel test								
1-6	2.72	PWB	2-7	3.05	PWB	3-6	2.00	PWB
1-7	2.00	PWB	2-8	2.32	PWB	3-7	2.66	PWB
1-8	3.14	PWB	2-9	0.40	S	3-8	2.98	PWB
1-9	1.90	PWB	2-10	0.60	S	3-9	1.60	PWB
1-10	2.35	PWB	2-11	0.80	S	3-10	1.78	PWB
AVG	2.42	—	—	1.43	—	—	2.20	—
Pull test								
1-1	>3.2	S	2-1	>3.2	S	3-1	>3.2	S
1-2	>3.2	S	2-2	>3.2	S	3-2	>3.2	S
1-3	>3.2	TBK	2-3	3.04	S	3-3	>3.2	S
1-4	>3.2	S	2-5	>3.2	S	3-4	>3.2	S
1-5	>3.2	S	2-6	>3.2	S	3-5	>3.2	S
AVG	>3.2	—	—	>3.17	—	—	>3.2	—
*PWB = Pad to PWB			S = Solder to termination			TBK = Termination broke		

Shear tests were attempted on several soldered terminations. However, a reliable means of pulling against the thin (0.009 in. thick), rounded termination could not be developed and the shear test was subsequently abandoned in favor of the pull test. Since the brass termination was found to be stronger than the solder joint (except for Term. No. 1-3) the pull strength test appears to be a viable alternative to the shear test.

HCC soldered connections

Fig. 9 shows the fillet configuration of typical HCC connections to PWB pads. RMA flux was used, and the connections were formed with a laser power of 8 W for 0.3 sec (2.4 Joules) without the use of additional solder beyond that already present on the solder-dipped HCC pads after hot belt flattening and that on the PWB after thick (0.0012 to 0.0018 in.) tin-lead plating. The most striking feature of these HCC connections is the size of the solder fillets, which is smaller than that usually obtained with solder paste. Fig. 10 shows one of these connections (second connection from left in Fig. 9) at high magnification. Another feature of these connections is their function as standoffs between the bottom surface of the HCC and the PWB. This serves a twofold purpose: the clearance allows easier entrance and exit of cleaning solvents; and (2) the standoff distance, 0.0044 in. on this particular HCC assembly, will allow longer temperature cycling life according to Fennimore (Ref. 3). Larger diameter HCC solder fillets are also desired; this was easily achieved by means of 8 W for 0.7 sec on HCCs that were laser soldered subsequent to these photographs.

Although these laser soldered HCC connections had visually acceptable solder fillets, they could not be visually inspected in accordance with MIL-STD-454, Requirement 5, Soldering, or with any other known government document, since none of these documents specifically address HCC or other leadless devices such as chip capacitors and chip resistors.

After visual inspection of these laser soldered connections on an 18-pad HCC device, a Hunter gauge was used to determine the shear strength of the soldered connections for the entire device. Total shear of the device from the PWB was obtained at an applied load of 18 lb against the five-pad side of the HCC, for an average shear strength of 1 lb per solder fillet. Figure 11 shows the solder fillets of Fig. 9 after shear test. The standoff height of these fillets is clearly shown, as is the direction of shear toward the lower left-hand corner of Fig. 11. Figure 12 shows the solder fillet of Fig. 10 after shear test. Note that the pad void shown in Fig. 10 is almost covered by displaced solder fillet material in Fig. 12, and that the dark area to the lower right of the pad void in Fig. 10 is not apparent in Fig. 12. Two other

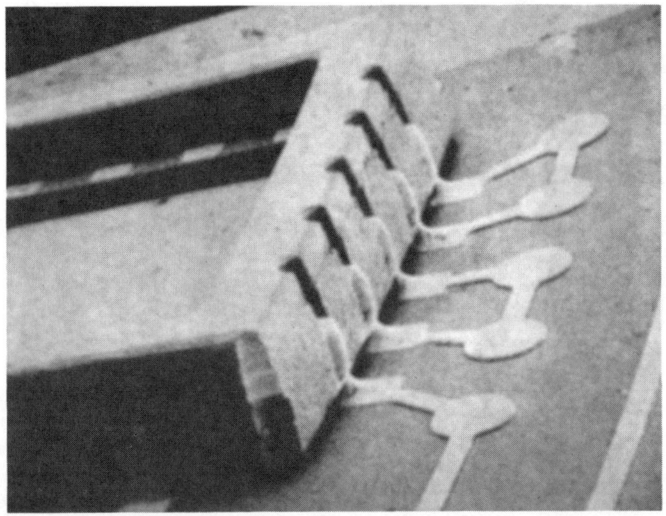

9. ***Laser-soldered connections*** *between a hermetic chip carrier and a substrate are shown in this photograph taken with an SEM at 30X.*

11. ***This SEM photograph*** *taken at 50X magnification shows the solder pads after a laser soldered HCC was removed during shear testing.*

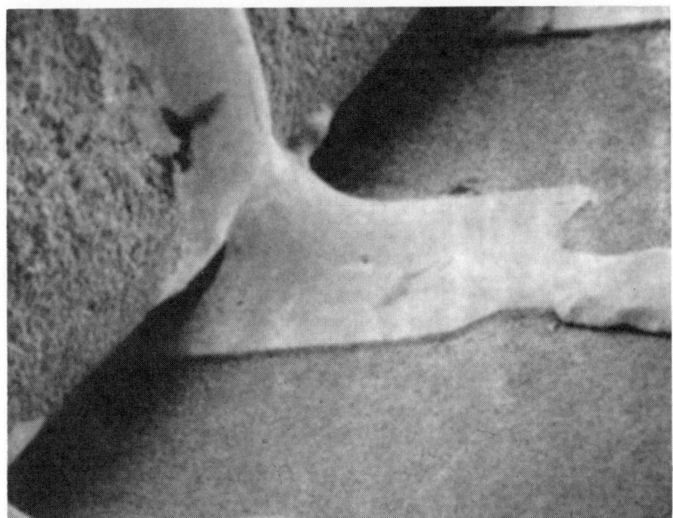

10. ***This closeup*** *of a laser-soldered connection was taken by an SEM at 200X magnification.*

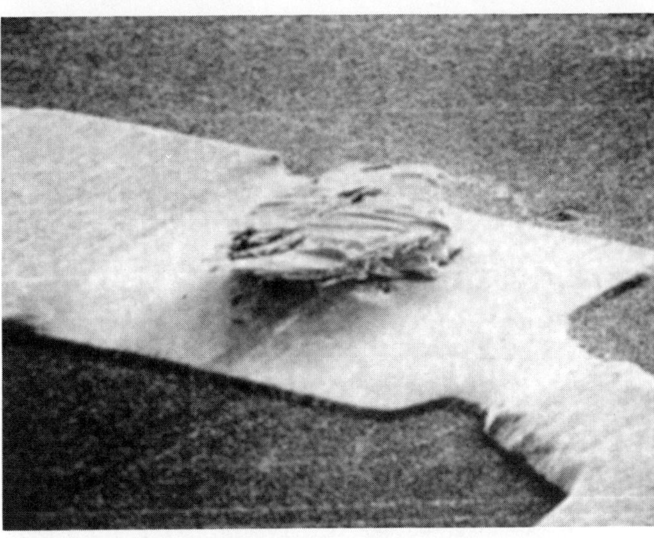

12. ***This 200X magnification*** *SEM photograph shows a laser-soldered connection after shear testing.*

noteworthy features of Fig. 12 should be mentioned: The solder melt zone on the 0.025-in. wide PWB pad is clearly outlined between unfused tin-lead plated areas, showing that the plating fused to a distance of about 0.005 in. beyond the pad and onto the 0.010-in. wide conductor line; and the curved line of demarcation between fused and unfused plating on the PWB pad is also visible to the left of the solder fillet, showing the raised, convex surface essential to fillet formation as described earlier. ■

References

1. Dixon, T., "Lasers Bring Precision to Electronics Manufacturing," *Electronic Packaging and Production* (March 1984)
2. Burns, F., and C. Zyetz, "Laser Microsoldering," *Electronic Packaging and Production* (May 1981)
3. Fennimore, John, "Hermetic Ceramic Chip Carrier Implementation," *Electronic Packaging and Production* (May 1981)
4. Woodruff, Bill, "The Leadless Chip Carrier," *Military Electronics/Countermeasures* (May 1981)
5. Tsantes, John, "Leadless Chip Carriers: The Answer to Pinout Profusion," *Electronic Business* (June 1981)
6. Leonard, Milton, "Packaging Technology for Tomorrow," *IC Update* (August 1981)
7. DeVore, John, "Solderability Defect Analysis," Soldering Technology Seminar, Naval Weapons Center — China Lake (February 1982)

Reprinted from *Annals of the CIRP*, Vol. 28/1/1979

Positioning System with Laser for Automatical Micro Spot Laser Welding

Akira Ono, Tadanori Komatsu, Naomi Hoshina (2) and Akira Kobayashi (1), Manufacturing Engineering Laboratory, Toshiba Corporation

The positioning system using He-Ne laser was newly developed for automatical micro spot YAG laser welding. Non contact, accurate and rapid positioning is necessary, to utilize advantage of high reliability and productivity of the high speed YAG laser welding operation for small electric parts. This positioning was attained by detecting the reflection beam from the edge of the part close to the welding spot with He-Ne laser beam scanned over the edge of the part. He-Ne laser beam incident and receiving angle of to the axis of the part are selected to an angle to reduce disturbance light reflection. After detecting the edge, the YAG laser welding optical system moves a given distance from the edge to the accurate welding spot, and then YAG laser irradiates the welding spot. The YAG laser welder with this positioning system is used as part of automatic assembly systems for massproduction of high quality electric device.

1. Introduction

The YAG laser welder is very useful for micro spot welding of small electric parts, having the advantage of excellent reliability and high productivity, and of preventing welding splash. In the electron tube assembly process, parts were welded by YAG laser after manual positioning using a microscope. The electron tube consists of many small parts, and has many welding spots. So an accurate and rapid positioning system is necessary for automatical YAG laser welding. This positioning system was attained by detecting a reflection beam from the edge of the parts close to the welding spot, using an He-Ne laser beam scanned over the edge of the part.

2. Position sensing method by He-Ne laser

An external view of the electron tube part and welding spot position are shown in Figs. 1(a) and (b). The electron tube parts consist of a number of electrodes, connection ribbons and isolators. The electrode has the shape of a small elliptical hat, the thickness of its brim is 0.2mm. Connection ribbon width is 1mm and it's thickness is 0.08mm. Both elements are made of stainless steel. The connection ribbon is connected to the top end of the electrode brim by YAG laser welder. Welding spot is on the mid-point of the brim thickness.

Figure 2 shows a cross section view of YAG laser welded spot. Molten zone size is about 0.5mm in diameter and nugget size is about 0.3mm. These value are larger than the electrode brim thichness. If a YAG laser beam center line does not coincide in position with the brim thickness mid-point, the laser beam makes a hole through the connection ribbon at the position where the brim does not exist. Such an electron tube part will be rejected. Tolerance of this coincidence is known to be less than 0.05mm from experimental values. The brim position fluctuation caused by loading and assemblage error is about ±0.2mm, so the development of an accurate positioning system is necessary. To have a high effect on productivity of the automatical YAG laser welder using this positioning system, 20μm position sensing accuracy and response time of less than 2ms are required for welding spot position sensing. A mechanical feeler is not useful, as it would interrupt the connection ribbon chuck movement, and has a response time of several tens of milli-seconds. As there are no marks on welding spot for position sensing, positioning systems by television pattern information are not available, and they are very expensive systems.

The positioning system using a He-Ne laser was newly developed. There is no need for extra marks on the welding spot in this system. This system senses the brim edge position instead of the welding spot by detecting the reflection beam from its edge. At preliminary experiment stage, an attempt of the edge sensing method by detecting the reflection beam from the edge was made, and the He-Ne laser beam irradiated the edge in normal direction to the surface as shown in Fig. 3(a). When the He-Ne laser beam scanned over the electron tube part, the laser beam scattered by not only the brim edge but also corner, electrode surface texture and so on. Therefore, the brim edge position could not sensed successfully. While, from the experimental result, it was found that the scattering reflection beams from the corner and the electrode surface texture were eliminated by selection of He-Ne laser beam incident angle θ and receiving angle ψ as in Fig. 3(b). As shown in Fig. 3(c), if the incident angle θ is larger than the given angle, the brim shadows the corner from He-Ne laser beam, and the corner is not irradiated by the laser beam. But the large incident angle causes deterioration in position sensing accuracy. The reflection beam scattered due to the fine surface texture decreases considerably with increasing of the angle ψ, but the reflection beam from the brim edge also decreases. So, θ=5° and ψ=50° were selected from the experimental results. In Fig. 3, "lens A" concentrates the He-Ne laser beam into a small spot, and "lens B" gathers the reflection beam.

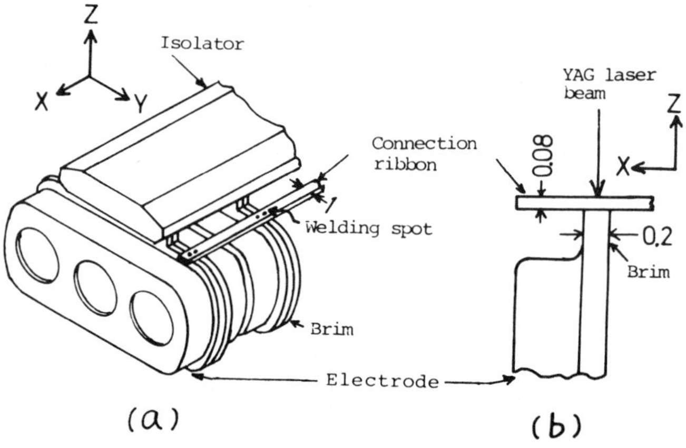

Fig.1 An external view of an electron tube part and welding spot position.

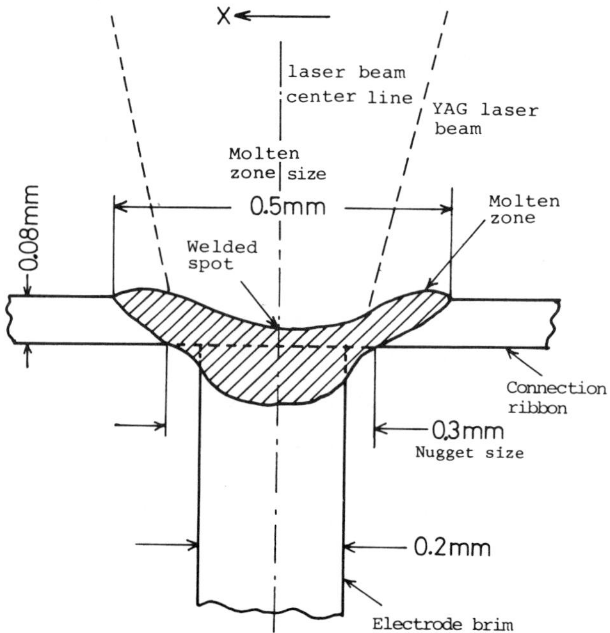

Fig.2 YAG laser welded spot cross section view.

Figure 4 shows the experimental apparatus for edge detection. The electrode is drawn exaggeratedly to explain easily. In this experiment, the electrode is set on a precise table and moved by a micrometer manually. The electrode position "x" is obtained from a micrometer scale. A microscope having forty magnification, is used to find the x=0 position where He-Ne laser beam irradiation spot coincides with bottom position of the brim edge. The He-Ne laser beam irradiates the brim, and the scattering reflection beam from the brim is detected by a photo-detector. Photo-current from the photo-detector is fed into a photo-detector amplifier. Output voltage "E" of the photo-detector amplifier is measured by a voltmeter.

Figure 5 shows the relationship between the electrode position x and the output voltage E. Five data are shown in Fig. 5. From these data it is found that x coordinate values of maximum voltage point fluctuate in wide range. This fluctuation was thought to be caused by variation in the brim edge profiles. This is revealed from Fig. 6. Figure 6 shows cross section profiles of several electrode brims. The lower edge has a keen and right-angled profile in Fig. 6(a). When the He-Ne laser beam irradiates the brim edge exactly, the output voltage E becomes maximum. Figure 6(b) shows round shape brim edge. In this case, the maximum E point is slightly above the bottom of the brim edge. Figure 6(c) shows the brim edge having burr. When the He-Ne laser beam irradiates burr, the reflection beam becomes maximum. This variation in brim edge profiles is caused by production process of the electrode. From curves in Fig. 5, position sensing accuracy by this maximum voltage method is expected around $\pm 20\mu m$. An electric circuit to find maximum voltage point is complicated, and requires a long response time. Therefore, we adopted the threshold voltage to define the detecting spot. This position sensing acuuracy is as small as that of the maximum point sensing method.

The positioning system requires capability of detecting the brim edge, in spite of electrode loading deviation of ± 0.2mm in Y and Z direction (See Fig. 1). This was attained by putting a cylindrical lens into the He-Ne laser beam path in front of lens A. The He-Ne laser beam is extended along the brim edge line at detecting spot. The position sensing accuracy was not deteriorated by using the cylindrical lens. The detected reflection laser beam is very weak, so the position sensing is influenced by illumination lamps. To reject stray light from the illumination lamps, red pass filter was put in front of the photo detector.

According to above results, a following position sensing method has been designed. An optical arrangement of the position sensing method is shown in Fig. 7. He-Ne laser output power is 1mW. "Lens A" has a focal length of 50mm, "lens B" focal length is 25mm. The "cylindrical lens" focal length is 200mm. The "aperture" has a 0.2mm diameter hole. The "photo-detector" is a silicon photo-diode. This optical system moves upward in x-axis direction, so He-Ne laser beam scans over the electrode. When the He-Ne laser beam irradiates the brim edge, the reflection beam feed into a photo-detector through the lens B and the aperture.

Sensing errors for this position sensing method are measured using the experimental apparatus shown in Fig. 4. This error was defined as the distance between the detecting spot and the bottom position of the brim edge. From measuring results for ten electrode samples, the doubled standard deviation $2\sigma=17\mu m$ is obtained. Position sensing accuracy was considered to be better than $\pm 20\mu m$ from the measuring results.

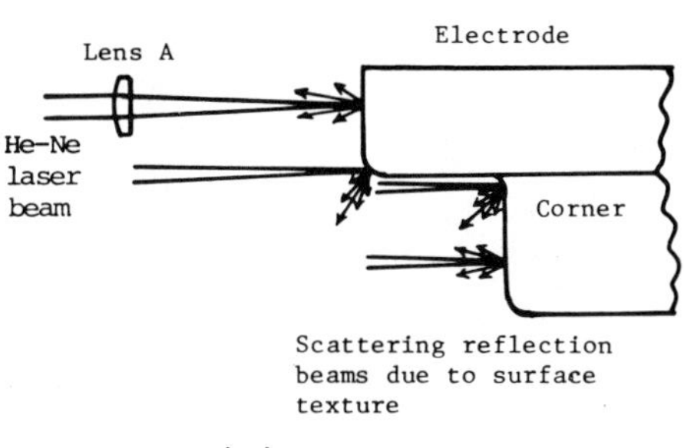

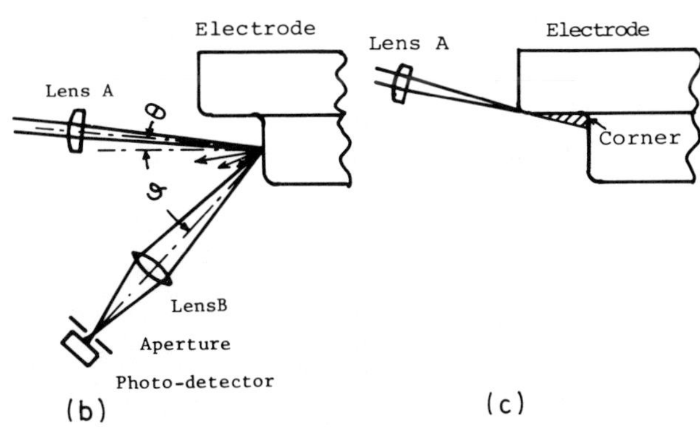

Fig.3 Examination for optical arrangements of the position sesnsing method

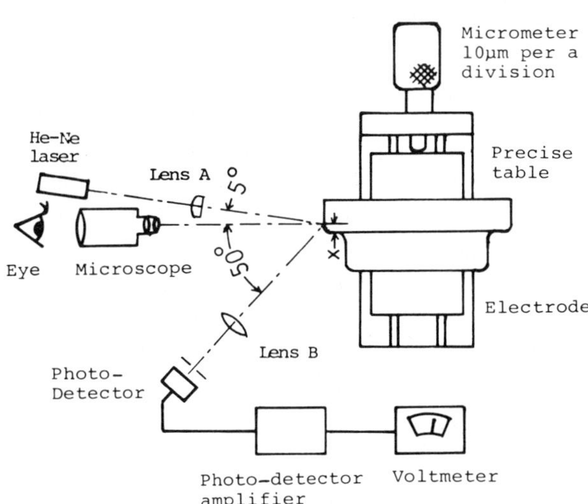

Fig.4 Experimental apparatus for edge detection

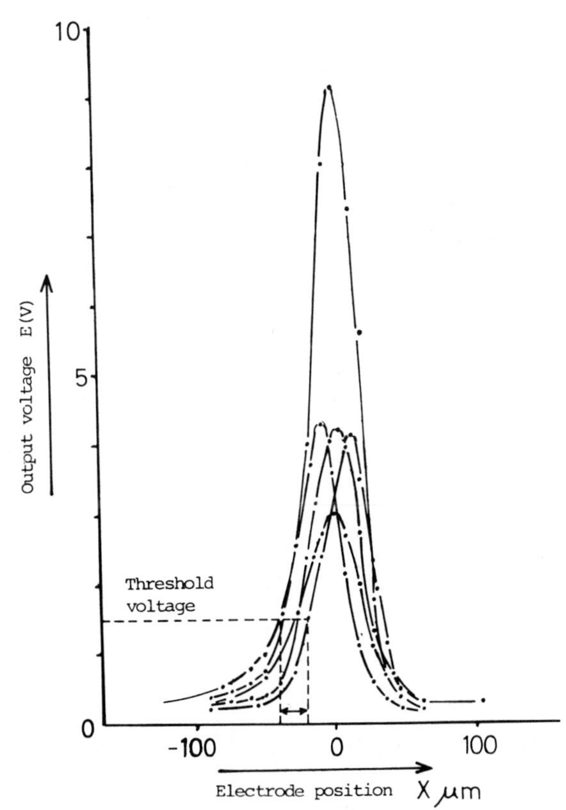

Fig.5 Relationship between electrode position and photo detector amplifier output voltage.

3. Positioning system for YAG laser welder

Figure 8[(1)] shows the developed positioning system using He-Ne laser position sensing method for automatical YAG laser welding system. Both optical apparatuses of the positioning system and the welder are on one working table. The table is driven upward by a ball screw and a stepping motor controlled by a controller unit. Both optical axes of YAG laser and He-Ne laser beams are made parallel to the table movement direction by "prism A" and "mirror A", and are led into the table So the beam positions on the optical elements do not vary with table movement. Then, YAG laser beam axis is folded into horizontal line by the "prism B", and led into the welding spot through "lens C". YAG laser beam is emitted only at welding time. The He-Ne laser beam axis forms an incident angle of 5° with horizontal line by "mirror B", and irradiates the detecting spot through lens B. The detecting spot is on the opposite side from the welding spot against the isolator. Both optical axes were set at same height on the electron tube parts. The electron tube part is loaded so that the brim edge line may be horizontal. When the He-Ne laser beam detects the brim edge, YAG laser beam optical axis is at the same brim edge too. The YAG laser optical axis is adjusted by using another He-Ne laser. After detecting the brim edge, the table needs to travel the distance from the brim edge to welding spot, that is a half of the brim thickness. Then, YAG laser beam is emitted, and the welding spot is irradiated.

Photo-current is fed into the detector circuit, and output signal from the detector circuit is fed into the controller unit. The controller unit controls stepping motor rotation and YAG laser emitting timing. Figure 9 shows a block diagram of the detector circuit. Photo current is amplified, and is compared with the threshold voltage. Output signal of the comparator is converted into the signal matched with controller unit input. The circuit has lms response time. Figure 10 shows a blockdiagram of the controller unit, and Fig. 11 shows the automatical YAG laser welding system operation. Its abscissa is defined as the working table position i.e. He-Ne laser position x on the electron tube part. Previously approximately designed position of the brim edge and the brim thickness are manually keyed into control unit from setting keys as a pulse number. One pulse corresponds to 20μm traveling. The gate is opened, and the detecting pulse can be accepted only at the time when the He-Ne laser beam is near the brim edge. This helps the controller prevent any control mistakes by noise signal caused by the reflection He-Ne laser beam from another position. As mentioned above, fluctuation in the brim real position is a ±0.2 mm value, so it is reasonable that the gate is opened at 0.5mm before the designed position. If the stepping motor is stopped suddenly, rotating slip can occur, so rotating speed is decreasing gradually from the gate opening time. Similarly, at start, the stepping motor rotating speed needs to increase gradually.

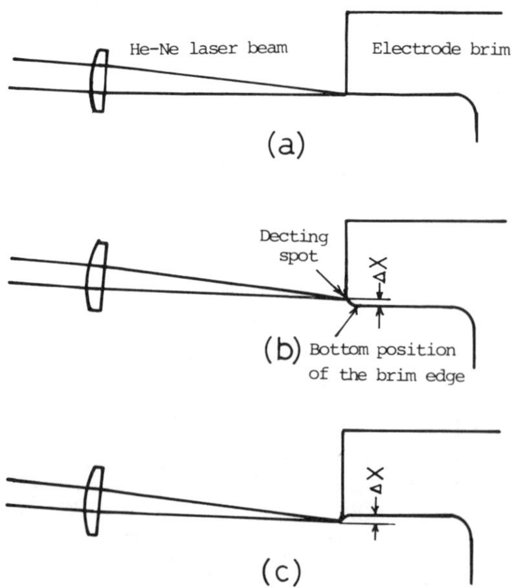

Fig.6 Cross section profiles of electrode brims.

The following, welding processes are explained with Fig. 11. After loading the electron tube part and the connection ribbon, the table starts to move upward. When the table arrives at a pre-set position that is 0.5mm below the designed position of the electrode brim, the controller opens the gate and decreases the stepping motor rotating speed gradually. When the He-Ne laser beam detects the actual position of the electrode brim, the detecting pulse is fed into the controller

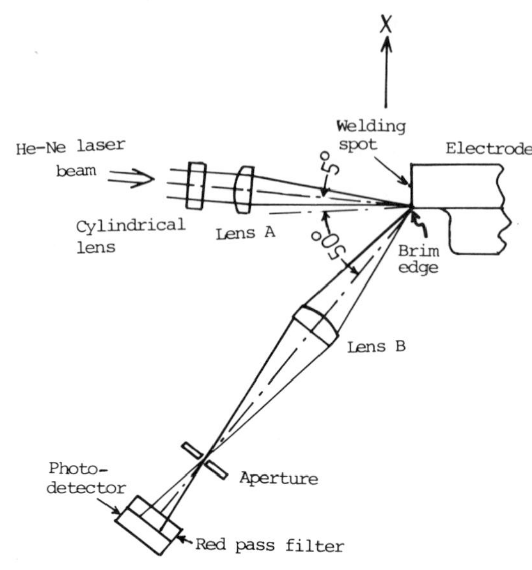

Fig.7 position sensing optical arrangement.

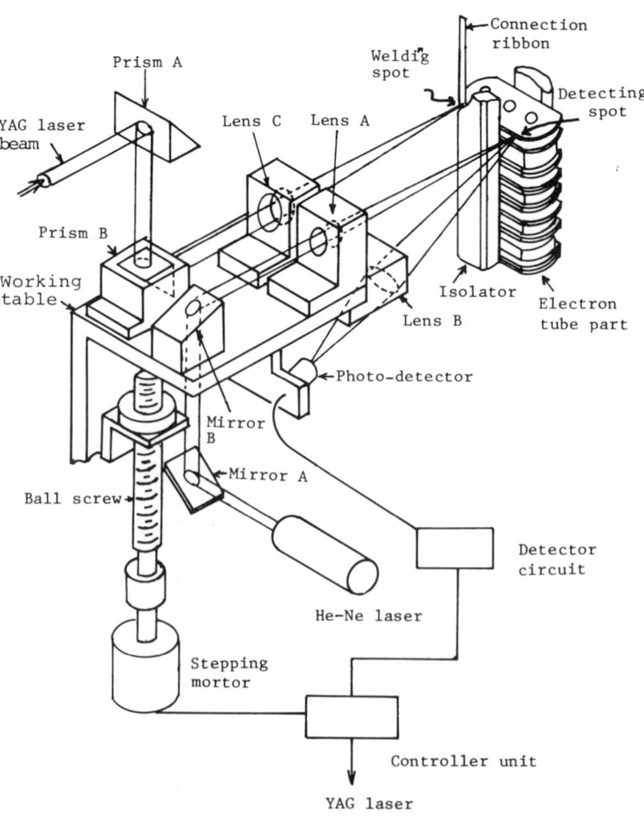

Fig.8 Positioning system for YAG laser welder

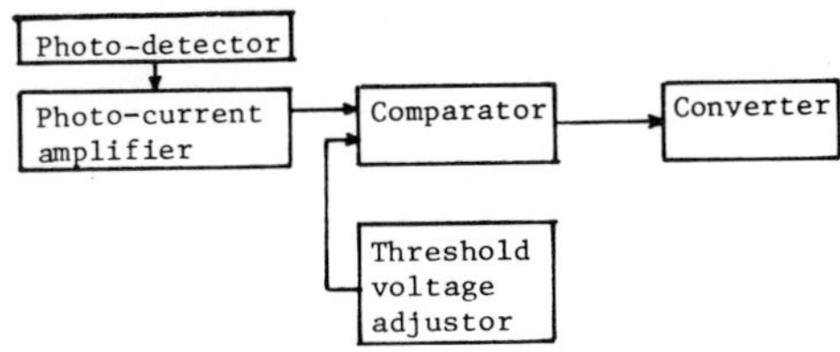

Fig.9 Photo-detector circuit block diagram

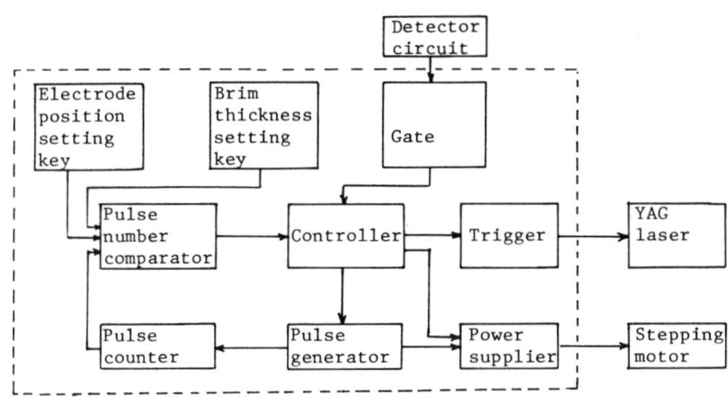

Fig.10 Controller unit blockdiagram

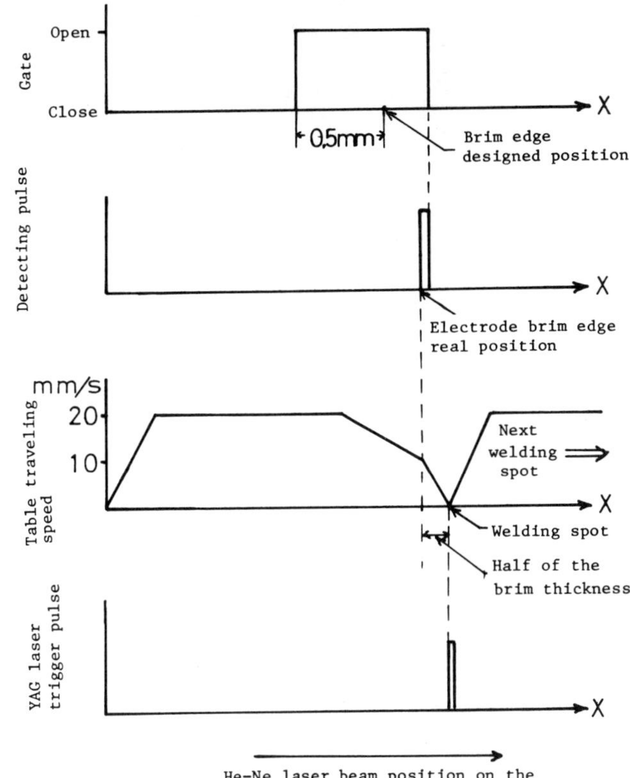

Fig.11 Automatical YAG laser welding system operation.

unit. The controller unit closes the gate, continues to move the table from the brim edge to the welding spot i.e; the distance of a half brim thickness. The table is stopped. Then YAG laser emits the beam by trigger pulse, and welds the connection ribbon into the electrode. After welding, the table start to move again the next welding spot.

The YAG laser welding system can weld up to six spots successively. In the electron tube assembly process, four of these systems are operating, but only one YAG laser and two He-Ne laser are used. The original laser beams from YAG laser and He-Ne laser are divided into four beams by half mirrors.

4. Conclusion

The positioning system was newly developed by detecting the reflection beam from the edge of the part with an He-Ne laser beam. Better than ±20μm positioning accuracy,1ms response time, and non-contact positioning were obtained. With this system, electrodes and connection ribbons could be welded automatically by YAG laser welder in electron tube assembly process. So, highly reliable electron tube parts have been produced on a large scale.

The YAG laser used is a LAY-603 laser, and The He-Ne laser is a LAG-3211A model. Both lasers were made by Toshiba Corporation.

(1) Patent pending

CHAPTER 8

VISION SYSTEMS IN ASSEMBLY

Reprinted from *Manufacturing Engineering*, April 1983

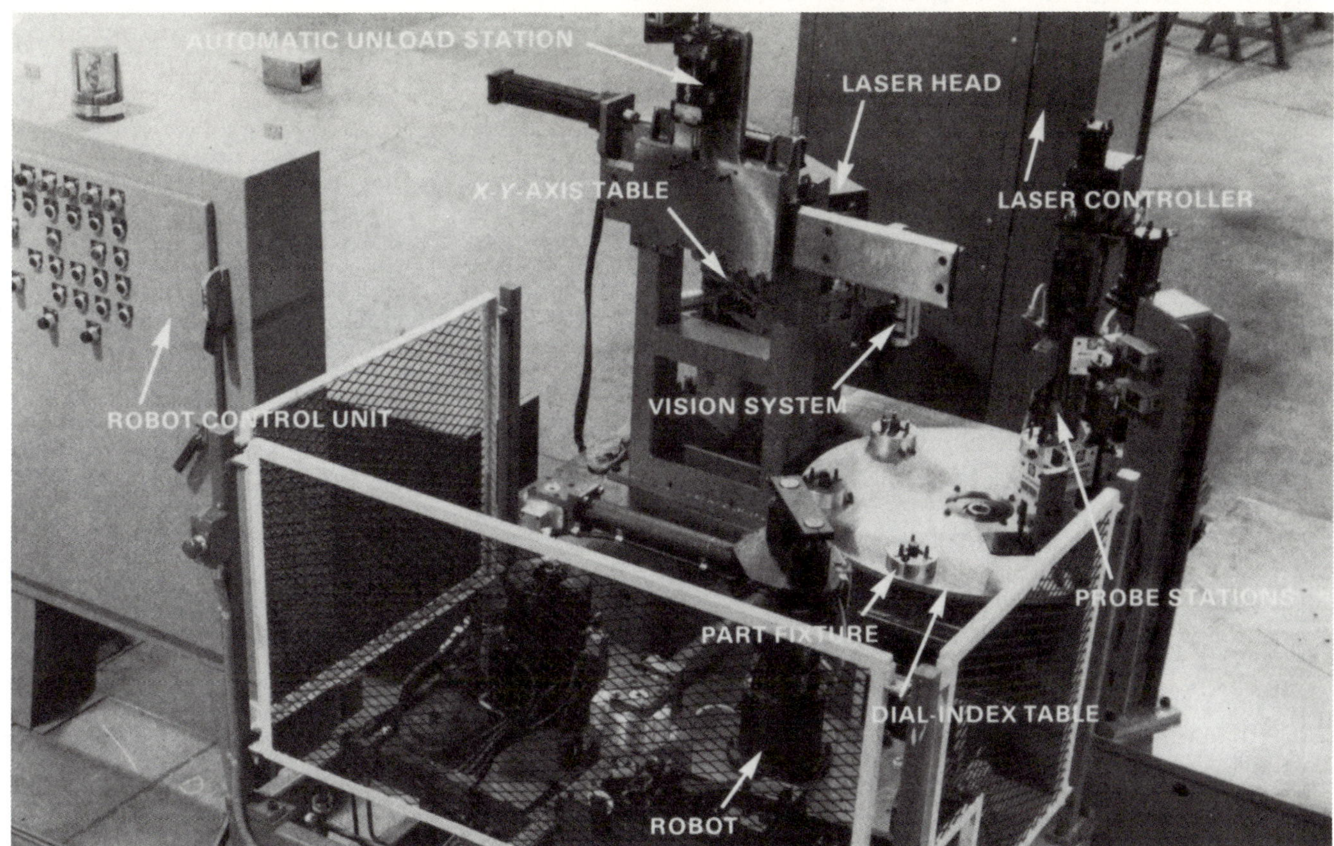

The automatic assembly system at Ford's Batavia, OH transmission plant is a six-station dial-index machine that combines laser welding with a vision system. It is expected that this approach will reduce the reject rate of planet-carriers to less than 5%.

Automatic Assembly: Making Use of Probes and Vision Systems

The use of probes and sensing systems is gaining in importance as the need increases to automatically verify the presence, position, and function of assembled components

HARRY WALDMAN
Senior Editor

LAST AUGUST, MANUFACTURING ENGINEERING took a look at automatic assembly. While it was stressed that no major breakthroughs in various assembly machine chassis appeared to be on the horizon, it was also felt that the next few years would see a growing recognition of the need to automatically verify the presence, position, and function of components that are to be inserted into assembly operations.

The development of probes and vision systems for use in directing the movement of other equipment is a step in this direction.

Cameras and Probes Aid Welding

Welding is growing in interest as a means of joining parts in automatic assembly: the percentage of welded joints continues to increase while that for screwed and riveted joints is decreasing. In laser welding, two parts are melted and then fused together.

Extending the precision and productivity of laser welding, the Ford Motor Company (Dearborn, MI) is installing a CNC laser system to automatically join a six-gear shaft to a planet-gear cage.

In the testing stage since November 1981, the system, developed by Letnan Industries, Inc. (Sterling Heights, MI), is a six-station dial-index machine (controlled by an Allen-Bradley PLC 220) that is automatically loaded by a robot to weld 200 planet-cages per hour. Set to run for one shift per day, it will replace a TIG welder that, each day, saw 250 carriers rejected out of a production lot of 1100 to 1200.

Impulse YAG Laser

The system matches an impulse YAG laser (*Figure* 1) with a vision system that directs the movements of the laser. It's hoped that this approach will greatly improve reliability — now at 70% — and reduce the reject rate to less than 5%. In this impulse laser (a Raytheon 501), the initial action of the beam is more intense than the rest of the pulse, causing greater melting of the metal at first. Once this occurs, the melted metal is pushed aside and a beam is then directed into

Using probes, these machines can detect malfunctions, eject incomplete assemblies, and sort good and bad parts.

the metal itself. For this 500-W laser, penetration occurs at a rate of 15 ipm (380 mm/min).

This approach to automatic assembly has a number of advantages. The fact that it's a noncontact method obviates the need for any external force. Second, the capability of making deep or shallow penetrations with equal ease is an added flexibility; and third, very localized, high-energy welds may be made on the most hard-to-reach places.

Sensor Capability

Following the automatic loading of a planet-carrier by a robot, two probes (Gilman Multisense) 3/8" (10.5 mm) in diameter, automatically check the depth-height relationships of six 9.5-mm shafts, three at a time, on the planet-cage surface. The matrix-type camera (Copperweld Opto-Sense) views the parts and locates the true center of each pin. This information is stored in memory and is used later for the actual welding.

The machine's laser-weld station works on the six pins; the laser head is mounted to an X-Y-axis table (controlled by an Allen-Bradley 8200 controller). The processor instructs the laser as to the appropriate positions for welding — within 0.003 to 0.004" (0.08 to 0.10 mm) of the center of each pin. The pins are not welded in any particular order because the machine is programmed to weld with the least amount of motion; the result is more camera scanning but less welding movement. Although the impulse welder's power rating is 500 W, in tests it has been operating very successfully at 250 W. It can even weld over welds.

Ford's new automatic assembly system may be a first in the industry.

While the cycle time for the welding is supposed to be not more than 18 seconds, tests indicate it's actually just over 16. And while the welds are guaranteed to withstand 1461 lb (6500 N) pushout, testing indicates that they actually withstand 2600 to 3000 lb (11 565 to 13 334 N) pushout. At the final station of the machine, an electromagnetic, three-position, pick-and-place device unloads the planet-carrier from the system.

First of Its Kind

According to the manufacturer, the special assembly system may be the first match of a laser with a scanning system in the industry. It's also been designed for adaptability. Sometime in the near future, Ford will be adding a second laser station, resulting in a configuration that can handle eight pins per planet-cage, with a laser-welding cycle time of 24 sec. And in this setup, each probe will be responsible for four pins.

The system was scheduled to go into full operation in early March at Ford's Batavia, Ohio transmission plant. It's long-term aim is to provide greater precision, minimal thermal distortion, production-cost savings, and high-yield productivity for Ford's spring 1983 in-

SENSING SYSTEMS FOR AUTOMATIC ASSEMBLY MACHINES

SENSING PROBES are mechanical devices which, in conjunction with limit switches, proximity detectors, or photoelectric switches, sense the presence, position, and other mechanical characteristics of an assembly as it progresses. These same probes, in combination with electrical contacts, diagnose the condition of parts.

Their main advantage, according to Automated Process, Inc. (Milwaukee, WI), a manufacturer of automated assembly machines, is that the assembly machine itself can detect malfunctions, eject incomplete assemblies, sort good and bad parts, and, in general, provide inspection and quality control functions at practically no cost.

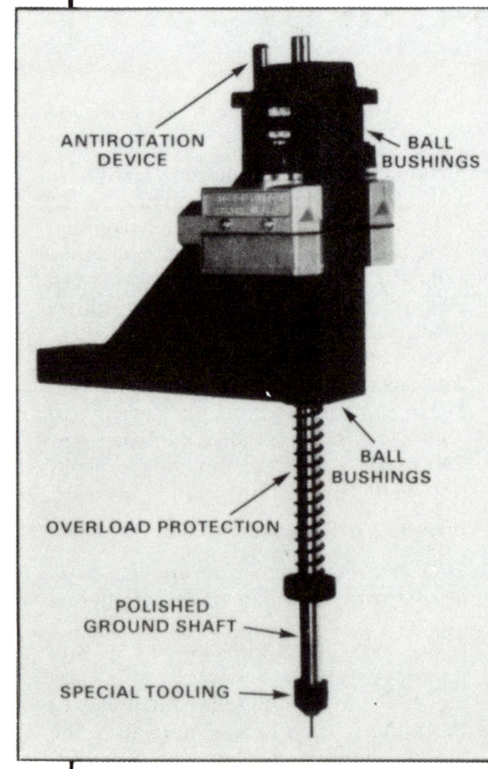

The diagnostic capability of a sensing probe is almost unlimited: it depends only upon the ability to incorporate the electrical and electronic components within the sensing probe and on the availability and accuracy of sensing components. In a typical automatic assembly, a properly placed part will result in one limit switch being actuated on a probe, and one remaining unactuated; an improperly placed part actuates both switches; and no part leaves both switches unactuated.

Normally used after each working station on an assembly machine, a sensing probe is brought into position with the product during the dwell portion of the machine cycle. A programmable controller usually looks at the electrical output of the sensing probe and accommodates a variety of machine functions as a result of that electrical output.

Typically, station malfunctions detected at any one nest lock out downstream stations whenever that nest is presented to them, eliminating further assembly work on already defective partial assemblies and preventing further damage in later processes.

For an occasional malfunction at a nest, the machine is not stopped, but it is programmed to stop whenever multiple malfunctions occur at a working station. An audio or visual alert may then be given. Machine uptime and throughput are improved, good parts are not wasted on a defective assembly, downstream damage is limited, incomplete and rejected assemblies are discarded, and the final product reflects properly assembled products. ■

The combination magnet-and-test machine combines five components (inset) and checks the quality of each assembly on an eight-station setup.

troduction: the ATX three-speed, front-wheel-drive transmission.

Assemble-Test Machine

By capitalizing on standard-machine dial movement and special station-to-station tooling, a new machine recently built by Newcor Bay City (Bay City, MI) uses cam-operated probes to verify the presence of parts. Without the aid of any operators, it produces a five-part unit that's needed in the functioning of submersible, electric, automotive-fuel-tank pump motors. (See *Figure* 2).

> This approach results in a 200% increase in today's production.

The assembly operation entails systematically aligning two curved magnets (unmagnetized at this stage) and two spring-type tracers that are between the magnets along the inside-diameter wall of a tube-shaped flux carrier. and then magnetizing the parts.

This combination magnet-assembly-and-test machine has eight working stations, arranged around a motor-driven, cam-operated, 36″ (914-mm) diameter dial plate. Also included is a reject switch, which prevents any work being done on parts that are unacceptable. Special hopper feeders automatically keep the stations supplied with parts that are to be assembled.

At three of the stations (2, 4, and 6), the cam-operated probes detect the presence of parts. At station 8, another probe and sensing device checks the magnetism of the appropriate parts.

The assembly cycle is as follows: A flux carrier is automatically fed to the dial tooling at station 1. At station 2, it is probed for presence and at station 3 two magnets are added. At 4, probes check for the presence of the magnets. At 5, two spring spacers are fed to the assembly and pressed into the flux carrier, securing the magnets in place. At station 6, probes check for spacer pressure and alignment before the magnetizing occurs at station 7. Finally at 8, the magnets are measured for magnetic strength.

At the unload station, a two-position ejector either removes good assemblies for storage or rejects them into another container, depending on the signal received. Completed assemblies are unloaded at the rate of 850 per hour. According to the user, this represents a 200% increase in production over a semimanual method of assembly.■

Presented at the SME 13th ISIR/Robots 7 Conference, April 1983

Recent Proliferation of Industrial Artificial Vision Applications

by Philippe Villers
Automatix Inc.

1. Introduction

The past year has seen a rapid proliferation of Artificial Vision systems in industry for increasingly diverse applications ranging from high-speed visual inspection of simple partsand visual servoing of arc welding robots, to robot guidance for semi-ordered bin-picking. This year also marked the commercial introduction of integrated units for robot control and artificial vision.

This paper will discuss the higher speed and higher performance of third generation Artificial Vision systems and a recent example of such a system. The third generation Artificial Vision systems are more versatile and can handle processing at increased rates -- over 10 parts per second for simple parts -- and can accept inputs from up to 16 different solid-state cameras while increasing available resolution from 256 x 256 picture elements to multiple 512 x 256 pictures, as well as recording of 64 shades of gray.

Artificial vision applications can be conveniently divided into visual inspection, robot position control, and robot visual servoing. Recent examples of the state of the art, as used in industry, are given for all three classes.

2. Autovision® 4 - A Third Generation Artificial Vision System

Shown in figure 1. is the Autovision 4, a third generation commercial artificial vision system. The first generation systems, which became available around 1980, used the feature extraction method developed in the 1970's at Stanford Research Institute (Reference 1) with a combination of specialized hardware for preprocessing and simple, but widely available, microprocessors for feature extraction and analysis. In the second generation, (Reference 2) microprocessor technology was used more widely including bit slice microprocessor front ends, with microcoded preprocessing instructions. The more powerful microprocessors with wide memory-addressing capabilities, (such as the Motorola 68000) made possible the practical addition of gray-scale capability, multiple cameras and an increasing sophistication in specialized applications.

The third generation, as illustrated by the Autovision 4, uses multiple, powerful microprocessors for more rapid preprocessing. Figure 2. is a block diagram of the Autovision 4.

Autovision 4 has the capability of handling sixteen cameras, and can use more resolution (up to 512 x 256 picture elements) to match improvement in commercially available, relatively low-cost, solid-state cameras. Both application programs and programs for specialized preprocessing, where required, can be easily written while retaining the ability to program at higher levels including the RAIL® user language.

The factory-hardened unit is packaged in a sealed cabinet to protect it from harsh industrial environments including fumes, corrosive atmospheres, and metallic dust as well as electronic noise.

Control inputs and outputs are through opto isolators. As compared to the second generation, Autovision II, the processing speed has been increased by a factor of two to five times.

Specifications of the Autovision 4 are summarized in Appendix A. The Autovision 4 preprocessing and vision processing capability are also available in an integrated configuration with a robot controller, the AI324™ controller shown in Figure 3, for robot guidance applications and visual inspection or servoing. Whereas the second generation Autovision II is limited to eight cameras, the Autovision 4 can support sixteen cameras. This can result in very low cost per station for applications requiring visual inspection at moderate speed. The cost per station when using multiple cameras can be as low as $3,000 to $5,500 depending on the camera used and therefore, are down to a small fraction of the annual cost of a single human inspector.

Artificial Vision applications are highly diversified and so ease of adapting artificial vision systems to many different applications is crucial. In the case of the Autovision 4 three levels of user programming are possible: 1. Application programming using the RAIL language at the Autovision 4, or remotely through off-line programming in this mode. A manufacturing engineer, who need not be a software engineer, can write programs for stand-alone operation or to be used in a menu-driven dialogue manner for use by the machine operator. 2. PASCAL language programming to provide analytical capabilities not currently available in the RAIL user langauge. 3. Specialized data preprocessing algorithms or techniques for novel applications, data formats, etc., which can be written in assembly language.

3. Artificial Vision Inspection Applications

The range of visual inspection applications has greatly increased as a result of recent advances in technology. The use for inspecting simple parts at high speed of up to 2,200 holes per minute, is shown in figure 4. This inspection of a 90 hole automotive stamping is in production use at a major automobile manufacturer. The inspecting system shuts down the stamping line system if any of the holes deemed critical are either missing or incorrect.

Representative of more sophisticated applications is the example shown in figure 5 for process control. In this case, an instrument cluster under test is inspected at high speed to determine the correctness of both analog outputs (dials) and digital outputs, such as the segments of a liquid crystal display and indicator lamps.

A similar application is an experimental application for a watch manufacturer who plans to use the vision system to command a special purpose fixture which on command depresses the proper watch buttons to set the correct time on newly manufactured digital watches. In this use, the artificial vision system observes the display, compares the results with its own real time clock, and signals the system when to stop cycling. The system will also verify the quality of the display during cycling.

Figure 6 pictures a form of process control using artificial vision. Here the system inspects gaps both between car doors and car bodies and car hoods and car bodies on an automotive assembly line. The artificial vision system takes a large number of three dimensional readings, typically 80 in 35 seconds, and then does a statistical analysis on the readings and communicates the information to earlier stations on the assembly line. If the process is drifting out of tolerance, it can be corrected before it results in faulty final assembly, in which the car-door-to-car-body gap is either too tight or too loose.

An example of an entirely different class of application is the use for rapid non-destructive testing for flaws in castings or forgings. This involves combining an artificial vision system with a classical magnetic dye penetrant and ultraviolet illumination technique. The forged automotive part shown in figure 7 is magnetized, immersed in the dye, and then placed under ultraviolet illumination and inspected by the Autovision system to find dye that has seeped into cracks. The high speed system can simultaneously inspect both sides of this forging at the rate of 60 per minute.

In the case of inspection -- where the average rate of defective parts is already below one in a hundred, and even more when it is below 1 in 1000 -- Artificial Vision becomes far more effective than the human inspector who must deal with the monotony problem of 999 cries of "wolf" followed by a single defect, a condition in which human performance is particularly poor.

One hundred percent inspection of simple parts to replace the implicit inspection by the human operator no longer present in robotic or other automated assembly tasks is another growing area. Figure 8 shows a Cybervision® III system and illustrates such an example, in this case a sophisticated application of a robotic assembly system for the batch insertion of keytops in keyboards (Reference 3). In this system, at its first station shown in figure 9, keytops are inserted -- using a series of vibratory bowl feeders -- in long tubes after being automatically inspected by the vision system which also 100% verifies the quality of the keytop labels (Reference 3). The second or Robotic Assembly station, shown in figure 10, can then proceed to insert the keys "knowing" that 100% good parts have been provided and that poorly labeled or wrongly labeled keys have been discarded.

An interesting and colorful application combining two classes of requirements is an experimental application to properly affix labels on champagne bottles. In this application, the customer wishes to ensure the alignment of the various labels on the champagne bottle and at the same time wishes to ensure that the bottle is filled to an exact fill level.

Another production line application, this time involving color discrimination using gray tones, comes from the pharmaceutical industry. A manufacturer of birth control pills wishes to ensure that each compartment of a transparent plastic package contains a pill and, further, that the pill in each compartment is of the correct type as determined by its color.

With the use of multiple cameras on a single artificial vision system, the cost per station often makes possible the combining of several inspection operations on an economical basis.

4. Robot Position Control

The Autovision 4 system is now being used in an increasingly broad range of visual servoing applications with the camera located either in the hand of the robot or at a fixed location. These applications occur particularly in connection with robot arc welding. In those cases where full-visual servoing, such as discussed in the next section, is not necessary; that is, where the problem is only lack of repeatability of the exact location of the part to be arc welded, the function of the Artificial Vision system is to determine, in three dimensions, the location and orientation of the part and to provide that information to the robot controller. Thus, a coordinate conversion is applied to ensure that the preprogrammed welding path is correctly executed on the shifted part. This technique, known as Vision Offset, permits welding of a less than ideally fixtured part or even, if desired, unfixtured parts.

A conventional assembly application is shown in Figure 11. This example, as demonstrated at Robots VI in 1982, shows the assembly of a small pulse transformer where the part-to-part variations are too great to permit successful assembly without individual correction for each part. In these kinds of applications, the vision system is also frequently used to verify the correctness of the part as well as its position and that of its mating surfaces.

In small-part electronic applications, as shown in figures 12 and 13, the same combination of visual inspection and determination of exact location - in this case pin location - is sometimes required to ensure successful insertion. Figure 14 shows the verification of relay cans to ensure that the correct part is being inserted, that it is not rotated 180° and, if necessary, to allow an offset to ensure that the pins, as observed by the camera, can be correctly inserted into the printed circuit board.

The use of vision position feedback in assembly is not limited to small assemblies. Figure 15, illustrates an experimental use in assembling a car door. In this application, there are a number of bolts in slots. The exact position of those bolts is determined at assembly stage by the fixture designed to ensure proper operation of the window mechanism. The solid-state camera, which inputs to the vision system, is located on the same arm as the torque wrench used to fasten the bolt. The vision system, using its gray-tone capability and its own ring of light, shown in figure 16, determines the exact position of the bolt head in its slot. The robot then proceeds to fasten the bolt with an average cycle time per bolt, including some relatively hard to access locations, of approximately three seconds.

A similar automotive application involving fastening of accessories on an engine block requires that the vision system locate the engine block itself (otherwise, the position of the block is not defined precisely enough). Signals from the vision system correct the robot command and allow the robot to tighten the fasteners with a torque wrench.

Figure 17 shows a Ford Motor Company example of the use of robot vision to guide the robot in bin-picking. Although the general problem of jumbled parts in a bin is the subject of much research work and is not yet practical, the example shown in figure 18, developed by Ford in Germany, shows what can be accomplished with a "semi-ordered" bin where the parts have been stacked in layers, and generally correctly oriented but where the exact position is not determined. In this Ford application, the vision system determines the exact location of each part so that the robot can grasp it. This example illustrates a class of applications in which bin picking is now feasible and economical by using semi-ordered bins instead of the more general random loaded case.

These industrial applications, as well as some experimental ones involving drilling and riveting in the aerospace industry, are examples of the use of vision to aid robots in large assembly tasks. As of this year, Artificial Vision systems are available as an integral portion of robot controllers such as the AI324 controller previously shown figure 3.

5. Robot Visual Servoing

Shown in figure 19 is a Robovision® II arc welding robot with optical seam tracking capability. This capability, first demonstrated late in 1981 at the Autofact IV Show, involves using a laser beam to illuminate the seam during the welding pass to correct for part mislocation, part inaccuracy, and thermal distortions of the part due to welding. This distortion is particularly troublesome in light gage material and it makes preprogrammed robot arc welding much more difficult.

Shown in figure 20 is a block diagram of this one-pass visual servo; and shown in figure 21 is a complex customer part used in the field trials of this system which has just recently reached the level of commercial use. The part is welded in about five minutes and includes approximately 80 inches of weld. The required tolerance on a 1/4" lap weld is 0.050" and the part-to-part variability is up to 0.250". With the vision system the position error is reduced to about 0.030".

6. Future Trends

It appears that the often predicted rapid growth of artificial vision systems is at last taking place. As the industry has expanded, the range of application in which the cost has now decisively passed the threshold of cost-effectivity is now substantial. The addition of gray-tone capability as well as increasing resolution and speed, have all contributed to this ability to deal with a variety of situations, while increasing integration with the robots themselves have made the robot guidance usage of artificial vision systems increasingly convenient and cost effective.

In this era of rapid expansion of artificial vision applications, it remains to be seen whether the United States will become the leading user of Artificial Vision technology or whether, as has happened in other fields of applied robotics, an early lead in U.S. usage is followed by more rapid acceptance and utilization in Japan.

In any case, it can been said with confidence that the next twelve to twenty-four months will bring a vast increase in the range, scope and effectivity of applications as, on the one hand robots increasingly receive the benefit of being sighted and, on the other hand Artificial Vision takes over more and more visual inspection tasks. These systems not only relieve the monotony and inattention of humans due to extreme boredom, but are used in those places where automation or hazardous conditions make human vision an unfeasible option, or where the rates of inspection are not compatible with human visual inspection.

REFERENCES

1. Gleason, Gerald J. and Gerald J. Agin, "A Modular Vision System for Sensor-Controlled Manipulation and Inspection", SRI International, Menlo Park, California, U.S.A., March, 1979.

2. Reinhold, A. and G.J. VanderBrug, "Robot Vision for Industry: The Autovision System", Robotics Age, Fall 1980, pp. 22-28.

3. VanderBrug, G.J.; Wilt, Donald; and Jim Davis, "Robotic Assembly of Keycaps to Keyboard Arrays", Robots 7, April, 1983.

Figure 1. Autovision® 4 Artificial Vision System

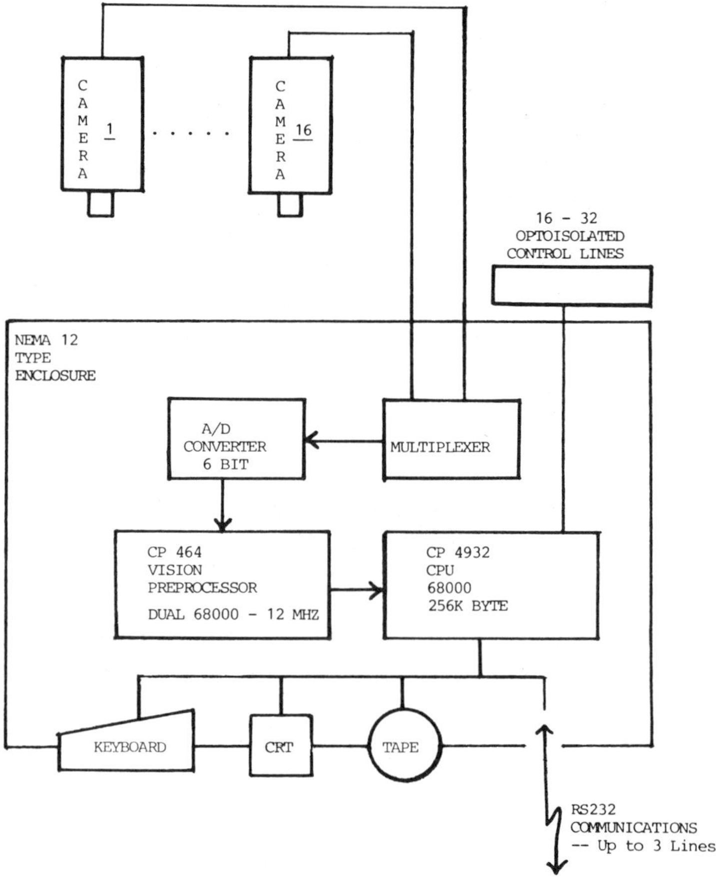

Figure 2. Block Diagram of Autovision 4

Figure 3. AI324[tm] Robot Controller

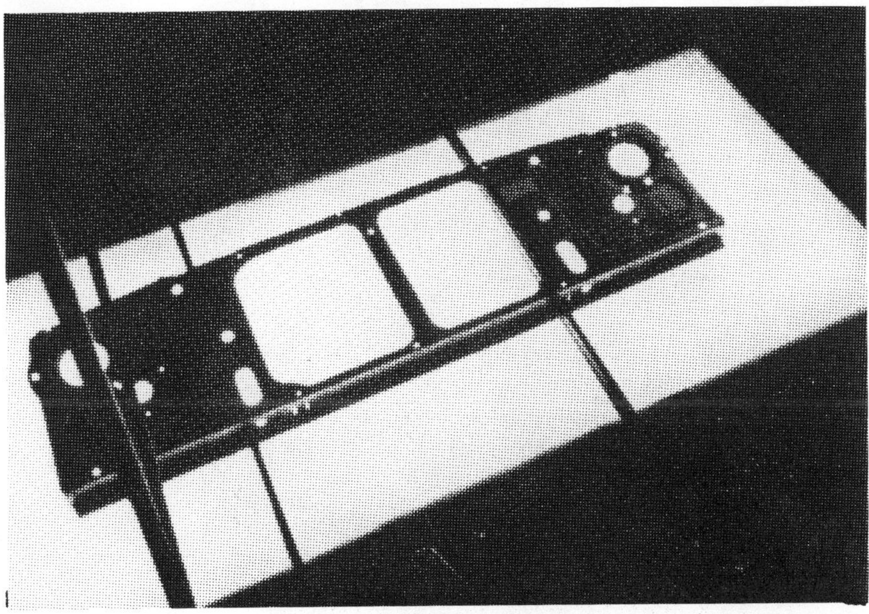

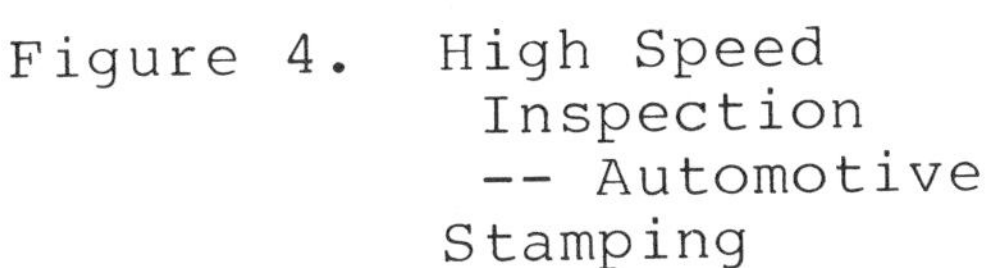

Figure 4. High Speed Inspection -- Automotive Stamping

Figure 5. Instrument Cluster Verification during Test

Figure 6. Gapsight -- Car/body Gap Inspection

Figure 7. AUTOVISION High Speed Flaw Inspection of Connecting Rods Using Ultraviolet Light

Figure 8. Cybervision® III Robot Assembly System

Figure 9. First Station of Keycap Assembly

Figure 10. Second Station of Keycap Assembly

Figure 11. Small Part Assembly Application - Pulse Transformer

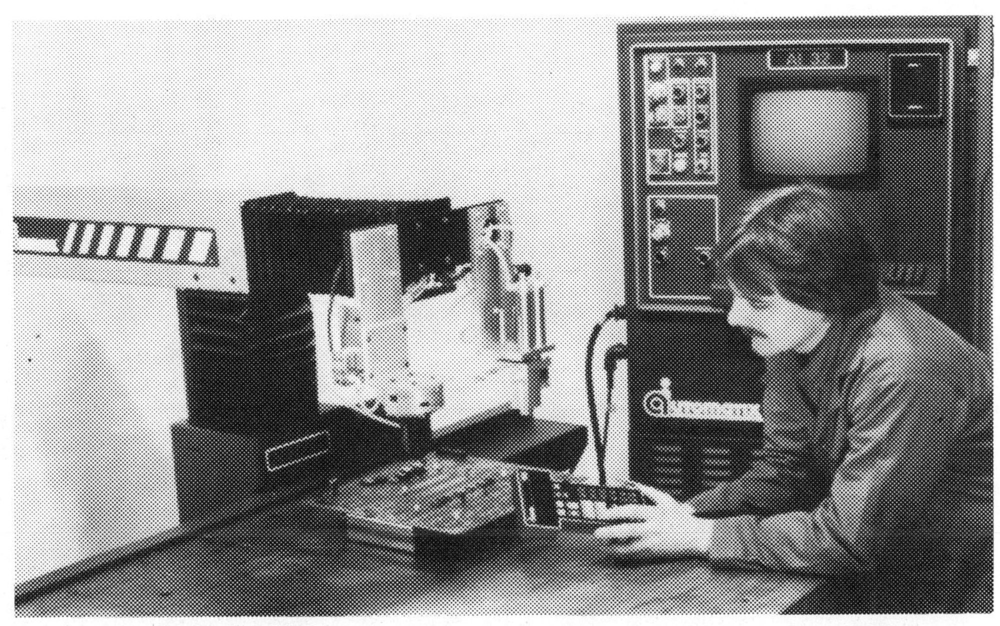

Figure 12. P.C. Board - Non Standard Component Application

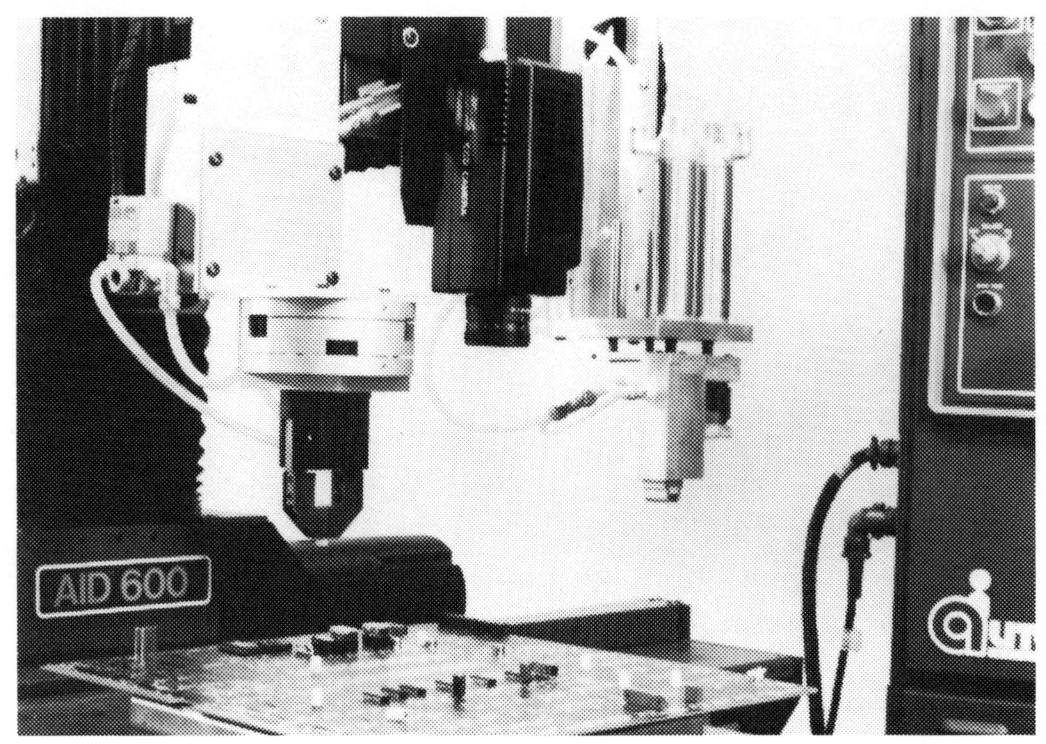

Figure 13. P.C. Board Insertion Using Vision

Figure 14. Verification of Relay Cans

Figure 15. Vision Guided Assembly - Car Door

Figure 16. Close-up - Car Door Assembly

Figure 17. Semi-ordered Bin-picking at Ford

Figure 18. Bin-picking - Close-up View

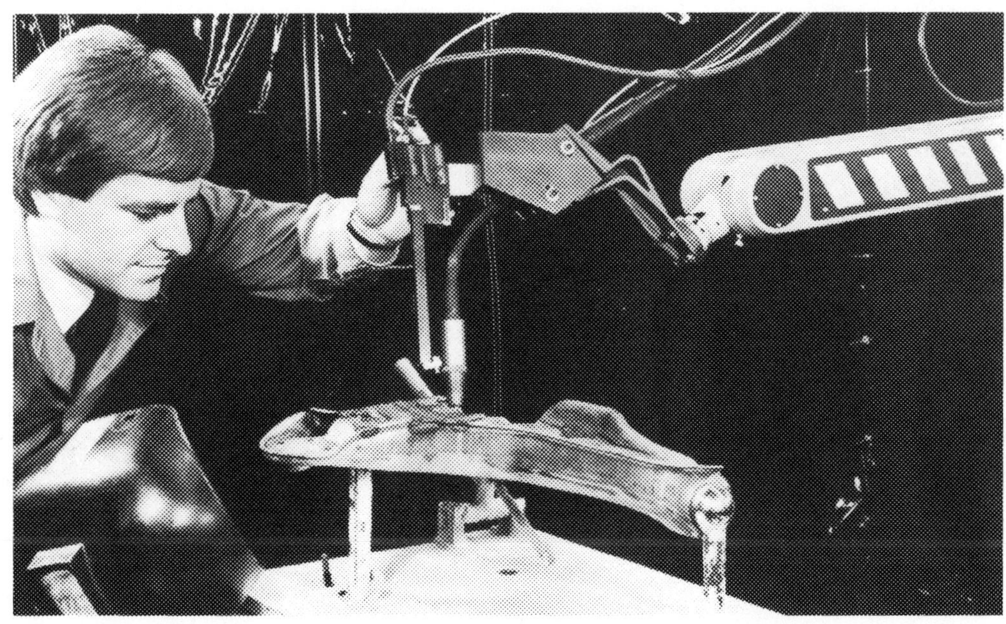

Figure 19. Robovision® IIA with Autovision Optical Seam Tracker

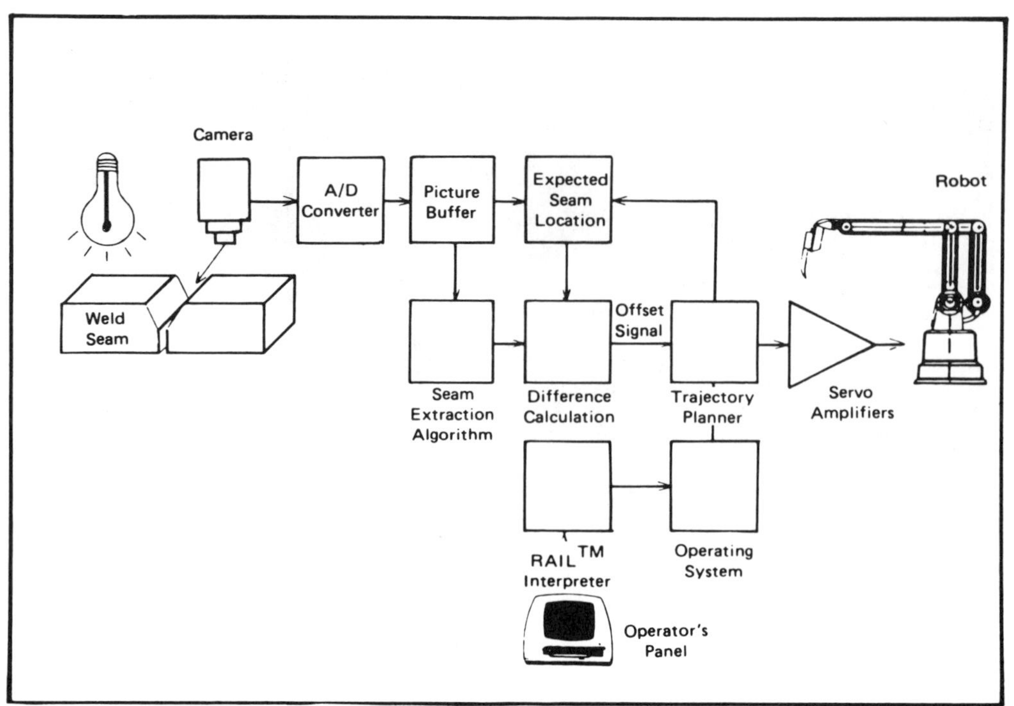

Fig. 20. Block Diagram of Robovision IIA with Seam Tracking

Figure 21. Complex Customer Part Welded with Optical Seam Tracker

AUTOMATIX INCORPORATED
AUTOVISION® 4 TECHNICAL SPECIFICATIONS

Appendix A

CP 464 Vision Processor
- o Flash lamp control for up to 16 cameras.
- o 64 levels of gray.
- o Frame storage for four 256 x 512 eight-bit pixels accessable by both microprocessor and CP4932. (512K bytes total)
- o Two-68000 at 12MHz each with 32KB static ram no wait states.
- o Magic box
- o Up to 4 boards

CP4932 Control Processor
- o 32-bit internal architecture with 16 general purpose 32-bit registers.
- o 256K bytes of read/write memory.
- o Expansion memory directly addressable up to 16 million bytes.
- o Dynamic control panel and keyboard - accepts operator commands and displays operator messages, diagnostic test results. programming data, video picture from camera, digitized video, and processed video. Pictures are displayed in either binary (black/white only) or gray scale.
- o Two RS-232 serial lines standard up to 4 additional lines available.
- o Sixteen configurable I/O lines connected to optoisolator mounting board. Three AC input modules (90 to 140 Vac) and five AC output modules (24 to 280 Vac, 3 A RMS) are standard, with sockets for eight more modules. An additional 16 lines are optionally available.
- o Optional dual cartridge tape drive and controller for up to 256K bytes of program or data per cartridge.
- o Auto reboot.
- o Ruggedized sealed enclosure with key lock control switch.
- o Non-volatile real time clock.

AI Vision Software License including:
- o Feature extract - display or output of 58 geometric features such as angular and linear displacement, area, moments of inertia and so on.
- o Autotraining - recognition by showing objects and decision based on nearest neighbor algorithm.
- o RAIL command language.
- o Diagnostic package.
- o pascal support package (optional).
- o Optional Gray Scale inspection package.

Physical Specifications:
- o Power: 100, 120, or 240 Vac strap selectable, 50 or 60Hz, 400W.
- o Temperature: 10°-45°C; Humidity: 0-90%, non-condensing.
- o Dimensions: (HxWxD) 60.7"x31"x21.5" (1540 mm x 790 mm x 460 mm)

Weight: 325 lbs. (148 kg)

Warranty:
- o 90 day, parts and labor, subject to terms and conditions.
- o Full service contract available thereafter.

Presented at the SME 13th ISIR/Robots 7 Conference, April 1983

Applications of Machine Vision to Parts Inspection and Machine Control in the Piece Part Manufacturing Industries

by Tim Pryor
Walt Pastorius
Diffracto Limited

Abstract

Machine vision has been heavily publicized because of its potentially crutial contribution to robotics. Basically the field appears to fall into four areas: inspection of piece parts and assemblies, recognition of objects for ID or orientation for open loop robot control, calibration of robots and closed loop guidance of robots or other machines.

To date, the vast majority of installed applications have been inspection. This results because of the relative ease of its application to industrial plants and processes and the current push toward improved quality and productivity.

Our company alone has nearly 1,000 cameras and other "machine vision" type sensors installed, in-line applications - primarily in the automotive industry. This paper reviews several of these applications and the various sensing technologies which have been applied to piece part inspection. An extrapolation is then made for robotic guidance applications on the same types of parts, in the same environment.

I Introduction

The term "machine vision" generally suggests "eyes" on robots, guiding them to do complex tasks. However, for the present, actual installations utilizing machine vision have been almost entirely in the area of electro-optical inspection.

Inspection is of considerable importance, occupying nearly 10% of the people and 5% of the machines in plants. Further, the new emphasis on quality clearly implies inspection, and automation using machine vision offers a chance to improve both quality and productivity simultaneously. Practically too, inspection has been first because it's easier to inspect a part than to handle or work on it. And too, the majority of present day robots do not have control systems which allow dynamic path changes as a result of sensor data, nor the programming support for such activities.

While employing essentially the same "machine vision" sensor technology, inspection applications to date have generally been concerned with parts whose locations and orientation are both known to a large degree. This is considerably simpler than the general free-for-all possibility of looking at randomly oriented parts.

There is another subtle difference. Most inspection applications are justified relative to either existing gages in the sense of dimensional inspection, or relative to visual inspectors, as in surface flaw detection. In general, therefore, the sensor unit employed has to be of relatively high resolution. In the robot guidance case, however, the sensor may have considerably lower resolution, suitable for certain simple tasks that are required.

Differences aside, the parts being sensed, the environment, the sensor, and computer technologies employed are very similar. This paper provides an overview of the many inspection applications installed by the author's company. Nearly 1,000 in-line camera and related electro-optical sensor installations are represented.

Inevitably, these applications are leading into generalized robot guidance in various forms and some projects in this area are also described.

II Rationale of Inspection

Typically, there are a large number of reasons why one would wish to automatically inspect a part. A table listing most of them is shown in Table I. Increasingly, 100% in-process or final inspection is required in order to meet ever higher quality goals simultaneously with higher rates of production and therefore increased productivity.

Due to the high rates and the 100% inspection requirements, it is becoming very difficult to do this in any other meaningful way but with automatic inspection. This is true both in the sense of dimensional gages, and in those replacing people in the visual context.

III Machine Vision Inspection Advantages

The electro-optical dimensional gage or inspection system has a large number of advantages over both its dimensional inspection competitors (commonly the LVDT or "electronic gage", and the air gage), and human visual inspection. The usual advantages are:

- NON-CONTACT SENSING
- HIGH SPEED OPERATION
- DIMENSIONAL ACCURACY (SHORT AND LONG TERM)
- ACCURACY RELATIVE TO HUMAN INSPECTION
- LARGE STAND-OFF
- ABILITY TO "SEE" IN SMALL OR RESTRICTED AREAS (snap ring grooves, turbine blades, leading edges etc.)

Relative to human visual inspection, electro-optical inspection also provides consistency of inspection, both in providing a true 100% check and in checking each part in a consistent manner. Neither is possible with subjective inspectors.

It has been our experience that the average totally automatic final inspection machine justifies in approximately 1.5 years on a 2 shift basis. The principle modes of saving are in the area of direct inspection labor costs and warranty repairs. Certain gages have been justified because of the safety critical aspect of the part without regard to direct cost savings.

IV Areas of Application

Electro-optical inspection in plants falls into the following general areas:

1. Inspection of part dimensions.

2. Inspection of parts for surface characteristics such as defects and surface microfinish.

3. Inspection of completed or semi-complete parts and assemblies for proper assembly and manufacturing operations. This includes checking presence of holes, threads, rollers, washers, etc.

V Measurement of Outer Diameters (ODs)

The most common technique for vision based piece part dimensional measurement is profile imaging, where one or more edge images of the part are scanned by a photodiode array or TV camera. This duplicates closely the typical shadowgraph or contour projector whose ground glass screen, in this case, has been effectively replaced by the image sensor.

The most common arrangement, and most accurate, is shown in figure 1. Light passing the edge of a part is imaged by a lens onto a photodiode array. A scan of the array determines the edge image location and thence the location of the part edges from which the dimension of the part can be determined. For large diameter parts, such as crankshaft journals, two such sensors, one for each edge, are generally used.

For piece part measurement, the imaging approach, particularly with the solid state diode array, has a great deal to recommend it. It is all-digital, with very high speed scan rates and high resolution. It also has a large range although not as large as the scanning laser technique which is the principal alternative.

Another advantage of the image scan type is that it can operate with stroboscopic illumination, allowing the part image to be "frozen" in space - essentially as if it had been dropped into a fixture. Units have been built, for example, with pulsed diode laser sources having 100 nanosecond bursts.

In most practical applications, photodiode arrays are vastly preferable to image tubes such as Vidicons, because of their dimensional stability and lack of "blooming".

Using a combination of optical and electronic magnification plus software, accuracy of practical, installed systems has reached 4 millionths of an inch (0.1 microns) with ranges as high as $+/-10^4$, the minimum resolvable dimension. These specifications exceed all other common automatic sensing units for high resolution piece part measurement, and have led to several gages of unique capability described below.

Figure 2 illustrates one of three gages for in-line inspection of engine valves. These gages each have 12 sensor units or "cameras" and inspect all pertinent dimensions of valves at rates of 4200/hr. The gage also provides for automatic visual detection of valve surface defects. Purchase of these gages was justified over conventional approaches principally because of high part rate and the virtual lack of maintenance and complex setup required with contact gages.

Figure 3 illustrates one of 6 pump vane gages for inspecting straightness, squareness, length, width and thickness profile of transmission pump vanes with one micrometer resolution. These machines operate on-the-fly at 7200/hr and sort vanes into 4 classes. A combination of 12 diode laser and gas laser light sources are used.

VI Measurement of Internal or Contoured Surfaces

Another principle dimensional check is of surfaces which cannot be seen in profile, such as gear or turbine blade contours, sheet metal body dimensions, female thread forms or bore inner diameters (IDs). For this purpose, the triangulation technique has proven of most value, using projected spots, lines or grids.

As shown in figure 4, triangulation (also variously called structured light, light section etc.) operates by projecting light from one angle and imaging the resultant spot on the surface from another. Image spot location on a position sensing detector, such as a photodiode array, provides a measure of surface location, from which part dimension, contour etc. can be accurately determined. Recent installations can achieve 20 microinches (0.5 um).

This type of measurement is much more difficult to accurately accomplish than profile (shadow) imaging because of the effects of surface reflectivity due to variations in color, roughness, texture and form. For example, on turbine blades, 100,000:1 variations in reflected light power have been noticed depending on part form and type.

One example of such triangulation based equipment is the 4 axis, dual triangulation sensor turbine blade contouring machine shown in figure 5. This machine has a total accuracy throughout its envelope of 2 um, and averages 6-8 points per second, even on the small trailing edge radii.

Figure 6 illustrates an automatic in-line gage using triangulation to determine camshaft casting bow and part line shift. Defective cams are rejected to prevent damage to expensive cam grinders.

A machine using four matrix arrays to contour sections of arc welds on automotive lower control arms such as those of figure 7, has also been provided. A complete inspection of 15 sections of data on each of 4 welds is accomplished in 2 seconds. Weld defects such as off joint, undercuts, porosity, poor fillets etc. can be determined. This same type of sensor can also be used to control a robot welding operation.

A semi-automatic gear gage, based on dual triangulation sensors is shown in figure 8. It inspects pitch line runout and tooth spacing errors on rear axle hypoid pinions. Total time to inspect all teeth on both drive and coast sides is 10 seconds including load/unload.

VIII Visual Inspection Replacement

The forgoing discussion has concentrated solely on measurement of dimension, in which case one compares the machine vision solution directly to the existing LVDT, air gages, etc. There is, however, another principle area of visual inspection of surface flaws, defective assemblies and the like, a large consumer of human labor.

In visual inspection, the present human standard is considerably lower than that presently achievable dimensionally by conventional gages. Inspectors called on for repeatitive examination are generally error-prone and much less than 100% effective.

Human acuity for spotting defective parts varies widely and is totally subjective. While formal visual inspection specifications defining the size of defects or whatever may be written, when it comes to the actual practice, accept or reject decisions often depend on how the inspector feels that day. There is seldom, if ever, actual measurement taking place. Rarely too, are accurate statistics as to defect types accumulated.

Not only is human inspection error-prone, it is also extremely expensive. Thus a large area of near term application of machine vision is in these areas since the justification is high. Added to this is the fact that most products which are visually inspected are items which, by their very nature, cause a failure directly (missing parts in assemblies, large surface defects which wear out break seals, bearings etc.).

VIII Automatic Surface Flaw Detection

Our company has been very active in automatic defect inspection applications of machine vision. The principle application has been inspection for surface defects such as pits, scratches, voids, tool marks, cracks and the like.

In one technique for performing such inspection, the part outer surface is illuminated and the image scanned, in this case, by a photodiode array. In the case of the cylindrical parts, rotation 360° past the line scan array provides inspection of the total part surface. Machines have been built in this manner to check valves, drums, wrist pins, brake discs, bearings, seals and the like.

A similar form of inspection can also be provided in internal bores using a circular scan. An automatic defect inspection machine of this type, shown in figure 9, is used for detecting defects in power steering housing bores, and can perform a complete inspection of these parts in less than 2 seconds. Another machine, one of 10, for pinion gear bores is shown in figure 10.

IX Inspection for Defective Assembly or Machine Operations

The second principle area of automatic visual inspection is inspection for defective assembly or machine operations. This is customarily done with solid state linear array or matrix cameras operating on the reflected part image, and thus bears most resemblance to the general machine vision problem. It is thus often called "pattern recognition".

Figure 11 illustrates one of two machines built to inspect for correct assembly of power steering worm assemblies. It utilizes two solid state TV cameras and linear arrays, coupled to two microcomputers. Part rate is 1500/hr.

Illustrated in figure 12 is an in-line machine to inspect threaded and machined holes in cylinder heads at 600/hr. It is the largest known electro-optical gage installed to date with 80 image sensors of various descriptions. Presence of each thread in every hole is checked plus blockage in all drilled oil holes.

Figure 13 illustrates one of six solid state TV camera based systems to check orientation of connecting rods on-the-fly and turn them over if incorrectly oriented. The purpose is to preclude downtime due to jamming on expensive transfer machines.

"Robot and Machine Vision"

What about guidance of robots? As previously pointed out, this is a natural extension of optical inspection. The easiest first step is to simply tell a robot (or other automation) what to do, based on sensing performed at an upstream or separate station.

For example, consider figure 14 which illustrates a camera system used with painting robots. From the image of a car body on-line, it determines the type of car in question and tells robots down the line what paint program to call up.

Similarly, the gage of figure 15 senses dimensions of parts made on an NC lathe and tells the robot whether to accept or reject as well as feeds back tool correction data to the machine.

Robot inspection is a natural outgrowth of the above activity in which robots move one or more sensor units programmably and flexibly to many points on a part. An example of this is the turbine blade inspection system shown in figure 5. Another is a unit presently under construction for a major automobile manufacturer for inspection of fender shape, outline of dimension and hole location.

In time, sensing of the types described herein, combined with additional computing power, will be used in real factory applications, to guide robots on dynamically changing paths. To allow widespread use, sensing will likely be based on 3 dimensional techniques operating independent of the part reflectivity. Backlit light tables and binary images commonly used in robot "vision" demonstrations, will be useful only in rare cases. Backlighting at a fixed range is the exception rather than the rule.

In closing, a word of caution needs to be voiced. Many first machine vision installations have been time consuming, involving substantial training of plant personnel in unfamiliar new technologies. In some plants, just getting someone to wipe a sensor window once a week is a major task.

Every effort, therefore, needs to be made to insure that trouble shooting and maintenance procedures are fully understood and documented, training programs implemented, and where possible, diagnostic aids and maintenance features incorporated. All these things add cost but can easily spell the difference between success and failure in plant.

Conclusion

Machine vision is a field whose time has come. Starting with insepction, "vision" will spread first to simple robot auxilliary tasks such as identification, and finally to dynamic guidance of the machines themselves.

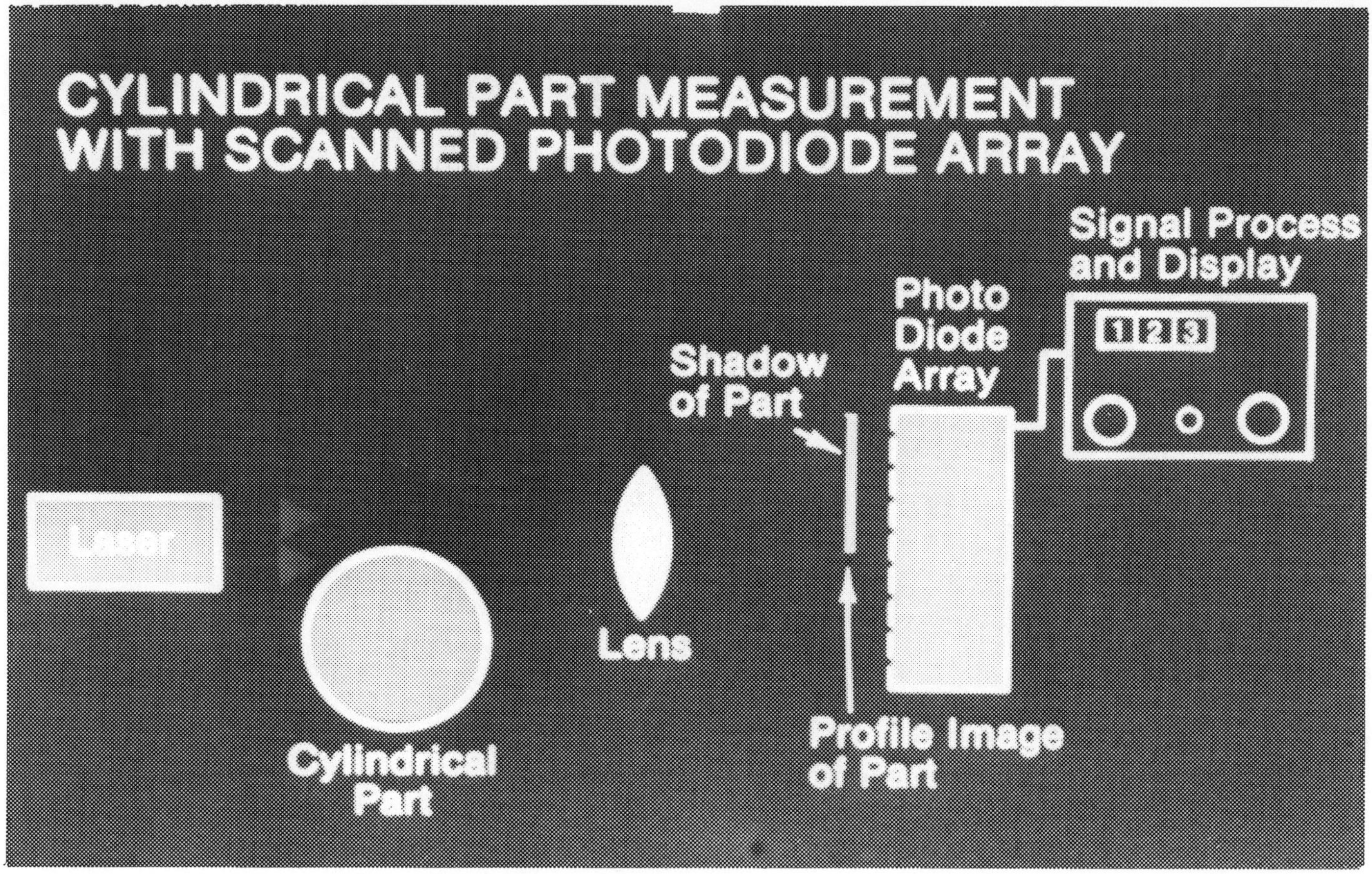

CYLINDRICAL PART MEASUREMENT
WITH SCANNED PHOTODIODE ARRAY

Signal Process
and Display

Photo
Diode
Array

Shadow
of Part

Laser

Lens

Cylindrical
Part

Profile Image
of Part

FIGURE 1

FIGURE 2

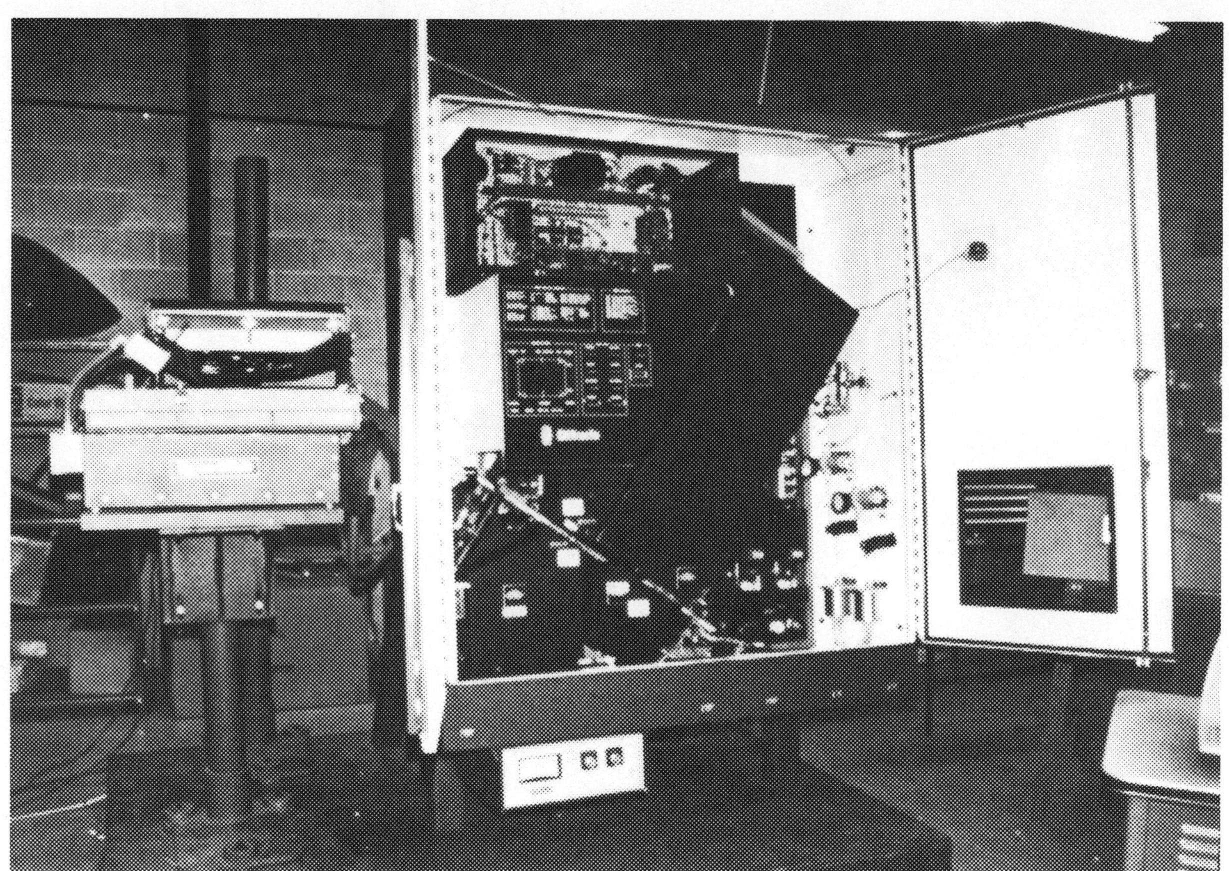

FIGURE 3

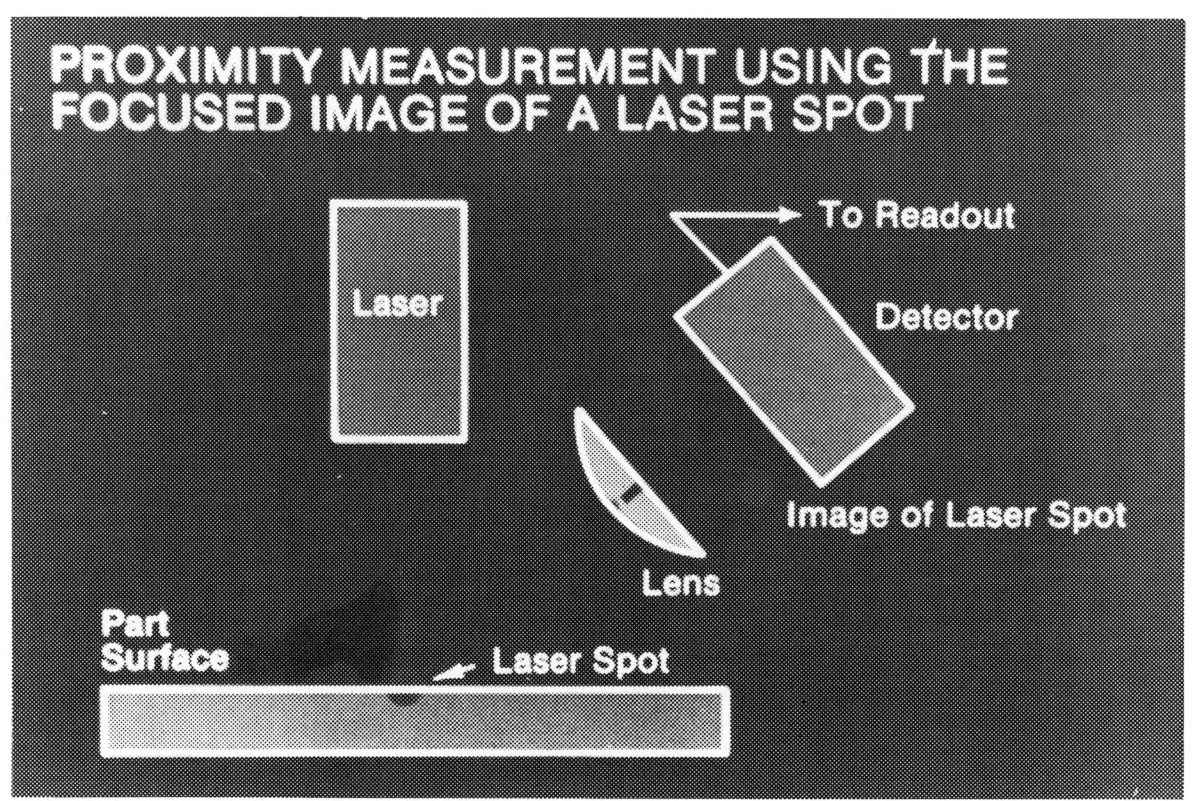

FIGURE 4

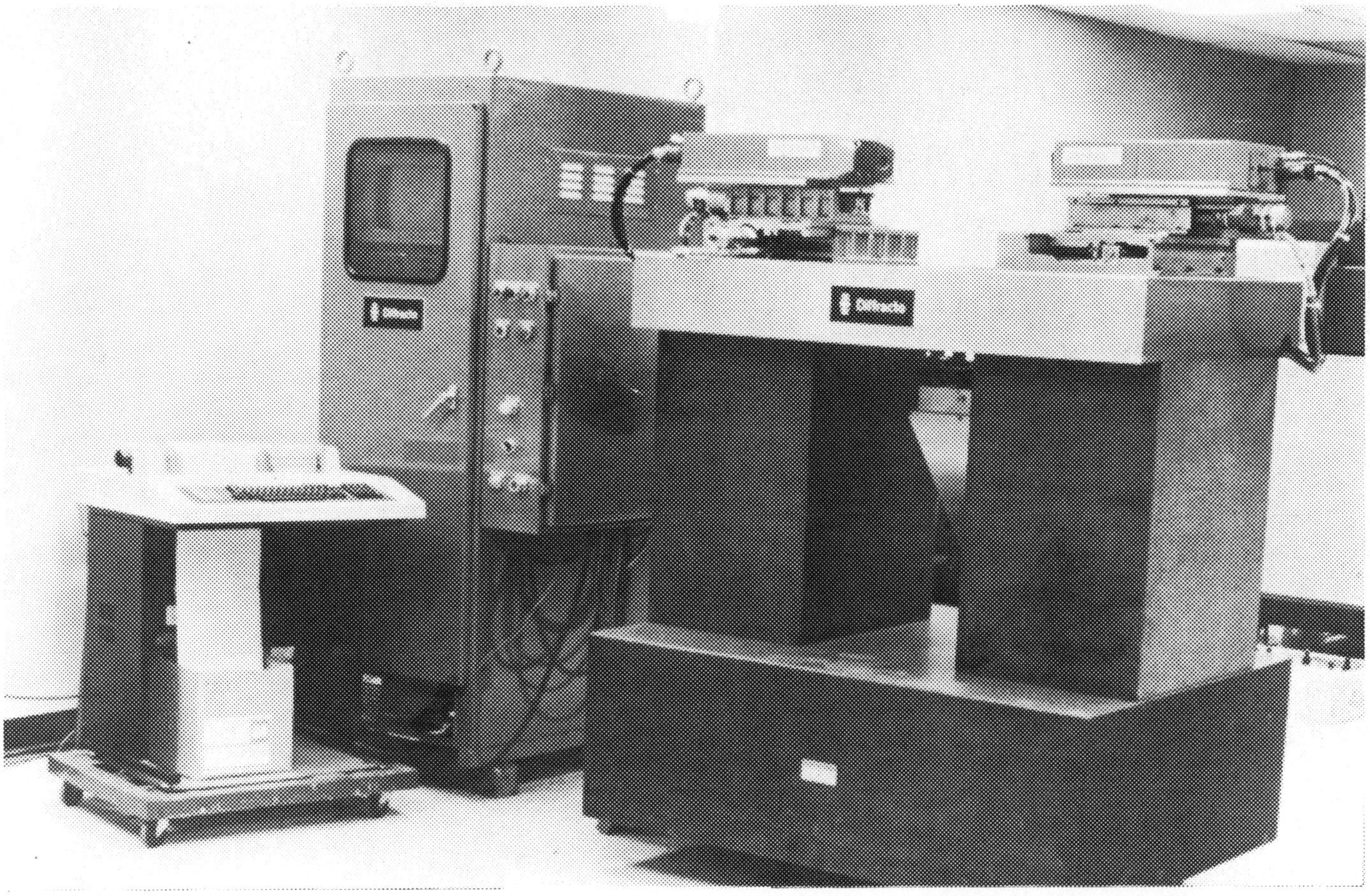

FIGURE 5

FIGURE 6

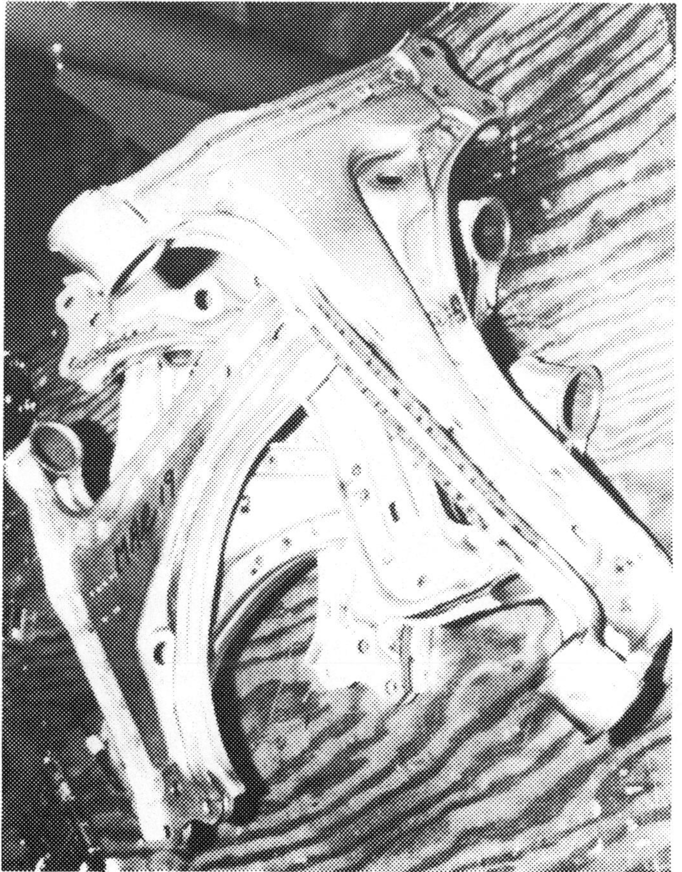

FIGURE 7

FIGURE 8

FIGURE 9

FIGURE 10

FIGURE 11

FIGURE 12

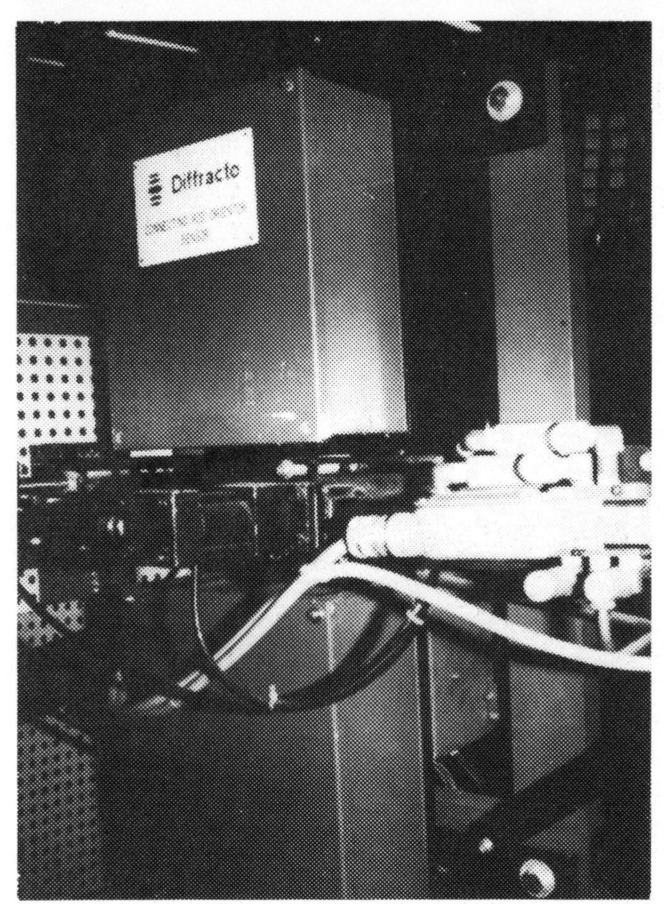

FIGURE 14

FIGURE 13

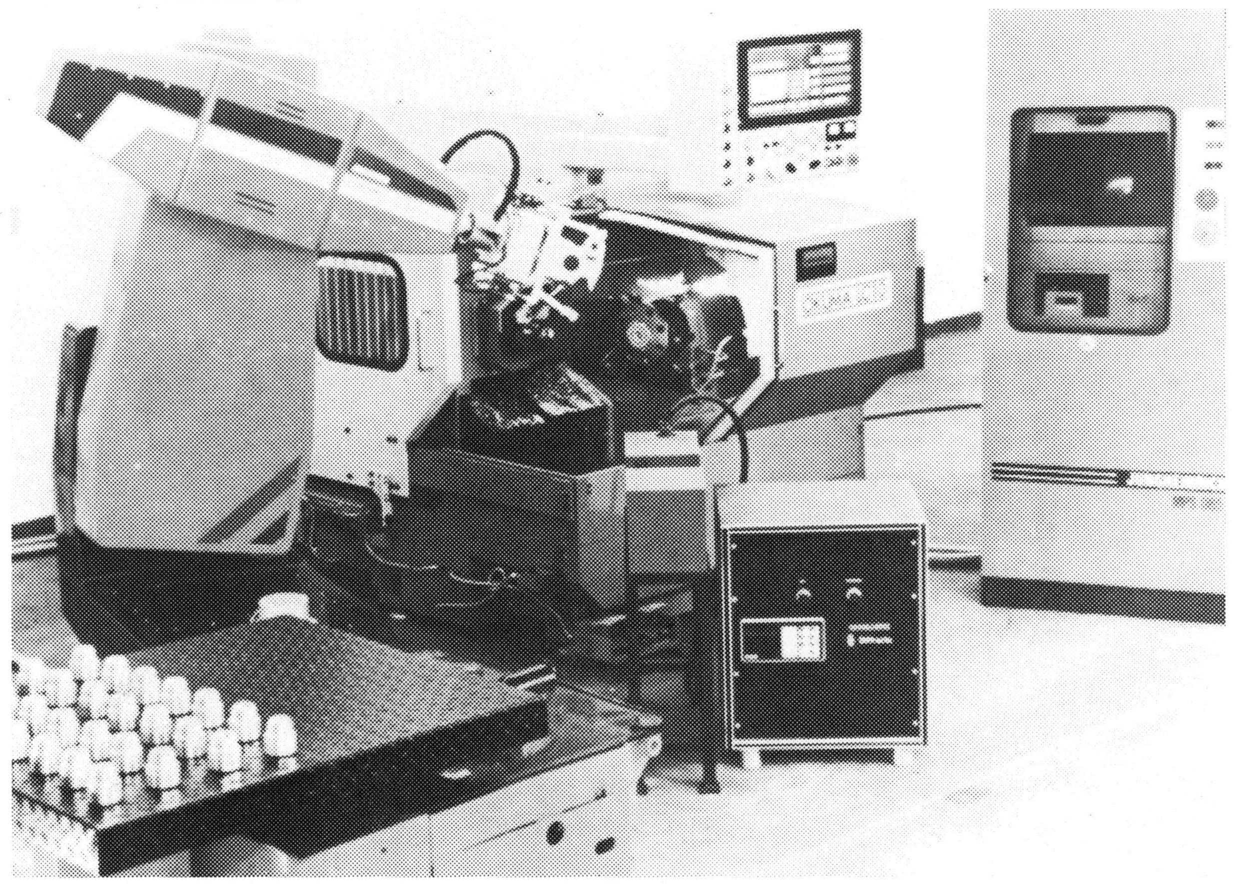

FIGURE 15

Presented at the SME Vision '86 Conference, June 1986

Performance Evaluation of Machine Vision Inspection of Assembled Printed Circuit Boards

by Jerry Kosner
Tellabs, Incorporated

INTRODUCTION

The success of every company depends on a consistent level of product quality. To maintain quality requires defined quality standards in a set of policies. All employees must then understand how the policies relate to their job function, and management must provide the tools required to measure their quality ouput against the predefined policies.

Machine vision is an effective tool to measure product quality against defined policies through process control, and it can be applied accross many industries. Tellabs, Inc. is the largest independent supplier of telecommuncation equipment, and this paper describes the integration of a turnkey, in line, 100%, wire side, machine vision inspection system for component lead through the hole verification before wave solder of assembled printed circuit boards. The assembly operation for the printed circuit boards is described, and defect repair costs are analyzed before and after the installation. The features, performance, and remaining work on the vision system are described as well as the frustrations, and trauma associated with a two year internal hardware software development effort. The move from the laboratory to the production floor was a very difficult 6 month learning experience.

PRODUCT QUALITY

The quality of a product encompasses many nebulous and intangible attributes that represent every characteristic of the product and the company that built it. From the quality level and the perceived real value of the product, the customer will make a buy decision between your product and your competitors. This decision will determine the ultimate success and profitability of your company. Included in the evaluation of quality are such diverse items as the color of text used in the software menu, the length of the leads on the back of the printed circuit board, and the cleanliness of the office rug when the customer comes to visit after the first spring thaw.

To maintain consistent product quality requires work and a significant amount of discipline. The first step is to define the quality level of the product and to describe the level in a set of policies. Examples for a company assembling printed circuit board products are lead lengths will not exceed 1.9 mm (0.075") from the wire side of the board, all components must be flush to 0.76 mm (0.030") above the component side of the board, all leads must have evidence of a component side fillets after solder, etc. Every employee of a company cares about the quality of their work, and they want to do the best job possible, but in many organizations the policies governing product quality remain undefined. After the policies required for quality are defined, each employee must be trained to understand how the policies apply to their function at the organization. The final step necessary for successful quality implementation, requires management to provide the tools required by the employees to measure their work output against the established policy. With the assault of imported foriegn products on US markets, management in many organizations has reacted without success by implementing quality circles and participative work group decision making. While implementing these techniques, they have totally ignored the three fundamentals of quality; define a set of policies, ensure that every employee understands how their function relates to these policies, and provide the tools necessary to measure the work output against the policies. Machine vision is an ideal tool to provide the process control required to ensure adequate and consistent product quality.

CIRCUIT BOARD ASSEMBLY OPERATION

Tellabs supplies 800 different products for the telecommunication and data communication end users, and the majority of the products are processed in lots of 100 to 500 boards. This manufacturing environment requires the use of flexible automation and quick set up, instead of continuous production runs and hard automation. Figure 1 illustrates a batch assembly operation flow chart and shows where the inspectors check the work of their respective work groups.

It is important to note that Figure 1 is broken up into two areas. Stage 1 is designated before wave solder, and assembly defects recognized and corrected in this area are ten times less costly to correct then errors found after wave solder in Stage 2, which is final assembly and functional test. The error repair rate is again multiplied by ten if the error is not found until the field. For a wire side assembly error found in Stage 1, the correction cost averages $.07, in Stage 2 of assembly the cost averages $.68, and for the field the cost is $6.80. Using the ten times guideline, it is clearly evident that process control for the assembly operation is required before wave solder. Therefore, the printed circuit board assembly process of Stage 1 will be described in detail, with only references made to Stage 2, which is included for clarity.

The first two work centers automatically insert and clinch axial components; resistors, capacitors, and diodes, and dual in line packages; integrated circuits. Inspectors 1 and 2 review the work of each station and correct any components jarred loose from the insertion of other components, knocked loose by the insertion head, or displaced by handling within the work center. With consistent parts, good boards, adequate setup and skilled operators, part placement is reliable, accurate, and secured by clinching.

The subsequent work center, hand insertion, inserts the radial parts, electrolytic capacitors, transistors, connectors and transformers and any other odd shaped parts not suitable for automatic insertion. The leads of the components placed in the boards by the hand insertion line are first trimmed and then bent to form the knees in component prep. An adequate clinch is also placed in the lead so that it will remain in the board during handling before wave solder. A pass along line of two to twelve people, inserts eight components each with a forty five second cycle time on the line. The supervisor, Inspector 3, of each line inspects the assembled board and reviews it for correct component placement and orientation. When assembly errors are found they are corrected, and feedback is given to the line personnel involved.

The final inspection before all assembly errors become permanently wave soldered is completed by Inspector 4. The actual implementation of Stage 1 pre wave is shown on Figure 2- Palletized PC Assembly: System Products and Figure 3- PC Assembly: Voice Frequency Products. There is a difference between the two systems which has caused significant process control problems. The palletized PC assembly for System Products uses a fixed, continuous conveyer to carry the boards from the hand insertion line to the wave solder. Although immediate feedback is provided to the hand insertion line, Inspector 4 sees three different assembled circuit boards in random patterns. A lower density, assembled, Type 18 printed circuit board inspected at station 4 is shown on Figure 4 component side and Figure 5 wire side. Each board has a maximum of 200 components and 2,860 holes to review in 20 seconds, an impossible task. Because of the random arrival of circuit boards, the inspection task has been split, component and wire side, and an additional Inspector 4 has been added. The benefit of instant feedback for the hand insertion line is definitely overshadowed by the drawback of random board arrival patterns into Inspector 4. Because of this process control problem, it is likely that the line will be returning to the batch orientation unless machine vision can be installed successfully.

The job of Inspector 4 is critical in maintaining low variance production costs, consistent product quality levels, and high profitability for the company. The inspector's job classification is three levels above the hand insertion personnel. Because of the pressure and the performance level required of Inspector 4, frequent job postings appear on the bulletin board as follows.
POSITION: Final Assembly Senior Inspector
DUTIES: Visually inspect PCB's to detect errors. Concentrated effort to be placed on missing components, correct polarities, and overall PCB appearance.
REQUIREMENTS: Ability and eagerness to follow directions. Ability to work independently. Minimum of one year final assembly experience or equivalent.

ASSEMBLY OPERATION PERFORMANCE

With all the policies related to quality defined, the personnel trained to understand how each policy related to their job function, and an additional Inspector 4 added, 75% of all circuit boards assembled and wave soldered had an average of 5.3 defects each. Figure 6 shows this data divided into 5 catagories; solder defects, lead trim defects, component side defects, wire side defects, and miscellaneous. The data was gathered by Inspector 6 while Inspectors 1, 2, 3, 4, and 5 were functioning during November and December of 1985. From this data the following general conclusions were reached.

Solder defects; shorts and bridges, excessive solder, insufficient solder, blow holes, fractured joints, and cold solder, account for the largest number of defects found, and these defects are the least costly to correct. Because of the critical process control required for the wave solder operation and the direct relationship on field reliability, automated process control using manual x-ray systems, laser thermographic systems, or machine vision are a candidate for implementation as soon as the technology is available in commercial systems.

A corporate wide zero lead trim program was implemented in August 1985, and the 8% lead trim defects resulted from depleting old stock not conforming to current specifications. With the zero lead trim program now in place, board scalping has been eliminated along with flagging. The integrity of the solder joint has also been improved.

Six percent of all assembly errors can be located using wire side machine vision inspection. Only 1% of all assembly errors can be located using component side inspection, which is significantly more difficult because parts must be identified. The six to one difference in error rates results from the consistency of the automatic insertion equipment and the orientation of the parts in the magazine.

Figure 7 breaks down the wire side defects into missing parts, lead in wrong hole, lead to short, and the lead not in the hole. As shown by the data, the largest percentage of defects can be easily found by wire side machine vision inspection. Repair costs are significant in wire side defect repair, because many times the part must first be unsoldered from the board and new part soldered in. The average cost for wire side defect repair was $.68 per defect. Summing the number of defects per month yields rework costs of $4573 or $.15 per board assembled for November and $2873 or $.14 per board assembled for December.

When reviewing the data, it is important to remember that November, December and January were the first three months the data was available. It took ten years to develop the discipline required to collect meaningful process data. With all the time required to launch new products, new technology, and new processes, it is often easy to overlook the improvements possible in present processes with improved discipline.

MACHINE VISION INSPECTION

Because of the high cost associated with reworking wire side defects after the wave and the technology available for in line, wire side, component lead through the hole verification before wave solder, it was concluded that a machine vision system would be designed and implemented. The system would be the tool with which the assembly personnel could measure their performance against the defined quality policies. Inspector 4 works as a team with the machine vision system inspecting only the component side of the board for reverse polarity, incorrect value, incorrect part, and poor workmanship. The machine vision system inspects the wire side of the circuit board for missing parts, legs bent under the component body, leads not through the board, and missing leads.

The features of the system are as follows.

* 100%, wire side, process verification of all assembly operations.
* Operates in line at wave line conveyer speeds up to 2.74 m/min (9 ft/min).
* Maximum board/panel size 457 mm (18") by 609 mm (24") and minimum board size 76 mm (3") square.
* Algorithm error rate less then .01% on lead through the hole inspection.
* Self teaching- the coordinates of the vision area are entered by the operator. The system learns from an empty board and an assembled golden board.
* No production set up required on a job previously run.
* Simple 2 key operator interface. No programming required.
* Color graphics assembly error display of the circuit board outline for on line board adjustment and repair.
* Buffer vision buffer conveyer makes an ideal workstation for Inspector 4.
* Hard copy travellers are provide for boards requiring major rework.

Figure 8 shows the IS32 optic ram mounted in the lense housing of the camera shuttle. A technical overview of the binary vision system was given last year during Vision 85 in the paper "Component Lead Through the Hole Verification Using 64K Dynamic Rams as Image Sensors".

MACHINE VISION PERFORMANCE

The path from the laboratory to the production floor for the machine vision system has been difficult but the preliminary results from limited production testing are encouraging. The schedule for full system implementation was August 1985, presently the implementation is 6 months behind. January 1986 was the first month the system was out of the laboratory for more than a single shift, and the limited engagement was only 5 days of 2 shift in line process control. During this production trial, knowledge was gained in writing bullet proof input output software routines, keeping production dust off the camera lenses, and designing easy to use operator interfaces.

With all the problems and design difficulties associated with the development, the results obtained in the production environment were very encouraging. Wire side defects were reduced from a constant 6% for November and December to 4% for January, a full 2% reduction. Along with the reduction in defects comes a significant reduction in wire side rework costs. Figure 9 summarizes the January data and shows that the rework costs for the month were reduced from $.14 per board assembled to $.09 per board assembled. Although there is no claim that the limited data is conclusive, it is the authors opinion that wire side defects can be eliminated from the post wave operation with an annual $40,000 savings in rework costs.

Currently the software is being toughened, and areas of the mechanics redesigned, correcting the deficiencies found in the first trial. A full 2 shift production month is planned for March, and the latest data will be presented at the conference.

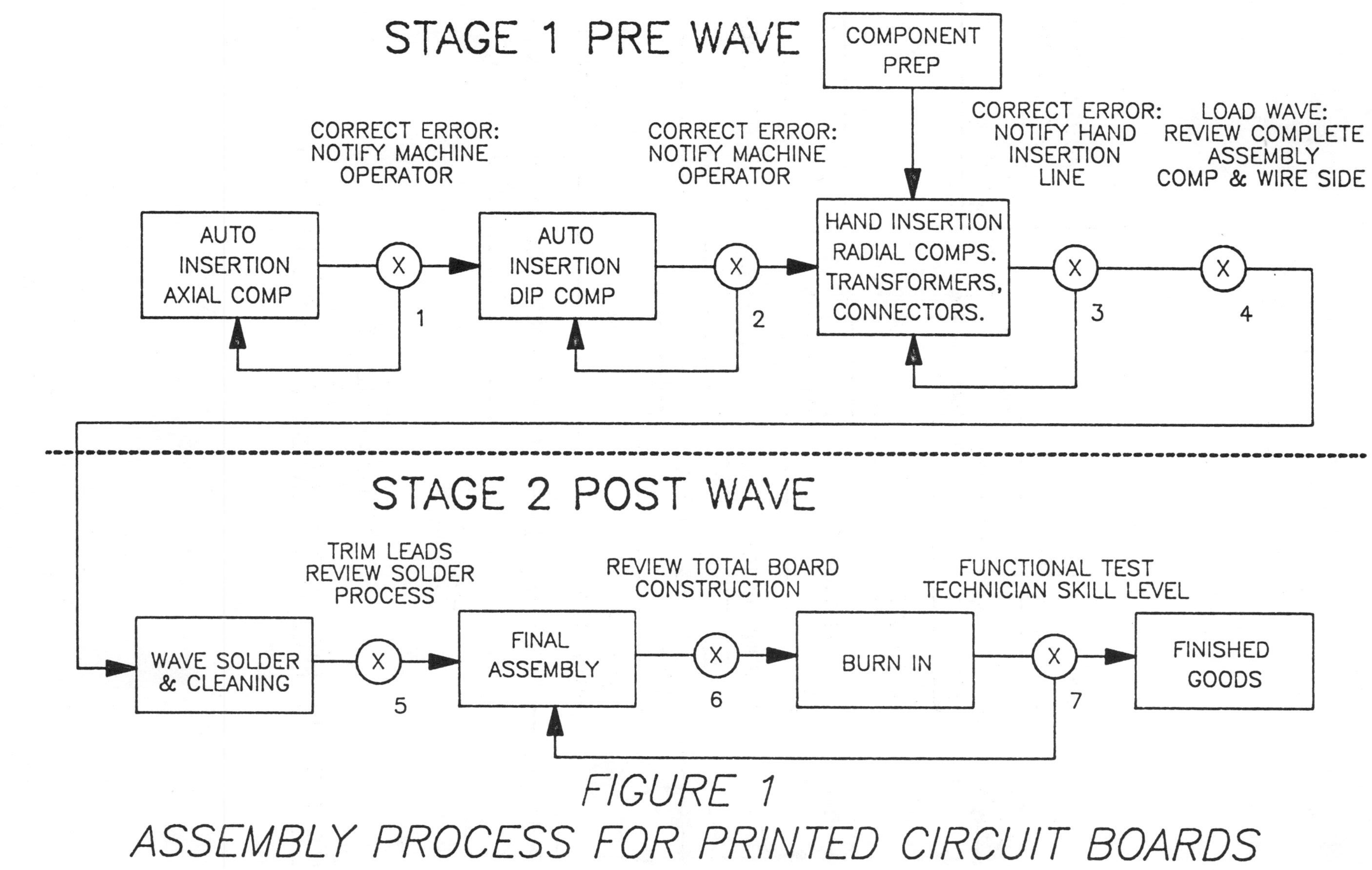

FIGURE 1
ASSEMBLY PROCESS FOR PRINTED CIRCUIT BOARDS

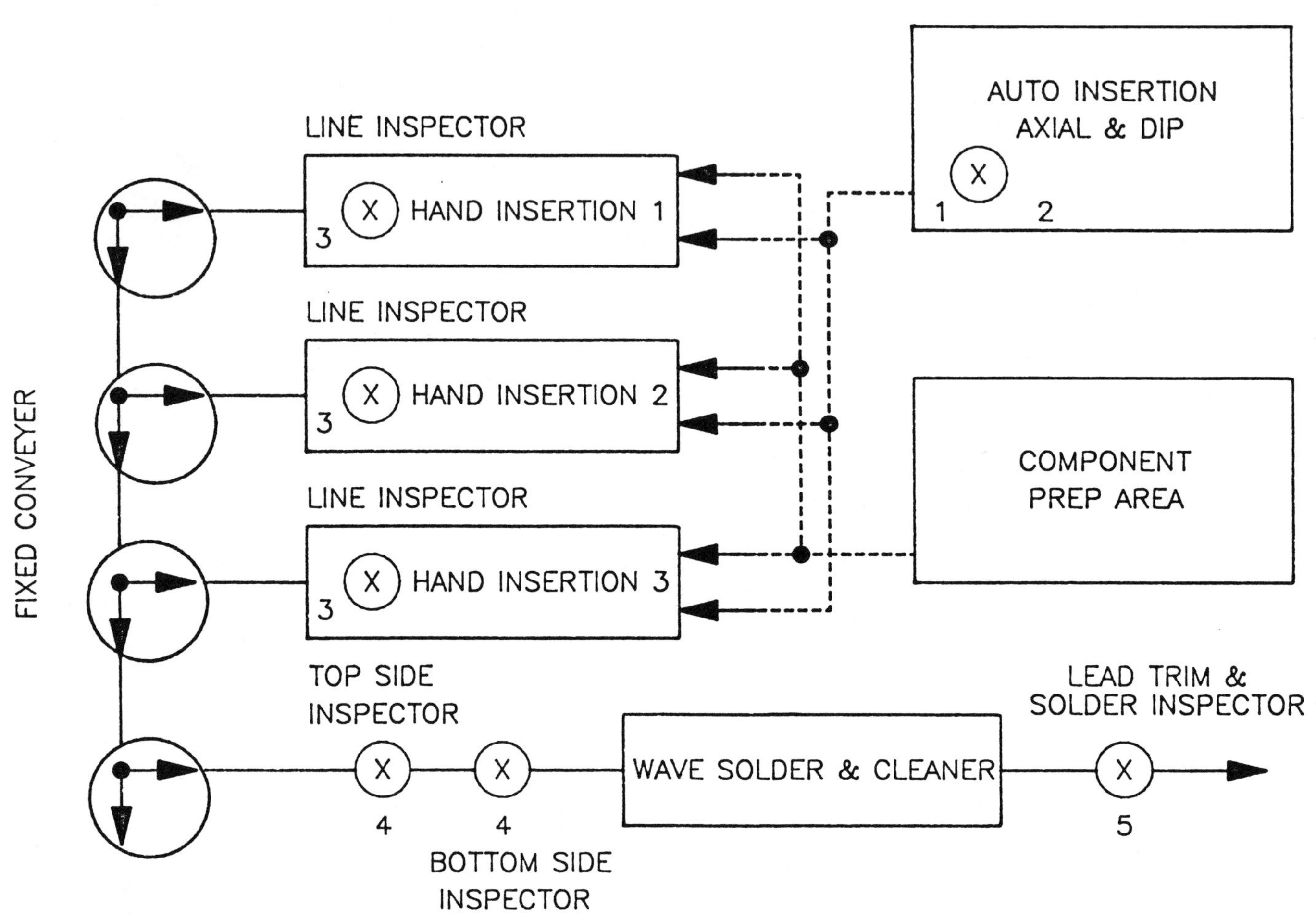

FIGURE 2

PALLETIZED PC ASSEMBLY: SYSTEM PRODUCTS

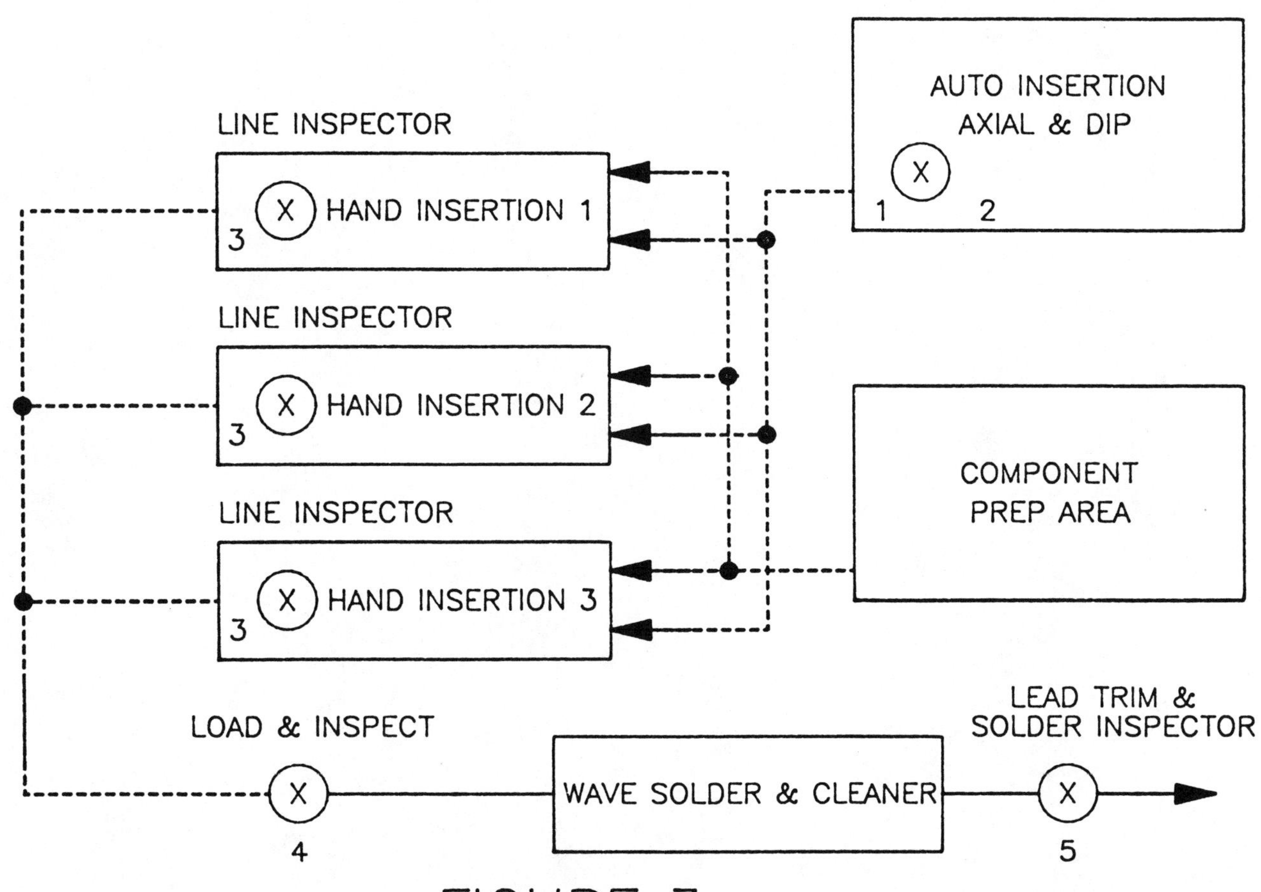

FIGURE 3

PC ASSEMBLY: VOICE FREQUENCY PRODUCTS

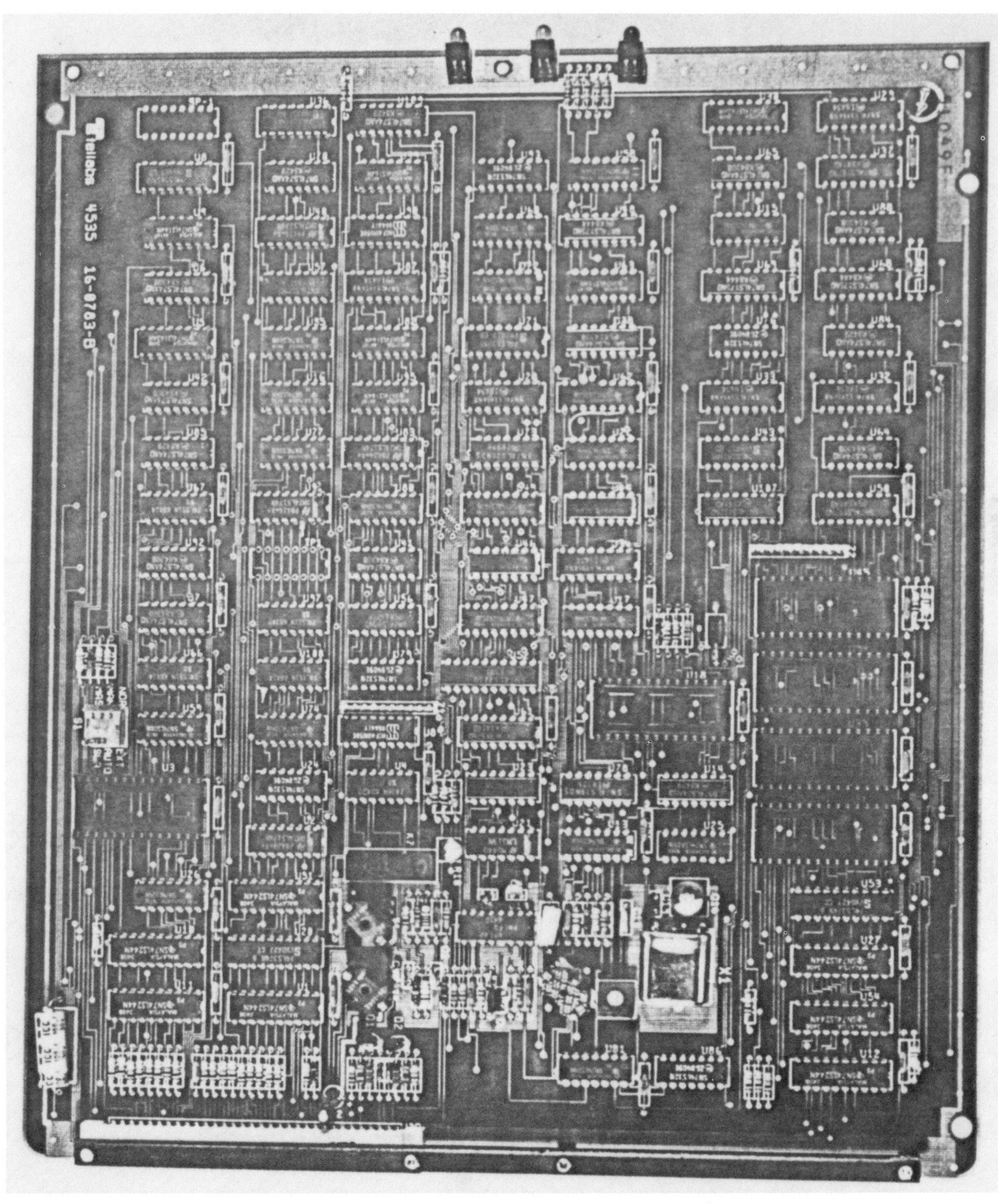

Figure 4 - Type 18 Component Side

Figure 5 - Type 18 Wire Side

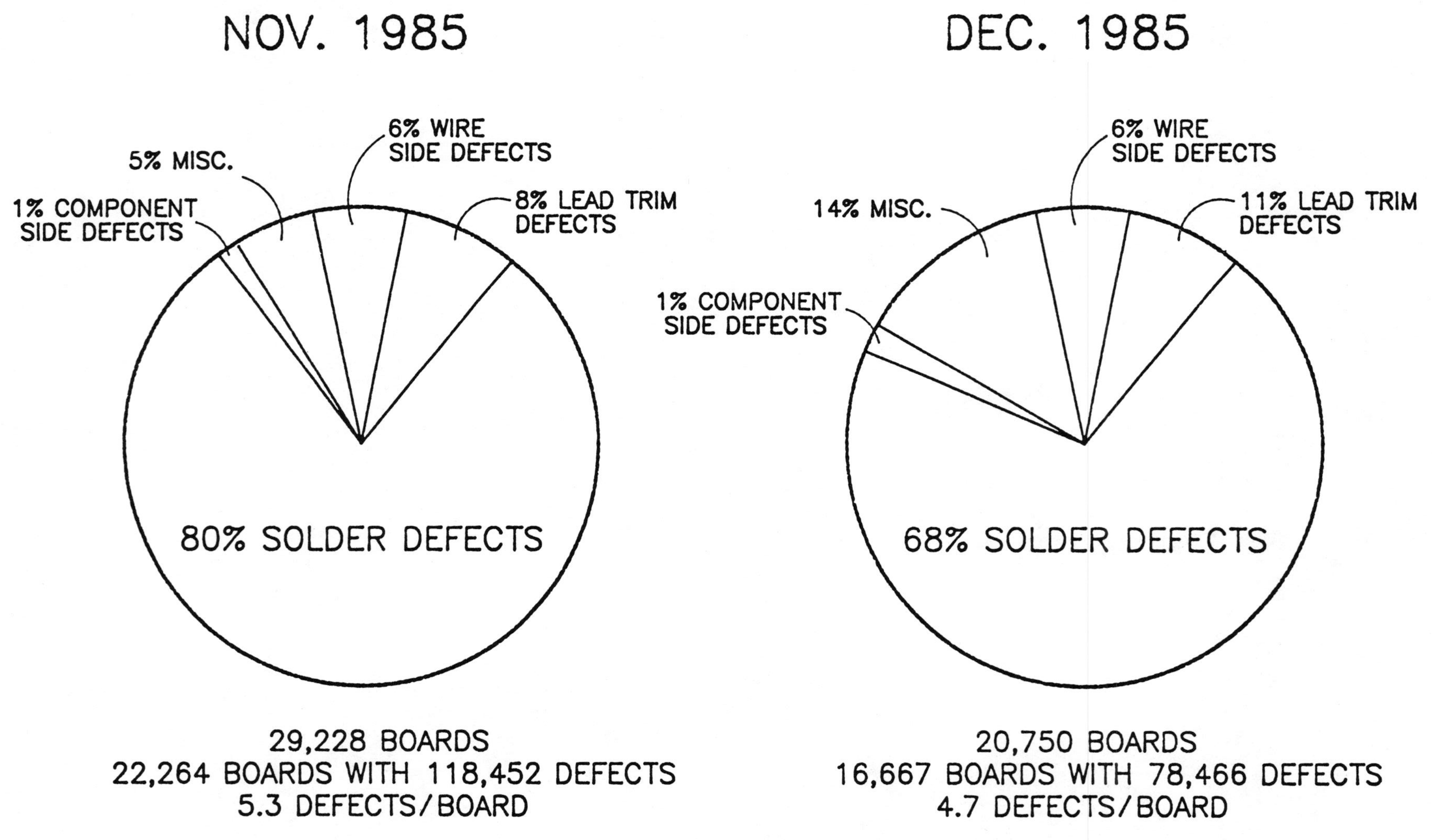

FIGURE 6 — PC BOARD DEFECTS #6 INSPECTION STATION

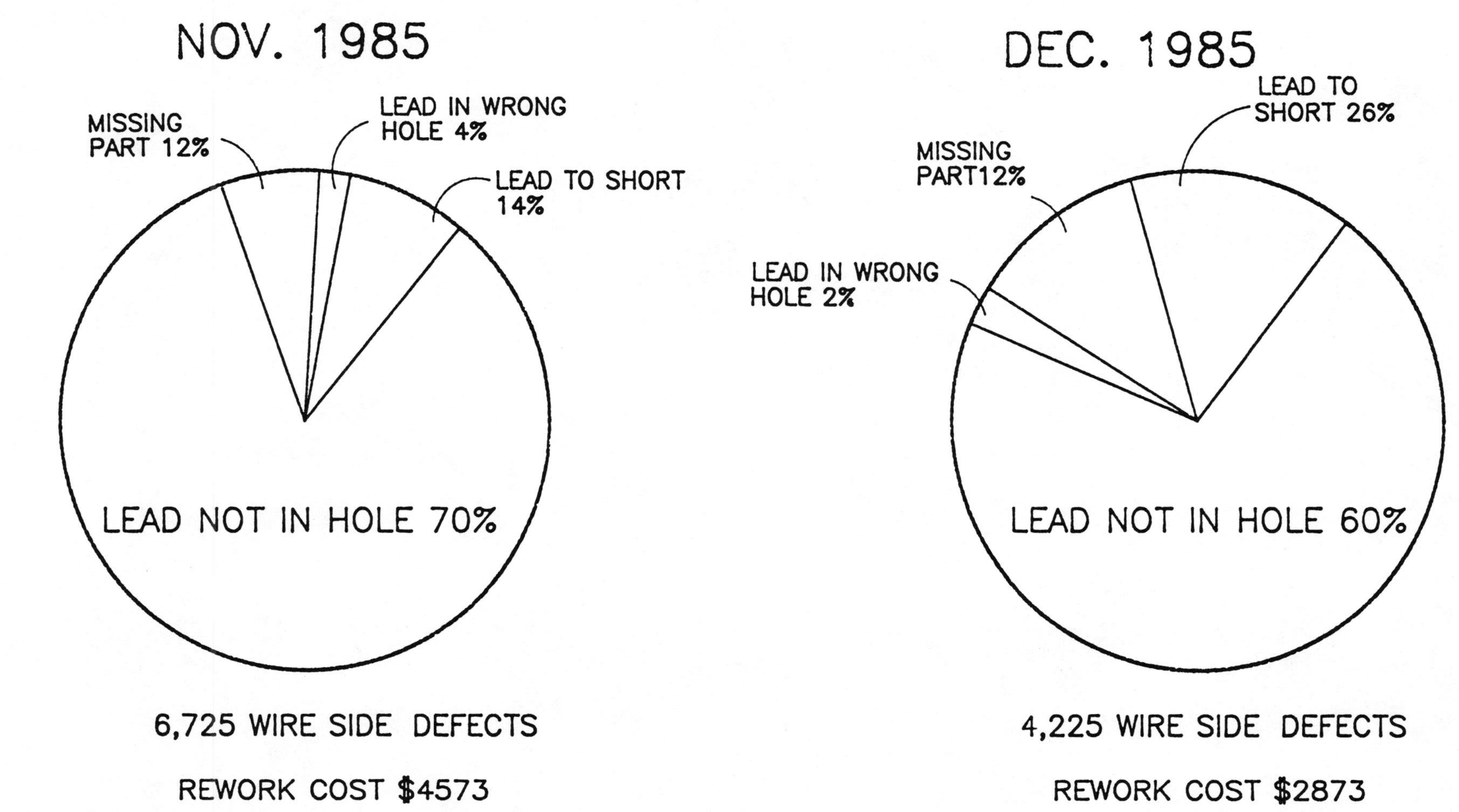

FIGURE 7 – WIRE SIDE DEFECTS #6 INSPECTION STATION

Figure 8 - Optic Ram and Lense Arrangement

JAN 1986

22,352 BOARDS.
16,540 BOARDS WITH 79,392 DEFECTS.
4.8 DEFECTS/BOARDS.

3175 WIRE SIDE DEFECTS.
REWORK COST $2159

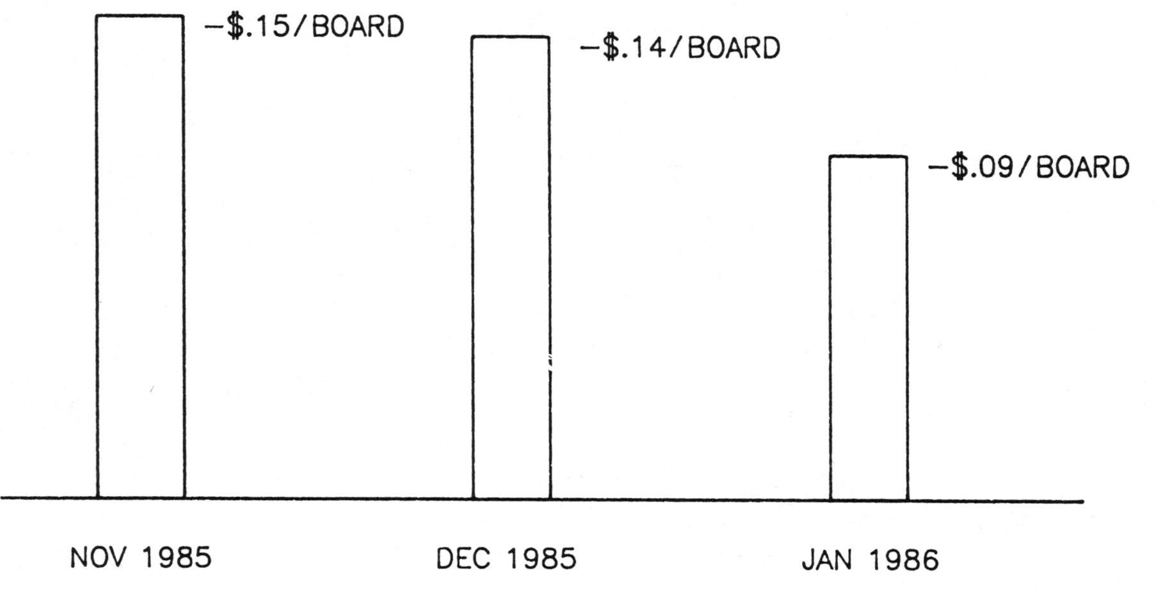

FIGURE 9

WIRE SIDE REWORK COST PER BOARD ASSEMBLED

CHAPTER 9

AUTOMATIC GAGING AND INSPECTION

Published as an SME technical paper, © 1976

Improved Productivity Through Inspection in Automatic Assembly

by Robert L. Douglas
Gilman Engineering & Manufacturing Co.

The evolution of inspection in assembly processes, starting with manual assembly and moving to automatic assembly is discussed. An example of selective assembly which depends on automatic inspection is presented.

Areas where productivity improvements can be made are discussed.

INTRODUCTION

Inspection has always been an important factor in the productivity of any assembly process. The physical characteristics of the components determine the acceptability of the completed assembly. Everyone has heard the story about the car which came off the assembly line with two doors on one side and one on the other and most of us have experienced the frustration of building a child's game or toy from a kit. Invariably, the kit came with erroneous assembly instructions or twice as many left-handed components as required without a single right-handed component. The productivity of the assembly process which produces "junk" is very low indeed.

The problems experienced by the assembler on Christmas Eve are not unique. Mass production systems experience the same productivity limiting problems. Incorrect components, shortage of components, damaged components, components which will not fit together, plague every assembly operation.

In manual systems, the inspection is accomplished by the ultimate inspection device------Man. The ingenuity of production people who have learned that a tap with a lead hammer at the key location or that inserting a component upside down will make an acceptable assembly is amazing.

In automatic assembly systems, that key manual inspector is not present, and this function must be filled by automatic inspection.

AUTOMATIC INSPECTION

An automatic machine receiving incorrect parts will produce reject parts cycle after cycle, hour after hour, all day until stopped. This is not zero productivity; this is negative productivity. Bad parts whether oversized, undersized, or damaged, usually cause misloads and jams in automatic equipment. A jammed feeder track can starve a machine and without adequate inspection following the loading operation, missing key components will not be detected. Therefore, a major factor in inspecting any component is to check the part presence in the assembly as close to the loading operation as possible. This is necessary because if other components are to be loaded over the missing one, they very often will cover up the fact that the first one is missing.

This type of inspection of incoming parts, parts coming to the machine, to determine their shape, orientation, color, size, function, is the first step of adding inspection to an automatic assembly process. However, the key to improving productivity is involved with the decision as to what action is to be taken when an inspection is not passed. Once again, at its simplest, the process can be stopped and the man alerted to the problem. He can then make a decision as to what to do with the faulty component. However, this slows up the entire assembly process and the productivity suffers. The technique that is widely employed is to isolate the faulty partial assembly and to add no further components to it. This partial assembly is then isolated in a reject chute or location somewhere on the machine. This allows for the part to be disassembled and its components recycled through the machine. In this way, the machine cycle remains constant with only one cycle of time lost to a reject assembly.

Let's examine in detail some options available when inspection identifies an unacceptable component or assembly. As already discussed, the machine can remember the reject condition and isolate the bad assembly. The bad assembly can then be reworked or discarded as justified by the cost of the components versus the cost to repair.

From a machine control standpoint, this requires machine memory to be added. The machine must know that a part at a given station is good or reject before it works on the part. From a mechanical standpoint, this requires that the stations can be "locked out" and prevented from operating on reject assemblies. It also requires a dual unload system for segregating good assemblies from rejects. Thus the cost of the inspection may be minimal within itself, but the cost to maintain a high level of productivity and still accommodate rejects can be substantial.

(For a discussion of machine memory and other required control features, see SME Technical Paper ADA-426, Programmable Controllers in Automatic Assembly, by R. L. Douglas).

To aid in diagnosing a major problem, inspection stations are often given a "sequential reject count and stop system". In this way, if tooling is broken or a part feeding track is jammed and good assemblies cannot be produced without manual intervention, the machine operator can be alerted. If three or any selected number of consecutive rejects are created, the machine is stopped and the operator alerted through the use of a warning light.

The discussion has been limited to inspection to insure that a part that should have been loaded is in fact in the assembly. Inspection of this type is common in most assembly systems.

Another problem is that associated with parts of the wrong size, shape or orientation finding their way into the feeding or loading system. When there is serious danger of this occurring, steps must be taken to prevent the part from being loaded. This is especially true if the faulty part affects the final function of the assembly. Finding this condition at final function testing can be very expensive. Not finding it until the part fails in the field can be extremely serious.

There are many ways of determining the condition of the parts as they are being fed to the machine. A check which is used includes forcing the part through a profile check and determining the orientation, size and shape. In other systems, gages are installed on feeding tracks to insure that the parts have exactly the correct size. In this way, most reject parts can be prevented from entering the machine feeding system at all.

If on occasion bad parts are put into the assembly and are detected later, a number of ways can be utilized to handle this condition. On non-synchronous machines, a repair loop can be introduced so that the part is diverted from the normal assembly process, manually repaired at an operator's pace on a repair loop, and re-introduced into this system.

The most sophisticated systems require a kind of selective assembly in order to produce satisfactory finished assemblies. Selective assembly is required when the parts making up an assembly must be matched, dimensionally or functionally, to produce a quality final assembly. For instance, a ten thousandths inch washer may fit into a given housing, whereas another housing may require a twelve thousandths inch washer. In this case, both the housing and the washers must be gaged to a high level of accuracy in order to make the desired selection. The following example demonstrates the flexibility of such a selective assembly system.

EXAMPLE OF SELECTIVE ASSEMBLY

The mechanism produced on this selective assembly system is a tightly dimensioned drive gear. The assembly, which can be produced in three different lengths, is made up of alternate gears and spacers to produce a kind of spiral gear drive. (See Figures 1, 2, 3 & 4)

In manual assembly, a Gear #1, referring to Figure 1, is loaded to a shaft and gaged for thickness. A decision is made as to whether a five thousandths or a seven thousandths washer is added in order to maintain the tolerance on the length of the assembly. The correct washer is added and Gear #2 is placed into the assembly. The assembly is again gaged and a decision is made as to whether a five or a seven thousandths washer is necessary and introduced into the system. Finally, Gear #3 is put onto the assembly and gaged and the same decision made as to the thickness of the washer required. This process is repeated time and time again up to 150 times to produce a necessary gear train. This particular part must be held to very close tolerances both for the overall length of the stack and the incremental length of the stack. (Figure 2)

As a manual assembly process, this is a slow, tedious assembly to work on. The necessity of constant gaging and selecting of components produces a high level of tedium to the operator and mistakes are very common. Even with an experienced operator, a complete assembly may take as much as a half an hour to produce. More serious than the time required to produce the assembly is the reject rate which approached 50 percent.

Gilman Engineering & Mfg. Co. proposed and provided a two machine system for producing this assembly. The first machine inspected the gears and loaded them into magazines according to the orientation of the key to the gear teeth (Figure 1). The thickness was determined by a LVDT gage and parts that exceeded the tolerance were automatically removed from the system. The acceptable gears were subsequently indexed to other stations on the machine where they were loaded into the correct magazine for the key gear orientation (Figure 2). These magazines of pre-inspected and selected gears were then available for loading into the final assembly machine.

The sequence of the final assembly machine is shown in Figure 3. An empty shaft was loaded at Station one and subsequently indexed to Station two, where it was gaged with a laser interferometer. A decision was made as to what subsequent component should be loaded, either a five thousandths or a seven thousandths washer, or Gear one, two, or three. The correct washer was loaded to the assembly, then Gear #1 was loaded to the assembly. The part was subsequently indexed around the machine back around to the gage. The decision was made as to whether a five or a seven thousandths washer was required on this pass. The correct washer was then loaded. The part then indexed to Station six where Gear #2 was loaded, indexed again to the gage where the decision was again made as to what washer to load. The part was then indexed to Station seven and Gear #3 was loaded. This process was continued until the complete stack up of parts had been loaded to the shaft. The laser interferometer and the press, which pressed the components to a final height, is shown in Figure 4.

The mini-computer controlling this system, interfaced to the laser interferometer and to all of the load stations, provided a number of features which resulted in a very productive machine. The first thing that the computer did when an empty shaft was loaded at Station one and subsequently indexed in Station two, was to determine if the zero reference for that particular fixture and dial plate location was correct. Each fixture location on the dial plate was given a reference zero dimension and was checked every time a new shaft was put into the system. In this way, the computer could ascertain that the shaft was in fact empty.

Because the computer continually knew what the actual height of the stack should be for a given number of gears and washers, a partial assembly could be introduced into the system at any time. In this way, a partial assembly could be introduced into Station one, indexed to the gage station, and the gage station could be used to determine what was the next correct component to put into the assembly.

The computer monitored both the overall stack height and the incremental height of the last three components going into the system. If a faulty component was introduced into the system and the overall tolerance or the incremental tolerance was not maintained, the computer printed out a message on the Teletype terminal and alerted the operator through the use of a red light. The machine did no further work on that assembly. At his leisure, the operator could return to the machine, ascertain that the part in fixture "X" was not being worked on, check the Teletype message which included the dimensions of the last three gaging operations, and determine how many parts to remove from the assembly. The operator then stopped the machine and removed the offending components, initiated the cycle by telling the computer that this was now a partial assembly, and the assembly was regaged and reinitialized into the system. The operator could remove only a few parts or could remove a large number of parts and the computer would always reinitialize at the gage station and pick up the right sequence to make a satisfactory assembly.

In the same way, when the assembly reached its final complete length, the computer printed out (Fixture "X" assembly complete) and alerted the operator with a horn. If the operator did not respond immediately, the part continued to cycle through the machine with no further work being performed on that assembly.

The operator could allow the machine to cycle until all fixtures had completed assemblies or he could immediately remove the one completed assembly and load an empty shaft and tell the computer it was now looking at an empty assembly.

Other features on the machine through the computer control, include a three consecutive reject count and stop. If a station made three consecutive rejects, then the machine was stopped and the operator alerted. This was a rather complicated feature to add because the computer had to remember that it was supposed to load Gear #1 and that it failed to load Gear #1 in order to produce a consecutive reject count feature.

In the same way, the computer had a "try-again" feature on rejects. If a given fixture required a Gear #2 and did not receive it, the computer would direct the machine to "try again" to load Gear #2. It would try to do this three times and if it failed to get Gear #2 after three tries, it would assume that there was something wrong with the shaft or with the assembly and stop work on the assembly.

When the computer received a signal that the part to be gaged was a new assembly or a partial assembly, the computer advanced the press slowly because it had no way of knowing how long a stack it was looking for. It pressed into the stack, stalled out and then referenced that dimension. The stalling of the cylinder was determined through the laser interferometer. The computer read the stack height and when three consecutive readings were determined over a period of 50 milliseconds, the computer determined that the press had been stalled.

One of the features available with computer control is production data. The computer on this machine printed out "on demand" and at the end of the shift, the production data for that period of time. Included in the report is the time of day and the number of cycles run, the number of completed assemblies, the number of stoppages due to problems within the machine, etc.

This type of selective assembly system would be impossible to implement without computer control and when the computer is introduced in this system, many side benefits are realized.

ADVANTAGES OF SELECTIVE ASSEMBLY

The advantages of selective assembly are as follows:

1. A machine can utilize parts with loose machining tolerance.

2. The machine can utilize a process with looser tolerance.

3. The machine can utilize a high percentage of parts with wide range of tolerance.

4. Rejects are reduced to a minimum by fitting components together to make adequate assemblies.

The looser machine tolerances of the parts is an important factor because the cost of components is directly proportional to the tolerance required in the manufacturing or machining process.

The selective assembly machine will accept parts coming to it with a wider range of tolerances and therefore the cost of the components can be reduced to a minimum.

Final functional inspection is very important in many assembly processes. If for instance, a shaft is bound up or an assembly leaks, or if it vibrates during functional testing, the assembly must be rejected. The range of functional tests that can be run on assemblies is as wide and diverse as the number of assemblies which can be assembled. A functional test is attractive because it usually tells the condition of a large number of components in the assembly without inspecting each of the single components.

CONCLUSION

Inspection is rapidly becoming a requirement in automatic assembly. In order to be cost effective; however, inspection must have some characteristics shown in Figure 5.

The inspection or gaging operation itself is only as effective as the "skill" of the machine. The machine must be fitted with the best sensors and tooling to make the desired measurements. Just as a manual inspector is only as good as his gages, micrometers and tools, a machine is only as good as the tranducers and sensors which it employs. Also, a skilled inspector makes optimum use of his tools. Similarly a machine designed and built by an experienced builder has the "skill" to perform precise inspection operations. Secondly, the inspector must be intelligent enough to make the correct decisions. The requirement for additional machine intelligence has made programmable control very widely used in automatic assembly systems. Programmable controllers, minicomputers, and micro-processors will continue to expand in usage. The buyer of automatic assembly systems should insure that his system is intelligent enough to make the correct decision. Thirdly, all of the inspection skill and intelligence is wasted if the required action cannot be accomplished. If the man lacks the physical dexterity, faculty, or support facility to take the required action, then his ability accomplishes nothing. Similarily, if the assembly machine lacks the tooling accuracy, flexibility, or reliability to satisfactorily perform, then precise inspection skill and superior intelligence is useless.

Inspection in automatic assembly can show dramatic increases in productivity by keeping the machine cycling efficiently, by reducing rejects, by providing a means of repairing rejects if they should occur and through selective assembly can reduce the cost by relaxing otherwise stringent components specifications.

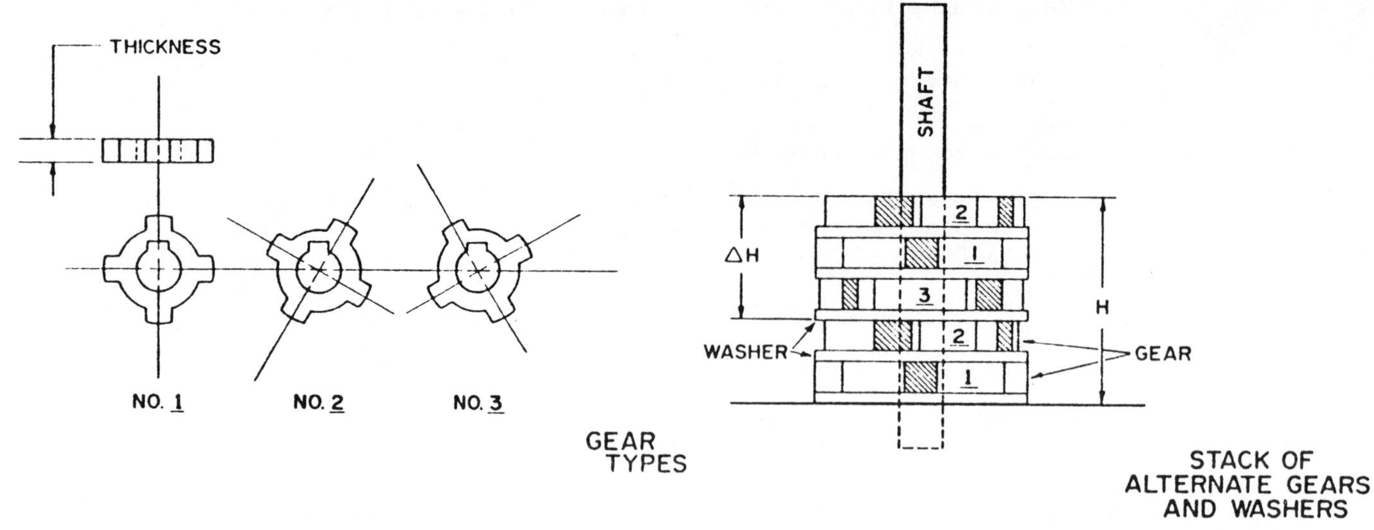

FIG. 1

FIG. 2

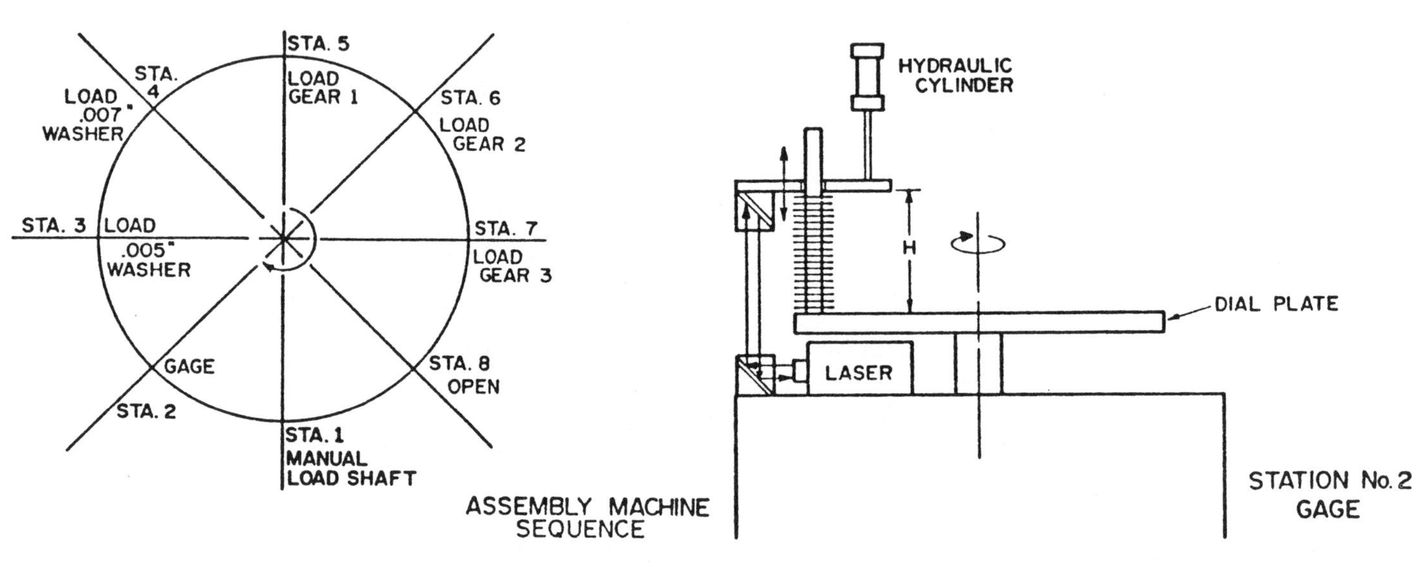

FIG. 3

FIG. 4

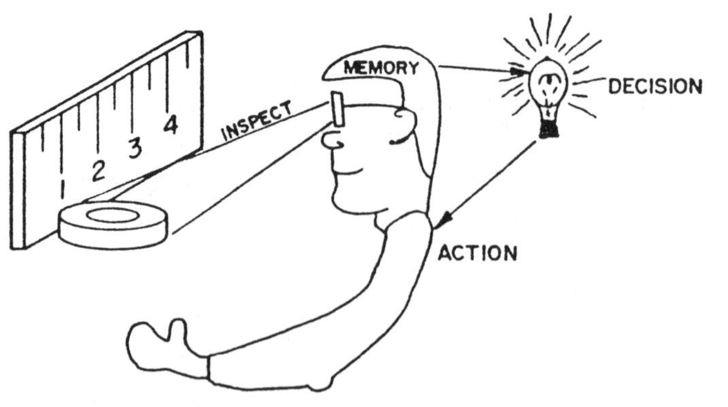

FIG. 5

Presented at the SME 1981 International Tool & Manufacturing Engineering Conference, April 1981

Automatic Gaging in the Automated Factory

by Cecil B. Henn
Ex-Cell-O Corporation

INTRODUCTION

The natural evolution of things is for a job to become easier to accomplish, requiring less time and usually resulting in higher quality. This is a tribute to the intelligence of man. This can be seen in the farm industry where today less than 4% of the population are on farms, yet we are well fed, with a generous surplus to sell to the rest of the world. Technological improvements on the farm freed 96% of the remaining population to do other things more to one's liking.

Manufacturing will follow the same path. Today a smaller percentage of people are involved in manufacturing and yet we are producing more of both quantity and quality than was dreamed of only 50 years ago.

From Henry Ford's first assembly line to today, the challenge goes forth. From Go No-Go plugs to electronic gages with computer control, to as far as we can imagine the changes keep coming. The state-of-the-art is already in place to advance even more rapidly in the next 15 years. Add to this what can happen with new developments in physics and technology and we have the makings of the automated factory as a very likely probability in the near future.

TODAY'S TECHNOLOGY IN TOMORROW'S AUTOMATED FACTORY

Today's advanced engine manufacturing process is a highly automated system and serves as a good example of present day technology that will evolve into tomorrow's completely automated factory. A brief look at existing gages involved in the automatic production of several related engine parts illustrates the point:

PISTON GAGE:

Automatically inspects all pertinent dimensions, specifically, the wrist pin bores and O.D. sizes. The O.D. size is automatically segregated into as many as nine classes of .0004 to .0005 inch (.01-.013 mm) per class. (Number of classes and size of class depends on factors peculiar to the manufacturer). Pistons are ink or metal stamped with size identification for later select fit to the engine cylinder bore. Pistons are also individually weighed to assure balance.

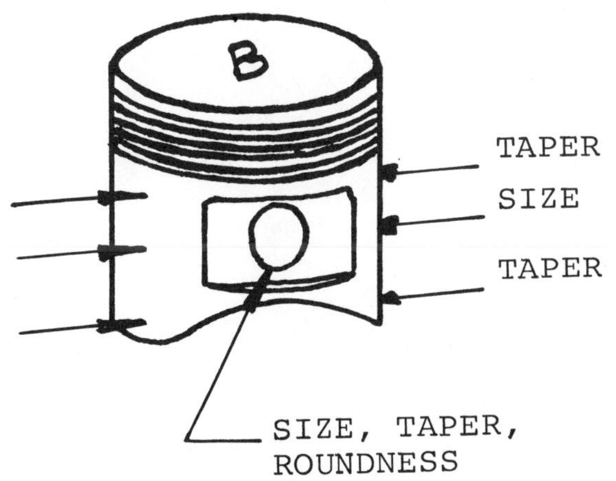

FIG.1. AUTOMATIC PISTON INSPECTION (SORT BY SIZE.)

WRIST PIN GAGE:

Automatically gages and segregates wrist pins into five size classifications for later select fit to the piston. Automatically rejects parts which are out-of-round or tapered in excess of .0001 inch (.0025 mm).

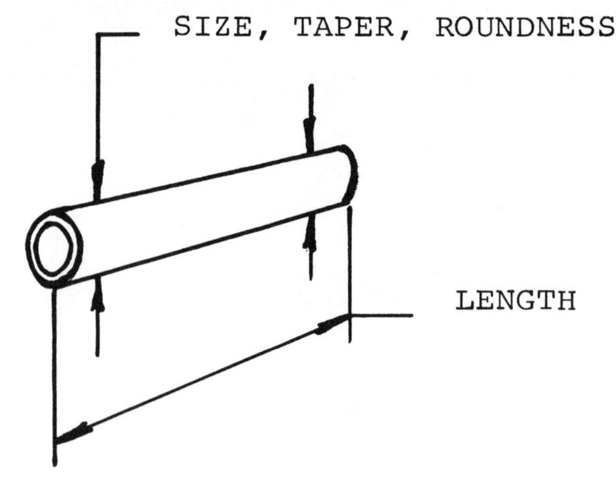

FIG. 2 AUTOMATIC WRIST PIN INSPECTION.
(SORT BY SIZE)

CONNECTING ROD GAGE:

Automatically gages and segregates rejects. Parts are inspected for size, taper, roundness, center distance, and the bend and twist relationship of holes, along with weight at each end for balance.

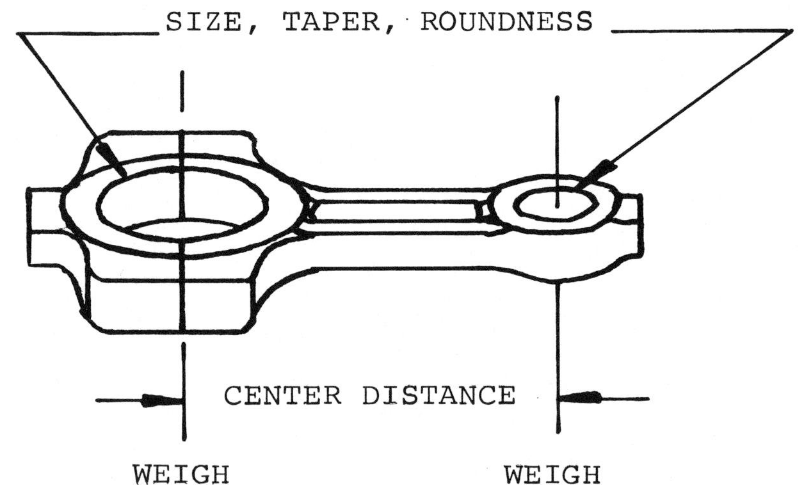

FIG. 3. AUTOMATIC CONNECTING ROD INSPECTION.

PISTON-PIN FITTING MACHINE:

Automatically selectively fits Pistons to Pins by size. Gages Piston pin hole and selects presorted pins to fit both ends for a predetermined clearance fit.

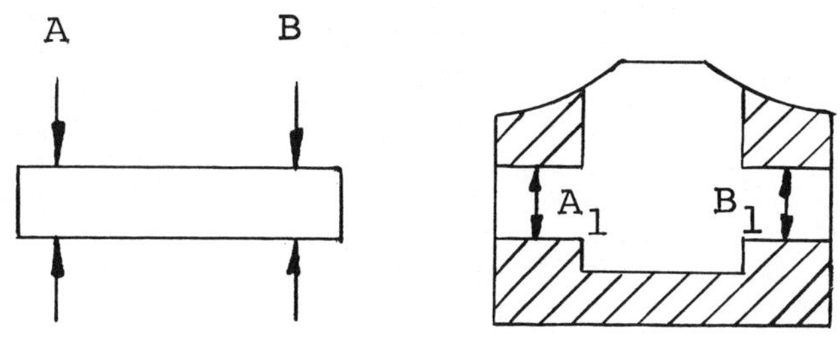

AUTOMATIC
FIG. 4. SELECT FIT PIN TO PISTON. BOTH ENDS OF PIN TO BORES.

PISTON-PIN-CONNECTING ROD ASSEMBLY MACHINE:

Automatically receives connecting rods and pistons (with pin fitted from previous machine), orients each for proper relationship to each other, induction heats the pin hole of the connecting rod to expand the hole, and assembles the connecting rod to the piston. Part is automatically gaged on the same machine for correct dimensional assembly and also "force" tested for tightness of fit of the connecting rod to the wrist pin. The machine automatically segregates good from reject assemblies.

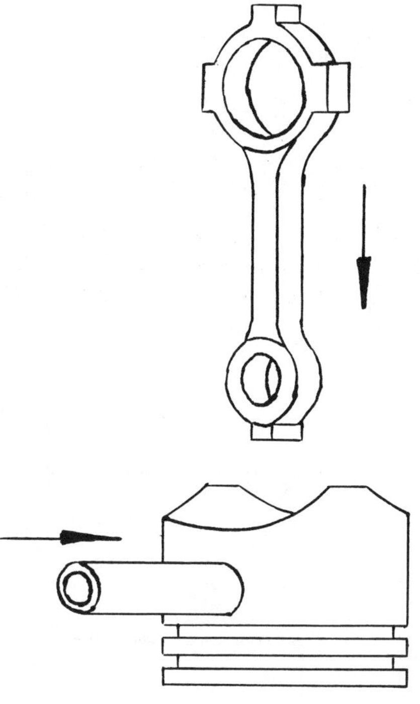

FIG. 5. AUTOMATIC ASSEMBLY OF PISTON, PIN AND ROD.

CYLINDER BORE GAGE:

Automatically gages cylinder bores for size, taper, and roundness, and classifies the bores into classes corresponding to the piston classifications. Bores are marked by ink or metal stamped for later match fitting to the piston.

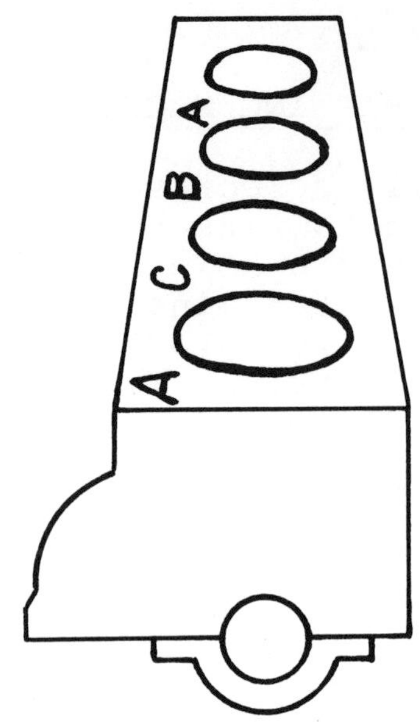

FIG. 6 CYLINDER BORE CLASSIFYING

Advanced features built into the six machines above and ready for tomorrow's factory are controls, sensors, and mechanisms which are necessary for the operation of unattended machines:

PROGRAMMABLE CONTROLLERS: Flexible and field proven, the industrial controllers are ready for complete automation. In addition to acting as a machine controller, they can communicate to one another and to a "Supervisor Controller", along with providing data storage and diagnostics.

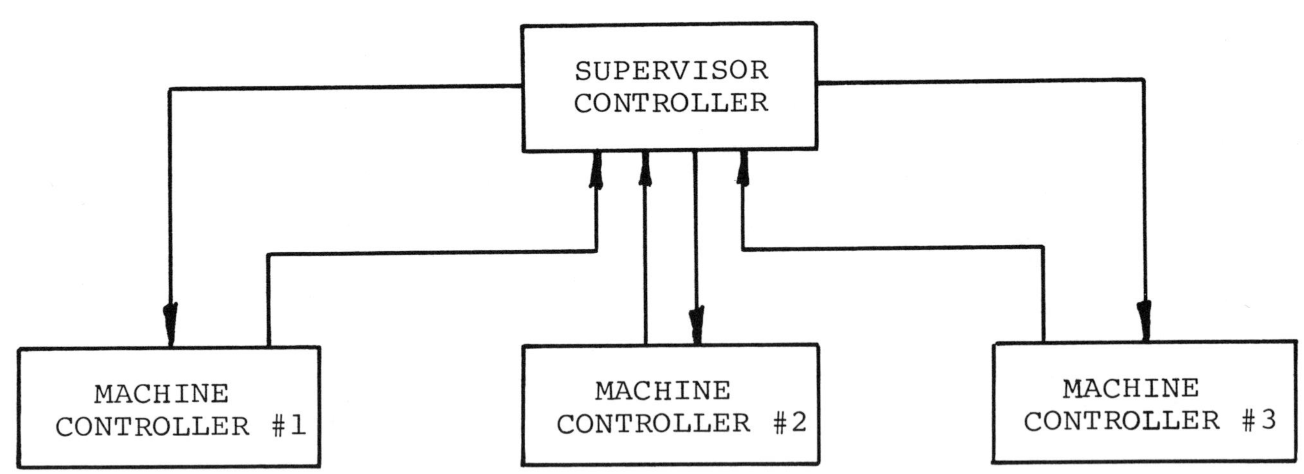

FIG. 7. MACHINE COMMUNICATIONS

AUTOMATIC CALIBRATION: A necessary feature for the automatic factory, this feature monitors the gage's calibration and automatically corrects for deviations. Built in masters are used for reference standards. The Gaging Arbor retracts into the Masters after inspecting the part.

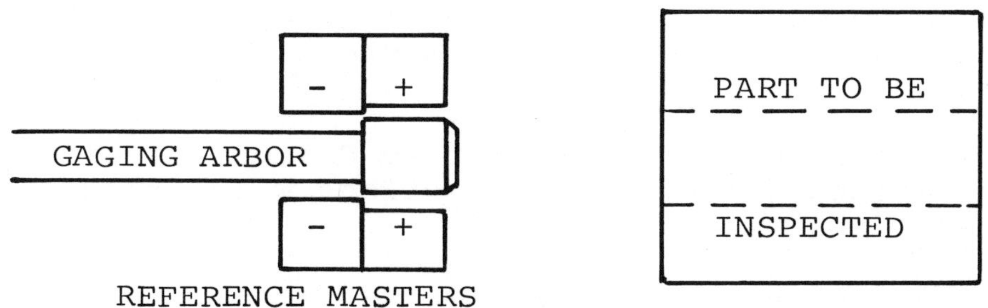

FIG. 8 AUTOMATIC CALIBRATION.

ENCODERS: Used to determine the position of machine elements such as the machine's cam shaft rotation, ball screw drives, turret indexers, etc. Encoders are the absolute, closed loop type, which do not lose position due to a loss of power to the encoder.

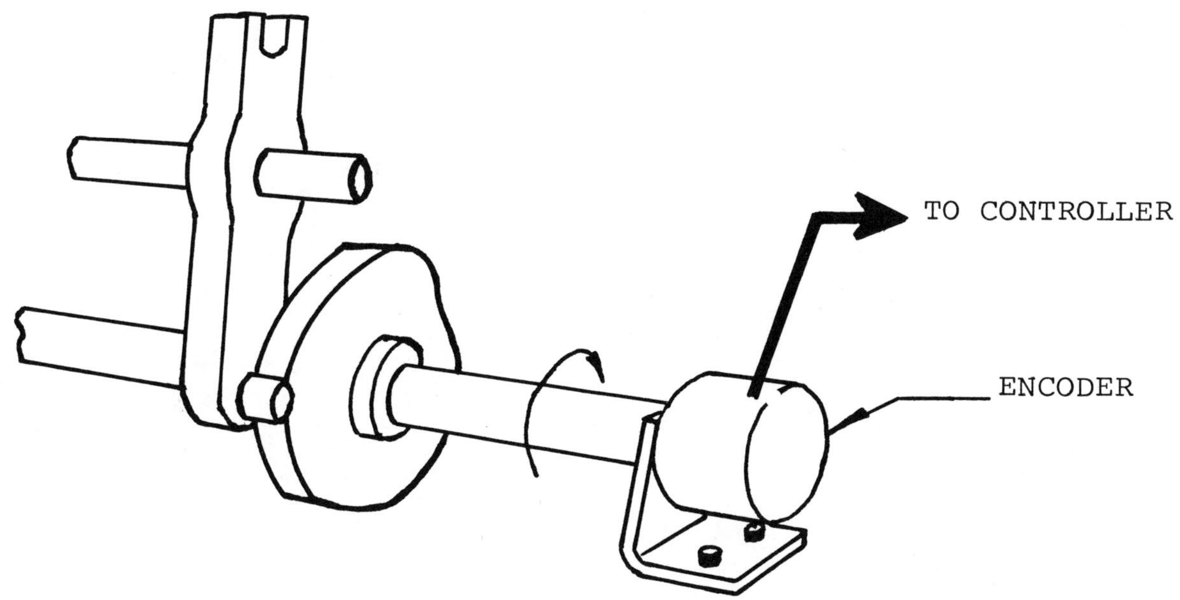

FIG. 9 Encoder used to determine angular position of Cam Shaft.

ELECTRONIC MOTOR DRIVES: Electronically controlled electric motors with absolute, closed loop feed back assures precision radial location of the motor output shaft and is able to move both clockwise and counterclockwise, reducing time required to locate turrets, etc.

SENSORS: Infra-red heat sensors monitor con rod temperature; Infra-red part presence detectors sense Connecting Rod and Piston locations and fiber optics are used in areas where limit switch mounting is difficult.

CAM DRIVE: Positive mechanical interlocked drives for high speed synchronous operation of many complex motions which require minimum maintenance and downtime.

JAM PROTECTION: Redundant jam protection for maximum protection against broken machine parts. A limit switch detects a jam and signals an emergency stop while a mechanical device "breaks away" without damage to the machine and resets itself on the return stroke if the electrical system fails.

THE AUTOMATED FACTORY

Timing is very important to the introduction of a new idea. It's just as sure a recipe for failure to have the right idea 50 years too soon as 5 years too late. For a new idea to become a reality, a series of events usually precede it and are woven into the fabric of the emerging idea. The wheel and engine were invented before the automobile.

Many things need to precede the automatic factory. Most of the required developments now exist in some form. The question is: "Is the idea of the automatic factory an idea whose time has come?" A completely automated machining line was built on the West Coast in 1958 utilizing flexible machining concepts and using computer control. It worked quite well, but it was ahead of its time, was scrapped out and discarded at considerable lost to the innovating company. It was not cost effective.

Today, the "wheel" that is needed to precede the automatic factory is well developed. It is the computer, well established throughout industry and business with thousands of people trained in its use. It is common place.

Robotics are rapidly reaching the same status, and of course depend upon the well developed microprocessor. The robot industry forecasts 50,000 robots by 1990. General Motors plans to use 14,000 of them.

The automatic factory of the future will probably be designed for manufacturing efficiency and the efficient use of energy, and not so much for human comfort. Less will be spent on rest rooms, lunch rooms, meeting rooms, parking lots, etc., due to having fewer people involved in its operation. The plant will be underground for lower energy costs and bomb proof. Heating and cooling is by solar systems. Without people, the temperature can be down to 40° F. in the winter and up to 110° F. in summer. Photovoltaic power to run lighting and computers will become more economical. (Now \$6/watt, the Energy Department's goal is \$0.70/watt by 1986.) Lighting will be dim by today's standards. Machine optics may see better in the dark and are only interested in vision into dedicated areas. Mobile robots can be guided by magnetic coordinate systems in the floor.

Production machines will be designed for quick replacement. Automatic anchoring to the floor through "T" slots and locating pins, battery packs for short term running and intelligence maintenance, no hard wiring interconnections, wireways, and conduits between machines. Long term power is received by microwave from the central power supply beamed to each unit. Intelligence between machine computers is by short range radio transmission or light beam for line of sight applications using multiplexing at the superfast scan rates of microprocessors. See Fig. 10.

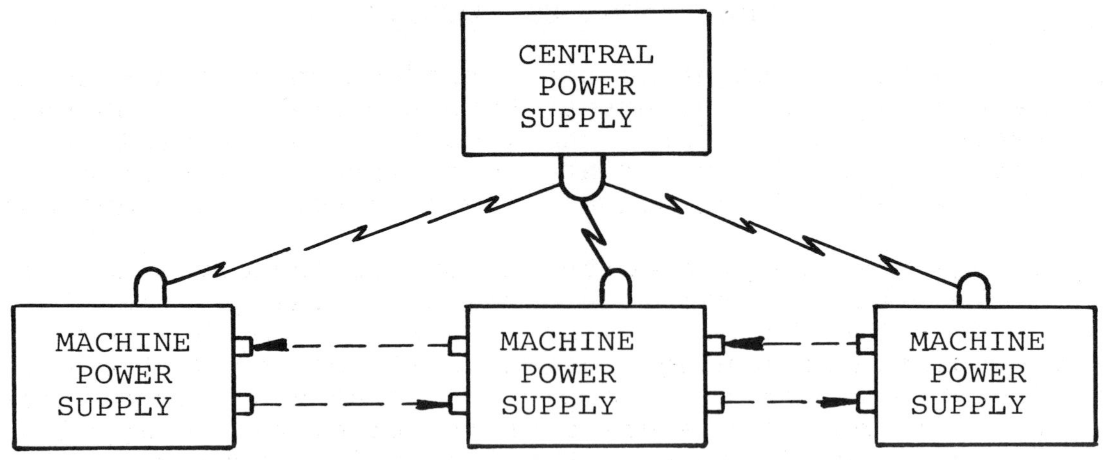

FIG. 10 Power Transmitted to Individual Machines by Micro-wave
Intelligence Transmitted between Machines by Multiplexing Light Beams

Gage heads, resolvers, encoders, limit switches and other sensing devices can have built in batteries charged by induction from an electrically charged atmosphere. This is particularly advantageous for flexible machining lines where gage heads need to be changed as often as tooling. Additionally, the lack of wire connections to sensors facilitates ease of automatic replacement by robots or tool changers.

THE END PRODUCT: We can expect to see the day of the throwaway appliance, perhaps even engines, as we now have with wrist watches, small appliances, etc., because the cost to repair exceeds the cost of replacement. As the throwaway concept gains momentum it will create an increasing demand for longer product life. Satisfing the demand for longer product life increases the demand for tighter tolerances, better wearing surfaces, and better material quality. If these demands are to become a reality without exorbitant costs it will be through a stabilization of products and the increased use of automatic manufacturing techniques.

INSPECTION REQUIRED: The requirement for a better end product will lead to intensive automatic inspection from the beginning. The automatic factory runs best when all forms of scrap are recognized in advance and discarded. (Garbage in, garbage out, GIGO.)

Parts will be packaged from suppliers for automatic handling and testing; mounted on pallets, belts, or suitable devices for automatic parts handling. Everything is designed for automatic parts handling by bulk feeders, conveyors, stack cranes, robots, etc.

Incoming parts are recognized by various means including numerical character recognition, bar codes, or physical size and shape recognition by optical sensors. After the part is recognized, the computer sends it to the appropriate testing machine and the part is accepted or rejected. If rejected, the receiving computer debits the account, reorders replacement, and pays the invoice for the amount accepted.

Because of the requirement for very high quality parts in the automatic system and the high cost of downtime, 100% inspection will become the rule on all critical items. (No more Russian Roulette with sampling inspection.)

100% automatic inspection will be required for Non Destructive Testing (NDT) of material quality, in-process gaging with automatic size control feedback, and increased amounts of selective fitting at assembly.

Much has been said about eliminating the need for final inspection machines and for selective fitting of mating parts through the use of better size control of the manufacturing process. It is this author's opinion that such thinking is not realistic for many applications. Machines wear, materials vary, cutting tools vary, temperature and humidity vary, and parts change shape when released from holding devices. And most of all, the desire to get the theoretical dimension we wanted, not some approximation of it. Automotive people who were happy with holding .001 inch tolerances 20 years ago today complain because they can't hold .0001 inch. Crankshafts and camshafts today are built to "master" tolerances on a production basis and yet they yearn for better control.

GAGING TOMORROW'S ENGINE: A desireable engine is well balanced, smooth, fuel efficient, and a high performer. The moving parts must be all the same weight to provide balance between pistons and the compression rates must be the same in all cylinders. See Fig. 11. This involves the weight matching or control of pistons, pins and rods along with the crankshaft and the dimensional control of the crankshaft throws, connecting rod length, pin to dome height of the piston, cylinder bore length as established by the head face location from the crankshaft bore, length of flange of the steel liners (if used), engine head cavity dimensions and valve seat dimensions. This, along with previously discussed gaging and classifying of Pistons, pins, rods and cylinders as is presently done, and the desire for ever better engines, leads one to believe the requirement for final inspection of piece parts and select fitting of mating parts will be a requirement in tomorrow's factory. The cost of select fit is less than the cost of holding tight tolerances in many cases.

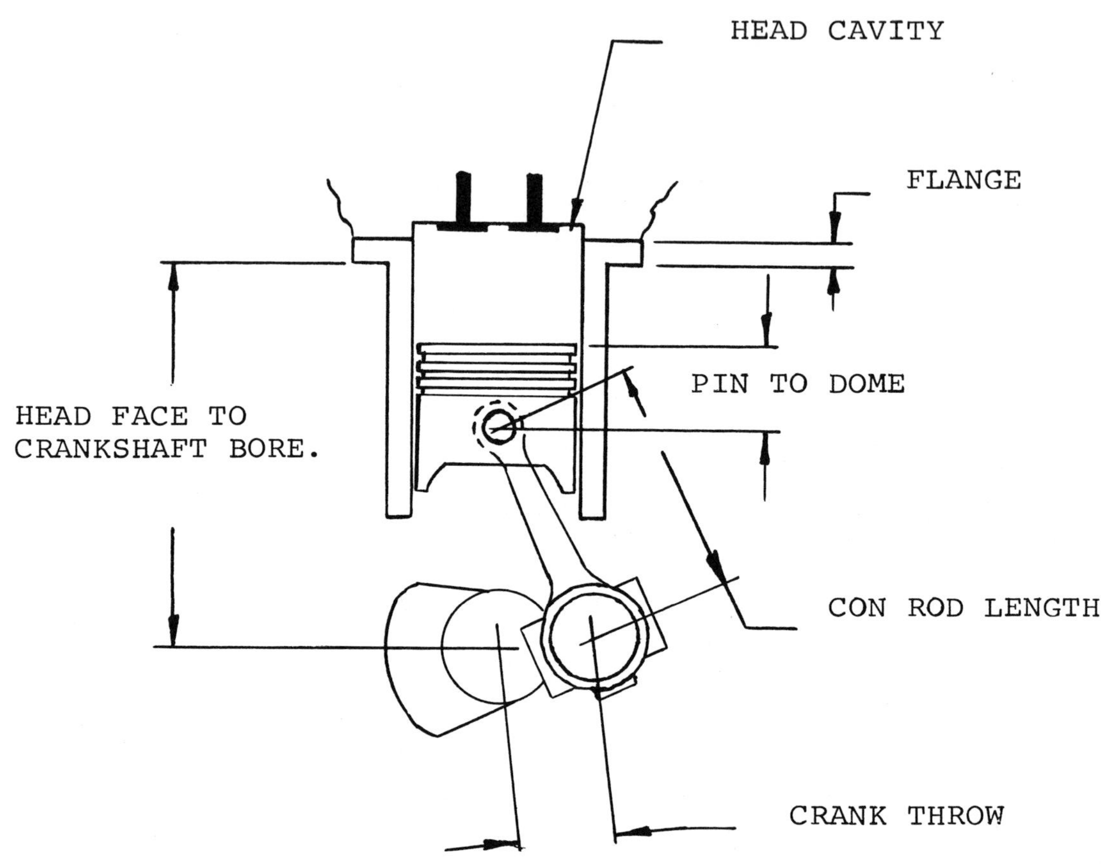

FIG. 11 DIMENSIONS WHICH AFFECT COMPRESSION RATIO.

SELECT FIT ASSEMBLY The desire for optimum fit demands selective fitting of mating parts. Tomorrow's factory will utilize many of today's techniques as discussed above where gaging and assembly are accomplished on the same machine as in the case of fitting wrist pins to pistons. In this case the random distribution of the process of both parts must match each other or an inventory problem develops.

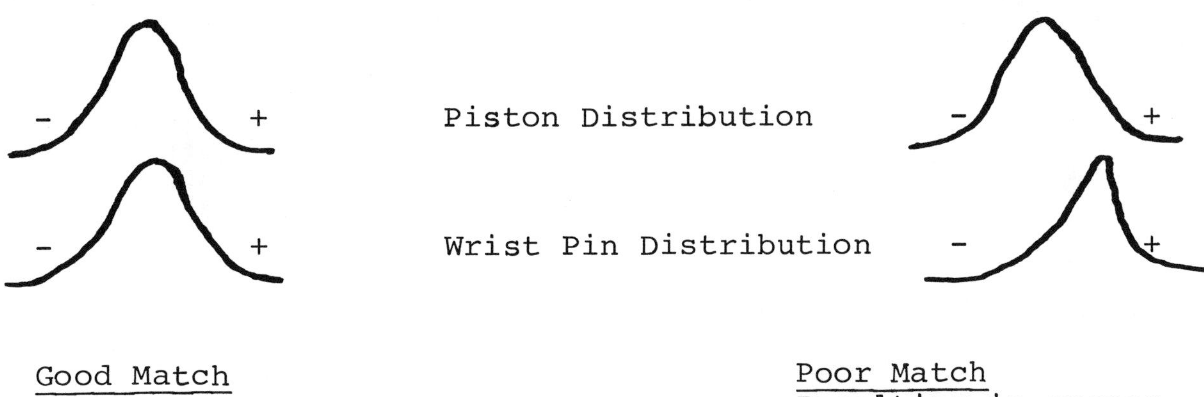

Good Match

Poor Match
Resulting in excess Small Pistons and Excess Large Pins

FIG. 12 MATCHING DISTRIBUTIONS OF MATING PARTS.

Another way is to size a hole while simultaneously gaging its mating part. The part being gaged controls the tool which is sizing the hole. This is currently being done in some companies producing valve bodies and spools; presently known as "match sizing".

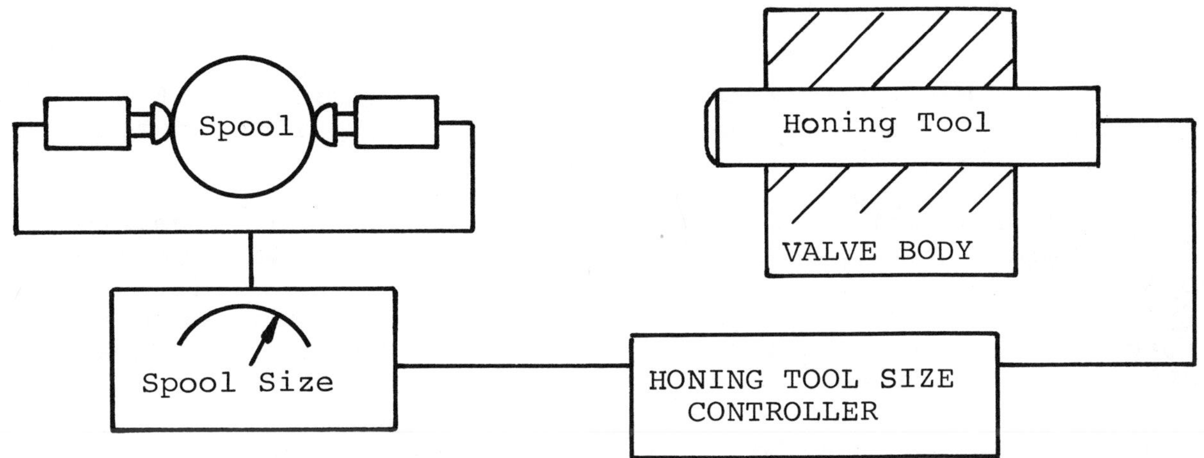

FIG. 13 Automatic Match Sizing. Spool Size Controls Stock Removal of Hole in Valve Body.

Perhaps a better way for tomorrow's factory would take advantage of the computer's growing power and flexibility, its data storage capacity, and increased used of sensors. Using the computer to its fullest, the machining process could be operated at its maximum speed, accepting its random misbehavior and size variations. Each part could then be gaged and serialized, with the gaged dimensions stored in the computer with its serial number. The serial number would be marked on the part in a manner suitable for machine reading, using character recognition as used on bank checks or bar code as used on grocery store items.

1349706

FIG. 14 AUTOMATIC SERIAL NUMBER READING.

As an example, an engine block arrives at final assembly, its serial number is read, the computer calls for the appropriate piston assemblies to fit the random sized bores whose sizes were stored in the computer and also dispenses the proper size main bearing inserts as called for by the mating crankshaft which was also serialized and stored in memory.

The computer would also have inventory responsibility. As inventory imbalances occur the supervisor controller directs the size control units to generate sizes to restore balance.

THE GAGING MACHINE: The "watchdog" to the automatic factory along with the microprocessor is the automatic gaging machine. It checks the prior machine's output, accepts, classifies or rejects its output and feeds back to the machine for tool correction and adjustment or stops the process if it can not be brought under control.

The gage, of course, must be in proper calibration at all times. This requires automatic calibration with built in reference masters in the gage to be certain of its accuracy.

Another watchdog feature of the gage will be to calculate the standard deviation (, Sigma) of the machining process on a continuous basis and to call for repair when the standard deviation goes beyond established limits. The gage's own performance is also monitored in the same way using its built in masters to check its own standard deviation.

When repair or replacement is indicated, the gage signals the supervisor controller which dispatches a redundant equipment to replace the ailing device. The crippled equipment is brought out of the production line to a repair area by a free running robot operating on a system of magnetic coordinate references in the floor, similar to VOLVO's "T" carriers presently in use in Sweden.

IN CLOSING: Most of the needs for an automatic factory are now field proven. Some of the items discussed above are known but as yet unproven in industrial applications. We stand today with respect to the automatic factory as we did in 1959 with respect to the space program which put a man on the moon in 1969. The automatic factory can become a reality soon if there are those who have the desire and the resources to do it, along with the vision to see a product that will be stable in the marketplace long enough to justify the cost.

More than likely, the completely automated factory will evolve over a longer time period than was required by the space program. The automatic factory must be justified on the basis of cost whereas the space program had no such constraints. The automated factory is a "Coming-to-be", guided, as it were, by Adam Smith's concept of the invisible hand.

Presented at the SME QualTest-I Conference, October 1982

Automatic Inspection Machines

By Alberto Imarisio
Marposs Gauges Corporation

INTRODUCTION

In high production manufacturing facilities, the use of automatic gauging machines to perform final inspection operations are an economical means to qualify parts for assembly or shipment. The variety of gauging machine designs are as varied as the vast array of parts produced. However, there are enough similarities between all gauging machines that a simple definition can be attempted.

A gauging machine automatically transports parts to one or more measuring stations so that they can be inspected for quality. This inspection is performed during the automatic cycle of the machine and does not involve direct human intervention to judge quality. Movement of the parts through the machine is accomplished by electro-mechanical, hydraulic or pneumatic mechanisms. The machine can be either manually or automatically loaded. Several measurements are usually taken, by this type of final inspection equipment, due to the great efficiency of their operation. In cases where full automation is incorporated into the cycle of the machine, some way of identifying good and bad parts must be accomplished. This may mean a marking station and/or the use of a reject chute to collect scrap parts.

In the past, the main purpose of an automatic gauging machine was simply to final inspect parts at the end of a manufacturing process. The present industry trend is to use them also for automatic inter-operational inspection for the purpose of realizing a direct process control. Inter-operation machines are normally simpler than final inspection machines. Fewer measurements are required because they are related to only one or two metal removal operations. In addition, most measurements are taken statically which simplifies the inspection process. By monitoring the quality output of a machine immediately after the operation is completed, direct process control can be achieved. If part size is approaching an out-of-tolerance limit, proper compensation feedback can be given. If size is out of tolerance, the machine can be stopped before any more scrap is produced. Electronic gauging machines are the most suitable means to achieve direct process control and overall quality control. This is because they can provide measurement output information which can be easily interfaced with statistical analyzers, printers and computers.

The individual components of either a final inspection or inter-operation gauging machine deserve a closer examination to appreciate their basic function and how they work together to make a complete system.

1. Part Transportation: Each different part requires a detailed study and engineering to assure efficient and safe movement through the machine.

2. Measuring Station: Depending upon the complexity of the gauging task and the physical characteristics of the part, one or more measuring stations are required to inspect the part.

3. Gauging Amplifier: Information received from the measuring station(s) is processed by the gauging amplifier and compared to a desired and predetermined standard.

4. Logic System: This system controls all of the machine movements while also keeping track of good and bad part conditions that were determined by the gauging amplifier.

5. Marking Station: Some way of identifying good and/or bad parts must be accomplished.

6. Reject Station: Bad parts can be moved out of the machine into a scrap chute to avoid any possibility of using them in a final assembly.

7. Segregation of Good Parts: In some instances there may be different classifications of good parts. Segregating by size can be an effective way in which to improve and facilitate assembly.

The development of microprocessor-based circuitry into the design of today's automatic gauging machine amplifiers has opened a whole new world of advantages to the user:

- Increased measuring capacity while reducing the physical size of the equipment.

- Precise and clearer measurement information is available to the operator because it is displayed as a written message on a screen.

- Easier resetting operations allow for faster and more economical checking procedures.

- Automatic or semi-automatic mastering of the system is easily performed.

- Self-diagnostic systems allow for easy and fast identification of faulty circuits reducing downtime and improving maintenance efficiency.

- Retooling is less expensive and faster because in most cases it will not require hardware changes, but only software reprogramming.

CASE STUDY

Considering all of the above factors, the realm of automatic gauging machines and their applications is indeed a broad area. The focus of this paper will examine one specific automatic gauging machine used as a final inspection system in an automotive plant. It represents the latest state of the art in automatic gauging machines.

Figure 1

This machine (Figure 1) performs the final inspection on crankshafts used in an automobile engine. A total of 85 measurements are taken simultaneously during a 16 second gauging cycle. These measurements are checked by two gauging stations. The first station gauges the five main bearings, post end, flange, pilot hole and thrust faces for a total of 57 measurements. The second station inspects each of the pin bearings for the remaining 28 measurements. All measurements are taken dynamically while the crankshaft is turned approximately two revolutions. Figure 2 shows a schematic diagram of the position of each gauging point and a list of measurements taken at each gauging station.

GAUGE STATION #1

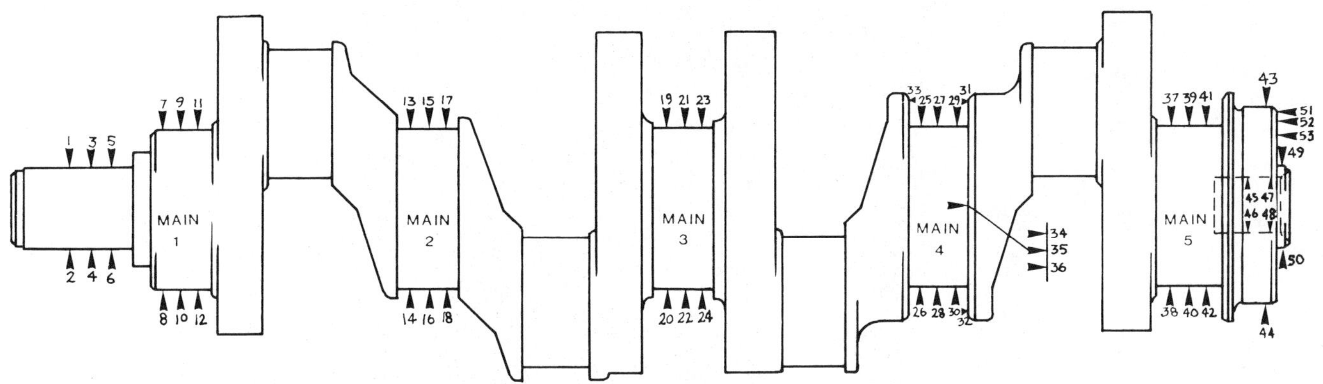

Measurements Performed:

Diameter, Maximum Diameter, Taper, Ovality and Straightness of Main Bearings 1 - 5; Post End, Gear Fit, Oil Seal, Hub Diameters; Bushing Bore Diameter, Taper and Concentricity to Main Bearings 1 & 5; Thrust Wall Width and Concentricity to Main Bearings 1 & 5; Total Indicated Runout of Mains 2, 3 & 4 with respect to Main Bearings 1 & 5, and overall T.I.R.

GAUGE STATION #2

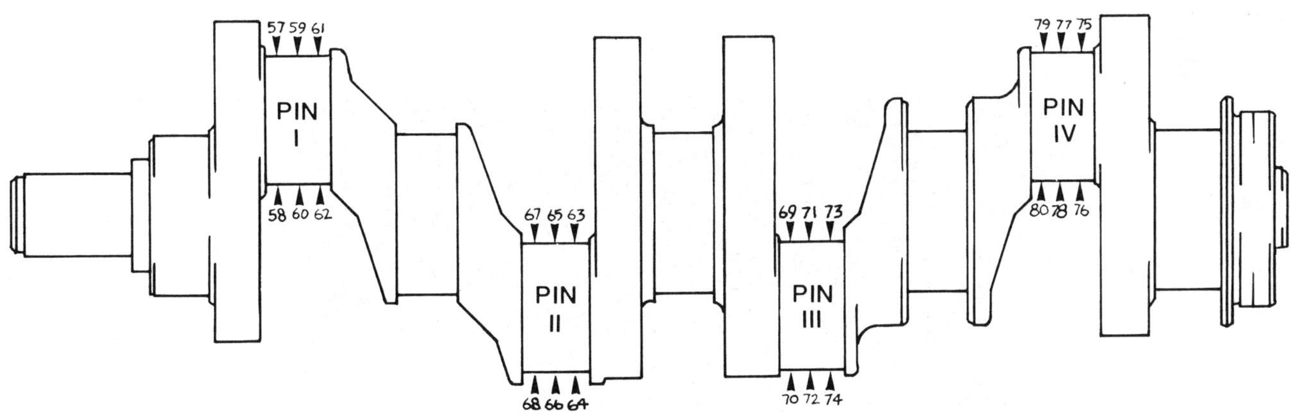

Measurements Performed:

Diameter, Maximum Diameter, Taper, Ovality and Straightness of Pin Bearings I - IV

Figure 2

Figure 3

Each part checked is loaded by the customer's automation system onto the gauging machine's lift and carry mechanism (Figure 3). After two idle stations, the part reaches the first gauging station. Here two centers lift the crankshaft from the nests of the lift and carry bars. A driver mechanism engages the end of the post end diameter while the frame supporting the gauging fixture is lowered onto the part. When each of the 53 contact gauging points are in proper position, the part is rotated and the gauging cycle begins (Figure 4). All measurement data from each of the contact points are sent to the microprocessor-based amplifier where they are stored in memory.

Figure 4

At the end of the second revolution the gauging fixture is raised, the driver disengaged and the centers opened so that the part rests again on the lift and carry nests.

Figure 5

After the crankshaft has been carried through the next five idle stations by the lift and carry mechanisms, it reaches the second gauging station (Figure 5). Although the gauging cycle is very similar to the one just described for the first station, the gauging structure is completely different because each gauging assembly related to an individual pin bearing is independent from the other three. This is due to the cranking movement necessary during the rotation of the part. Again the part is rotated for approximately two revolutions, and measurement data is sent to the gauging amplifier.

Figure 6

Figure 7

After an idle station, the following two are marking stations. Here selective number of colored dots are stamped onto the part to identify the size class of each main bearing (Figure 6), or each pin bearing (Figure 7). Only good parts are marked.

Following the marking stations, parts judged to be out of tolerance that must be scrapped or reworked, are taken out of the main stream of the machine by a reject station (Figure 8).

Figure 8

In the reject station crankshafts are lifted and turned 90° by a lift-turntable mechanism which unloads them onto a reject chute magazine. Good parts are allowed to continue to the last station of the gauging machine where the customer's automation unloads them.

The brain of the measuring system is the gauging amplifier which is located in a separate cabinet on top of the pushbutton console (Figure 1). All signals coming from the 77 contact gauging points are fed into a multiple analog amplifier where all of the dynamic measurement computations like T.I.R. or minimum and maximum diameters are performed. The resulting information is then fed into a microprocessor-based, amplifier-controller. Here the results of all 85 measurements are compared with each related set of tolerance limits. This measurement data is also memorized in order to control the machine's marking and reject stations so that they can be activated at the proper time and sequence. Due to the transfer sequence, each part is gauged and marked/rejected at different times within the machine's cycle. In fact, the amplifier-controller must keep track of several parts which move simultaneously through the machine.

The microprocessor amplifier is equipped with interface circuitry so that it can be connected with a Marposs Statistical Analyzer and an industrial printer. In this way, complete documentation of part quality and statistical evaluation of production can be accomplished. These devices are not permanently connected to the automatic gauging machine. They are conveniently mobile so that they can also be utilized and plugged into other inspection machines within the plant.

DESIGN CONSIDERATIONS

This condensed description of an automatic crankshaft gauging machine shows that complex multiple measurements can be confined into a relatively small space with many advantages. The nature of those measurements, the close tolerance required and the conditions that this machine must work in (a normal manufacturing environment) require very special engineering and design techniques. A close-up examination of some of these special features will provide a clearer understanding of how it was designed to work in a difficult production environment while also limiting the inevitable possibility of human error.

The structure of the machine was engineered to completely separate the part transportation system from the gauging structures. In fact, the base of the hydraulically-operated, lift-and-carry system has floor anchors which are completely independent from those supporting the gauging structures. In addition, special antivibration mounts are used on each of the supporting legs of the gauging structures. Designing the machine in this way helps to insulate the gauging fixture from any possible vibrations from the rest of the machine and the surrounding environment. This will help to ensure that any vibration which may influence gauging accuracy is eliminated.

To avoid any possibility of the machine mechanically going out of sequence, special design considerations were incorporated. The sequence of part lifting, clamping between centers, the gauge fixtures moving down and up, as well as the rotation operation are all linked by a single control mechanism. All movements of the gauge structures in each gauging station are cam operated and powered by a single electric motor reducer. Using this engineering technique avoids any out-of-sequence mishaps.

The centers used to hold the part during gauging rotation are used only for mechanical positioning. They are not used by the gauging system as reference axis. The part axis in fact is electronically computed using the gauging points on main bearing #1 and #5. In this way, the accuracy of the measurement results and gauge repeatability are not affected by small particles present on the part holding centers or from any mechanical wear. The "floating" capability of the measuring points will more than compensate for these error conditions.

A full complement of protective guards are used on all exposed areas of the machine. Not only do they protect the safety of the operator and bystanders, but they also ensure the integrity of the gauging operation. The protective guards help to ensure that a part cannot be loaded into the middle of the machine during the automatic cycle. In fact, if a guard is opened during the normal operation of the machine, it stops running and the "manual" mode is automatically triggered. The manual mode can also be called for by the operator when the mastering procedure of the gauge stations are performed. Safety of hands and arms during this procedure are assured and meet all government regulations.

Normally, due to the stability and linearity of transducers and circuitry used by Marposs, only a single mean size master would be required. In this case, the customer desired to use both a minimum and maximum master, so that he could check the high and low tolerances. The minimum master is used for the automatic mastering operation. In this situation a failsafe method has to be employed so that the machine will not master itself to the wrong master or to a part. After the minimum master is in gauging position and the machine put into the mastering mode, all dimensions are automatically zeroed. If by mistake the operator calls for automatic mastering after he placed the maximum master into the machine, the amplifier screen will display a warning that the wrong master is in place and the mastering command will be refused. Similarly, if a crankshaft is in the gauging station and the mastering command is called for, it too will be refused. The machine has special sensing devices that communicate to it exactly what kind of master or part it is inspecting. Using this failsafe method during mastering, the customer is assured that the gauging machine is properly set up for his final inspection operation.

Another typical human error the machine is equipped to recognize is if the part is mounted backwards or improperly seated on the transfer nests. This is true both at the loading station and the gauging stations. One can imagine the physical damage that would occur if the measuring fixture were to try to gauge a part in the wrong orientation. Not only would it be expensive to repair, but the time lost on the machine would not be acceptable.

A frequent occurrence in a production environment is an unground or oversized crankshaft coming down the line with finished parts. Again, to avoid damage to the gauging fixture a number of safety devices are used to monitor these kinds of conditions. All of the mechanical or gauging devices "entering" the part are springloaded into their proper position instead of being positively pushed. The main gauge frame, which supports most of the gauging points, has a template-like guard which is used for oversized part protection. If excessive force is needed to reach the correct gauging position, it will automatically recoil away from the part before any damage can occur. A safety switch is immediately triggered which shuts the machine down so that the obstruction can be manually removed.

If for any reason too many rejected parts are fed into the reject chute magazine, a full magazine switch will stop the machine so that the operator can check the condition causing the rejects.

OPERATIONAL CONSIDERATIONS

Although this machine is truly an automatic gauging device intended to inspect approximately 225 parts per hour, the operator may want to visually monitor the measurements related to a particular part. In order to do this he must first switch the machine into the manual cycle mode. He can now scan the various measurements of the part, on the amplifiers display screen. This can be accomplished even if the gauge fixture has just been retracted. The memory of the amplifier retains the information automatically.

Should a more comprehensive study be needed of several dimensions for many consecutive parts, a different and much more efficient method is available. A Marposs Statistical Analyzer can be plugged into the machine. This device will automatically gather the necessary data while the machine proceeds in its normal production mode. The statistical analyzer not only gathers the measurement data faster than other methods, but it is much more reliable and accurate. This is achieved by the efficient and direct communication between the two microprocessor-based systems.

The versatility of the microprocessor-based gauging amplifier-controller becomes more evident when a system malfunction needs investigation. A built-in diagnostic self-test mode can be used to pinpoint any electronic failures. Simply, should any portion of the circuit be malfunctioning, the amplifier screen will display which card needs to be replaced. Repairs can be done promptly and downtime for the machine will be minimal.

SUMMARY

Recalling the content of this paper we can say that automatic gauging machines can be divided into two main categories; final inspection and inter-operational. All final inspection gauging machines automatically transport the parts to be inspected to one or more measuring stations. A number of individual components are working together to make a complete system. Today's state of the art takes full advantage of modern microprocessor-based circuitry.

In our Case Study we examined a specific automatic gauging machine used for the final inspection of an automobile crankshaft.

- A total of 85 measurements were taken at a production rate of 225 parts per hour.

- Two gauging stations were used and the part was dynamically gauged during rotation.

- Marking was provided as a classification method.

- A reject station segregates reworkable or bad parts from the good crankshafts.

- It is possible to connect a statistical analyzer and printer to the machine.

- Special design characteristics are necessary to allow a gauging machine to perform at high standards in a manufacturing environment.

- Safety of operators and system integrity must be considered.

- Easy maintenance procedures are accomplished because of the use of modern microprocessor-based gauging amplifiers.

With careful attention to the design and function of an automatic gauging machine, it is possible to bring the accuracy of the quality control lab right into the manufacturing environment. However to do this, a close cooperation among manufacturing and quality control departments within the plant and the designer of modern gauging machines must be maintained. Only this combination will be able to succeed and bring quality control where it is needed most: on the plant floor where parts are manufactured.

Reprinted from *MACHINE DESIGN*, May 6, 1982

THE TECHNOLOGY OF AUTOMATIC GAGING

HARRY M. ABRAHAM, Jr.
Vice President, Engineering
AA Gage Inc.
Detroit, Mich.

How gaging probes are actuated

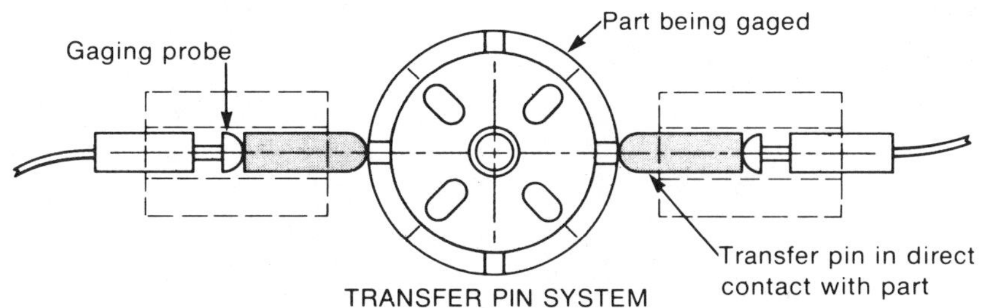

TRANSFER PIN SYSTEM

Whenever feasible, gaging probes are designed to be actuated by an auxiliary probe, or transfer pin, that functions in tandem with the gaging probe. The tip of the transfer pin is machined to the exact shape of the gaging probe.

This technique greatly increases the service life of the relatively expensive probes by eliminating direct contact with the parts being checked. The protection is especially important in automatic-gaging systems where advancing parts would be sliding against or ramming into the probes. The auxiliary tips also help ensure precise linear movement of the probe plunger with respect to the part being gaged.

When the part configuration or the means of conveyance limits the space for probe placement, the probes can be made retractable. This method holds the probe heads back until the part is in gaging position, preventing possible wear or damage to the probes.

Another type of probe actuation uses rocker-arm components to transfer movement from the datum through one or more pivot points to the gaging probe. The ratio of movement from the surface being gaged to the probe plunger can be varied through the mechanical linkage, amplifying the measured tolerance for greater accuracy.

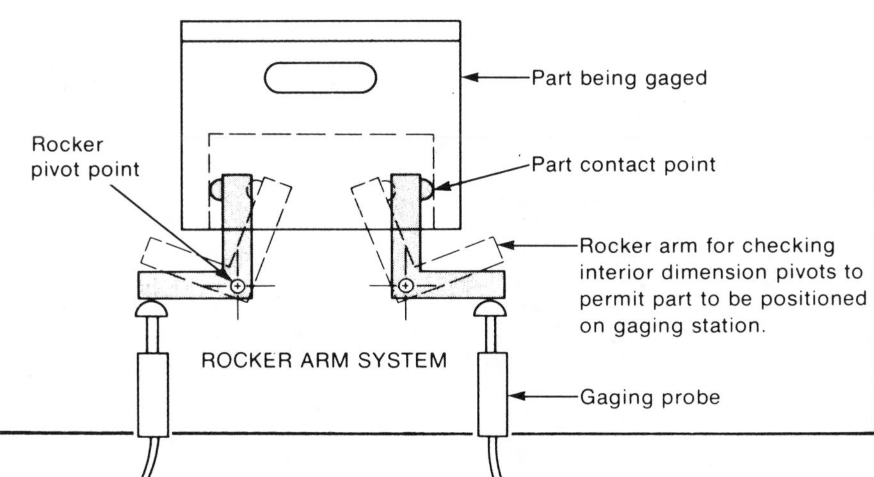

ROCKER ARM SYSTEM

Spurred by product liability and escalating labor costs, more companies are turning to automation for dimensional inspection of manufactured parts. Initital investment in equipment can be substantial, but prudent design planning can keep costs reasonable. The trick is to specify inspection of only essential features.

DIMENSIONAL inspection can be one of the most critical operations in a manufacturing plant. Usually, dimensions are checked manually with micrometers, scales, gages, comparators, or coordinate-measuring equipment.

Inspectors are not infallible, however, and fatigue or carelessness can cause errors. Out-of-tolerance parts may be sent to the assembly line, or the product may fail in service. Manual inspection is also slow and expensive, often contributing to a weak competitive position, especially in selling against foreign-made products.

Thus, for reasons of product safety, manufacturing economy, eliminating the human decision, and legal protection, automation is increasingly being viewed as a practical approach to checking and documenting the dimensional

How a typical automatic gaging system works

Dimensional buildup in a tractor track consisting of 150 or more links could, if the parts are out of tolerance, cause the assembled track to be too short or too long. Thus, the individual links for tractors are 100% inspected by automatic gaging equipment.

This part, one of four sizes of forged-steel track links in production, is gaged directly downstream from the production machining system. The automatic gaging machine, using electronic gaging, inspects eight dimensions on the link, compares them against established tolerance limits, and makes a decision as to whether all measurements are in tolerance, out of tolerance, or trending toward out of tolerance.

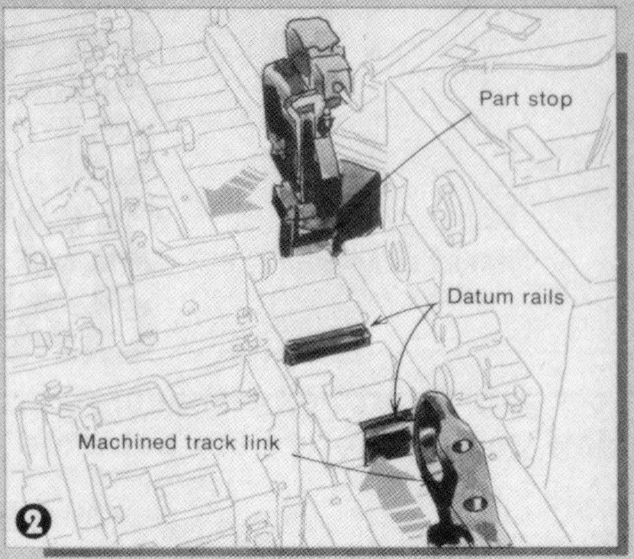

The part is released to the gaging station by a reciprocating gate (not shown), which restrains any additional links waiting to be gaged. As the link approaches the gaging area on the inclined conveyor, a bar moves out, stopping the part at the gaging station. The conveyor rolls then drop down, allowing the part to rest on the precision datum rails.

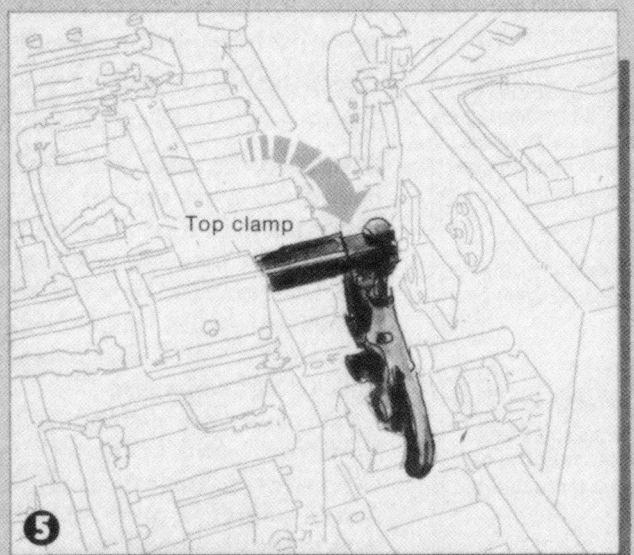

A clamp swings down and presses onto the top of the part, ensuring that it is properly seated against the base datum rails. The part is now "staged" and is ready to be checked. Before the gage heads are activated, however, a compressed-air blast from the gage spindles cleans the part of chips and coolant.

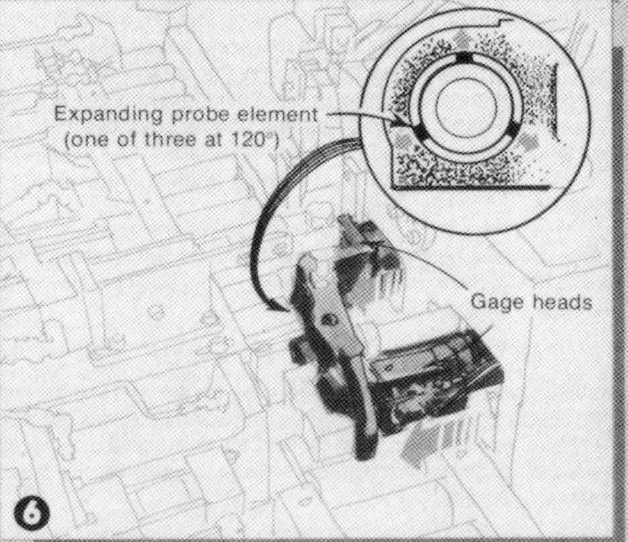

The gage heads, which can "float" in both the vertical and horizontal directions, then advance into the bores of the part. The heads are designed so there is no rubbing contact as they move into the part. After they are in position, the probe elements (similar to a machine-tool chuck) expand to contact the diameters to be checked. With this one gaging operation, a total of eight checks are made and compared against set limits, and a decision is rendered. Features checked are:

- 1 and 2: diameters of both through holes
- 3 and 4: diameter and depth of the counterbore
- 5: radius of the counterbore
- 6 and 7: height of both holes above the base rail
- 8: location of the holes with respect to each other

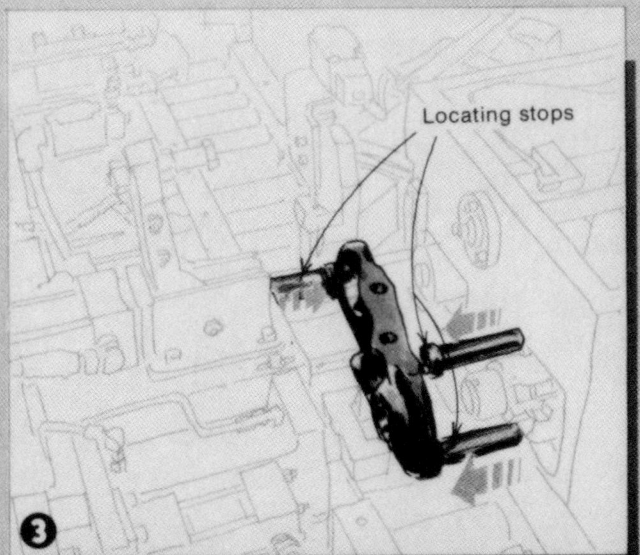

Three locating stops move out to a fixed position to establish a plane. The stop bar is retracted.

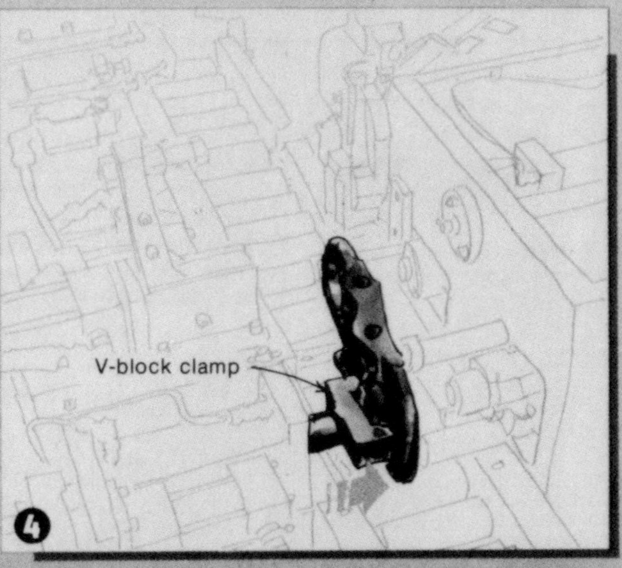

A V-block clamp advances out from the opposite side, securing the link in the plane of the three locating stops. The clamp also adjusts the part into the proper fore-and-aft position.

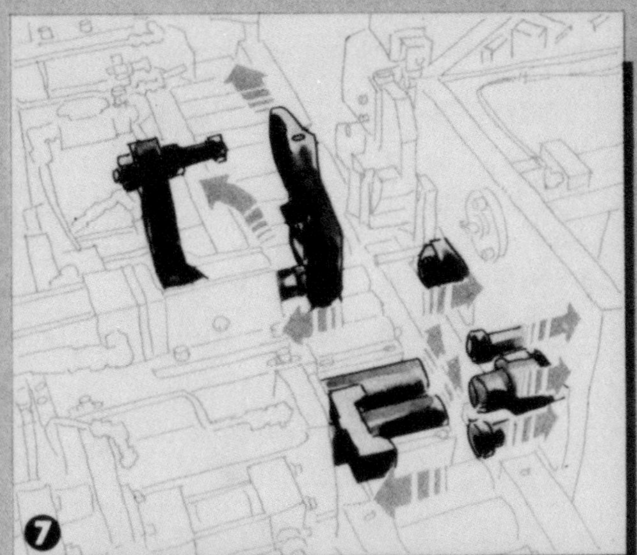

After the gaging operation, the gage heads, locating stops, and clamps are retracted, the conveyor rolls are raised, and the link moves along. If all features are within tolerance, the conveyor feeds straight ahead to either an assembly or storage area.

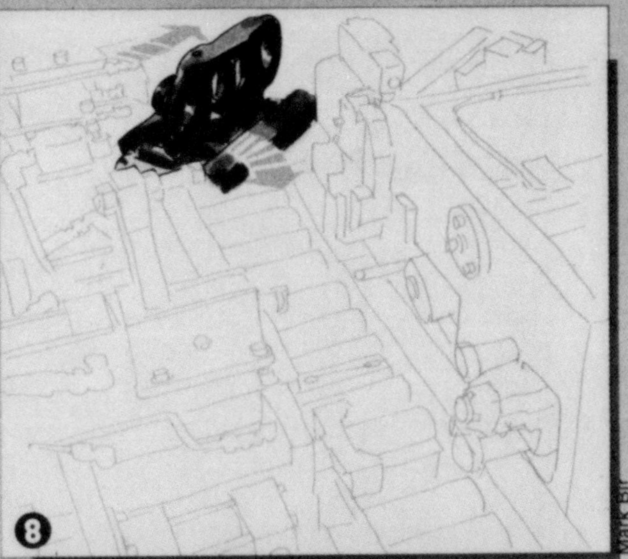

Mark Bir

If the link has one or more of the checks out of tolerance, a section of the conveyor shunts the link off to another area where it is checked manually for possible salvage.

characteristics of mass-produced parts.

In most cases, the design and configuration of fabricated parts are determined entirely by functional, assembly, or manufacturing requirements. In others, however, some options exist as to the location,

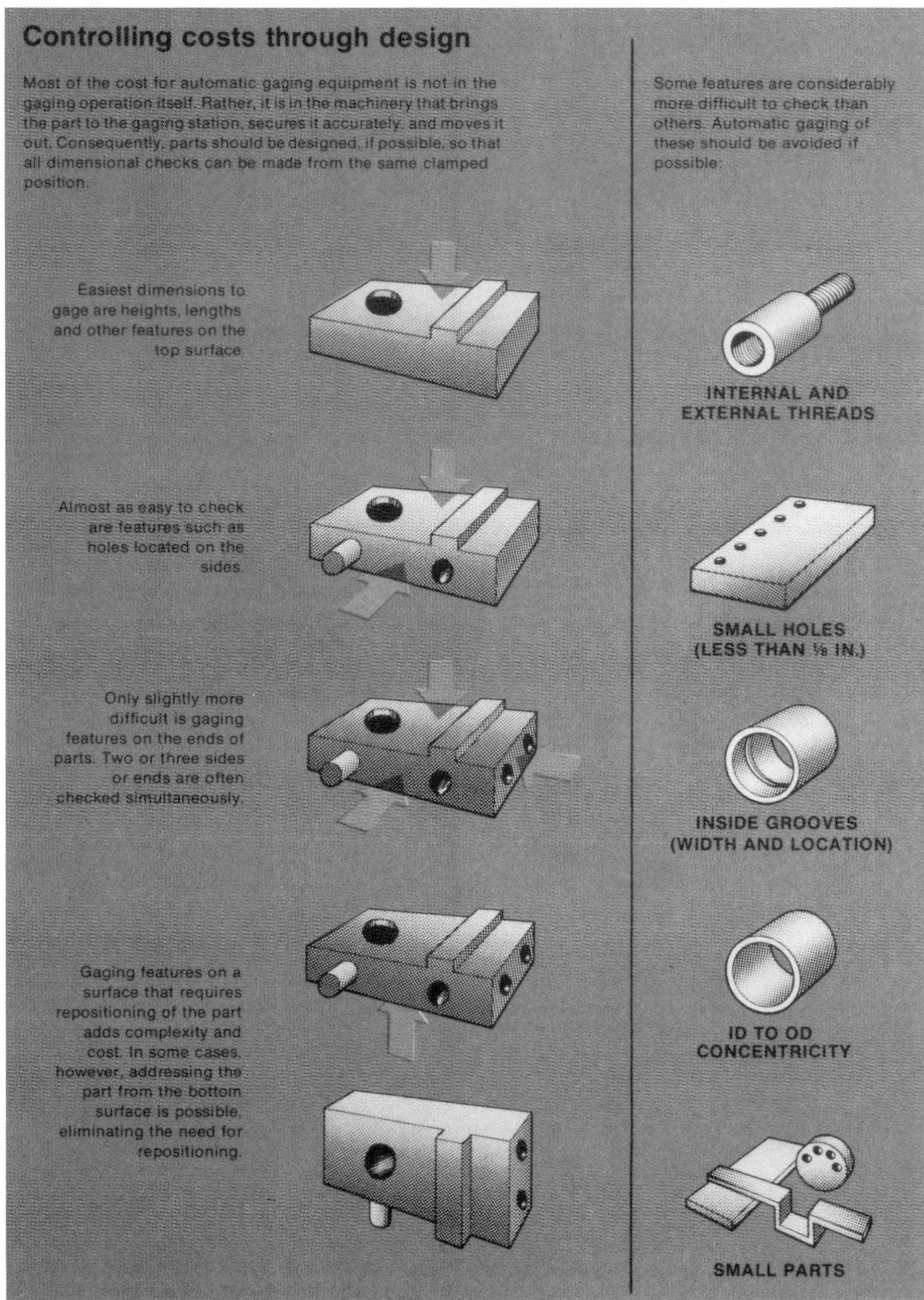

size, or shape of part features. If such parts are destined for automatic inspection, chances are that economies in inspection-equipment needs can be made in the design stage.

By far the largest use of automatic inspection equipment has been in checking automotive production such as rear-axle assemblies, wheel spindles, brake parts, and collapsible steering-column components. Other high-production parts that warrant automatic inspection are, for example, track links for tractors and relay-core spools for communication equipment.

Gaging in action

Automatic inspection—or automatic gaging as it is called—almost always requires custom-built equipment. Although certain standard components and instrumentation are common to a number of installations, each system involves individual engineering, assembly, and break-in, or "debugging," attention. Thus, even a relatively small gaging system—one that checks only one or two features on a part—may cost in the range of $40,000 to $50,000. From there, the cost of automatic gaging can escalate into the area of $500,000 for a highly sophisticated system. Software cost alone for one of the more complex machines can exceed $100,000.

Interestingly, a large portion of the machine cost is usually not in the gaging system at all but in the means necessary to bring the parts to the gage and to move them out after inspection.

The principles of automatic gaging are simple. A part is brought to the gaging station by any of a number of methods and clamped so that its significant features can be measured (usually from a machined datum surface). Sensing probes then move into or around the part to measure designated features, and electronic responses compare the gaged values to standards. The part is then moved out of the gaging station, often through one of several trap doors or chutes, depending on how the machine is programmed. The exit paths may be simply "acceptable," "salvagable," and "scrap" or, in cases involving selective assembly, for example, parts may be segregated according to the size of a given critical feature.

Staging the part: Before a part can be gaged, it must be delivered to the gaging station in proper orientation. This may be done manually or by means of a vibrator bowl or conveyor. Some of the newer systems use robots for this function. Robots are not particularly fast, however, nor are they especially accurate; moderately priced robots cannot repeatedly locate parts closer than about 0.060 to 0.070 in. Most systems, robot-loaded or otherwise, require a locating device such as a tapered clamp or a hole in the part to position the part precisely.

Taking the measurements: After the part is positioned and clamped, various sensing probes, pins, and expanding-chuck devices move into and over the part to measure the significant features and relay the data to the control instrumentation. The measuring probes need not touch the part directly, however. The delicate probe tips are usually protected from impact and wear from the moving part by transfer pins and other intermediate devices.

Another method used to protect delicate probes from damage involves a two-station arrangement. At the first station, a simple plug moves into, for example, a hole in the part to verify the presence of a hole of approximately the right size. The part then moves to the second station where a measuring probe checks the exact dimension. If the plug, or "functional probe," detects no hole or one considerably undersize, the part is rejected before it reaches the gage station. Alternatively, the conveyor line may be shut down automatically until the problem upstream—for example, a broken or worn tool—is corrected. This type of probe protection is used most commonly where the gaging equipment is directly downstream from the machining operation.

Other than functional probes, there are two other types of gaging probes: electronic and pneumatic. The primary considerations in choosing between electronic and pneumatic logic are the size and configuration of the feature being checked.

Electronic probes are readily adaptable to most part configurations except inside diameters under ⅝ in. Pneumatic probes can be used to measure any inside diameter larger than the probe itself—about ⅛ in. in diameter is the smallest.

A typical pneumatic gaging spindle has a central air passage and diametrically opposed air jets. It gages a hole diameter based on the amount of clearance between the spindle and the wall of the hole. The change in air-flow velocity is interpreted as a dimension on the indicator. Air gaging, the slower of the two types, limits the gaging speed of a system to about 1,800 parts per hour. Electronic gaging can handle parts at speeds three or more times faster.

Another important consideration is the type of control logic to be used. In potentially hazardous areas, air logic is usually favored to minimize or eliminate the use of electricity.

Other factors are the dimensional tolerance required, indicator resolution, and amplification. For example, gaging tolerances of 0.0001 in. requires electronic means. And for very high amplification, air systems become unstable. Only through electronic techniques can high resolutions be utilized with any degree of stability. Based on these criteria, the system could well use a combination of pneumatic and electronic probes.

Speed of gaging is also affected by the closeness of tolerances being measured and by whether the part is gaged while at rest or "on the fly." Parts on a continuously moving conveyor can be gaged at a faster rate than those that are checked while stationary but not to the same degree of accuracy. For measurements involving a tolerance of one ten-thousandth of an inch, even the at-rest parts require a second or two of "stabilization" time for the moving components in the gaging station to recover from any flexing or vibration. Only then can acceptable readings be taken.

Sorting the parts: The exit path of the part is determined by automatic comparison of measured features with the standards programmed into the controlling instrumentation. Some systems are designed to recognize only a simple pass/fail gaging result; others are more sophisticated. For example, an additional notification of "trending toward out of tolerance," can be incorporated into a program to warn an operator of tool wear that will soon require attention.

Other types of systems sort parts according to size to simplify selective assembly. One such system, which gages telephone-relay cores at 3,600 parts per hour, segregates the parts into five sizes according to length. The five groups—differing by only 0.0002 in.—are then guided into separate bins.

The mechanics of moving the parts after gaging can be handled by gravity roller conveyor—often with a bypass section for out-of-tolerance parts—or by various arrangements of chutes and trap doors controlled by the machine program.

Another variation of an automatic gaging system has been applied to checking a stacked-tolerance assembly of components for an automotive differential. Such assemblies are often specified to be manufactured so that the sum only of the maximum-material values equals the desired total length. The difference, then, for most assemblies, is made up by adding a shim member (usually a washer-shaped part) to reach the desired length. An automatic gaging machine is used to measure each assembly and determine the exact shim thickness needed. The value appears as a digital readout on the panel, and the operator then adds a shim of the indicated size.

Design recommendations

The principal design recommendation applying to designers of parts to be gaged automatically is to try not to do too much. The tendency to specify "check all dimensions to tolerances indicated" must be tempered with reality. Such a specification would almost always result in an unnecessarily complex, expensive piece of equipment. Responsible builders of gaging machines will, in fact, discourage this type of order and ask that only the essential dimensions be checked automatically—and preferably those that can be reached from a single positioning of the part. With some parts, this may involve only one or two dimensions; others may require eight or ten checks.

A recommended goal is to gage the features that could cause perhaps 90% of the problems (in assembly or safety, for example) rather than to increase the complexity—and cost—of the machine two or three-fold to reach near 100%. The increased complexity may lead to the machine being down 50% of the time. The simpler the system, the higher its reliability.

In general, the features that are easiest to gage are lengths, heights, holes larger than ⅝ in. in diameter (both diameter and depth), hole location, and runout. Features that are difficult—and hence expensive—to gage include threads, inside groove locations, concentricity of an inside to an outside diameter, and holes having diameters smaller than ⅛ in. In addition, small parts are often difficult to orient or to hold for gaging. Finally, parts must be clean, free from burrs in critical areas and, particularly for checking tolerances of one ten-thousandth of an inch or finer, must be stabilized at a known temperature at which the tolerances are to apply.

Automatic inspection is not limited to the checking of dimensions only. For example, valve seats in automotive engine heads can be checked for hardness and for the presence of cracks or other flaws. This inspection involves electronic instrumentation using an eddy-current response to verify hardness or to detect discontinuities.

Another characteristic that can be inspected automatically is porosity of vessel-type parts. Here, the part is sealed and filled with a fluid under pressure. Pressure readings taken before and after a suitable dwell period are then compared to detect any leakage.

CHAPTER 10

ASSEMBLY IN AN AUTOMATED FACTORY

Presented at the SME 13th ISIR/Robots 7 Conference, April 1983

The Unattended Factory FANUC's New Flexibility Automated Manufacturing Plant Using Industrial Robots

by Douglas F. Urbaniak
GMF Robotics Corporation

Ever since the word "robots" crept into our vocabulary in the mid-sixties, people have been talking about the unmanned factory -- a factory where, at the push of a button, everything operates automatically; where robots are involved in machining, assembly and inspection operations; and in fact, in the supervision of other robots. As managers and engineers strive to improve quality and increase productivity by demanding and developing flexible manufacturing systems that utilize minimal or no manpower, we come closer and closer to the truly unmanned factory. Before that day arrives, there needs to be additional developments in the areas of sensors -- robots that can see, hear, feel and touch; improvements in product design and system engineering; and finally, more sophisticated control systems that will allow machines and robots to communicate almost instantaneously.

FANUC's new flexibly automated manufacturing plant at the base of Mt. Fuji is the closest thing to an unmanned factory in the world today. The plant was completed in August of 1982. It is a two-story building with 8,000 square meters of manufacturing space on each floor. There are approximately 60 people and 101 robots working in the plant to produce 10,000 motors per month. Currently, there are over 40 different kinds of AC spindle motors, AC servo motors, and DC servo motors manufactured at the motor assembly plant in lots ranging from 20 to 1,000 pieces.

On the first floor of the motor assembly plant, where all the machining of parts is done, there are 60 machining cells, supported by 52 robots, which carry out loading and unloading assignments. On the second floor, where the motors are assembled on four different assembly lines, there are 25 assembly cells and 49 robots.

Connecting the two floors is an automatic warehouse located at the rear of the building. The automatic warehouse is served by unmanned carriers which transport workpieces, finished parts and assembled units to and from the work stations on both floors. Parts machined on the first floor are temporarily stored in the warehouse, then are gradually sent out according to the assembly schedule on the second floor. When the motors have been totally assembled on the second floor, they are carried by unmanned carriers to the automatic packing machine; the packed motors are then sent to the automatic warehouse for temporary storage.

First Floor Machining Operations

On the first floor, robots load and unload parts to and from turning, grinding, tapping, drilling, boring and facing machines. These machining cells are designed to work unattended 24 hours a day. Parts are brought to and from the machining operations on standard pallets by the unmanned carriers. Each pallet is capable of holding from 1 to 1,000 parts depending on its shape. Robots pick up and return workpieces to and from the standard pallets. When the pallet is full, it is returned to the automatic warehouse for subsequent use on the assembly floor. The key to unattended manufacturing is equipment and part reliability. Parts must be uniform and consistent, and all dimensions must be within the tolerances specified. Material quality must be consistent. All equipment and tooling must be designed and manufactured to specifications for automatic manufacturing. Defective workpieces must be identified and removed. Equipment and tooling must be maintained and changed at the appropriate intervals.

Unmanned operations are possible only if these conditions are met.

Machining Sequence

Robots 1 through 8 -- load shafts to finish grinders.

Robots 9 through 13 -- load rough shaft castings to turning machines.

Robot 14 -- loads rotors to a turning machine.

Robots 15 and 16 -- load machines that put in the key way and drill holes in the end of the shaft.

Robot 17 -- loads shafts to a turning machine.

Robots 18 through 21 -- load end covers to vertical drilling machine where holes are drilled and tapped.

Robot 22 -- loads brake housings to a vertical drilling and tapping machine.

Robot 23 -- loads the front housing to a drilling machine.

Robot 24 -- loads brush end covers to a horizontal boring and facing machine.

Robots 25 and 26 -- load brush end covers to a horizontal drilling, tapping, reaming and facing machine.

Robot 27 -- loads the pulse coder flange to a horizontal drilling machine.

Robots 28 through 31 -- load front flange castings for machining of the end face.

Robot 32 -- loads pulse coders to a facing machine.

Robot 33 -- loads pulse coders to a grinding machine.

Robot 34 -- loads brake housings to a turning machine.

Robot 35 -- loads balancing rings to a turning machine.

Robots 36 and 37 -- load brush holder flange castings to facing machines.

Robots 38 through 43 -- load brush holder flange castings to turning machines.

Robots 44 through 51 -- load end covers to turning machines.

Robot 52 -- loads rotors to a key slot machines.

Second Floor Assembly Operations

On the second floor -- the assembly floor -- there are four different assembly lines. Each line is made up of several assembly cells, which are arranged on both sides of the unmanned carrier path. Each cell has several robots combined with peripheral devices such as presses for pressing bearings, nut runners for screwing, and other machines.

The unmanned carrier brings pallets to the assembly cells and parts are removed for assembly. When the work at a given assembly cell is completed, the partially assembled motor is sent to the next cell by a workpiece feeder.

Although 49 robots on the assembly floor work the 25 assembly cells, such jobs as wiring, installing detectors and so on, are still too difficult for robots. These jobs are done by workers.

Currently, the second floor operates only 8 hours a day. After the motors are totally assembled, they are inspected automatically on an inspection machine and finally taken by unmanned carriers to the automatic packing machine and then to the automatic warehouse for temporary storage.

Motor Assembly Sequence

Robots A and B are used for loading and unloading the magnets and pole shoe in a bonding operation.

Robot C handles the motor shell through a cleaning and heating operation and then places it over the magnets for bonding.

Robot D unloads the shell with the bonding magnets to a pallet.

Robot E handles the rotor through a turning operation, slotting operation, and then to a deburring operation.

Robot F takes the rotor from the previous operation and loads them into a press that presses the bearings on the end of the rotor and places it on an in-feed conveyor for the next assembly sequence.

Robot G loads the front flange to a machine that puts some bonding material on the flange and transfers it to a machine that presses the oil seal in place.

Robot H takes the rotor and places it into the front flange where the front flange and the rotor assembly are pressed together.

Robot I takes the rotor front flange assembly and places it on another transfer conveyor.

Robot J gets the shell with the magnets from a pallet and places it on a checking fixture -- and takes it off the checking fixture and places it over the rotor on the transfer conveyor.

Robot K picks up the brush holder flange assembly from a pallet and places it onto the rotor front flange and shell assembly on a conveyor.

Robot L inserts the four long bolts and screws the bolts to the front flange.

Robot M places a lock washer and a nut over the four bolts which secures the assembly.

Robot N picks up a spacer and places it over a fixture, and Robot K takes the brush holder flange and places it onto the fixture where the spacer is secured.

Robot K also takes the secured assembly from the previous transfer conveyor and places it onto another conveyor that transfers it down to a manual testing and inspection area. The pulse encoder and cover are added to the assembly here.

Robot O loads the rotors to an automatic testing station.

Several of the assembly robots utilize automatic hand changers -- a device which permits the robot to use several hands or end effectors.

Central Computer Monitor

All machining and assembly operations are monitored by a central computer. The central control room has a big display board, which shows the status of each machining and assembly cell. There is one colored graphics terminal that displays the machine running time. It can display the status of daily and monthly production. There are three TV screens, which are attached to cameras that are monitoring various operations on the plant floor. These cameras are extremely important for unmanned or unattended operations.

Summary

FANUC's engineers have done an excellent job utilizing the latest robotic, computer and sensor system technology as they planned and developed the Fuji Motor Manufacturing facility. Before FANUC built the new motor assembly plant, 108 people and 32 robots manufactured 6,000 motors per month. At the new factory, 60 people and 101 robots manufacture 10,000 motors per month, resulting in a three-fold productivity improvement. FANUC's Fuji Motor Assembly Plant is not a totally unmanned plant, but it offers a glimpse into the future.

Reprinted courtesy of B. Kuttner
Computer Tool and Die Systems, Inc.

Approaching the Factory of the Future

by B. Kuttner
Computer Tool and Die Systems, Inc.

Introduction

Much has been said and written about the Factory of the Future[1]; a second industrial revolution has been heralded with the introduction of microprocessors into manufacturing; complete automation is promised to be immenent with the proliferation of robots through the industry.[2] If we take the Factory of the Future to mean complete automation, then it should be stated at the outset that its achievement can only be asymptotic: it can be approached, but never fully achieved; some form of maintenance and programming will be with us for ever, even in the most sophisticated robot-operated installations. Markets change, orders change, designs change and machinery wears out; all this requires human intervention to support and maintain.

But the Factory of the Future will be automated not simply in its mechanical operations (in many ways the simplest aspects of a factory to automate) but also in its coordination: in order processing, inventory control, tool and fixture deployment, process planning, scheduling, inspection, shipping. By far the greatest economies to be achieved lie in this coordination aspect since over 75% of a product's turnaround time is absorbed in waiting on machining, assembly or shipping.[3] The potential increase in coordination efficiency can give small and medium sized shops the same economies as mass production manufacturers. It will also enlarge the scope for subcontracting in almost every phase of manufacturing, including casting and forging, by converting product design specifications automatically into tooling requirements and machine instructions, sending this information electronically into the job shop.

Given that a fully automated manufacturing system can only be approached and never fully achieved (or purchased), what is the best means of approach? How should a company producing small or medium runs best set about developing their manufacturing system? Three approaches will be laid out in this paper. Although not mutually exclusive, there may be a pay-off in emphasizing one approach over the others or applying them in sequence in a given installation. The criteria for making this decision will be discussed below.

Approaches

How can the economies offered by the integration of Computer Aided Engineering and Computer Aided Manufacturing be acheived? Basically there are three approaches available, as indicated in Table I.

APPROACH	DESCRIPTION	ELEMENTS
1. System Architecture Design	Top-Down	Solid Modeling Manufacturing Control
2. Hardware Integration	Horizontal	Robot Cell Communications Network
3. Applications Integration	Bottom-Up	Applications Data Base

TABLE I

The distinctions between the approaches lie mainly in the emphasis given to certain key components. All involve computers, programmable control systems, data bases and, of course, applications. However, the strategy that is chosen in putting these components together can have a profound impact on long and short term return on investment. For instance, there is a trade-off between wholesale, complete integration of all shop floor equipment, and the demands of near term applications. The pursuit of complete integration may prove Quixotic in some instances and result in little or no benefit to the company for many years; on the other hand, total preoccupation with immediate applications of CAD/CAM technology may delay integration of the manufacturing system and its accompanying benefits indefinitely.

An analogy may be made here with data processing in general.[4] Initially, computer systems were essentially hardware plus collections of applications. This was essential in proving the utility of the technology. However, every installation had to "reinvent the wheel" to perform common calculations such as square root, integration, etc. Operating systems were quickly developed that came already equipped with utilities to perform these functions, together with compilers for higher level languages that made use of them. Next we saw networks of computers connecting geographically dispersed installations, making computing power from one location available to the other nodes in the network. These were primarily large, time-sharing systems. As a user you would still essentially "do your thing" in terms of applications, taking advantage of the languages and utilities offered on the system. With the advent of data-base technology, this trend was taken a step further, whereby applications were tied together (and in some cases even generated) by the data base management system. This means that users in different parts of a company, say Order Entry and Accounting, get their own perspective of the self-same data, and changes are communicated instantly between them [5] using low cost mini or microcomputers.

The Factory of the Future may not follow the same course of development as commercial data processing. It may be possible to skip some steps; it may also be desirable not to. The remainder of this article will explore the options available in the manufacturing system development process.

System Architecture Design

For the purpose of this article, Computer Aided Engineering and Computer Aided Manufacturing tools are regarded as more than just gadgets that speed up existing processes; they are regarded, in addition, as development tools by which the next stage of manufacturing automation may be reached. Thus the principle benefit of Computer Aided Design systems is taken to be not the speed and ease with which layouts may be generated and altered, but rather the *symbolic* representation of product geometry that this technology provides. Armed with this representation, we have a machine-readable basis for part and mechanism analysis.

With the internal representation or mathematical model maintained by the computer system, product geometry can be manipulated and analysed, e.g. for interference checking, without having to redescribe the problem to the computer. Similarly, Numerical Control provides a symbolic representation of the machining process (in addition to actually guiding the machine tool). With this representation, it is possible to obtain tooling and machining requirements for process planning of similar products. In fact, with the machining process definable using NC programming, and the design process captured with CAD technology, there is only one additional step required to integrate all the discrete engineering and manufacturing data available: a higher level language that can adequately describe the manufacturing environment.[6]

The Systems Architecture or "Top-Down" approach to manufacturing system development demands such a formal representation of the manufacturing environment. It requires that the overall process, from design through manufacturing, be defined mathematically, as the CAD and NC substeps are today, so that the symbols can be manipulated to optimize the processes they represent. Without this overall symbolic representation, the architecture of the manufacturing system cannot be adequately specified and the situation is comparable to an architect trying to define a structure without blue-prints.

In order to achieve this depth of description, standardized tooling/ fixturing notation and procedures must be developed. Likewise, NC-type descriptions of casting, forging, fastening operations must be developed.

To get an idea of the way this approach would work in practice, at least at the highest level, consider the scheduling regime depicted in schematic outline in Figure 1. The external events feed into a production planning process that results in the formulation of a master schedule, which in turn is transformed into production and assembly schedules, depending on inventory levels of components required by bills of materials for the products. Using the top-down approach, these requirements would be automatically broken down into subprocesses and machine instructions. Only one level of this break-down will be considered here, to give the reader an idea of the notational tools currently available. The hierarchical nature of the process should be noted.

Following the conventions of the SIMULA programming language,[7] two <u>classes</u> of objects are defined, one for work orders, and another for the master schedule itself. The work order class contains a field for part number and a pointer (<u>Ref</u>) to the next work order; the master schedule class contains pointers to the first and last work orders. If we assume that an algorithm is available for sorting the work orders into a sequence that meets corporate objectives (e.g. maximizing revenue) then the program looks like that represented in Figure 2. This process would be continued to finer and finer levels of detail, until each work center has been assigned its work load, Figure 3. This complete allocation of resources defines a single sequential state of the system.

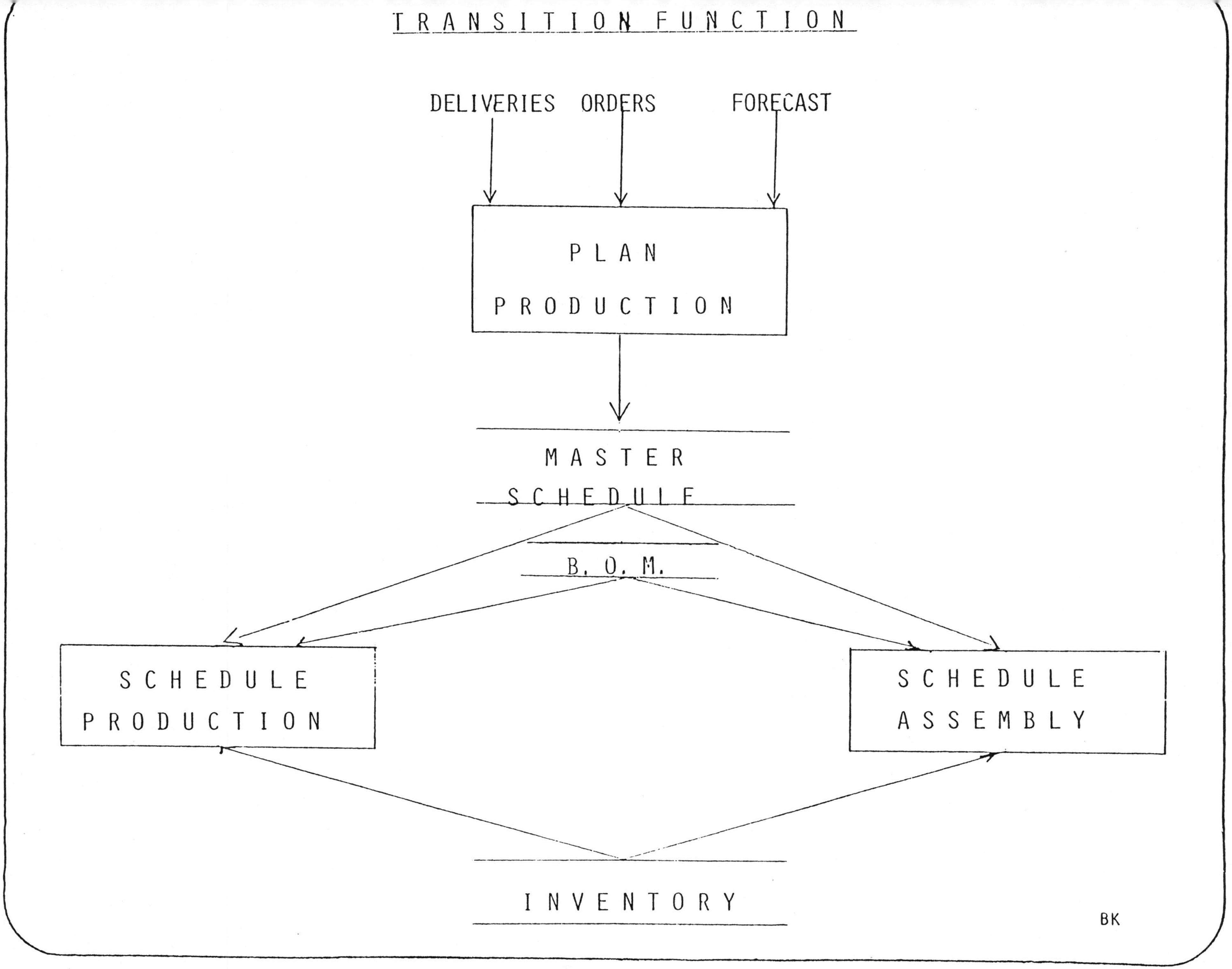

TRANSITION FUNCTION

DELIVERIES
ORDERS
FORECAST
PLAN PRODUCTION
MASTER SCHEDULE
B. O. M.
SCHEDULE PRODUCTION
SCHEDULE ASSEMBLY
INVENTORY
BK

FIGURE 1

MASTER SCHEDULING

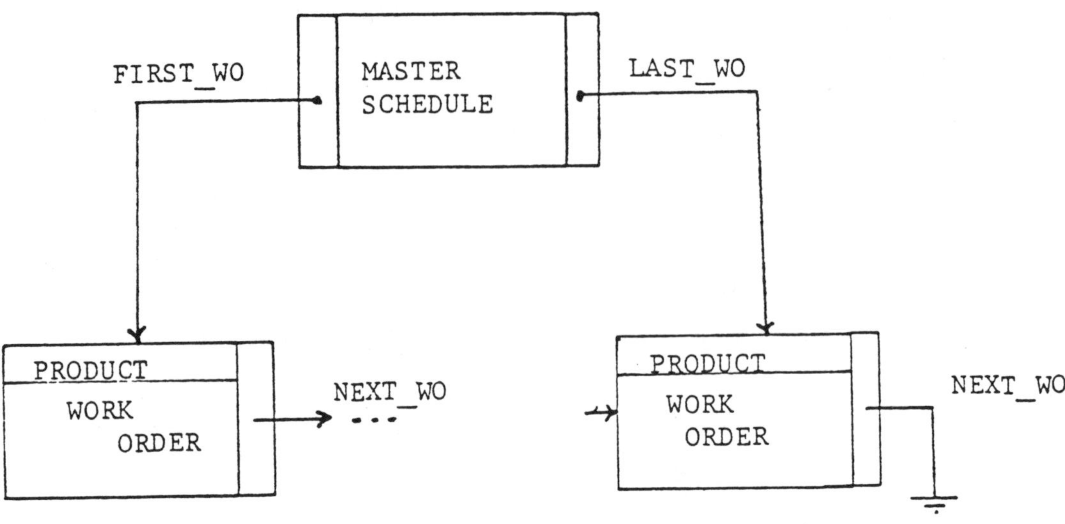


```
CLASS       WORK_ORDER (PRODUCT); INTEGER PRODUCT;
            BEGIN   REF  (WORK_ORDER) NEXT_WO;
            END;

CLASS       MASTER_SCHEDULE;
            BEGIN   REF  (WORK_ORDER) FIRST_WO, LAST_WO;
            COMMENT SET PRIORITIES AND DUE DATES ON ORDERS BY
                    SORTING THEM WITH SOME ALGORITHM;
                    SORT (FIRST_WO, LAST_WO) ;
            END     MASTER-SCHEDULE;
```

BK

FIGURE 2

FIRST_WO
MASTER
SCHED
LAST_WO
FIRST_TASK
WO
NEXT_WO
LAST-TASK
FIRST_TASK
WO
NEXT_WO
LAST_TASK
TASK
NEXT_
TASK
TASK
I
TASK
NEXT_
TASK
TASK
I
TSK
TSK
JOB
NEXT
JOB
NEXT
FIRST JOB
LAST JOB
WRK_CTR [I]
INSTANCE OF PRODUCTION
SCHEDULING DATA STRUCTURE
BK

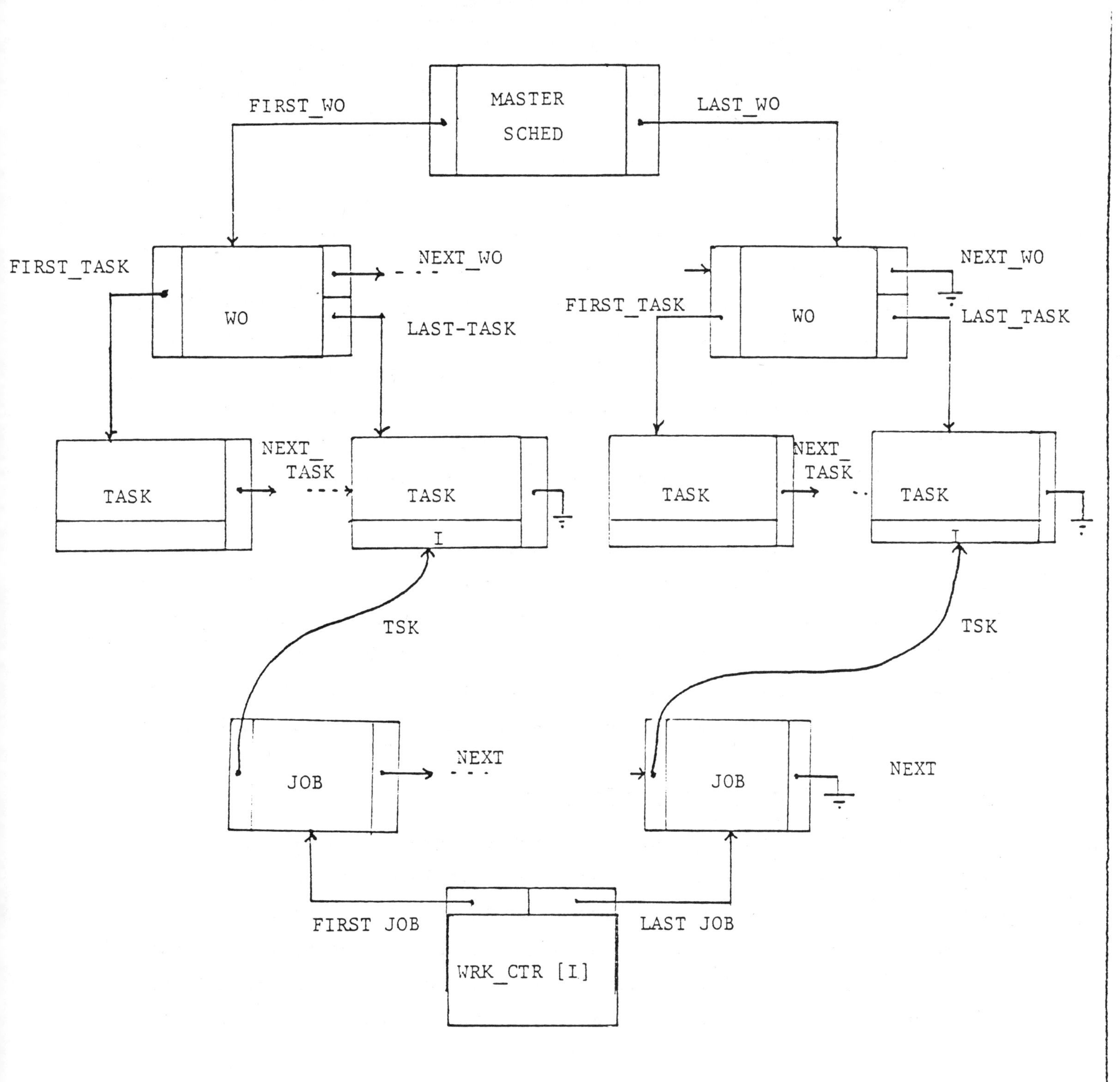

FIGURE 3

To actually set this system in motion, the output function must be defined, taking the sequential states, and transforming them into output values. This is the NC-like language that directs all the mechanisms involved, transforming product geometry defined in the design process into machine instructions for execution on the shop floor. The motion of every axis, or degree of freedom in the entire system is specified in this language. But first the product must be specified in a manner that can be translated directly into this output language.

Computer Aided Design systems employing solid modeling techniques are seen as a major step forward in achieving the aim of transforming product geometry directly into machine instructions.[8] In these systems, a mathematical model of the product is maintained in three dimensional form (unlike the two dimensional model of conventional drafting practices). Three dimensional shapes (e.g. spheres, cones, cubes) are combined or substracted to give the desired product geometry. Now since a tool path is nothing but a subtractive process defined by the volume of material removed from the stock by the tool, it is easy to see how this form of representation lends itself to the automatic generation of NC programs.[9] Similarly, mathematical models are under development to describe casting and forming operations, using finite element analysis of the shapes to be produced to obtain simulations of fluid flow and heat exchange.[10] Combining these into the unified notation recommended above will provide the possibility of programming an entire factory in a single, structured system.

Advantages/Disadvantages of Top-Down Approach

The principal advantage of the manufacturing system architecture design (top-down) approach, is that it guarantees complete integration of Computer Aided Engineering and Manufacturing (in fact without complete integration it will not work). This means that full automation, in the sense defined at the start of this article is possible. Automatic optimization of the manufacturing process (i.e. maximizing throughput) is also possible.

The disadvantages of this approach are:

Difficult technology transfer

The work process implied by this approach does not fit that employed in any existing facility, so only very sophisticated users will benefit, intially; and human factors are not taken into account at all. Trained personnel to implement such a system are not readily available.

Subprocesses not defined

Most of the subprocesses involved in the output function are not yet fully defined. For example, machine set-up and tooling, and mass forming operations have not been well-defined in any symbolic notation.

Despite these disadvantages, the top-down approach holds out great promise as it forces the user to "think-through" the operations to be performed and, if commercial data processing is any indication of the path to be trodden, the benefits in terms of reduced maintenance and other overhead, of this sytem design process are enormous.[11]

Hardware Integration

A second line of approach toward the Factory of the Future being actively pursued (amongst others, by General Motors[12]) is the Hardware Integration approach. The approach requires the assembly of building blocks called "robot cells" or "islands of automation" into a manufacturing configuration that suits the user's needs. Each cell or island would consist of a machine tool, one or more robot arms, inspection devices and a "bus interface unit" to connect it to the rest of the network.

Each cell is programmed independently of the other cells in the network, but sends and receives information to and from the other cells or Computer Aided Engineering system and host computer. The flexibility of the system derives from this broad-band, "any-to-any" communication that makes it possible, for instance, to pick up and process shop floor monitoring information instantaneously and divert material from a cell that is having a problem or is overloaded, to another, equivalent cell, dynamically. However, it should be noted that simply getting the cells "talking" to each other does not in itself accomplish any coordination function; the necessary software still has to be written.

The hardware integration approach is described as "horizontal" because each cell behaves as an island unto itself and, except for cycle start and emergency stop conditions, the inner workings of the cell are a black box to the other cells in the system--unlike the top-down approach which is hierarchical in its command structure. The intent of the horizontal approach is to devise a "layer" of communications software that will approximate an operating system for a general purpose computer, to monitor the performance of an entire factory of cells, with each cell's application being written off-line but executed in synchronization by the operating system. The paradigm of distributed data processing may find application here, although much development work remains to be performed.

Advantages/Disadvantages of Horizontal Approach

The principal advantage of the Hardware Integration approach is the possibility of rapid introduction. It promises incremental growth and flexibility, the user simply adding new cells as the need arises. In addition, the user determines and programs applications procedurally, obviating the need for the higher level language called for in the top-down approach.

The disadvantages of this approach are:

Limited Optimization Potential

The separate islands of automation are, as their name implies, independent entities under local control with only coordination potential between them. Thus if a cell requires a complex fixturing operation, it may not be able to take advantage of another cell's existing facilities to perform the operation (without human intervention).

Lack of Standardization

This feature of the hardware integration approach is a disadvantage from the end user's point of view, but a lucrative advantage from the vendor's point of view. At present, and for the forseeable future, the formats and protocols of the devices to be connected differ from manufacturer to manufacturer. Each has its own proprietary idiosyncrasies buried in its code. This implies a permanent need for post-processors or translators to communicate effectively between components made by different manufacturers, much as is the case today in the NC environment when programming in a language other than the machine language of the NC control.

Applications Integration

The third line of approach to the Factory of the Future, one that is being pursued in any case, whether its practitioners are aware of it or not, is the Applications Integration or "bottom-up" approach. In this approach, applications are developed in a more or less *ad hoc* manner and integrated after the fact. Despite its haphazard appearance, this approach may prove to be the most appropriate at the present stage of development for many installations. The Applications Integration approach has four phases: User Identification, Problem Definition, Problem Solution and Solution Integration; so it is essentially a problem-solving (and sometimes a "fire-fighting") approach. As an analyst employing this approach, one looks around a company for potential users in large and small shops, among management, salaried and hourly personnel. The problems to be grappled with include:

- Barriers to productivity improvement
- Imprecise engineering techniques (rules of thumb)
- Quality control
- Health, safety,ergonomic issues

Once identified, the solutions will involve software applications modules, either developed or acquired, to solve the problem at hand, plus a user-interface tailored to the specific needs and skill level of the user environment, as well as the necessary interfaces to other existing or future modules.

It is this last point, the analysis of interface requirements, that gives this activity the legitimacy of a methodology or approach. The intent is to produce not a collection of incompatible applications, but a solid array of modules to be employed as utilities once the system starts to "gel". Perhaps the most important aspect of the bottom-up development process, as far as the user is concerned, is the human/machine interface. This is what the user sees and it is a matter of indifference to him if behind it lies the most sophisticated 32 bit microprocessor or a cage of chipmunks on a treadmill, so long as the results are satisfactory. In order to achieve satisfactory results using this approach, the decision rules used by today's skilled designers and machinists must be captured and embodied in the system. In practice this means the provision of a "user-friendly" front end to the application that will prompt the user for key parameters. In this way, the criteria and decision rules generally found in standards guides and the user's own memory, will gradually be transferred to the system itself.

The data base in which the decision rules are deposited is a crucial element in the bottom up approach. It is the "glue" that holds the disparate applications together and allows for the progressive accumulation of experience and skill. It should accummulate optimal solutions found for various product shapes and associated processes, for reuse in a group technology procedure[13] (which in turn requires an efficient classification and coding scheme). The data base must also supply the "hooks" to other components in the system, in either the design or manufacturing phase, so that a single master file can be maintained for each product during all stages of its life cycle, so that changes are made in one place only and are reflected throughout the system: in design, manufacturing, quality control, invoicing, etc.

Finally, this software must be tied to the progressive introduction of appropriate policies and procedures to ensure, for instance, the gradual standardization of tooling, fixturing and set-up procedures, or the control of design changes to preserve integrity of the data base. These procedures will allow other applications modules (e.g. Group Technology and Process Planning) to take advantage of the information captured.

Advantages/Disadvantages of the Bottom-Up Approach

The advantages of the Applications Integration approach are:

User Acceptance

The bottom-up approach responds to immediate, recognized needs. It provides the user with tools to solve existing problems and gives him a sense of ownership in the tools produced, by utilizing and enhancing existing skills.

Procedural Data Base

The data base will be customized to the user's operations and products, and will expand as applications are developed. It is not necessary to know in advance all the information that will be required once the system is fully integrated.

The disadvantages to this approach are similar to those already experienced in the data processing industry: integration may be delayed by incompatibility between applications, requiring an expensive "rewrite" of the system downstream. Secondly, the specific users may not have the capital to finance the development of a fully integrated system individually.

Discussion

Given the advantages and disadvantages of the approaches discussed above, the reader will no doubt have already concluded that the best solution is to pursue all three approaches in parallel. How such a combined approach would be pursued depends on the resources of the user. A small shop would find the quickest return on investment by following the bottom-up approach for the most part, with gradual addition of hardware integration as the need arises. A larger, multi-departmental shop should consider the hardware integration approach as its first priority, making sure that applications, as they are developed, are accomodated on the chosen hardware vehicles and that they fit, at least conceptually, into an overall system design. The pure top-down approach appears to be somewhat extravagant at this time for all but the largest manufacturers (i.e. automotive and aerospace).

Conclusion

Within the next decade, Computer Aided Engineering will be integrated with Computer Aided Manufacturing in some sectors of the industry: the electronic communication of design information and the automatic generation of machine instructions will be common-place. The installations that position themselves now, by choosing an appropriate development strategy, will be the only ones to benefit from this technology as the transition from traditional techniques to full integration is comparable to the leap from horseback to space travel: the skills employed in the one technology simply do not apply in the other. That is the reason to start now in choosing a manufacturing system development approach, and to incorporate it into long term planning. System development demands long term commitment to a coherent strategy.

FOOTNOTES

1. Grierson, D., interview, Iron Age, (December,81).

2. Smith, D.N., Colwell, L.V., and Colding, B., A Delphi Forecast of Manufacturing Technology: Manufacturing Systems Materials Removal, (SME, Dearborn, 77), and "Surveys Reveal Robot Population and Trends" in Robotics Today, p. 79, (Feb., 82).

3. Merchant, M.E., "The Coming of the Automatic Factory" in Manufacturing Engineering, (Mar., 80).

4. Hurd, C.C., "Early IBM Computers", Annals of the History of Computing, 3,2, (April, 81).

5. Fry, J.P. and Sibley, E.A., "Evolution of Database Management Systems", Computing Surveys, 8, 1, (1976).

6. Wesley, M.A., Lozano-Perez, T., Lieberman, L.I., Lavin, M.A., and Grossman, D.D., "A Geometric Modeling System for Automated Mechanical Assembly", in IBM Journal of Research and Development, 24, 64, (1980).

7. Birtwistle, G.M., et al, Simula Begin, (Stockholm, 1977), and Palme, J., "Uses of the Simula Process Concept", in Software, 12, 1, (1982)

8. Meyers, W., "An Industrial Perspective on Solid Modeling" IEEE Computer Graphics, 2, 2 (March, 1982).

9. Requicha, A.A.G. and Voelcker, H.B., "Solid Modeling: A Historical Summary and Contemporary Assessment," ibid.

10. Semiatin, S.L., and Lahoti, G.D., "The Forging of Metals", in Scientific American, 245, 2 (Aug, 1981).

11. Brooks, F.P., The Mythical Man Month: Essays on Software Engineering, (Addison Wesley, 1975), and
Constantine, C.C. and Yourdon, E., Structured Design: Fundamentals of a Discipline of Computer and System Design, (Prentice Hall, 1979).

12. See MAP Protocol Reference.

13. Arn, E.A., Group Technology: an Integrated Planning and Implementation Concept for Small and Medium Batch Production (Springer Verlag, 1975).

Presented at the SME Autofact 4 Conference, November 1982

Automatic Storage and Retrieval in the Automated Factory

by Darrell B. Searls
Jervis B. Webb Company

Automatic storage and retrieval (AS/RS) is a key element in the computer directed material handling system that will be needed to make computer integrated manufacturing and the automated factory a reality. The automated factory implies the continuous controlled movement of material to and through the factory processes. Many of these processes have been highly automated for some time, giving rise to the concept of "islands of automation". The automated factory integrates these islands of automation with a material handling system that provides for the scheduled flow of material through the factory with continuous feedback of information to regulate and adjust all of the factory elements to achieve maximum productivity with the expenditure of a minimum of resources. This results in:

- Reduced cycle time
- Reduced inventories
- Smaller physical plant
- Better overall utilization of capital and labor

Large production units combine processes with different operating rates and reliability and, in the event of production problems, provision must be made to store material between processes until the total system can be brought back into equilibrium. Automated storage and retrieval, then, is required in the automated factory wherever buffering between processes is necessary and wherever there may be an unscheduled pause in the flow of material. The AS/RS Product Section of the Material Handling Institute has completed a study to define the place of AS/RS in the automated factory which I have excerpted for this presentation. First let me define AS/RS as including automated storage, automated transportation, automated material identification and tracking equipment, and real time computer control of inventory. Six major processes exist in most manufacturing facilities with material handling serving as a vital link which enables each one to interface with the others. They are:

- Receiving
- Inspection
- Picking
- Manufacturing
- Assembly
- Shipping

The objective is to reduce costly inventories to their lowest possible levels while maintaining an uninterrupted work flow through the six factory processes.

A great deal has been done to automate and improve productivity as it pertains to the equipment available to design the product, to produce the parts, to ship, to package and protect the materials, to aid in picking and to load and unload shipping vans and rail cars. Those developments not only help to produce materials of high quality workmanship but they greatly increase the quantity of goods produced. The common link to each of these processes is material handling and inventory control. The increased production capability of many of these processes has rendered conventional material handling, storage and inventory control techniques ineffective and incapable of supporting sophisticated modern manufacturing processes. There is much to be gained by skillfully integrating factory islands of automation into a smooth running, harmonious system. AS/RS systems provide that common link and have become the core of the automated factory as a result of these attributes:

1. Space efficient storage of materials for each function.
2. High speed input/output of materials to/from storage.
3. High speed controlled material transportation from process to process.
4. Real time material identification and tracking capability.
5. Real time inventory control of all materials in storage or in transit.

An automated factory evolves as internal processes and functions are automated to some degree. Automated factories are seldom achieved in one sweeping action. There are five typical and identifiable characteristics common to the automated factory.

1. <u>Data Automation</u>

   Conventional methods of manually gathering, recording, checking, correcting and updating data are too slow, costly, inaccurate and leave out the most important common element of the automated factory: material tracking. Production scheduling, inventory control and material requirements planning, to name a few, rely on the availability of accurate, real time data.

   In the automated factory, entry and sensing equipment collect data through keyboard terminals, magnetic wands, scanners, limit switches and other means of material tracking. Paperwork is reduced and data timeliness and realiability are greatly improved.

   Capturing and transmitting data in real time is vital to the functioning of automated machines and equipment and is essential to timely, informed decision making by management.

2. <u>Networking Controls</u>

   Automatic control of equipment is well established in today's factories. But in the automated factory, equipment must interface with and transmit data directly to other facets of the total operation such as storage and transportation systems as well as with management through a control network. In addition to providing essential information among major stations, an instantaneous communications network also provides an important degree of administrative control and capacity to monitor the entire operation. Such administrative controls, therefore, tie all the functional elements together into a cohesive operational whole.

3. <u>Production Automation</u>

   Production automation includes the use of automatic machines that do everything from automated storage and retrieval to welding, die casting, stacking, painting, assembly and inspection. The capability to interface with other automated systems has made these automated production machines an integral part of the automated factory. Flexible machining centers are being serviced by AS/RS systems that can feed machine tools and store the materials between operations.

4. Flexibility

With approximately 75% of all manufacturing in the United States now classified in the batch or job shop category, the element of flexibility in the automated factory is most desirable.

The ability to make rapid changes in factory processes has become a primary justification for installing automated equipment control systems. Flexibility also extends to the capability of isolating machines when they require service or maintenance, rerouting material flow and avoid disruption of the entire system, and the effective handling of material surges between work stations.

5. Automated Material Handling

The final key element in the automated factory is the automation of the physical handling and control of materials. According to many factory automation experts, it is the most vital characteristic of the automated factory because it is the common interface to all functions and processes. It is also a fruitful area for productivity improvements since most of the time that material is in the plant, it is being handled or stored.

From the time material is received until the finished product is shipped, it may be picked up, moved, stored, moved again, worked on and handled dozens of times. Manual, semi-automatic and fully automatic equipment will come into contact with the material. Each piece of handling equipment must interface with other equipment, and at each step, the material must be tracked and controlled accurately in real time.

Todays AS/RS concept have become so advanced that they match or exceed the most sophisticated manufacturing machines and equipment. There are innovative solutions to virtually every material handling requirement and a standardization in equipment and installation by system suppliers has helped make AS/RS systems highly cost effective.

The integration of AS/RS systems as an element of an automatic factory requires careful planning and will depend on whether a new facility is being planned which would allow the design form to follow the logically arranged layout of processes and facilities, or whether the automated factory concept is to be applied to existing facilities which would require carefully phased planning with integration of a few elements at one time so that production could continue. This "phasing in" approach is most easily assimilated by both labor and management and assures favorable acceptance and minimal production interruption. The following guidelines are offered for implementing an automated factory project:

- A completed master plan is essential before changing any of the material handling system elements.

- The system is an integration of independent elements which can be installed as convenience, economics and justification permit.

- Product control is fundamental to any system, even the existing one. Obtain control as you incorporate elements, and make every successive system element continue this control.

- AS/RS provides positive control over stock movement and storage and should be an early system element to capture product control and to provide the common link for future automation elements.

- Accurate, timely data is essential to benefit from product control. Real time reporting is necessary for positive inventory control.

An integrated system is the consolidation of system elements, brought together by the development of reliable interfaces and centralized system control.

A "top down" examination of the entire operation reveals the areas for consolidation, elimination, or expansion and a fresh look at the need for present departmental structures.

Once the "new facility is configured, a careful, step-by-step "bottom up" development of equipment, interfaces and controls will contribute to a successful and profitable automated facility.

The "building blocks" of the automated factory exist today as reliable, proven equipment. Such subsystems have provided many cost-effective benefits as "stand alone" solutions to localized problems. Similarly, computerized control of factory data is an established and successful achievement.

Receiving

The operational difficulties in many conventional receiving areas are generally well known. Material staging areas often are choked with incoming shipments as they await destination assignments. Material is often handled several times before it is dispatched, increasing the probability of product damage, loss or mis-assignment. Material checkers, industrial truck drivers and other personnel make the receiving function labor intensive. Paperwork is shuffled from one point to another, delaying delivery of materials to the point of need. Real-time control of inventory in these conventional receiving areas seems to be impossible to achieve.

In an automated receiving area, material is managed and controlled precisely from the moment it arrives on the dock -- and sometimes even before its actual arrival. The expanded capacity of the automated system can handle virtually any planned volume which eliminates the need for large staging areas. As material is loaded onto the dock, it is quickly identified to a computer using a keyboard terminal, light pen, magnetic wand or other automatic identification device.

The computer, which is linked to other Material Requirements Planning (MRP) data banks, verifies the identification and quantity of the incoming material against material orders. The information control system has the ability to then direct the material to storage, inspection, order picking, manufacturing, assembly, or distribution as appropriate. The computer may direct that a move ticket be printed for attachment to the material at the receiving dock.

The material moves rapidly out of the receiving area via an automated transportation system such as a conveyor, in-floor tow-line, automated guided vehicle or other means. This systematized process greatly reduces the number of handling operations, the duration of each process and the frequency of handling.

Materials destined for storage move to the automated high-rise storage system which typically occupies less than one-third of the floor space required by conventional floor storage. The system computer has already assigned a storage location by storage aisle and storage rack opening as the material arrives. An S/R machine picks up the load, stores it in the assigned location, and verifies to the computer that the load has been correctly stored. Precision S/R machines and equipment assure product protection through precise handling while the system protects the material from unauthorized removal by restricting access and through absolute inventory accountability.

Inspection

Management information system can determine the need for and size of lot samples to be sent to quality inspection. There is no need, therefore, to retain the entire shipment in the receiving area while awaiting disposition. The material can be sampled, sent to quality control, and the remaining shipment moved directly to storage all in one operation. The material will be quarantined in the computer and cannot be accessed until released by quality control. This not only alleviates congestion in the receiving area, but also assures that materials will not be used until cleared through the inspection process.

When inspection is ready to check the material, the sample is retrieved, the material checked, and the computer notified that the material is ready for release. If material is found to be unacceptable, it can be held in the AS/RS until shipping is ready to return it to the vendor, at which point the AS/RS system automatically directs it to the shipping dock.

By integrating AS/RS into the inspection process, use of time is maximized by the efficient flow of materials. There is no longer a need for a large staging area in the inspection department, releasing space for manufacturing or other activities. Productivity is increased because duplicate handling has been eliminated and automated equipment moves and stores the material.

Picking

The picking activity is labor intensive in the typical factory, whether in support of inspection, fabrication, assembly or to support shipping in less than unit loads.

Conventional order picking operations normally require larger numbers of people to be engaged in repetative tasks which allow a high margin for error. In addition, close supervision is difficult because the pickers must, of necessity, be scattered throughout the storage areas. In addition to actually picking the order, they must also update inventory records, deliver the order and return for their next assignment. Going to and from the picking areas creates additional unproductive time and encourages haste which, in turn, often encourages errors.

Keeping all the items readily accessible to pickers necessitates storage within a relatively limited reach. Floor space must be utilized for storage rather than manufacturing or other processing operations. Comfortable environmental conditions such as heating/air conditioning and lighting must be maintained for the comfort level of the employee, not of the product. In an AS/RS system, however, between 30 and 40 percent of the floor space usually can be recovered for other operations. And the costly environmental conditions can be revised to achieve significant additional savings. This will greatly increase the utilization of the cube, effect substantial financial savings and increase overall plant productivity.

Automated storage/retrieval systems can reduce the time required to pick materials by bringing the material to the picker. An S/R machine can either take the person to the materials for efficient picking or it can direct the S/R machine to retrieve the material from picking stations in front of the storage aisle or via conveyor or transporter to a picking area away from the storage system. The system computer can be designed to direct the picker via a CRT or ticket printer or both to pick the material and send it to a designated area. The number of configurations of picking methods is almost unlimited.

Because picking is done at a well lighted, human engineered work station, picking errors are reduced and picking rates dramatically increased. In some cases, picking rates have been increased from 10 to 15 per hour in a conventional operation to 240 line items picked per hour. These increased picking rates have sharply reduced manpower requirements, turn-over and training costs. Supervision is effective since pickers stay in one place and need not wander through the storage aisles.

When a pick is completed, the picker signals the system computer via his control panel that the task is accomplished and the computer automatically updates the inventory records in real-time, thereby eliminating the human error factor often associated with the completion of paperwork by the picker. Also, it gives management timely information. The accuracy of inventory records maintained by the system computer can be checked periodically without a physical count of all inventory in the system.

An automatic cycle inventory (ACI) check can be performed by the computer, directing random sample counts which are compared with the computer records. Real-time inventory control, combined with the speed and accuracy of material picking, contributes directly to the reduction of back-up or "safety" stocks. Management can now rely on the validity and timeliness of inventory records.

Manufacturing

Inefficiencies in material handling and control on the manufacturing floor are often the root of productivity problems. Variable path transporters, principally industrial trucks, hand trucks and carts, require the use of up to 40 percent of the floor space for travel and turn-around. To assure that expensive machine tools and equipment have a high operating uptime, additional floor space is used to queue up plenty of materials at each work station.

In addition, space is used to store materials between manufacturing operations. This work-in-process storage often exceeds the space in which the associated productive work is performed. With materials stacked in work station queues and work-in-process areas, it is difficult to maintain accurate control, protect material from damage or loss, or to determine the real-time status of material, data which are vital to an effective production material control and scheduling system.

Automated storage/retrieval systems, long proven effective in warehousing operations, are solving the problems associated with in-process storage, in or near the manufacturing area. In some plants, automated S/R systems directly service manufacturing by providing access to materials through the storage rack openings facing the work stations. Although the S/R machines are captive in the storage aisles, their independent vertical and horizontal motions allow access to any one of a large set of locations on either side of the storage aisle. Thus, the S/R machine behaves like a random path vehicle and can bring materials to work stations as needed. Automated Storage/Retrieval Systems may be located adjacent to the manufacturing area and materials delivered via in-floor tow carts, automatic conveyors or automatic guided vehicles in situations where direct interfacing of AS/RS storage and work stations is not practical. The concentration of storage at or near the processing site does more than save space. It provides a virtually limitless working queue at each process. The storage space may be "partitioned" into any configuration needed, and the partitioning may be changed by changing computer software rather than changing physical facilities or moving machinery. Thus, a production machine is not limited by the queue in front of it, thereby increasing the productivity of the capital equipment while eliminating floor storage of material.

Automated S/R systems are central to the success of real-time control of materials in the manufacturing area. When combined with automatic identification systems and with automatic transportation systems, such as automatic conveyors or automatic guided vehicles. Material can thus be tracked and controlled in real-time throughout the process.

Assembly

Raw material goes through multiple processes prior to becoming part of a finished product. Material stored in its lowest manufacturing state is the least costly as there is a sharp use in value when labor is added. The sharpest rise in value generally occurs in the assembly operation. An unplanned shutdown of an assembly line due to slow delivery of materials or mis-picked parts is very costly. To hedge against shutdowns, materials or assemblies are often stored in large quantities near the assembly line, taking up valuable space while increasing the probability of damage or loss. It is here that AS/RS systems can provide efficient storage, transportation and real-time control to keep inventories at a minimum and maximize productivity. In high speed assembly operations, the material handling function is typically labor intensive. AS/RS can improve the productivity of existing manpower through controlled transportation of materials. Automated horizontal transportation systems also reduce the potential for material damage that occurs through conventional material handling methods. In assembly operations where there are a variety of products or models being produced there are frequent assembly-line changeovers making real-time control provided by AS/RS essential.

Storing parts, materials and assemblies in an automated storage/retrieval system will reduce the amount of material needed at a given time along the assembly line. Production control management can initiate the assembly process with the knowledge that materials are available in the storage system and can be accessed quickly and accurately when needed. There is no need to stockpile materials in excess of that day's or that shift's requirements. Materials needed to meet production assembly schedules can be quarantined in the computer. This also prevents excess quantities of materials from being horded which may be needed in other areas. Material kept in the AS/RS system is protected from damage or loss which may occur as it sits for long periods in the assembly area.

Shipping

At this stage of the manufacturing process product damage is more costly than at any other time since all the value has been added to the raw materials. Unfortunately, product damage seems to occur more often as it is moved from manufacturing, assembly and packaging into finished goods storage; then out of storage to order build up, onto shipping docks and into trucks or rail cars. In most large finished goods storage warehouses, a third or more of the storage space is lost to aisles for the movement of industrial trucks and people. Areas are often set aside for building up orders. Sometimes, orders cannot be filled on time because inventory records show items to be in storage, yet they cannot be found. Even in conventional operations, there is usually more control and supervision in the assembly area than in the shipping operations.

In an AS/RS system, materials and finished products move from the assembly line or manufacturing and, where appropriate, through packaging and automatic palletizing, into the shipping/warehouse area where they are identified to the computer. According to shipping schedules, they are sent directly to the shipping dock or to storage via an automated transportation system. The high-density AS/RS system effectively utilizes floor space and maintains accurate inventory control over all items in storage at all times. To fill an order, the computer directs the S/R machines to bring out the items in a predetermined sequence for the efficient loading of trucks or rail cars. The speed and accuracy of an AS/RS system reduces the floor space requirements for staging orders prior to shipment. This single step will eliminate the practice of accumulating orders ready for shipment on the floor of the warehouse or the shipping dock.

The installation of an automated storage/retrieval system in the warehouse/shipping area can improve productivity significantly, reduce damage and can improve safety. The resultant manpower reduction and reduction of conventional transportation equipment also will improve the level of job satisfaction of the personnel in the shipping area.

Implementation -- The Team Approach

The ancient Chinese had a saying to the effect that the longest journey starts with a single step. A material handling and storage application also needs an initiating event, and we strongly suggest that this event be the appointment of a person within your organization, and reporting at a high level in your corporate structure, as project team coordinator whom can be assigned the responsibility for collecting the data, identifying the scope of the problem and leading the implementation team. This coordinator should be totally familiar with your company's interdepartmental relationships since his primary function is to communicate with the heads of the departments and with outside sources such as vendors, architects and government agencies. While the person selected may not have established contacts outside of your company, it would be helpful if he/she is acquainted with your department managers and key executives and enjoys a good working relationship with them.

The next person assigned to the project team should be a systems analyst. The project team coordinator should approach your systems manager and request that a person with a good computer systems background and a good working knowledge of your data processing and paperwork procedures be assigned to the project team for the life of the project. Most material handling systems installed today are controlled by computers that become a part of your overall computer hierarchy and are, consequently, inserted in your paperwork stream, making the systems person an invaluable member of the project team.

These two persons, the project team coordinator and the systems analyst, need to sit down together and do three things: look at the "big picture"; list the tasks to be completed; and identify the other members needed for the project team. The first two, looking at the big picture and identifying the tasks to be completed, more or less automatically identify the areas of expertise that will be required by the team and will thus simplify the selection of its other members.

Let me explain to you what we mean when we say that your project team needs to "look at the big picture". Your project team should consider your company in its entirety. All of its employees, all of its suppliers, all of its customers, all of its buildings, machines, forktrucks, its business strategy, all taken together, represent a number of elements united by regular interaction which is the definition of a system. This "system" produces a profit by purchasing raw materials, converting them into products that serve a useful purpose in our society, shipping them to customers, receiving a fair price in return so the process can be started over again, and all the while keeping hundreds of records so the shareholders and the various governmental bodies can be compensated. In this context, material handling becomes a subsystem of the complex company system, and this relationship must be kept in mind by the project team. The fact that material handling is interrelated to receiving and shipping, production control, inventory control, manufacturing, purchasing, accounting, plant engineering, plant maintenance and data processing suggests that all these departments should be represented at one time or another on the project team.

The project team is to be charged with no less responsibility than the analysis of your company's total operation and the development of a plan to achieve cost objective for material handling that are consistent with the strategic plan for the company.

A study of this scope and depth will produce a report that will have tremendous impact on the future of your company. If it is cursory and shallow, it will be rejected and you will have lost the opportunity for action now to improve your productivity and profitability or, worse, if it recommends sweeping changes that promise great improvements, you risk the possibility of implementing a plan that may be flawed because the project team lacked the full range of expertise necessary to ensure success.

This suggests a cooperative effort -- a team that combines your people with their expert knowledge of your business and its problems and material handling systems people with their knowledge of material handling.

A cooperative effort requires a commitment on the part of both parties to provide well-qualified persons from varied backgrounds and disciplines. Well-qualified people in any company are a scarce and valuable resource, and their assignment must be made carefully by those in the executive suite for them to be utilized efficiently and accepted by other departments.

To initiate a joint effort on a material handling project, we suggest a negotiated contract with specific rights for either of the parties to terminate the contract at the end of the Feasibility Study or the Design Phase.

This method of procurement has several practical advantages. You get added strength for your project team near the beginning of the project; you get a tailor-made system from a supplier who has gained insight and intimate understanding of your needs; if trade secrets are involved in your operations, you get greater confidentiality; and since certain design/construct areas can be overlapped, you get your system and its payback sooner. We might add that you will probably save money because every available option will be discussed with you and those that do not result in a benefit that justifies their cost will be discarded.

Involving so many members of your organization from the outset has many downstream advantages. You will be preparing members of your organization to accept a system that represents a change in procedure and greater job enrichment.

Through participation, your managers will develop a pride of ownership that will help ensure the success of the installation. And you will find that an in-house study such as this will pay for itself again and again through a heightened awareness of just what is going on in your operation.

Selecting a partner for such an important project is one of the critical decisions to be made and you will want to check that partner's credentials very carefully.

Summary

Automated Storage/Retrieval Systems are operating successfully in manufacturing centers, improving productivity and profitability through effective handling, storage, movement and control of a wide variety of materials.

The elements of the automated factory are all available and in operation today. But the elements must be integrated effectively to realize the full potential of each element. The key to factory integration is the automation of the material handling and control function. Without this vital link, the automated factory is simply not possible.

The automated factory will not be devoid of humans and their irreplaceable capabilities. By bringing together high technology and human engineering, automated factories will enhance and improve working conditions for people at all levels. Heavy physical demands can be removed from the workplace by precision, automated machines while materials are tracked and controlled in real-time with a speed and accuracy no humans can match.

Through an evolutionary process, the automated factory is becoming a reality. The phasing in of automated systems has proven to be a sound process, economically and humanistically. But a master plan is essential before implementing any element.

We have suggested a cooperative approach to achieve the automated factory. This will mean changes in the traditional supplier/user relationship. Traditional approaches may be stifling creativity and contributing to high costs. Design and implementation will be shared by members of a project team composed of the user, the supplier, and the consultants. The traditional compartmentalized approach with arm's length relationships does not permit the intimate relations required to successfully produce integrated systems. Suppliers will be chosen much earlier in the chain of events and will contribute their expertise as equal partners on the project team. The automated factory is a concept whose time has come. By working together, we can secure its benefits for our society.

INDEX

A

B

C

D

E

F

G

H

I

J

K

L

M

N

O

P

Q

R

S

T

U

V

W

X

Y